Contributions to Statistics

The series **Contributions to Statistics** contains publications in theoretical and applied statistics, including for example applications in medical statistics, biometrics, econometrics and computational statistics. These publications are primarily monographs and multiple author works containing new research results, but conference and congress reports are also considered.

Apart from the contribution to scientific progress presented, it is a notable characteristic of the series that publishing time is very short, permitting authors and editors to present their results without delay.

More information about this series at http://www.springer.com/series/2912

Katarzyna Filipiak · Augustyn Markiewicz ·
Dietrich von Rosen
Editors

Multivariate, Multilinear and Mixed Linear Models

Springer

Editors
Katarzyna Filipiak
Institute of Mathematics
Poznań University of Technology
Poznań, Poland

Augustyn Markiewicz
Department of Mathematical and Statistical Methods
Poznań University of Life Sciences
Poznań, Poland

Dietrich von Rosen
Department of Energy and Technology
Swedish University of Agricultural Sciences
Uppsala, Sweden

ISSN 1431-1968
Contributions to Statistics
ISBN 978-3-030-75493-8 ISBN 978-3-030-75494-5 (eBook)
https://doi.org/10.1007/978-3-030-75494-5

Mathematics Subject Classification (2010): 62H12, 62H15, 15A03, 15B52, 62K05, 62J99, 15A18, 15A27

This Springer imprint is published by the registered company Springer Nature Switzerland AG
The registered company address is: Gewerbestrasse 11, 6330 Cham, Switzerland

To Prof. Tadeusz Caliński

Preface

In 2009, the birth of a workshop series on multivariate models and mixed linear models took place, as the result of an earlier collaboration between Dietrich von Rosen, Augustyn Markiewicz and Katarzyna Filipiak. The first meeting was organized as a Polish-Swedish research group on Multivariate Linear Models and Designs of Experiments. Since then (till November 2020), 20 international research group meetings have taken place with 68 participants attending at least one event. To celebrate 10 years of successful work, a conference was organized in May 2019 and a book project was initiated to spread some of the achievements which have been generated.

Now the book project is finished, and it contains mainly the overviews of selected results on multivariate and mixed linear models. It consists of 12 articles including 27 authors from all over the world, where 17 have participated at least one time in a workshop meeting. Several of the authors have repeatedly joined the meetings for many years.

Our idea is to guide, in a detailed manner, the readers along the topics related to the estimation and testing of multivariate/mixed linear model parameters. Therefore, the book starts with multivariate distributions theory, then under various multivariate models identification and testing of covariance structures and means is considered, and it finishes with estimation in mixed linear models and their transformations. The readership is expected to be researchers and Ph.D. students interested in multivariate linear and mixed linear models.

During research meetings also, many different topics on multivariate and mixed models were studied and published in regular journals. The full list of articles inspired by discussions within the meetings is available at the end of the book.

All the meetings have been run at the Mathematics Research and Conference Center in Bedlewo, Poland, which is headed by the Institute of Mathematics of the Polish Academy of Sciences. We are very grateful for the hospitality of the Conference Center and highly appreciate their professionalism. Moreover, we would like to thank Stefan Banach International Mathematical Center and Institute of Mathematics of the Polish Academy of Sciences, for all the financial support which has been received over the years, including the recent conference. Now we could focus on the content of the meetings and without the support, it would have been likely that the series would have been interrupted.

We would also like to thank all the contributors who accepted the invitation to create this book. Since there are so many authors, we can imagine that to follow guidelines and where the readership is expected to have a mixed background, it must have taken a lot of energy when preparing the articles.

The help by the Springer team and the support we have received is also highly appreciated.

The workshop series will continue to run, nowadays with two meetings per year, and we warmly welcome old and new participants. The topics considered during the meetings are also within the theme of related events, such as The International Conference on Trends and Perspectives in Linear Statistical Inference, LinStat (https://linstat2020.science.upjs.sk/history.php), or International Workshop on Matrices and Statistics, IWMS (https://webpages.tuni.fi/uta_statistics/iwms/), co-organized by participants of our meetings. There are still many important challenging problems to be solved within the area of multivariate and mixed linear models.

Będlewo, Poland
February 2021

Katarzyna Filipiak
Augustyn Markiewicz
Dietrich von Rosen

Contents

Contributors

Syed Ejaz Ahmed Department of Mathematics & Statistics, Brock University, St. Catharines, ON, Canada

Dursun Aydın Department of Statistics, Mugla Sitki Kocman University, Mentese, Turkey

Carlos A. Coelho Mathematics Department (DM) and Centro de Matemática e Aplicações (CMA), NOVA School of Science and Technology, NOVA University of Lisbon, Caparica, Portugal

Xiangzhao Cui Department of Mathematics, Honghe University, Mengzi, China

Katarzyna Filipiak Institute of Mathematics, Poznań University of Technology, Poznań, Poland

Miguel Fonseca Centro de Matemática e Aplicações (CMA) and Departamento de Matemática, Universidade Nova de Lisboa, Lisbon, Portugal

Stephen J. Haslett Research School of Finance, Actuarial Studies and Statistics, The Australian National University, Canberra, Australia;
Centre for Public Health Research, Massey University, Wellington, New Zealand;
School of Fundamental Sciences, Massey University, Palmerston North, New Zealand

Jarkko Isotalo Faculty of Information Technology and Communication Sciences, Tampere University, Tampere, Finland

Radosław Kala Department of Mathematical and Statistical Methods, Poznań University of Life Sciences, Poznań, Poland

Daniel Klein Faculty of Science, Institute of Mathematics, P. J. Šafárik University, Košice, Slovakia

Tõnu Kollo Institute of Mathematics and Statistics, University of Tartu, Tartu, Estonia

Arkadiusz Kozioł Faculty of Mathematics, Computer Science and Econometrics, University of Zielona Góra, Zielona Góra, Poland

Chun Li Department of Mathematics, Honghe University, Mengzi, China

Augustyn Markiewicz Department of Mathematical and Statistical Methods, Poznań University of Life Sciences, Poznań, Poland

Monika Mokrzycka Institute of Plant Genetics, Polish Academy of Sciences, Poznań, Poland

Jianxin Pan Department of Mathematics, The University of Manchester, Manchester, UK

Jolanta Pielaszkiewicz Linköping University, Linköping, Sweden; Stockholm University, Stockholm, Sweden

Simo Puntanen Faculty of Information Technology and Communication Sciences, Tampere University, Tampere, Finland

Anuradha Roy Department of Management Science and Statistics, The University of Texas at San Antonio, San Antonio, TX, USA

Akimichi Takemura Faculty of Data Science, Shiga University, Shiga, Japan

Dietrich von Rosen Linköping University, Linköping, Sweden; Swedish University of Agricultural Sciences, Uppsala, Sweden

Ersin Yılmaz Department of Statistics, Mugla Sitki Kocman University, Mentese, Turkey

Li Zeng Department of Mathematics, Honghe University, Mengzi, China

Ivan Žežula Institute of Mathematics, Faculty of Science, P. J. Šafárik University, Košice, Slovakia

Defei Zhang Department of Mathematics, Honghe University, Mengzi, China

Jine Zhao Department of Mathematics, Honghe University, Mengzi, China

Roman Zmyślony Faculty of Mathematics, Computer Science and Econometrics, University of Zielona Góra, Zielona Góra, Poland

Chapter 1
Holonomic Gradient Method for Multivariate Distribution Theory

Akimichi Takemura

Abstract In 2011, we introduced the Holonomic Gradient Method (HGM) as a new approach to multivariate distribution theory. It is a general method applicable when the density function $f(x, \theta)$ is "holonomic" in the random variables x as well as the parameters θ. Since the multivariate normal density is holonomic, HGM is applicable to the whole classical distribution theory based on the multivariate normal distribution. In fact since 2011, we applied HGM to many problems of classical multivariate distribution theory and showed that some difficult distributional problems, such as those involving hypergeometric functions of a matrix argument, can be successfully treated by HGM. In this article, we discuss the origin, the basic theory and some applications of HGM to multivariate distribution theory.

1.1 Introduction

In Nakayama et al. [22], we introduced the holonomic gradient method and applied it to the Fisher–Bingham distribution on the sphere. Mathematically, HGM is based on the theory of D-modules and holonomic functions. Actually, Nobuki Takayama has been long suggesting the applicability of the theory of D-modules to statistics, as seen in a comprehensive treatment of the background theory of HGM in Chap. 5 of Hibi [6]. However, statisticians including me did not understand his intentions. This changed during a tutorial seminar in September of 2009, where statisticians and algebraists gathered together and gave introductory talks related to algebraic statistics. From a simple example discussed by Nobuki Takayama, which I will discuss in the next section, I realized that the theory of D-modules can be applied to multivariate distribution theory.

In Japan, there was a long tradition of deep mathematical research of the theory of D-modules by Masaki Kashiwara and his colleagues. Masaki Kashiwara won the Kyoto Prize and the Chern Medal of the International Mathematical Union in 2018 for "establishing the theory of D-modules, thereby playing a decisive role in the

A. Takemura (✉)
Faculty of Data Science, Shiga University, Shiga 522-8522, Japan
e-mail: Akimichi.Takemura@gmail.com

K. Filipiak et al. (eds.), *Multivariate, Multilinear and Mixed Linear Models*, Contributions to Statistics, https://doi.org/10.1007/978-3-030-75494-5_1

creation and development of algebraic analysis". Also, there was an active research on the algebraic algorithms for problems for D-modules in Japan, conducted by Toshinori Oaku, Nobuki Takayama and other people. See, for example, Oaku [24, 25] and Saito et al. [27]. On the other hand, in Japanese statistics, there is a strong tradition of research on multivariate distribution theory. Hence, the invention of HGM provided interesting research topics both for mathematicians and statisticians and led to active research on HGM in Japan. The list of papers on HGM maintained by Takayama [32] lists more than 35 papers on HGM.

In statistics, we very often need parameterized integrals. If the density function $f(x, \theta)$ is not normalized, then it is important to evaluate the normalizing constant

$$Z(\theta) = \int f(x, \theta) dx, \tag{1.1}$$

not only for a particular value of θ but also for many values of θ in the parameter space Θ. Note that in some textbooks, $1/Z(\theta)$ is called the normalizing constant. Here, we call $Z(\theta)$ the normalizing constant, because as discussed in the next section the holonomicity of a function is not preserved by division. Even if the density function is normalized, if we want to evaluate a power function of a testing procedure, then we need the probability of the rejection region R

$$P_\theta(R) = \int_{x \in R} f(x, \theta) dx \tag{1.2}$$

for many values of θ of the alternative hypothesis. HGM is particularly useful for these problems, because it evaluates $Z(\theta)$ or $P_\theta(R)$ by numerically solving a system of differential equations in θ, thereby evaluating these functions along a path in the parameter space. Note that Monte Carlo simulation is a general method for evaluating an integral, but usually simulations have to be repeated for different values of θ. Also, usually the numerical accuracy of the Monte Carlo method is about 4 significant digits, whereas by HGM we usually achieve about 10 digit accuracy.

The advantage of HGM is that it is a general methodology based on the theory of D-modules and holonomic functions, and hence is applicable to the whole multivariate distribution theory based on the multivariate normal distribution. Also, it is usually numerically very accurate. A disadvantage of HGM is that the derivation of the Pfaffian system from a given set of differential equations is often very heavy.

The objectives of this paper are to discuss the origin of HGM, its basic theory and its applications to multivariate distribution theory. In the subsequent sections, we describe the origin of HGM, see Sect. 1.2, as well as give definitions and properties of holonomic functions; see Sect. 1.3. Then we discuss various applications of HGM. In Sect. 1.4, we consider the evaluation of the hypergeometric function of a matrix argument. In Sect. 1.5, we consider the evaluation of probabilities of some regions under multivariate normality, and in Sect. 1.6 we discuss some distribution problems in wireless communication. We discuss some other applications of HGM and present concluding remarks in Sect. 1.7.

1.2 Origin of HGM

Here, we discuss a particular problem, which was given by Nobuki Takayama in a tutorial seminar in the September of 2009. Consider the following problem.

> **Problem**: Obtain a differential equation satisfied by $A(x)$, which is defined by
>
> $$A(x) = \int_0^\infty e^{-t-xt^3} dt, \qquad x > 0. \tag{1.3}$$
>
> The answer to this problem is as follows.
>
> **Answer**: $A(x)$ satisfies the following differential equation:
>
> $$27x^3 A''(x) + 54x^2 A'(x) + (6x + 1)A(x) = 1. \tag{1.4}$$

Let us consider the statistical meaning of this problem. Actually, $A(x)$ is the normalizing constant of the unnormalized probability density e^{-xt^3}, with respect to the base measure $e^{-t}dt$, for $t > 0$. Here, x is the parameter of the distribution and the random variable is t. It is an example of an exponential family with the sufficient statistic t^3. As explained after (1.1), we call $A(x)$ (rather than $1/A(x)$) the normalizing constant, because $A(x)$ is holonomic as explained below.

Actually, it is often not easy to realize that $A(x)$ is a normalizing constant. If the problem was given in the following notation, where (t, x) is replaced by (x, θ), it would have been immediately obvious that $A(\theta)$ is the normalizing constant:

$$A(\theta) = \int_0^\infty e^{-x-\theta x^3} dx, \qquad \theta > 0. \tag{1.5}$$

This example shows that such trivial things as the differences in notational conventions might hinder collaborations between researchers of different fields. Based on (1.5), we can define a probability density function $f(x, \theta)$, $x > 0$, by

$$f(x, \theta) = \frac{1}{A(\theta)} e^{-x-\theta x^3}.$$

In this distribution, the normalizing constant $A(\theta)$ does not have an explicit expression, but the differential equation (1.4) is useful from a statistical viewpoint. Note that the second derivative of $A(\theta)$ corresponds to the Fisher information. Hence by (1.4), we can immediately evaluate the Fisher information of the distribution if we compute $A(\theta)$ and its derivative $A'(\theta)$.

1.3 Definition and Properties of Holonomic Functions

In this section, we give the definition and properties of holonomic functions. The notion and properties of holonomic functions were developed by Zeilberger [35]. Note that $A(x)$ in the previous section satisfies the differential equation in (1.4). We consider here a linear differential equation since the equation is linear in the derivatives of $A(x)$, but the coefficients are (e.g., $54x^2$) polynomials (or more generally rational functions) in x. Hence, $A(x)$ is a holonomic function in a single variable x following Definition 1.1.

Definition 1.1 A univariate smooth function $f(x)$ is holonomic if it satisfies a linear differential equation with rational function coefficients, i.e.,

$$f^{(k)}(x) + h_{k-1}(x)f^{(k-1)}(x) + \cdots + h_1(x)f'(x) + h_0(x)f(x) = 0, \tag{1.6}$$

where $h_i(x), i \in \{0, 1, \dots, k-1\}$, are rational functions in x.

In this definition, the right-hand side of (1.6) is 0 and the differential equation is called *homogeneous*. The right-hand side of (1.4) is 1 and it is called *inhomogeneous*. From the viewpoint of numerical computation by HGM, the inhomogeneous case can be treated as in the homogeneous case and in the following we consider the homogeneous case.

By using a "companion matrix", we can write (1.6) in a matrix form as

$$\frac{d}{dx}\begin{pmatrix} f(x) \\ f'(x) \\ \vdots \\ f^{(k-1)}(x) \end{pmatrix} = \begin{pmatrix} 0 & 1 & 0 & \dots & 0 \\ 0 & 0 & 1 & \dots & 0 \\ \vdots & \vdots & \vdots & \ddots & \vdots \\ -h_0(x) & -h_1(x) & -h_2(x) & \dots & -h_{k-1}(x) \end{pmatrix} \begin{pmatrix} f(x) \\ f'(x) \\ \vdots \\ f^{(k-1)}(x) \end{pmatrix}. \tag{1.7}$$

We call (1.7) the Pfaffian system for $f(x)$. The Pfaffian system is used for numerically solving the differential equation.

We now give the definition of holonomic function in several variables.

Definition 1.2 A smooth function $f(x_1, \dots, x_n)$ is holonomic if it satisfies a linear differential equation with rational function coefficients in each argument (fixing the other arguments), i.e.,

$$\begin{aligned}
&\sum_{i=0}^{k_1} h_i^1(x_1,\ldots,x_n)\partial_{x_1}^i f(x_1,\ldots,x_n) = 0,\\
&\ldots \\
&\sum_{i=0}^{k_n} h_i^n(x_1,\ldots,x_n)\partial_{x_n}^i f(x_1,\ldots,x_n) = 0,
\end{aligned} \tag{1.8}$$

where the coefficients $h_i^j(x_1,\ldots,x_n)$ are rational functions in $x_1,\ldots,x_n$.

In this homogeneous case, we can set the highest coefficients to 1

$$h_{k_1}^1(x_1,\ldots,x_n) = \cdots = h_{k_n}^n(x_1,\ldots,x_n) = 1,$$

without loss of generality, by dividing the jth equation by the highest coefficient $h_{k_j}^j(x_1,\ldots,x_n)$, $j \in \{1,\ldots,n\}$, because $h_i^j/h_{k_j}^j$ remains to be a rational function in $x_1,\ldots,x_n$.

In the multivariate case, the Pfaffian system becomes a bit more complicated than in (1.7). Consider the case $n = 2$ and $k_1 = k_2 = 2$. Then we have

$$\partial_x^2 f(x,y) + h_1^1(x,y)\partial_x f(x,y) + h_0^1(x,y)f(x,y) = 0, \tag{1.9}$$
$$\partial_y^2 f(x,y) + h_1^2(x,y)\partial_y f(x,y) + h_0^2(x,y)f(x,y) = 0. \tag{1.10}$$

By differentiating (1.9) with respect to y, we have

$$\begin{aligned}
\partial_x^2\partial_y f(x,y) = &-(\partial_y h_1^1(x,y))\partial_x f(x,y) - h_1^1(x,y)\partial_x\partial_y f(x,y)\\
&-(\partial_y h_0^1(x,y))f(x,y) - h_0^1(x,y)\partial_y f(x,y).
\end{aligned}$$

Now consider differentiating the vector $(f(x,y), \partial_x f(x,y), \partial_y f(x,y), \partial_x\partial_y f(x,y))'$ by x. Then we obtain

$$\begin{aligned}
&\partial_x \begin{pmatrix} f(x,y)\\ \partial_x f(x,y)\\ \partial_y f(x,y)\\ \partial_x\partial_y f(x,y)) \end{pmatrix}\\
&= \begin{pmatrix} 0 & 1 & 0 & 0\\ -h_0^1(x,y) & -h_1^1(x,y) & 0 & 0\\ 0 & 0 & 0 & 1\\ -\partial_y h_0^1(x,y) & -\partial_y h_1^1(x,y) & -h_0^1(x,y) & -h_1^1(x,y) \end{pmatrix} \begin{pmatrix} f(x,y)\\ \partial_x f(x,y)\\ \partial_y f(x,y)\\ \partial_x\partial_y f(x,y) \end{pmatrix}.
\end{aligned} \tag{1.11}$$

Similarly, by differentiating (1.10) by y, we have

$$\partial_y \begin{pmatrix} f(x,y) \\ \partial_x f(x,y) \\ \partial_y f(x,y) \\ \partial_x\partial_y f(x,y)) \end{pmatrix}$$
$$= \begin{pmatrix} 0 & 0 & 1 & 0 \\ 0 & 0 & 0 & 1 \\ -h_0^2(x,y) & 0 & -h_1^2(x,y) & 0 \\ -\partial_x h_0^2(x,y) & -h_0^2(x,y) & -\partial_x h_1^2(x,y) & -h_1^2(x,y) \end{pmatrix} \begin{pmatrix} f(x,y) \\ \partial_x f(x,y) \\ \partial_y f(x,y) \\ \partial_x\partial_y f(x,y) \end{pmatrix}. \tag{1.12}$$

The Pfaffian system for $f(x, y)$ consists of (1.11) and (1.12).

Now consider the case $n = 2$ and general k_1 and k_2. Suppose we want to compute a higher order derivative $\partial_x^{r_1}\partial_y^{r_2} f(x, y)$ of $f(x, y)$. If $r_1 \geq k_1$ or $r_2 \geq k_2$, then we can repeatedly apply (1.8) for $x = x_1$ or $y = x_2$ and reduce the order of differentiation. This implies that we only need to (numerically) keep $\partial_x^i\partial_y^j f(x, y)$, $i \in \{0, \ldots, k_1 - 1\}$, $j \in \{0, \ldots, k_2 - 1\}$, in memory. Then we can always compute the value of higher order derivatives as needed. In particular, let $\boldsymbol{f}$ denote the vector consisting of $\partial_x^i\partial_y^j f(x, y)$, $i \in \{0, \ldots, k_1 - 1\}$, $j \in \{0, \ldots, k_2 - 1\}$. Then as in (1.11) and (1.12), we can obtain two square matrices $\mathbf{P}_1$ and $\mathbf{P}_2$ consisting of rational function elements in x, y and of size $(k_1 - 1)(k_2 - 1)$, such that

$$\partial_x \boldsymbol{f} = \mathbf{P}_1 \boldsymbol{f}, \quad \partial_y \boldsymbol{f} = \mathbf{P}_2 \boldsymbol{f}.$$

Usually a holonomic f satisfies more differential equations than given in (1.8). Then we can further reduce the number of lower order derivatives to be kept in memory. The set of lower order derivatives, which we need to keep in memory, is the set of "standard monomials". The number of standard monomials is called the holonomic rank. In the case of $n = 2$, the set of standard monomials looks like Fig. 1.1.

Given a set of differential equations, the set of standard monomials and the Pfaffian system can be obtained by the Gröbner basis computation in D-modules. In this sense, the algebraic computation of the Gröbner basis is an important element of HGM.

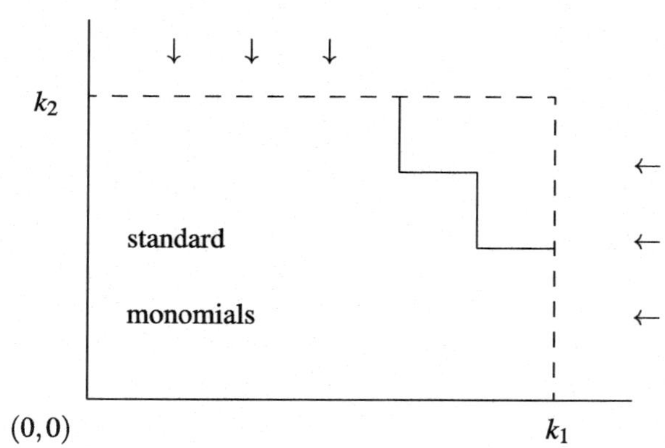

Fig. 1.1 The set of standard monomials

We have given the definition of a holonomic function. Now the question is how large is the class of holonomic functions. The basic properties of holonomic functions, together with appropriate algorithms, were given by Zeilberger [35]. We give the following listing of holonomic functions:

1. polynomials and rational function are holonomic;
2. exp(rational function) and log(rational function) are holonomic;
3. if f and g are holonomic, then $f+g$ and $f \times g$ are holonomic;
4. if $f(x_1, \dots, x_m)$ is holonomic, then $\int f(x_1, \dots, x_m) dx_m$ is holonomic;
5. the restriction of a holonomic f on a linear subspace $f(x_1, \dots, x_{m-1}, c)$ is holonomic.

By property 3., the set of holonomic functions is closed with respect to addition and multiplication. Hence it forms a ring. Note that $1/f(x)$ may not be holonomic even if f is. An example is $f(x) = \cos x$, which is holonomic but $1/\cos x$ is not holonomic. By property 4., the holonomicity is preserved under "marginalization", which is very important to statistics, because the normalizing constant $Z(\theta)$ in (1.1) is holonomic in θ if the density function $f(x, \theta)$ is holonomic jointly in x and θ.

The holonomicity can be extended to discontinuous functions, in particular to the indicator function

$$I_S(x) = \begin{cases} 1 & \text{if } x \in S, \\ 0 & \text{otherwise,} \end{cases}$$

where S is a semialgebraic set, which is defined by polynomial equalities and inequalities. Then

$$P_\theta(S) = \int_S f(x, \theta) dx = \int I_S(x) f(x, \theta) dx$$

in (1.2) is holonomic in θ.

An interesting example of this is given by Oaku [26]. He considers a parameterized integral

$$v(t) = \int_{x^3 \geq y^2} e^{-t(x^2+y^2)} dxdy$$

and shows that it satisfies the following differential equation

$$[216t^4\partial_t^4 + (32t^4 + 1836t^3)\partial_t^3 + (224t^3 + 3594t^2)\partial_t^2 + (326t^2 + 1371t)\partial_t + 70t + 15]v(t) = 0.$$

Note that $v(t)$ corresponds to the probability $P(X^3 \geq Y^2)$, where X and Y are independent standard normal random variables. It seems that the exact evaluation of this kind of integral is only possible by techniques associated with HGM.

There are three steps of HGM: (1) Pfaffian, (2) initial value and (3) numerical integration. So far we have discussed the first step of HGM. The second step of HGM is to evaluate the initial values for the standard monomials at a convenient initial point. Note that the differential equation does not give any information on the initial value, and we need to obtain initial values by some other method. Usually, we use an infinite series expansion for the evaluation of the initial values. The third step of HGM is to apply some numerical integration method, such as the Runge–Kutta method for solving the Pfaffian system numerically.

1.4 HGM for the Hypergeometric Function of a Matrix Argument

For the development of HGM, an application of HGM to a hypergeometric function ${}_1F_1$ of a matrix argument in Hashiguchi et al. [3] was an important milestone, because the numerical evaluation of ${}_1F_1$ with a matrix argument was a classical difficult problem. ${}_1F_1$ with a matrix argument appears in the cumulative distribution of the largest root of the central Wishart matrix with a general covariance matrix $\mathbf{\Sigma}$. James [7] and Constantine [1] and other people developed the theory of zonal polynomials and the infinite series expansions of hypergeometric functions of a matrix argument in terms of zonal polynomials. A general background of zonal polynomials is given in Takemura [34]. The theory of zonal polynomials is mathematically deep and leads to many important theoretical studies. However, from a practical viewpoint, there is a formidable combinatorial difficulty in evaluating zonal polynomials, and the numerical evaluation of the hypergeometric function of a matrix argument remained a long-standing difficult problem. The expansion in terms of zonal polynomials can be considered as the Taylor expansion around the origin, and the convergence of the series becomes slow as the elements of the matrix argument become large.

Later, Muirhead [20] derived partial differential equations satisfied by ${}_1F_1$, but he did not use the differential equations for the numerical evaluation of ${}_1F_1$. From the viewpoint of HGM, Muirhead's partial differential equations give a Gröbner basis in a straightforward manner and HGM can be implemented based on them. In Hashiguchi et al. [3], we confirmed that HGM works well up to dimension 20 and for large $\mathbf{\Sigma}$, for which the expansion in terms of zonal polynomials cannot be evaluated due to combinatorial difficulties associated with zonal polynomials of high degrees.

The following partial differential equations for ${}_1F_1(a; b; \mathbf{Y})$ were derived by Muirhead [20].

Theorem 1.1 (Muirhead [20], Th. 5.1] Muirhead [21], Th. 7.5.6) *The hypergeometric function* $F = {}_1F_1(a; c; \mathbf{Y})$ *of a diagonal matrix argument* $\mathbf{Y} = \mathrm{diag}\,(y_1, \ldots, y_m)$ *is the unique solution of the following set of m partial differential equations:*

$$y_i\, \partial_{y_i}^2 F + \left\{ c - \frac{m-1}{2} - y_i + \frac{1}{2} \sum_{j=1, j\neq i}^{m} \frac{y_i}{y_i - y_j} \right\} \partial_{y_i} F$$
$$- \frac{1}{2} \sum_{j=1, j\neq i}^{m} \frac{y_j}{y_i - y_j} \partial_{y_j} F - aF = 0, \qquad i \in \{1, \ldots, m\}, \qquad (1.13)$$

subject to the conditions that F is symmetric in $y_1, \ldots, y_m$ *and F is analytic at* $\mathbf{Y} = \mathbf{0}$, $F(\mathbf{0}) = 1$.

Muirhead's differential equations correspond to those in Definition 1.2 with $k_i = 2, i \in \{1, \ldots, m\}$. Hence as in Fig. 1.1, the set of standard monomials consists of all the square-free derivatives of F:

$$\boldsymbol{f} = \{\partial_{y_1}^{i_1} \ldots \partial_{y_m}^{i_m} F \mid i_1, \ldots, i_m \in \{0, 1\}\}.$$

As the second step of HGM, we need initial values of $\boldsymbol{f}$. We obtained the initial values by appropriately modifying the algorithm given by Koev and Edelman [8] for the derivatives of F.

One important point of (1.13) is that it requires $y_i \neq y_j$ for all $i \neq j$. Muirhead's partial differential equations are valid only when the roots of $\mathbf{Y}$ are all distinct, and the region of multiple roots forms the singularity of the partial differential equations. On the other hand, ${}_1F_1$ is an entire function and there is no singularity of the function itself. This example shows that the differential equation satisfied by a function may have a singularity at the point where the function itself is not singular. From the numerical viewpoint, the singularity of the differential equation poses a serious problem for HGM, because the path of the integral involved in HGM cannot pass the singularity. In Hashiguchi et al. [3], we used L'Hôpital's rule to obtain differential equations for the region with multiple roots from Muirhead's differential equations. At the same time, we multiplied (1.13) by $\prod_{j \neq i}(y_i - y_j)$ and used the restriction algorithm for the Weyl algebra to show that both methods give the same result. Later, we learned a remarkable mathematical fact: for $m = 4$ the ideal generated by differential equations with polynomial coefficients obtained from (1.13) after multiplication by $\prod_{j \neq i}(y_i - y_j)$ is not a holonomic ideal and the restriction algorithm does not work Kondo [9]. Hence from a mathematical viewpoint, Muirhead's differential equations have different properties for $m = 2, 3$ and $m \geq 4$. The idea of L'Hôpital's rule was further developed in Noro [23].

In Hashiguchi et al. [4], we extended our results on ${}_1F_1$ to ${}_2F_1$, which appears in the power function of Roy's maximum root test of equality of two covariance matrices.

1.5 HGM for Evaluation of Probability of Some Regions Under Multivariate Normality

The density function $f(\mathbf{x}, \boldsymbol{\mu}, \boldsymbol{\Sigma})$ of the d-dimensional multivariate normal distribution $N_d(\boldsymbol{\mu}, \boldsymbol{\Sigma})$ is holonomic in $(\mathbf{x}, \boldsymbol{\mu}, \boldsymbol{\Sigma})$, because $\exp(-(\mathbf{x} - \boldsymbol{\mu})'\boldsymbol{\Sigma}^{-1}(\mathbf{x} - \boldsymbol{\mu})/2)$ is holonomic by properties 1. and 2. in Sect. 1.3. Therefore, probability $P_{\mu,\Sigma}(S)$ of a semialgebraic set S is holonomic in $(\boldsymbol{\mu}, \boldsymbol{\Sigma})$. Hence, it is of interest to derive the differential equations satisfied by $P_{\mu,\Sigma}(S)$ and use them to evaluate the probability.

First we consider the problem of orthant probability $P_{\mu,\Sigma}(S)$ for

$$S = \{(x_1, \ldots, x_d) \mid x_1 \geq 0, \ldots, x_d \geq 0\},$$

Table 1.1 Comparison with one-dimensional integral ($d = 10$)

Correlation	Dunnett	HGM	Error
0.00	0.00097656	0.00097656	0.00000000
0.10	0.00658647	0.00658647	0.00000000
0.25	0.02660319	0.02660319	0.00000000
0.50	0.09090909	0.09090909	0.00000000

Table 1.2 Average computational time of orthant probabilities (in seconds)

Dim	Miwa	HGM
5	0.002	0.018
6	0.011	0.075
7	0.075	0.207
8	0.615	0.659
9	5.637	2.140
10	57.012	6.521
11	636.207	19.427
12	NA	62.947

which has been studied by many authors. Koyama and Takemura [14] applied HGM to the orthant probability. If $\boldsymbol{\mu} = \mathbf{0}$, then the differential equations obtained in Koyama and Takemura [14] correspond to the classical Schläfli formula on the relation between hyperplane angles and the volume of a polyhedron. HGM applied for the orthant probability is numerically very accurate and fast. The numerical accuracy of HGM can be seen in the special case of the orthant probability of the "equicorrelation" case, for which the orthant probability reduces to a one-dimensional numerical integration (Dunnett and Sobel [2]) and can be computed with high accuracy. By comparing the one-dimensional integral ("Dunnett") and HGM in Table 1.1, we see that HGM has more than 8 digit accuracy. The numerical accuracy of HGM is very good, which is an advantage of HGM.

Concerning the computational time, compared to the Miwa method in Miwa et al. [19], HGM becomes faster as the dimensionality grows (see Table 1.2).

Next, we consider the probability of a ball of radius r

$$S = B_r = \{(x_1, \ldots, x_d) \mid x_1^2 + \cdots + x_d^2 \le r^2\},$$

which was studied in Koyama and Takemura [15].

$$P(B_r) = \int_{B_r} \frac{1}{(2\pi)^{d/2}|\boldsymbol{\Sigma}|^{1/2}} \exp\left(-\frac{1}{2}(\mathbf{x} - \boldsymbol{\mu})'\boldsymbol{\Sigma}^{-1}(\mathbf{x} - \boldsymbol{\mu})\right) d\mathbf{x}. \quad (1.14)$$

For this problem, we can use the result for the Fisher–Bingham integral on the sphere of radius r

$$S^{d-1}(r) = \{(x_1, \dots, x_d) \mid x_1^2 + \cdots + x_d^2 = r^2\}.$$

The Fisher–Bingham integral on $S^{d-1}(r)$ is defined by

$$f(\lambda, \tau, r) = \int_{S^{d-1}(r)} \exp\left(\sum_{i=1}^{d} \lambda_i t_i^2 + \sum_{i=1}^{d} \tau_i t_i\right) dt,$$

where dt is the volume element on $S^{d-1}(r)$ with

$$\int_{S^{d-1}(r)} dt = r^{d-1} S_{d-1}, \quad S_{d-1} = \mathrm{Vol}(S^{d-1}(1)) = \frac{2\pi^{d/2}}{\Gamma(d/2)}.$$

The integral in (1.14) for the infinitesimal interval

$$r < (x_1^2 + \cdots + x_d^2)^{1/2} < r + dr$$

is the Fisher–Bingham integral. In statistical interpretation, the conditional distribution of $\mathbf{X}$ given $\|\mathbf{X}\| = r$ is the Fisher–Bingham distribution.

By rotation in (1.14), we can assume that $\mathbf{\Sigma}$ is a diagonal matrix, that is $\mathbf{\Sigma} = \mathrm{diag}\,(\sigma_{11}, \dots, \sigma_{dd})$ without loss of generality. Hence $\|\mathbf{X}\|^2 = X_1^2 + \cdots + X_d^2$ is the weighted sum of independent noncentral chi-square random variables. The weights are σ_{ii} and the noncentralities are μ_i^2/σ_{ii}, $i \in \{1, \dots, d\}$.

Let

$$\mathbf{\Sigma} = \mathrm{diag}(\sigma_1^2, \dots, \sigma_d^2), \qquad \boldsymbol{\mu} = (\mu_1, \dots, \mu_d)'.$$

By change of parameters

$$\lambda_i = -\frac{1}{2\sigma_i^2}, \qquad \tau_i = \frac{\mu_i}{\sigma_i^2},$$

the ball probability is written as

$$P(B_r) = \frac{\prod_{i=1}^{d} \sqrt{-\lambda_i}}{\pi^{d/2}} \exp\left(\frac{1}{4} \sum_{i=1}^{d} \frac{\tau_i^2}{\lambda_i}\right) \int_0^r f(\lambda, \tau, s) ds,$$

where $f(\lambda, \tau, s)$ is the Fisher–Bingham integral. Let

$$\boldsymbol{f} = \left(\frac{\partial f}{\partial \tau_1}, \dots, \frac{\partial f}{\partial \tau_d}, \frac{\partial f}{\partial \lambda_1}, \dots, \frac{\partial f}{\partial \lambda_d}\right)'.$$

Then $\boldsymbol{f}$ satisfies the Pfaffian for differentiation by r

$$\partial_r \boldsymbol{f} = \mathbf{P}_r \boldsymbol{f}.$$

The $2d \times 2d$ matrix $\mathbf{P}_r$ can be written as

$$\mathbf{P}_r = \frac{1}{r} \begin{pmatrix} 2r^2\lambda_1 + 1 & \mathbf{0} & \tau_1 & \cdots & \tau_1 \\ & \ddots & \vdots & & \vdots \\ \mathbf{0} & 2r^2\lambda_d + 1 & \tau_d & \cdots & \tau_d \\ r^2\tau_1 & \mathbf{0} & 2r^2\lambda_1 + 2 & & \mathbf{1} \\ & \ddots & & \ddots & \\ \mathbf{0} & r^2\tau_d & \mathbf{1} & & 2r^2\lambda_d + 2 \end{pmatrix},$$

where $\mathbf{0}$ denotes a triangular off-diagonal block of 0's and $\mathbf{1}$ denotes a triangular off-diagonal block of 1's. Note that

$$f(\lambda, \tau, r) = r^2 \times \left(\frac{\partial f}{\partial \lambda_1} + \cdots + \frac{\partial f}{\partial \lambda_d} \right),$$

because $t_1^2 + \cdots + t_d^2 = r^2$ on $S^{d-1}(r)$. Hence we do not need $f(\lambda, \tau, r)$ as an element of $\boldsymbol{f}$.

We can check the accuracy of HGM against the asymptotic Laplace approximation of $f(\lambda, \tau, r)$. Suppose $\sigma_1 > \sigma_2 \geq \cdots \geq \sigma_d$, i.e., $\lambda_1 > \lambda_2 \geq \cdots \geq \lambda_d$. Then as $r \to \infty$

$$\frac{f(\lambda, \tau, r)}{\tilde{f}(\lambda, \tau, r)} \to 1,$$

where

$$\tilde{f}(\lambda, \tau, r) = (e^{r\tau_1} + e^{-r\tau_1}) \exp\left(r^2\lambda_1 - \sum_{i=2}^{d} \frac{\tau_i^2}{4(\lambda_i - \lambda_1)} \right) \pi^{(d-1)/2} \frac{1}{\prod_{i=2}^{d} (\lambda_1 - \lambda_i)^{1/2}}.$$

Asymptotic Laplace approximations of derivatives of $f(\lambda, \tau, r)$ appearing in $\boldsymbol{f}$ are also given in Koyama and Takemura [15].

We show a numerical example. Let $d = 3$ and

$$\sigma_1 = 3.0, \qquad \sigma_2 = 2.0, \qquad \sigma_3 = 1.0,$$
$$\mu_1 = 0.01, \qquad \mu_2 = 0.02, \qquad \mu_3 = 0.03,$$

i.e.,

$$\lambda_1 = -0.0555556, \qquad \lambda_2 = -0.125, \qquad \lambda_3 = -0.5,$$
$$\tau_1 = 0.00111111, \qquad \tau_2 = 0.005, \qquad \tau_3 = 0.03.$$

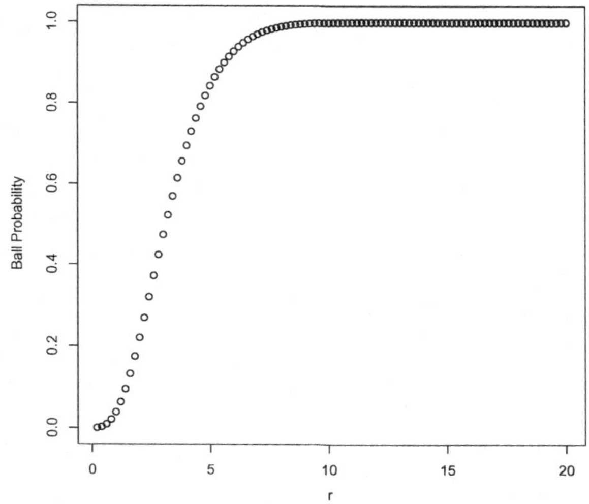

Fig. 1.2 CDF of $\|\mathbf{X}\|^2$

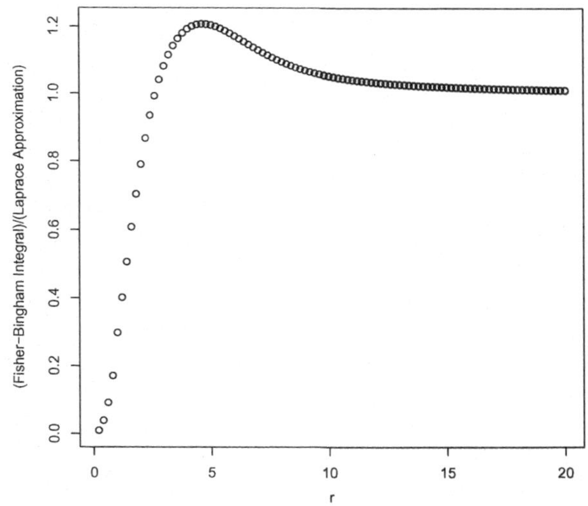

Fig. 1.3 Ratio of FB and Laplace approximation

As expected, the CDF converges to 1 as r increases; see Fig. 1.2. Similarly, the ratio of the Fisher–Bingham integral to its Laplace approximation converges to 1 for $r \to \infty$; see Fig. 1.3. This convergence of the ratio to 1 demonstrates the numerical accuracy of HGM, which is an advantage of HGM.

1.6 Application to Problems in Wireless Communication

In the area of wireless communication, complex normal distribution and complex Wishart distribution (e.g., James [7, Sect. 8]) are used for statistical performance evaluation of communication protocols. In particular for multiple-input multiple-output (MIMOs), the noncentral complex Wishart distribution plays an essential role. In collaboration with Constantin Siriteanu, we obtained some new distributional results. In the complex case, the Schur polynomial takes the place of the zonal polynomial (which is easier to handle). Moreover, the hypergeometric function becomes easier. However, as in the real case, the series expansion in terms of the Schur polynomials does not work numerically when the argument becomes large.

Siriteanu et al. [30] applied HGM to the zero-forcing protocol of MIMO by deriving the Pfaffian system by hand. In the subsequent work (Siriteanu et al. [31]), the derivation of the Pfaffian system was too hard for hand calculation, and we used the software developed by Koutschan [10–12]. The resulting Pfaffian looks complicated but we confirmed that HGM works well both in accuracy and speed. A disadvantage of HGM is that the derivation of the Pfaffian is often heavy and it is sometimes necessary to use algebraic software.

1.7 Some Other Applications and Developments of HGM

In our first paper on HGM (Nakayama et al. [22]), we applied it to directional statistics and there were more developments of HGM for directional statistics. Koyama et al. [16, 17], Koyama [13] improved the results in Nakayama et al. [22] and showed that HGM works reliably in high dimensions. Sei and Kume [28] considered the Bingham distribution, which is a special case of the Fisher–Bingham distribution and gave the explicit form of the Pfaffian system needed for HGM, including the case of multiple roots. In Sei et al. [29], HGM for maximum likelihood estimation of the Fisher distribution on the real orthogonal group was studied.

Marumo et al. [18] discussed the differential equations satisfied by the characteristic function and the density function of the sum of cubes of n i.i.d. standard normal random variables and applied HGM to the density function. Hayakawa and Takemura [5] proposed an exponential family of distributions extending the original exercise problem in (1.3) and discussed maximum likelihood estimation by HGM. Takayama et al. [33] studied properties of the normalizing constant and the maximum likelihood estimation of generalized hypergeometric distribution over two-way contingency tables from the HGM perspective.

HGM is the common meeting ground of statistics, D-module theory and numerical integration. Hence, researchers with various backgrounds can study and improve HGM. Difficulties of HGM, e.g., the singularity of the differential equations and the selection of the path of numerical integration, remain to be solved.

Acknowledgements We thank two referees for very detailed and useful comments. This work was supported by JSPS KAKENHI Grant Number 18H04092.

References

1. Constantine, A.G.: Some non-central distribution problems in multivariate analysis. Ann. Math. Stat. **34**, 1270–1285 (1963)
2. Dunnett, C.W., Sobel, M.: Approximations to the probability integral and certain percentage points of a multivariate analogue of Student's t-distribution. Biometrika **42**, 258–260 (1955)
3. Hashiguchi, H., Numata, Y., Takayama, N., Takemura, A.: The holonomic gradient method for the distribution function of the largest root of a Wishart matrix. J. Multivariate Anal. **117**, 296–312 (2013)
4. Hashiguchi, H., Takayama, N., Takemura, A.: Distribution of the ratio of two Wishart matrices and cumulative probability evaluation by the holonomic gradient method. J. Multivariate Anal. **165**, 270–278 (2018)
5. Hayakawa, J., Takemura, A.: Estimation of exponential-polynomial distribution by holonomic gradient descent. Comm. Stat. Theory Methods **45**, 6860–6882 (2016)
6. Hibi, T.: Gröbner Bases—Statistica and Software Systems. Springer, Japan, Tokyo (2013)
7. James, A.T.: Distributions of matrix variates and latent roots derived from normal samples. Ann. Math. Stat. **35**, 475–501 (1964)
8. Koev, P., Edelman, A.: The efficient evaluation of the hypergeometric function of a matrix argument. Math. Comp. **75**, 833–846 (2006)
9. Kondo, T.: On a holonomic system of partial differential equations satisfied by the matrix ${}_1F_1$ (in Japanese). Kobe University (2013)
10. Koutschan, C.: Holonomic Functions package for Mathematica (2017) http://www.risc.jku.at/research/combinat/software
11. Koutschan, C.: Advanced applications of the holonomic systems approach. PhD thesis, Research Institute for Symbolic Computation (RISC), Johannes Kepler University, Linz, Austria (2009)
12. Koutschan, C.: HolonomicFunctions (user's guide). Technical Report 10-01, RISC Report Series, Johannes Kepler University, Linz, Austria (2010)
13. Koyama, T.: A holonomic ideal which annihilates the Fisher-Bingham integral. Funkcial. Ekvac. **56**, 51–61 (2013)
14. Koyama, T., Takemura, A.: Calculation of orthant probabilities by the holonomic gradient method. Jpn. J. Ind. Appl. Math. **32**, 187–204 (2015)
15. Koyama, T., Takemura, A.: Holonomic gradient method for distribution function of a weighted sum of noncentral chi-square random variables. Comput. Stat. **31**, 1645–1659 (2016)
16. Koyama, T., Nakayama, H., Nishiyama, K., Takayama, N.: The holonomic rank of the Fisher-Bingham system of differential equations. J. Pure Appl. Algebra **218**, 2060–2071 (2014)
17. Koyama, T., Nakayama, H., Nishiyama, K., Takayama, N.: Holonomic gradient descent for the Fisher-Bingham distribution on the d-dimensional sphere. Comput. Stat. **29**, 661–683 (2014)
18. Marumo, N., Oaku, T., Takemura, A.: Properties of powers of functions satisfying second-order linear differential equations with applications to statistics. Jpn. J. Ind. Appl. Math. **32**, 553–572 (2015)
19. Miwa, T., Hayter, A.J., Kuriki, S.: The evaluation of general non-centred orthant probabilities. J. R. Stat. Soc. Ser. B. Stat. Methodol. **65**, 223–234 (2003)
20. Muirhead, R.J.: Systems of partial differential equations for hypergeometric functions of matrix argument. Ann. Math. Stat. **41**, 991–1001 (1970) ISSN 0003-4851
21. Muirhead, R.J.: Aspects of Multivariate Statistical Theory. John Wiley & Sons Inc., New York (1982)
22. Nakayama, H., Nishiyama, K., Noro, M., Ohara, K., Sei, T., Takayama, N., Takemura, A.: Holonomic gradient descent and its application to the Fisher-Bingham integral. Adv. Appl. Math. **47**, 639–658 (2011)
23. Noro, M.: System of partial differential equations for the hypergeometric function ${}_1F_1$ of a matrix argument on diagonal regions. In: ISSAC'16 Proceedings of the ACM on International Symposium on Symbolic and Algebraic Computation, pp. 381–388 (2016)

24. Oaku, T.: Algorithms for b-functions, restrictions, and algebraic local cohomology groups of D-modules. Adv. Appl. Math. **19**, 61–105 (1997)
25. Oaku, T.: D-modules and Computational Mathematics (in Japanese). Asakura Shoten (2002)
26. Oaku, T.: Algorithms for integrals of holonomic functions over domains defined by polynomial inequalities. J. Symbolic Comput. **50**, 1–27 (2013)
27. Saito, M., Sturmfels, B., Takayama, N.: Gröbner Deformations of Hypergeometric Differential Equations. Springer-Verlag, Berlin (2000)
28. Sei, T., Kume, A.: Calculating the normalising constant of the Bingham distribution on the sphere using the holonomic gradient method. Stat. Comput. **25**, 321–332 (2013)
29. Sei, T., Shibata, H., Takemura, A., Ohara, K., Takayama, N.: Properties and applications of Fisher distribution on the rotation group. J. Multivariate Anal. **116**, 440–455 (2013)
30. Siriteanu, C., Blostein, S.D., Takemura, A., Shin, H., Yousefi, S., Kuriki, S.: Exact MIMO zero-forcing detection analysis for transmit-correlated Rician fading. IEEE Trans. Wirel. Commun. **13**, 1514–1527 (2014)
31. Siriteanu, C., Takemura, A., Koutschan, C., Kurikia, S., Richards, D.S.P., Shin, H.: Exact ZF analysis and computer-algebra-aided evaluation in rank-1 LoS Rician fading. IEEE Trans. Wirel. Commun. **15**, 5245–5259 (2016)
32. Takayama, N.: List of papers related to HGM; http://www.math.kobe-u.ac.jp/OpenXM/Math/hgm/ref-hgm.html
33. Takayama, N., Kuriki, S., Takemura, A.: A-hypergeometric distributions and Newton polytopes. Adv. Appl. Math. **77**, 441–436 (2018)
34. Takemura, A.: Zonal Polynomials. Institute of Mathematical Statistics, Hayward, CA (1984)
35. Zeilberger, D.: A holonomic systems approach to special functions identities. J. Comput. Appl. Math. **32**, 321–368 (1990)

Chapter 2
From Normality to Skewed Multivariate Distributions: A Personal View

Tõnu Kollo

Abstract An overview of the development of multivariate distributions' theory is given starting with T. W. Anderson's monograph [1]. In the book, multivariate analysis methods were presented in the compact matrix language for the first time. The methods were developed for normally distributed populations. Ten years later, the class of elliptically contoured distributions (from now on, elliptical distributions) was described by Kelker [22]. Elliptical distributions formed a new class of symmetric multivariate distributions which included the normal distribution. To model skewed data, the theory of multivariate Edgeworth type expansions was developed in the 1980s. Ten years later, A. Azzalini and his colleagues introduced multivariate skew elliptical distributions by transforming symmetric elliptical distributions with an additional parameter vector. In the 1990s, copula theory became important in modeling multivariate data. With copulas, it is possible to build multivariate distributions with a given set of marginals and a stipulated correlation structure. These models were intensively developed from the second half of the 1980s through the 1990s. Special attention has been paid by several authors to the Gaussian and t copulas, as well as to skew normal and skew t copulas.

2.1 Introduction

In Anderson [1], multivariate analysis for a normal population is presented in a compact and elegant way. Main results from the literature were collected, combined with his own contributions and presented in matrix form. Together with C. R. Rao's book on statistical inference (Rao [45]), these two monographs became a set at hand for all people who worked on data analysis and developed multivariate methods. Let us now introduce some basic notation and notions.

T. Kollo (✉)
Institute of Mathematics and Statistics, University of Tartu, 50090 Tartu, Estonia
e-mail: Tonu.Kollo@ut.ee

K. Filipiak et al. (eds.), *Multivariate, Multilinear and Mixed Linear Models*, Contributions to Statistics, https://doi.org/10.1007/978-3-030-75494-5_2

When treating multivariate distributions, we shall use notions such as the vec-*operator, commutation matrix, Kronecker product* and *matrix derivative.*
For deeper insight into this technique, we refer to Magnus and Neudecker [36] or Kollo and von Rosen [29].

If $\mathbf{A}$ is a $p \times n$ matrix, $\mathbf{A} = (\mathbf{a}_1, \ldots, \mathbf{a}_n)$, the $pn \times 1$ vector $\mathrm{vec}\,\mathbf{A}$ is formed by stacking the columns $\mathbf{a}_i$ under each other; that is

$$\mathrm{vec}\,\mathbf{A} = \begin{pmatrix} \mathbf{a}_1 \\ \vdots \\ \mathbf{a}_n \end{pmatrix}.$$

Commutation matrix $\mathbf{K}_{p,n}$ is an orthogonal $pn \times pn$ partitioned matrix consisting of $n \times p$ blocks with the property

$$\mathbf{K}_{p,n}\mathrm{vec}\,\mathbf{A} = \mathrm{vec}\,\mathbf{A}',$$

where $\mathbf{A}$ is a $p \times n$-matrix.

The Kronecker product $\mathbf{A} \otimes \mathbf{B}$ of a $p \times q$ matrix $\mathbf{A}$ and an $m \times n$ matrix $\mathbf{B}$ is a $pm \times qn$ partitioned matrix consisting of the $m \times n$ blocks

$$\mathbf{A} \otimes \mathbf{B} = [a_{ij}\mathbf{B}], \quad i \in \{1, \ldots, p\}, \; j \in \{1, \ldots, q\}.$$

The key notion in deriving asymptotic expansions is the matrix derivative. The expressions in Sect. 2.1 and later on are obtained by using the derivative defined in Definition 2.1.

Definition 2.1 Let the elements of $\mathbf{Y} \in \mathbb{R}^{r\times s}$ be functions of the elements of $\mathbf{X} \in \mathbb{R}^{p\times q}$. The matrix $\dfrac{d\mathbf{Y}}{d\mathbf{X}} \in \mathbb{R}^{pq\times rs}$ is called matrix derivative of $\mathbf{Y}$ by $\mathbf{X}$ in a set A, if the partial derivatives $\frac{dy_{kl}}{dx_{ij}}$ exist, are continuous in A and

$$\frac{d\mathbf{Y}}{d\mathbf{X}} = \frac{d}{d\mathrm{vec}\,\mathbf{X}}\mathrm{vec}'\mathbf{Y},$$

where

$$\frac{d}{d\mathrm{vec}\,\mathbf{X}} = \left(\frac{\partial}{\partial x_{11}}, \ldots, \frac{\partial}{\partial x_{p1}}, \frac{\partial}{\partial x_{12}}, \ldots, \frac{\partial}{\partial x_{p2}}, \ldots, \frac{\partial}{\partial x_{1q}}, \ldots, \frac{\partial}{\partial x_{pq}}\right)'.$$

The higher order derivatives are defined iteratively

$$\frac{d^k\mathbf{Y}}{d\mathbf{X}^k} = \frac{d}{d\mathbf{X}}\left(\frac{d^{k-1}\mathbf{Y}}{d\mathbf{X}^{k-1}}\right), \quad k \in \{2, 3, \ldots\}.$$

Let $\mathbf{x}$ be a random p-vector. The characteristic function of $\mathbf{x}$ is defined as an expectation

$$\varphi_{\mathbf{x}}(\mathbf{t}) = \mathrm{E}\left(e^{\mathrm{i}\mathbf{t}'\mathbf{x}}\right).$$

The cumulant function $\psi_{\mathbf{x}}(\mathbf{t})$ is defined as

$$\psi_{\mathbf{x}}(\mathbf{t}) = \ln \varphi_{\mathbf{x}}(\mathbf{t}).$$

When the absolute mixed moments $\mathrm{E}\left(\left|X_1^{k_1} X_2^{k_2} \times \cdots \times X_p^{k_p}\right|\right)$ of order k exist, where $k = \sum_{i=1}^{p} k_i$, the characteristic function is k-times differentiable and the moments of $\mathbf{x}$ up to the order k we get by differentiation

$$m_k(\mathbf{x}) = \frac{1}{\mathrm{i}^k} \frac{d^k \varphi_{\mathbf{x}}(\mathbf{t})}{d\mathbf{t}^k}\bigg|_{\mathbf{t}=\mathbf{0}}.$$

We obtain the cumulants $c_k(\mathbf{x})$ by differentiating the cumulant function

$$c_k(\mathbf{x}) = \frac{1}{\mathrm{i}^k} \frac{d^k \psi_{\mathbf{x}}(\mathbf{t})}{d\mathbf{t}^k}\bigg|_{\mathbf{t}=\mathbf{0}}.$$

The first three moments have the following representation:

$$\begin{aligned}
m_1(\mathbf{x}) &= \mathrm{E}\,(\mathbf{x}), \\
m_2(\mathbf{x}) &= \mathrm{E}\left(\mathbf{x} \otimes \mathbf{x}'\right), \\
m_3(\mathbf{x}) &= \mathrm{E}\left(\mathbf{x} \otimes \mathbf{x}' \otimes \mathbf{x}\right).
\end{aligned}$$

The corresponding central moments are

$$\begin{aligned}
\overline{m}_2(\mathbf{x}) &= \mathrm{E}\left[(\mathbf{x} - \mathrm{E}\,(\mathbf{x})) \otimes (\mathbf{x} - \mathrm{E}\,(\mathbf{x}))'\right] = \mathrm{D}\,(\mathbf{x}) = c_2(\mathbf{x}), \\
\overline{m}_3(\mathbf{x}) &= \mathrm{E}\left[(\mathbf{x} - \mathrm{E}\,(\mathbf{x})) \otimes (\mathbf{x} - \mathrm{E}\,(\mathbf{x}))' \otimes (\mathbf{x} - \mathrm{E}\,(\mathbf{x}))\right] = c_3(\mathbf{x}).
\end{aligned}$$

Let us consider p independent standard normal variables X_i that form a p-vector with the marginal density functions

$$f_{X_i}(x_i) = \frac{1}{\sqrt{2\pi}} \exp\left(-\frac{x_i^2}{2}\right), \quad i \in \{1, \ldots, p\}.$$

Then, if also all possible linear combinations of the X_i have a univariate normal distribution, $\mathbf{x}$ has a standard p-variate normal distribution, $\mathbf{x} \sim N_p(\mathbf{0}, \mathbf{I}_p)$, where $\mathbf{I}_p$ denotes the $p \times p$ identity matrix. Let

$$\mathbf{y} = \boldsymbol{\mu} + \mathbf{A}\mathbf{x}, \quad \mathbf{A} : p \times p,$$

then $\mathbf{y} \sim N_p(\boldsymbol{\mu}, \boldsymbol{\Sigma})$ with $\boldsymbol{\Sigma} = \mathbf{A}\mathbf{A}'$.

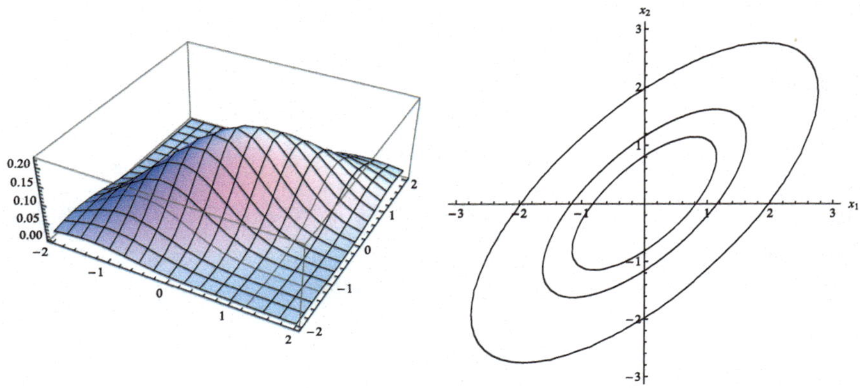

Fig. 2.1 The density function and contour plot of $N_2(0, \mathbf{\Sigma})$, $\mathbf{\Sigma} = \begin{pmatrix} 1 & 0.7 \\ 0.7 & 1 \end{pmatrix}$

If the determinant $|\mathbf{A}| \neq 0$, then $\mathbf{y}$ has the density function

$$f_{\mathbf{y}}(\mathbf{y}) = \frac{1}{\sqrt{2\pi}^p} |\mathbf{\Sigma}|^{-\frac{1}{2}} \exp\left[-\frac{(\mathbf{y} - \boldsymbol{\mu})' \mathbf{\Sigma}^{-1} (\mathbf{y} - \boldsymbol{\mu})}{2} \right].$$

The density of the multivariate normal distribution is axial symmetric where the matrix $\mathbf{\Sigma}$ determines its shape. In the bivariate case, this is illustrated in Fig. 2.1.

In fact, much of the further development of the multivariate distributions' theory is related to the normal distribution.

Unfortunately, real data often violate the assumption about symmetricity. In Fig. 2.2, data from the Institute of Geography, University of Tartu, presents measurements of concentration of nitrogen on x-axis and soluble phosphorus on y-axis in soil. The obtained empirical distribution is extremely skewed.

While in the univariate case several methods for transforming data to the normal law exist, in multivariate situation things are not that simple. The class of elliptical distributions is the first generalization of the normal distribution and includes the class of the normal distributions. In Sect. 2.2, a short outline of elliptical distributions is given. For modeling skewed data, the theory of Edgeworth type expansions was developed which initially was based on transforming the normal distribution. Section 2.3 gives an overview of how to approximate an unknown multivariate density function of interest with a simpler (often the normal) density via Edgeworth type expansions. In Sect. 2.4, skew elliptical distributions are considered with special attention to the skew normal and skew t distribution. Here also multivariate skewness and kurtosis characteristics are considered. Section 2.5 presents a short summary on the asymmetric Laplace distribution, and in Sect. 2.6 the copula theory is considered with special attention to the Gaussian and t copulas, the skew normal and skew t copulas. The property of tail dependence is also discussed here. Finally, we summarize the paper with a short conclusion and present a list of references.

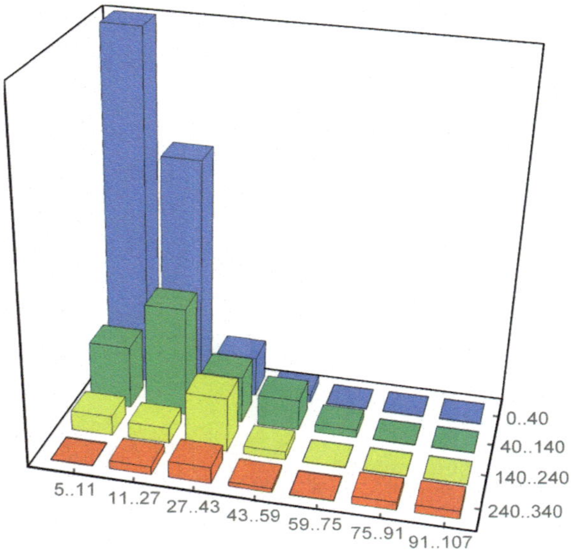

Fig. 2.2 Empirical distribution of concentration of phosphorus and nitrogen in soil

2.2 Elliptical Distributions

As pointed out in Fang et al. [13], the topic can be tracked back to Maxwell [39] but modern study of elliptical distributions starts from 1960s with the description of the family of elliptical distributions in Kelker [22]. The first books on the topic were written by Kariya and Sinha [21], Fang and Zhang [14] and Fang et al. [13]. Also in Muirhead [42], an overview is given and in Anderson [3] the methods of multivariate analysis are applied to elliptical distributions. In the class of elliptical distributions, a *spherical distribution* has the same role as the standard normal distribution $N_p(\mathbf{0}, \mathbf{I}_p)$ in the family of multivariate normal distributions $N_p(\boldsymbol{\mu}, \boldsymbol{\Sigma})$.

Definition 2.2 A random p-vector $\mathbf{x}$ is spherically distributed if $\mathbf{x}$ and $\boldsymbol{\Gamma}'\mathbf{x}$ have the same distribution for all orthogonal matrices $\boldsymbol{\Gamma} : p \times p$.

The following necessary and sufficient conditions hold.
A random vector $\mathbf{x}$ is spherically distributed iff

$$\varphi_{\mathbf{x}}\left(\boldsymbol{\Gamma}'\mathbf{t}\right) = \varphi_{\mathbf{x}}(\mathbf{t})$$

or

$$\varphi_{\mathbf{x}}(\mathbf{t}) = \phi(\mathbf{t}'\mathbf{t}),$$

where $\varphi_{\mathbf{x}}(\mathbf{t})$ is its characteristic function and $\phi(\cdot)$ is a scalar function. Table 2.1 due to Jensen [20], given in Fang et al. [13], presents a list of widely used classes of continuous spherical distributions where c denotes the normalizing constant.

Table 2.1 Some classes of p-dimensional continuous spherical distributions

Distribution	Density $f(\mathbf{x})$ or characteristic function $\varphi(\mathbf{t})$
Kotz type	$f(\mathbf{x}) = c(\mathbf{x}'\mathbf{x})^{N-1} \exp\left[-r(\mathbf{x}'\mathbf{x})^s\right],\ r, s > 0,\ 2N + p > 2$
Standard Normal	$f(\mathbf{x}) = c \exp\left(-\frac{1}{2}\mathbf{x}'\mathbf{x}\right)$
Pearson VII type	$f(\mathbf{x}) = c\left(1 + \frac{\mathbf{x}'\mathbf{x}}{s}\right)^{-N},\ s > 0,\ N > \frac{p}{2}$
t-distribution	$f(\mathbf{x}) = c\left(1 + \frac{\mathbf{x}'\mathbf{x}}{s}\right)^{\frac{p+m}{2}},\ s > 0,\ m \in \mathbb{N}$
Cauchy	$f(\mathbf{x}) = c\left(1 + \frac{\mathbf{x}'\mathbf{x}}{s}\right)^{\frac{p+1}{2}}, s > 0,$
Pearson II type	$f(\mathbf{x}) = c\left(1 - \mathbf{x}'\mathbf{x}\right)^m,\ m > 0$
Logistic	$f(\mathbf{x}) = c\frac{\exp(-\mathbf{x}'\mathbf{x})}{(1+\exp[-\mathbf{x}'\mathbf{x})]^2}$
Bessel	$f(\mathbf{x}) = c\left(\frac{\lVert\mathbf{x}\rVert}{\beta}\right)^{\alpha} K_{\alpha}\left(\frac{\lVert\mathbf{x}\rVert}{\beta}\right),\quad \alpha > -\frac{p}{2},\quad \beta > 0,$
	K_α is a modified Bessel function of the 3rd kind
Scale mixture	$f(\mathbf{x}) = c\int_0^\infty t^{-\frac{p}{2}} \exp\left(\frac{\mathbf{x}'\mathbf{x}}{2t}\right) dG(t)$
Stable laws	$\varphi(\mathbf{t}) = \exp\left[(\mathbf{t}'\mathbf{t})^{\frac{\alpha}{2}}\right] \quad 0 < \alpha \leq 2, \quad r < 0$
Multiuniform	$\varphi(\mathbf{t}) = {}_0F_1\left(\frac{p}{2}; -\frac{1}{4} \parallel \mathbf{t} \parallel^2\right),\quad {}_0F_1(\cdot)$ is a hypergeometric function

For the modified Bessel function of the 3rd kind in the density of the Bessel distribution, we refer to Kotz et al. [31, p. 315], and ${}_0F_1(\cdot)$ is a special case of the generalized hypergeometric function in the expression of the characteristic function of the multiuniform distribution (Muirhead [42, p. 20–21], for instance).

Definition 2.3 A p-vector $\mathbf{x}$ is elliptically distributed with location parameter $\boldsymbol{\mu}$: $p \times 1$ and scale parameter $\mathbf{V}$: $p \times p$ if

$$\mathbf{x} = \boldsymbol{\mu} + \mathbf{A}\mathbf{y},$$

where $\mathbf{y}$ is spherically distributed and $\mathbf{A} : p \times k$, $\mathbf{A}\mathbf{A}' = \mathbf{V}$ with rank $\mathrm{r}(\mathbf{A}) = k$.

Then

$$\varphi_{\mathbf{x}}(\mathbf{t}) = \exp(i\mathbf{t}'\boldsymbol{\mu})\phi(\mathbf{t}'\mathbf{V}\mathbf{t}).$$

When $\mathbf{V} : p \times p$ is of full rank and $\mathbf{x}$ is a continuous random vector, the density is

$$f_{\mathbf{x}}(\mathbf{x}) = c_p|\mathbf{V}|^{-1/2} g\left[(\mathbf{x} - \boldsymbol{\mu})'\mathbf{V}^{-1}(\mathbf{x} - \boldsymbol{\mu})\right]$$

for some non-negative function $g(\cdot)$.

This generalization seems broad but some properties of elliptical distributions are unexpectedly restrictive. For instance, the second and fourth moments and cumulants differ from the corresponding values of the normal distribution by a constant which depends only on the function $\phi(\cdot)$. So, any mixed fourth-order cumulant of the coordinates of $\mathbf{x}$ can be presented in the form

$$c_4(X_i X_j X_l X_m) = \kappa c_4(Y_i Y_j Y_l Y_m),$$

where $\mathbf{y} = (Y_1 \ldots, Y_p)' \sim N_p(\boldsymbol{\mu}, \boldsymbol{\Sigma})$ and

$$\kappa = \frac{\phi''(0) - (\phi'(0))^2}{(\phi'(0))^2}.$$

Elliptical distributions were successfully used in robustness studies, and they introduced the possibility to use other than normal error distributions in multivariate models. Nevertheless, they all are axisymmetrical and hence do not take into account skewness. One can ask how to transform a symmetric multivariate distribution into a non-symmetrical version. This task was solved by introducing the multivariate Edgeworth type expansions.

2.3 Multivariate Edgeworth Type Expansions

Here, we address a question regarding the method of transforming a symmetric continuous elliptical distribution into a skewed distribution. The idea is to present a complicated skewed multivariate density or distribution function of interest as a series expansion through a simpler multivariate distribution. In the univariate case, this idea has been realized by Edgeworth expansions and Gram-Charlier series. For an overview, the interested reader is referred to Cramér [11], for instance.

Let $\mathbf{x}$ and $\mathbf{y}$ be two p-vectors with densities $f_{\mathbf{x}}(\mathbf{x})$ and $f_{\mathbf{y}}(\mathbf{x})$. Our aim is to express a complicated density $f_{\mathbf{y}}(\mathbf{x})$ through a simpler one, $f_{\mathbf{x}}(\mathbf{x})$.

In the univariate case, Cornish and Fisher [8] gave a principal solution and applied this when $X \sim N(0, 1)$. The density $f_Y(x)$ was expressed as a series expansion

$$f_Y(x) = f_X(x) + a_1 {f_X}^{(1)}(x) + a_2 {f_X}^{(2)}(x) + a_3 {f_X}^{(3)}(x) + \ldots,$$

where constants a_k are expressions from differences of products of the corresponding cumulants of Y and X up to the order k so that sum of the orders of the products of the cumulants equals k and ${f_X}^{(k)}(x)$ is the kth derivative of $f_X(x)$. When $X \sim N(0, 1)$, the series is called Edgeworth expansion.

For $X \sim N(0, 1)$, the derivatives of the density are expressed through the Hermite polynomials $h_k(x)$

$$\frac{d^k f_{N(0,1)}(x)}{dx^k} = (-1)^k f_{N(0,1)}(x) h_k(x).$$

Then the density of Y can be given as

$$f_Y(x) = f_{N(0,1)}(x)\,[1 - a_1 h_1(x) + a_2 h_2(x) - a_3 h_3(x) + \ldots].$$

If the terms in the expansion will be in diminishing order and we would like to use the few first terms as an approximation of the density function of Y, a complication arises. As the terms are polynomials, the expression on the right-hand side can have negative values in the tail area. Even if the shape of $f_Y(x)$ is approximated quite well, we cannot trust the behavior of the approximation on the tail. This makes Edgeworth expansions not suitable for applications where the approximation of quantiles on the far tail is of special interest. We refer to this construction as Edgeworth type expansion when X has some other than normal distribution, both in univariate and multivariate cases. Use of a skewed distribution in the role of X seems natural while approximating a skewed distribution. This has been done in many applications. For example, Gerber [17] approximated claim size distribution through a Γ-distribution, Hall [18] approximated a sum of skewed random variables via a chi-square distribution, etc.

Finney [15] was the first to consider the relation between densities of random vectors $\mathbf{x}$ and $\mathbf{y}$. When $\mathbf{x} \sim N_p(\mathbf{0}, \boldsymbol{\Sigma})$, we get a multivariate Edgeworth type expansion in matrix notation (Traat [50]).

Theorem 2.1 *Let* $\mathbf{x}$ *and* $\mathbf{y}$ *be p-variate continuous random vectors with* $\mathbf{x} \sim N_p(\mathbf{0}, \boldsymbol{\Sigma})$. *Then*

$$\begin{aligned} f_{\mathbf{y}}(\mathbf{x}) = f_{\mathbf{x}}(\mathbf{x})\Big\{&1 + (\mathrm{E}\,(\mathbf{y}))' H_1(\mathbf{x}, \boldsymbol{\Sigma}) + \\ &+ \frac{1}{2}\mathrm{vec}'\left[\mathrm{D}\,(\mathbf{y}) - \mathrm{D}\,(\mathbf{x}) + \mathrm{E}\,(\mathbf{x})\mathrm{E}\,(\mathbf{y})'\right] \mathrm{vec}\, H_2(\mathbf{x}, \boldsymbol{\Sigma}) \\ &+ \frac{1}{6}\left\{\mathrm{vec}'\left[\overline{m}_3(\mathbf{y}) + 3\mathrm{vec}'(\mathrm{D}\,(\mathbf{y}) - \mathrm{D}\,(\mathbf{x})) \otimes \mathrm{E}\,(\mathbf{y})\right] + (\mathrm{E}\,(\mathbf{y}))'^{\otimes 3}\right\} \mathrm{vec}\, H_3(\mathbf{x}, \boldsymbol{\Sigma}) \\ &+ \ldots\Big\}, \end{aligned}$$

where $H_i(\mathbf{x}, \boldsymbol{\Sigma})$, $i \in \{1, 2, 3\}$, *are the multivariate Hermite polynomials*

$$H_0(\mathbf{x}, \boldsymbol{\Sigma}) = 1, \qquad H_1(\mathbf{x}, \boldsymbol{\Sigma}) = \boldsymbol{\Sigma}^{-1}\mathbf{x}, \qquad H_2(\mathbf{x}, \boldsymbol{\Sigma}) = \boldsymbol{\Sigma}^{-1}\mathbf{x}\mathbf{x}'\boldsymbol{\Sigma}^{-1} - \boldsymbol{\Sigma}^{-1},$$
$$H_3(\mathbf{x}, \boldsymbol{\Sigma}) = (\boldsymbol{\Sigma}^{-1}\mathbf{x})^{\otimes 2}\mathbf{x}'\boldsymbol{\Sigma}^{-1} - \mathrm{vec}\,\boldsymbol{\Sigma}^{-1}\mathbf{x}'\boldsymbol{\Sigma}^{-1} - \boldsymbol{\Sigma}^{-1} \otimes (\boldsymbol{\Sigma}^{-1}\mathbf{x}) - (\boldsymbol{\Sigma}^{-1}\mathbf{x}) \otimes \boldsymbol{\Sigma}^{-1}.$$

Multivariate Hermite polynomials are obtained by differentiating the density function

$$\frac{d^k f_{\mathbf{x}}(\mathbf{x})}{d\mathbf{x}^k} = (-1)^k H_k(\mathbf{x}, \boldsymbol{\Sigma}) f_{\mathbf{x}}(\mathbf{x}), \quad k \in \{0, 1, \ldots\}$$

When can an Edgeworth type expansion be used for density approximation?

Let $\mathbf{y}$ be a p-dimensional statistic with unknown distribution. The sample cumulants of $\mathbf{y}$ depend on the sample size n. Assume that the cumulants $c_i(\mathbf{y})$ depend on n as follows:

$$\begin{aligned}
c_1(\mathbf{y}) &= n^{-\frac{1}{2}}\boldsymbol{\Gamma}_1(\mathbf{y}) + o(n^{-1});\\
c_2(\mathbf{y}) &= \mathbf{K}_2(\mathbf{y}) + n^{-1}\boldsymbol{\Gamma}_2(\mathbf{y}) + o(n^{-1});\\
c_3(\mathbf{y}) &= n^{-\frac{1}{2}}\boldsymbol{\Gamma}_3(\mathbf{y}) + o(n^{-1});\\
c_j(\mathbf{y}) &= o(n^{-1}), \quad j \geq 4,
\end{aligned}$$

where $\mathbf{K}_2(\mathbf{y})$ and $\boldsymbol{\Gamma}_i(\mathbf{y})$ depend on the underlying distribution but not on n. These assumptions are fulfilled in the important practical situation when statistic $\mathbf{y}$ is a function of the sample mean or/and sample covariance matrix. Then the first three terms in the expansion give us an approximation of $f_{\mathbf{y}}(\mathbf{x})$ of order $n^{-\frac{1}{2}}$. This result can be straightforwardly carried over to random matrices after vectorization

$$\mathbf{X} \rightarrow \operatorname{vec}\mathbf{X}.$$

This gave a possibility to approximate an unknown matrix density via matrix normal distribution or a mixture of normal distributions. When approximating the distribution of a square matrix via Wishart distribution, in Kollo and von Rosen [27] generalization of Laguerre polynomials was introduced in matrix form.

All these expansions were derived for the case when both random quantities (variables, vectors and matrices) have the same dimension.

Generalization of the above results is needed for the approximation of smaller dimensional variables through higher dimensional ones. For example, we want to approximate the distribution of trace or determinant or eigenvalues, eigenvectors through the distribution of the matrix whose functions they are. In Kollo and von Rosen [28], a density relation between different dimensional random variables was derived. The result is presented as Corollary 3.3.1.1 in Kollo and von Rosen [29, p. 333], we present it here as Theorem 2.2.

Theorem 2.2 *Let* $\mathbf{y}$ *be a* p*-vector,* $\mathbf{x}$ *an* r*-vector,* $\mathbf{P} : p \times r$ *of rank* $\mathrm{r}(\mathbf{P}) = p$, $\mathbf{A} : r \times r$ *positive definite and* $(\mathbf{P}')^o : r \times (r-p)$ *of rank* $\mathrm{r}((\mathbf{P}')^o) = r - p$*. Then, the density* $f_{\mathbf{y}}$ *can be presented through* $f_{\mathbf{x}}$ *as*

$$\begin{aligned}
f_{\mathbf{y}}(\mathbf{y}_0) = |\mathbf{A}|^{\frac{1}{2}}|\mathbf{P}\mathbf{A}^{-1}\mathbf{P}'|^{\frac{1}{2}}(2\pi)^{\frac{1}{2}(r-p)}\Big\{ & f_{\mathbf{x}}(\mathbf{x}_0) - (\mathbf{M}_0 + \mathbf{M}_1)'\operatorname{vec}\mathbf{f}_{\mathbf{x}}^1(\mathbf{x}_0)\\
& + \tfrac{1}{2}((\mathbf{M}_0 + \mathbf{M}_1)'^{\otimes 2} + \operatorname{vec}'\mathbf{M}_2)\operatorname{vec}\mathbf{f}_{\mathbf{x}}^2(\mathbf{x}_0)\\
& - \tfrac{1}{6}\{\operatorname{vec}'\mathbf{M}_3 + (\mathbf{M}_0 + \mathbf{M}_1)'^{\otimes 3} + \operatorname{vec}'\mathbf{M}_2 \otimes (\mathbf{M}_0 + \mathbf{M}_1)'\\
& \times (\mathbf{I}_{r^3} + \mathbf{I}_r \otimes \mathbf{K}_{r,r} + \mathbf{K}_{r^2,r})\}\operatorname{vec}\mathbf{f}_{\mathbf{x}}^3(\mathbf{x}_0) + \dots \Big\},
\end{aligned}$$

where $\mathbf{f}_{\mathbf{x}}^k(\mathbf{x}_0)$ *is the kth matrix derivative of* $f_{\mathbf{x}}$ *at* $\mathbf{x}_0$,

$$\begin{aligned}
\mathbf{M}_0 =& \mathbf{x}_0 - \mathbf{P}'\mathbf{y}_0,\\
\mathbf{M}_1 =& \mathbf{P}'c_1(\mathbf{y}) - c_1(\mathbf{x}) = \mathbf{P}'\mathrm{E}\,(\mathbf{y}) - \mathrm{E}\,(\mathbf{x}),\\
\mathbf{M}_2 =& \mathbf{P}'c_2(\mathbf{y})\mathbf{P} - c_2(\mathbf{x}) + \mathbf{Q} = \mathbf{P}'\mathrm{D}\,(\mathbf{y})\mathbf{P} - \mathrm{D}\,(\mathbf{x}) + \mathbf{Q},\\
\mathbf{M}_3 =& \mathbf{P}'c_3(\mathbf{y})\mathbf{P}^{\otimes 2} - c_3(\mathbf{x}),\\
\mathbf{Q} =& \mathbf{A}\mathbf{P}'^{o}\left(\mathbf{P}'^{o\prime}\mathbf{A}\mathbf{P}'^{o}\right)^{-1}\mathbf{P}'^{o\prime}\mathbf{A},
\end{aligned}$$

and $\mathbf{P}'^{o}$ is any full rank matrix which spans the orthogonal complement of the column space of $\mathbf{P}'$.

This is a general relation between two density functions which enables to approximate distributions of determinant and eigenvalues, eigenvectors of the sample covariance and correlation matrices through the Wishart distribution or multivariate normal distribution. Most often in the role of $f_{\mathbf{x}}(\mathbf{x})$, we have the density of the normal distribution $N_r(\mathbf{0}, \mathbf{\Sigma})$. Take $\mathbf{x}_0 = \mathbf{P}'\mathbf{y}_0$ which implies that $\mathbf{M}_0 = \mathbf{0}$, and we reach the following expansion (Kollo and von Rosen [29, Th. 3.3.5, p. 339]).

Theorem 2.3 *Let $\mathbf{x} \sim N_r(\mathbf{0}, \mathbf{\Sigma})$, where $\mathbf{\Sigma}$ is positive definite, λ_i, $i \in \{1, \dots, r\}$, be the eigenvalues of $\mathbf{\Sigma}$ and $\mathbf{V}$ the matrix of corresponding eigenvalue-normed eigenvectors. Let $m_2(\mathbf{y})$ be non-singular and $m_2(\mathbf{y})^{\frac{1}{2}}$ any square root of $m_2(\mathbf{y})$. Then*

$$\begin{aligned}
f_{\mathbf{y}}(\mathbf{y}_0) =& |\mathbf{\Sigma}|^{\frac{1}{2}}|m_2(\mathbf{y})|^{-\frac{1}{2}}(2\pi)^{\frac{1}{2}(r-p)} f_{N_r(\mathbf{0},\mathbf{\Sigma})}(\mathbf{x}_0)\\
&\times \left[1 + \mathrm{E}\,(\mathbf{y})'\mathbf{P}\mathbf{H}_1(\mathbf{x}_0, \mathbf{\Sigma}) + \tfrac{1}{6}\mathrm{vec}'\left(\mathbf{P}'c_3(\mathbf{y})\mathbf{P}^{\otimes 2} - 2\mathbf{M}_1^{\otimes 3}\right)\mathrm{vec}\,\mathbf{H}_3(\mathbf{x}_0, \mathbf{\Sigma}) + \dots\right]
\end{aligned}$$

where $\mathbf{x}_0 = \mathbf{P}'\mathbf{y}_0$,

$$\mathbf{P} = \left[m_2(\mathbf{y})\right]^{-\frac{1}{2}}\mathbf{V}'$$

and multivariate Hermite polynomials $\mathbf{H}_i(\mathbf{x}_0, \mathbf{\Sigma})$, $i \in \{1, 3\}$, are defined in (3.1) and (3.3), respectively. If λ_i, $i \in \{1, \dots, p\}$, comprise the p largest eigenvalues of $\mathbf{\Sigma}$, the expansion is optimal in the sense of the norm $\| \mathbf{Q} \| = \sqrt{\mathrm{Tr}\,(\mathbf{Q}\mathbf{Q}')}$.

When approximating unknown distributions of statistics, very seldom we are able to find error bounds for the approximation and one has to rely on simulations to get information about the asymptotic behavior of the statistic of interest. Sometimes, there may arise some unexpected surprises using the above given approximations. When approximating the distribution of the location parameter $\mathbf{B}$ in the Growth Curve model

$$\mathbf{X} = \mathbf{ABC} + \mathbf{E},$$

in Kollo et al. [26] it came out that the first three terms in the Edgeworth type expansion form a density as the mixture of a multivariate normal distribution and a Kotz distribution where the weights are determined by dimensions of the matrices $\mathbf{A}$ and $\mathbf{C}$.

2.4 Skew Elliptical Distributions

Breakthrough in skewed data models was made in the second half of the 90s. Already in 1985, Adelchi Azzalini introduced univariate skew normal distribution (Azzalini [4]), but it took 10 years to carry over the idea to multivariate distributions (Azzalini and Dalla Valle [7]).

Definition 2.4 A continuous random p-vector has p-variate skew normal distribution $SN_p(\boldsymbol{\mu}, \boldsymbol{\Sigma}, \boldsymbol{\alpha})$ when its density function is of the form

$$f_{p,SN}(\mathbf{x}, \boldsymbol{\mu}, \boldsymbol{\Sigma}, \boldsymbol{\alpha}) = 2 f_{N_p(\boldsymbol{\mu},\boldsymbol{\Sigma})}(\mathbf{x}) \Phi(\boldsymbol{\alpha}'(\mathbf{x} - \boldsymbol{\mu})),$$

where $\boldsymbol{\Sigma} : p \times p$ is the positive definite scale parameter, $\boldsymbol{\mu} \in \mathbb{R}^p$ the location parameter, $\boldsymbol{\alpha} \in \mathbb{R}^p$ the shape parameter and $\Phi(\cdot)$ is the distribution function of $N(0, 1)$.

The idea of constructing skewed distributions from elliptical ones attracted many statisticians, and many generalizations have been made since 1996. In Genton [16], the results and applications of multivariate skew normal distribution were presented in a collective monograph. In 2014, a monograph Azzalini and Capitanio [6] was published with a comprehensive list of references on the topic. In Fig. 2.3, an illustrative graph of the density function is presented in the bivariate case.

Different parameterizations and their estimation are examined in Käärik et al. [33].

Estimation
Unfortunately, the maximum likelihood estimates cannot be obtained in explicit form. These estimates can be found iteratively but sometimes the iteration process converges to a local extremum. Therefore, one is interested in finding good initial values of the parameters.

Fortunately, the moment generating function has a simple form (Azzalini and Capitanio [6, Sect. 5.1.2]):

$$M(\mathbf{t}) = 2e^{\frac{1}{2}\mathbf{t}'\boldsymbol{\mu} + \mathbf{t}'\boldsymbol{\Sigma}\mathbf{t}} \cdot \Phi\left(\frac{\boldsymbol{\alpha}'\boldsymbol{\Sigma}\mathbf{t}}{\sqrt{1 + \boldsymbol{\alpha}'\boldsymbol{\Sigma}\boldsymbol{\alpha}}}\right)$$

and moments can be found by differentiating $M(\mathbf{t})$.

The first 4 central moments are

$$\mathrm{E}(\mathbf{x}) = \boldsymbol{\mu} + \sqrt{\frac{2}{\pi}}\boldsymbol{\delta}, \qquad \mathrm{D}(\mathbf{x}) = \boldsymbol{\Sigma} - \frac{2}{\pi}\boldsymbol{\delta}\boldsymbol{\delta}',$$

$$\overline{m}_3(\mathbf{x}) = \sqrt{\frac{2}{\pi}}\left(\frac{4}{\pi} - 1\right)\boldsymbol{\delta} \otimes \boldsymbol{\delta}' \otimes \boldsymbol{\delta},$$

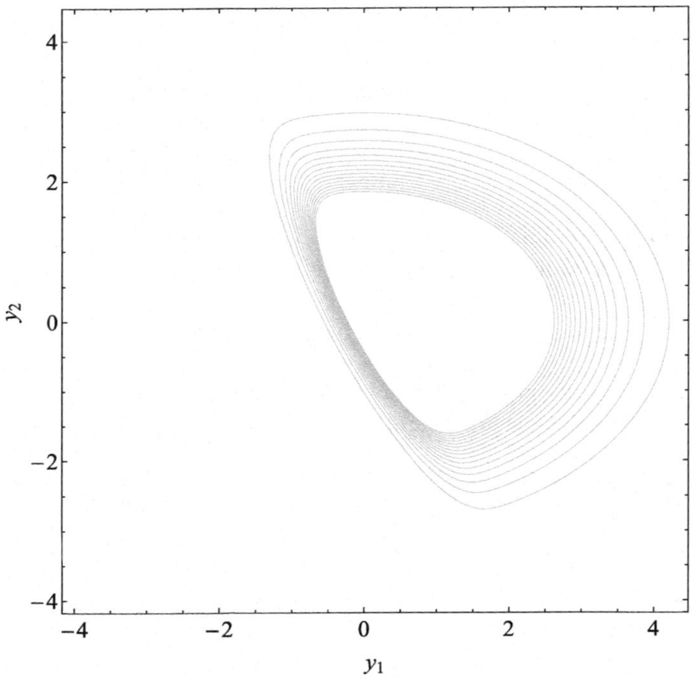

Fig. 2.3 Contour plot of a bivariate skew normal density, $\boldsymbol{\mu} = \mathbf{0}, \boldsymbol{\Sigma} = \begin{pmatrix} 2 & 0 \\ 0 & 1 \end{pmatrix}$, $\alpha = (4, 2)'$

$$\begin{aligned}\overline{m}_4(\mathbf{x}) &= \frac{8}{\pi}\left(1 - \frac{3}{\pi}\right)\boldsymbol{\delta} \otimes \boldsymbol{\delta}' \otimes \boldsymbol{\delta} \otimes \boldsymbol{\delta}' \\ &+ (\mathbf{I}_{p^2} + \mathbf{K}_{p,p})\,[\mathrm{D}\,(\mathbf{x}) \otimes \mathrm{D}\,(\mathbf{x})] + \mathrm{vec}\,[\mathrm{D}\,(\mathbf{x})]\,\mathrm{vec}'\,[\mathrm{D}\,(\mathbf{x})]\end{aligned}$$

where

$$\boldsymbol{\delta} = \frac{\boldsymbol{\Sigma\alpha}}{\sqrt{1 + \boldsymbol{\alpha}'\boldsymbol{\Sigma\alpha}}}.$$

Naturally there arises a question of how to measure skewness and kurtosis of a multivariate skewed distribution? Several scalar characteristics have been proposed; the list includes Mardia [38], Malkovich and Afifi [37], Isogai [19], Srivastava [49] and Song [48]. From these, the skewness measures of Malkovich and Afifi and Mardia are equivalent and Srivastava's skewness measure is not affine invariant. In several papers, the skewness measure has been introduced for constructing normality test (Isogai [19], Malkovich and Afifi [37], Srivastava [49] and Song [48]). Most often, Mardia's characteristics are calculated.

Let $\mathbf{x}$ be a random p-vector with $\mathrm{E}\,(\mathbf{x}) = \boldsymbol{\mu}$, $\mathrm{D}\,(\mathbf{x}) = \boldsymbol{\Sigma}$, and $\mathbf{x}_1,\ \mathbf{x}_2$ be two independent copies of $\mathbf{x}$. Mardia [38] introduced the skewness measure $\beta_{1,p}(\mathbf{x})$ and the kurtosis measure $\beta_{2,p}(\mathbf{x})$ as

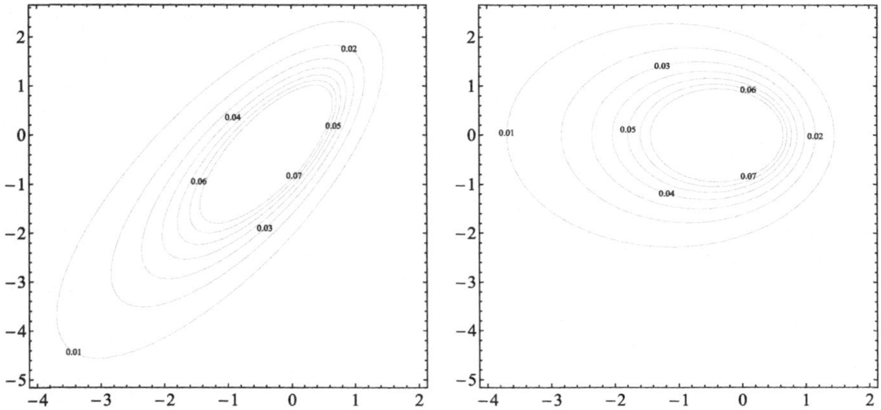

$\boldsymbol{\mu} = (-1, -1)', \boldsymbol{\Sigma} = \begin{pmatrix} 1.5 & 1.5 \\ 1.5 & 3 \end{pmatrix}$ $\boldsymbol{\mu} = (-1, 0)', \boldsymbol{\Sigma} = \begin{pmatrix} 1.5 & 0 \\ 0 & 1.5 \end{pmatrix}$

Fig. 2.4 Contour plots of two bivariate asymmetric Laplace distributions, both with $\beta_{1,2} = 5.63$, $\beta_{2,2} = 20$

$$\beta_{1,p}(\mathbf{x}) = \mathrm{E}\,[(\mathbf{x}_1 - \boldsymbol{\mu})'\boldsymbol{\Sigma}^{-1}(\mathbf{x}_2 - \boldsymbol{\mu})]^3,$$
$$\beta_{2,p}(\mathbf{x}) = \mathrm{E}\,[(\mathbf{x} - \boldsymbol{\mu})'\boldsymbol{\Sigma}^{-1}(\mathbf{x} - \boldsymbol{\mu})]^2.$$

Unfortunately, they can have the same numerical values for distributions with different shapes. In Fig. 2.4, two different asymmetric bivariate Laplace distributions (see Sect. 5) are shown with the same value of the skewness characteristic $\beta_{1,2}(\mathbf{x})$ and kurtosis $\beta_{2,2}(\mathbf{x})$.

Móri et al. [41] defined a vector $\mathbf{s}(\mathbf{x})$ to measure multivariate skewness of a random vector $\mathbf{x}$

$$\mathbf{s}(\mathbf{x}) = E(\|\ \mathbf{y}\ \|^2\ \mathbf{y})$$

and multivariare kurtosis as a $p \times p$-matrix

$$\mathbf{K}(\mathbf{x}) = E(\mathbf{y}\mathbf{y}'\mathbf{y}\mathbf{y}') - (p+2)\mathbf{I}_p,$$

where

$$\mathbf{y} = \boldsymbol{\Sigma}^{-1/2}(\mathbf{x} - \boldsymbol{\mu})$$

and $\boldsymbol{\Sigma}^{1/2}$ stands for a symmetric square root of $\boldsymbol{\Sigma}$. Not all third- and fourth-order mixed moments were taken into account in these definitions. In Kollo [23], all third- and fourth-order mixed moments were used in definitions of skewness vector and kurtosis matrix.

We define the star product of matrices due to its use in the coming definitions.

Definition 2.5 (MacRae [35]) Let **A** be an $m \times n$ matrix and $\mathbf{B} : mr \times ns$ partitioned matrix, which consists of $r \times s$ blocks $\mathbf{B}_{ij}, i \in \{1, \ldots, m\}, j \in \{1, \ldots n\}$. The star product $\mathbf{A} \star \mathbf{B}$ of matrices **A** and **B** is the $r \times s$ matrix

$$\mathbf{A} \star \mathbf{B} = \sum_{i=1}^{m} \sum_{j=1}^{n} a_{ij} \mathbf{B}_{ij}.$$

For example, $\mathbf{A} \star (\mathbf{A} \otimes \mathbf{A}) = \sum_{i=1}^{m} \sum_{j=1}^{n} a_{ij}^2 \mathbf{A}$.

Using the star product, we can transform the $p^2 \times p$ matrix of the third moment of **y** into the p-vector of skewness

$$\mathbf{b}(\mathbf{x}) = \mathbf{J}_p \star m_3(\mathbf{y}),$$

where $\mathbf{J}_p$ is a $p \times p$ matrix of ones.

The $p^2 \times p^2$ matrix of the fourth-order mixed moments is transformed into the $p \times p$ kurtosis matrix

$$\mathbf{K}(\mathbf{x}) = \mathbf{J}_p \star m_4(\mathbf{y}),$$

where $\mathbf{y} = \boldsymbol{\Sigma}^{-1/2}(\mathbf{x} - \boldsymbol{\mu})$.

Applying the star product to the matrix of the sample third moment, it is possible to find the parameter estimates of $SN_p(\boldsymbol{\mu}, \boldsymbol{\Sigma}, \boldsymbol{\alpha})$

$$\widehat{\boldsymbol{\mu}} = \overline{\mathbf{x}} - \sqrt{\frac{2}{\pi}} \widehat{\boldsymbol{\delta}},$$

$$\widehat{\boldsymbol{\Sigma}} = \mathbf{S} + \frac{2}{\pi} \widehat{\boldsymbol{\delta}\boldsymbol{\delta}'},$$

$$\widehat{\boldsymbol{\alpha}} = \frac{\left(\mathbf{S} + \frac{2}{\pi}\widehat{\boldsymbol{\delta}\boldsymbol{\delta}'}\right)^{-1} \widehat{\boldsymbol{\delta}}}{\sqrt{1 - \widehat{\boldsymbol{\delta}}' \left(\mathbf{S} + \frac{2}{\pi}\widehat{\boldsymbol{\delta}\boldsymbol{\delta}'}\right)^{-1} \widehat{\boldsymbol{\delta}}}},$$

where

$$\widehat{\boldsymbol{\delta}} = \mathbf{J}_p \star \widehat{m_3}(\mathbf{x}) \frac{\pi \widehat{H}^{2/3}}{(2(4 - \pi)^2)^{1/3}}, \quad \widehat{H} = \sum_{i=1}^{p^2} \sum_{j=1}^{p} (\widehat{m_3}(\mathbf{x}))_{ij}$$

and $\overline{\mathbf{x}}$ and **S** are the sample mean and the sample covariance matrix, respectively.

The multivariate skew t distribution has been of special interest because of heavier tails than the normal (skew normal) distribution. In the monograph Kotz and Nadarajah [32], a systematic overview is given on the multivariate t distribution as well as its skewed versions. Different constructions of skewed t distributions are based on the density function of the multivariate t distribution:

$$f_{p,\nu}(\mathbf{x}, \boldsymbol{\mu}, \boldsymbol{\Sigma}) = \frac{\Gamma((\nu+p)/2)}{(\pi\nu)^{p/2}\Gamma(\nu/2)|\boldsymbol{\Sigma}|^{1/2}}\left[1 + \frac{1}{\nu}(\mathbf{x}-\boldsymbol{\mu})'\boldsymbol{\Sigma}^{-1}(\mathbf{x}-\boldsymbol{\mu})\right]^{-(\nu+p)/2}, \tag{2.1}$$

where ν denotes degrees of freedom, $\boldsymbol{\mu}$ is the mean and $\boldsymbol{\Sigma} : p \times p$ the scale matrix, $\mathrm{D}(\mathbf{x}) = \frac{\nu}{\nu-2}\boldsymbol{\Sigma}$.

Azzalini and Capitanio [5] defined a multivariate skew t distribution with the density

$$f_{p,\nu}(\mathbf{x}, \boldsymbol{\mu}, \boldsymbol{\Sigma}, \boldsymbol{\alpha}) = 2 f_{p,\nu}(\mathbf{x}, \boldsymbol{\mu}, \boldsymbol{\Sigma}) T_{1,\nu+p}\left(\sqrt{\frac{\nu+p}{Q+\nu}}\boldsymbol{\alpha}'\boldsymbol{\Sigma}_d^{-\frac{1}{2}}(\mathbf{x}-\boldsymbol{\mu})\right), \tag{2.2}$$

where $f_{p,\nu}(\mathbf{x}, \boldsymbol{\mu}, \boldsymbol{\Sigma})$ is the multivariate t density in (2.1), $Q = (\mathbf{x}-\boldsymbol{\mu})'\boldsymbol{\Sigma}^{-1}(\mathbf{x}-\boldsymbol{\mu})$ and $\boldsymbol{\Sigma}_d^{-\frac{1}{2}}$ is the diagonal matrix with the diagonal elements $\frac{1}{\sqrt{\sigma_{ii}}}$. Here $T_{1,\nu+p}(\cdot)$ denotes the distribution function of the univariate t distribution with $\nu + p$ degrees of freedom. In this case,

$$\mathbf{x} = \boldsymbol{\mu} + \sqrt{\nu}\frac{\mathbf{y}}{\sqrt{V}},$$

where V is chi-square distributed with ν degrees of freedom and is independent of the skew normal vector $\mathbf{y}$ with density

$$f(\mathbf{y}) = 2 f_{N_p(\mathbf{0},\boldsymbol{\Sigma})}(\mathbf{y})\Phi(\boldsymbol{\alpha}'\boldsymbol{\Sigma}_d^{-\frac{1}{2}}\mathbf{y}).$$

Contour plot of a bivariate skew t density function is presented in Fig. 2.5.

As the distribution of the t distributed random vector is determined by the product of two independent variables, the skew normal vector $\mathbf{y}$ and the inverse chi-distributed random variable $\frac{1}{\sqrt{V}}$, the moments of $\mathbf{x}$ are the products of the corresponding moments of $\mathbf{y}$ and $\frac{1}{\sqrt{V}}$. Estimation of the parameters of the skew t distribution by the method of moments can be done using the sample moments of these distributions similar to the parameter estimation of the skew normal distribution.

2.5 Asymmetric Laplace Distribution

Symmetric p-variate Laplace distribution was introduced in Anderson [2] via characteristic function

$$\varphi(\mathbf{t}) = \frac{1}{1 + \frac{1}{2}\mathbf{t}'\boldsymbol{\Sigma}\mathbf{t}}$$

where $\boldsymbol{\Sigma} : p \times p$ is the positive definite scale parameter.

Asymmetric Laplace distribution was studied in several papers by T. J. Kozubowski with colleagues at the end of the 1990s; in the monograph Kotz et al. [31], the distribution corresponding to the characteristic function

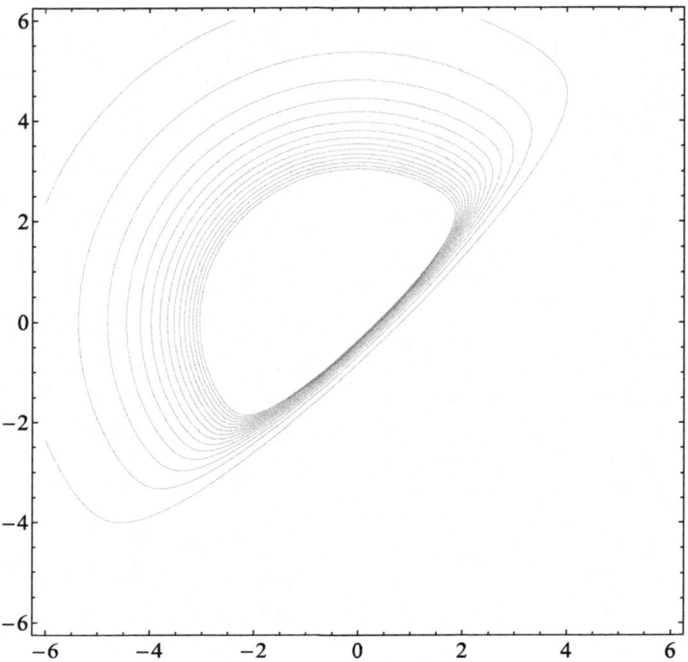

Fig. 2.5 Contour plot of a bivariate skew t density function with parameters $\nu = 2$, $\mathbf{\Sigma} = \mathbf{I}_2$ and $\boldsymbol{\alpha} = (-5, 5)'$

$$\varphi(\mathbf{t}) = \frac{1}{1 - i\mathbf{t}'\boldsymbol{\mu} + \frac{1}{2}\mathbf{t}'\mathbf{\Sigma}\mathbf{t}}$$

is thoroughly studied. Here p-vector $\boldsymbol{\mu}$ stands for the shape parameter (regulates location and skewness) and positive definite $p \times p$ matrix $\mathbf{\Sigma}$ is the scale parameter. The distribution has a heavier tail area than normal distribution but still the moments exist.

Shifted Laplace distribution was introduced in Visk [51] via characteristic function. A random vector $\mathbf{y}$ has shifted asymmetric Laplace distribution when

$$\varphi_{\mathbf{y}}(\mathbf{t}) = \varphi_{\mathbf{x}+\mathbf{a}}(\mathbf{t}) = \frac{e^{i\mathbf{t}'\mathbf{a}}}{1 - i\mathbf{t}'\boldsymbol{\mu} + \frac{1}{2}\mathbf{t}'\mathbf{\Sigma}\mathbf{t}},$$

where $\mathbf{x}$ has asymmetric Laplace distribution and p-vector $\mathbf{a}$ is the shift parameter.

Differentiation of $\varphi_{\mathbf{y}}(\mathbf{t})$ gives us the first moments

$$\mathrm{E}(\mathbf{y}) = \mathbf{a} + \boldsymbol{\mu}, \qquad \mathrm{D}(\mathbf{y}) = \mathbf{\Sigma} + \boldsymbol{\mu}\boldsymbol{\mu}',$$

$$\overline{m}_3(\mathbf{y}) = 2\boldsymbol{\mu} \otimes \boldsymbol{\mu}\boldsymbol{\mu}' + \operatorname{vec}\mathbf{\Sigma} \cdot \boldsymbol{\mu}' + \boldsymbol{\mu} \otimes \mathbf{\Sigma} + \mathbf{\Sigma} \otimes \boldsymbol{\mu}.$$

The estimates of all three parameters of the shifted asymmetric Laplace distribution were found by the method of moments in Visk [51] using the star product of matrices as it was done for the estimation of the parameters of the skew normal distribution.

2.6 Copulas

Abe Sklar introduced the notion *copula* in 1959 (Sklar [46]) and for more than 40 years, the development of copula theory remained a topic of interest in measure theory. Only in the 1990s, statisticians discovered that it gives a new possibility for constructing data models in statistical analysis. Copulas can be considered as multivariate distribution functions with uniformly distributed marginals in the unit cube which take into account the dependence between the variables. Copulas give us a principally new possibility to construct statistical models with differently distributed marginals and given correlation structure between the margins (for applications in finance, see Cherubini et al. [10], in environmental studies Chen and Guo [9], for example). In 1999, the popular book (Nelsen [43]) of R. B. Nelsen appeared. From the mathematical point of view, a bivariate copula is a transformation from $\mathbb{R}^2 \to [0, 1] \times [0, 1]$, which satisfies certain requirements. These requirements guarantee that the following result holds (Sklar [46]).

Theorem 2.4 (Sklar's theorem) *Let H be a joint distribution function with margins F and G. Then there exists a copula C such that for all $x, y \in \mathbb{R}$,*

$$H(x, y) = C(F(x), G(y)).$$

If F and G are continuous, then C is unique, otherwise, C is uniquely determined on the range $RanF \times RanG$. Conversely, if C is a copula and F and G are distribution functions, then the function H defined above is a joint distribution function with margins F and G.

Here $RanF$ denotes the set of all values of the distribution function F in the unit interval $[0, 1]$. In other words, a copula is a joint distribution function with uniformly distributed (possibly dependent) marginals. If a bivariate copula $C(u, v)$ is differentiable, the copula density is defined as

$$c(u, v) = \frac{\partial^2 C(u, v)}{\partial u \partial v}.$$

Multivariate copulas and their densities are defined in a similar way.

If a p-variate copula $C(u_1, \ldots, u_p), u_i \in [0, 1]$, is differentiable by all arguments, then the copula density is defined as

$$c(u_1, \ldots, u_p) = \frac{\partial^p C(u_1, \ldots, u_p)}{\partial u_1 \ldots \partial u_p}.$$

Theorem 2.5 (Sklar's theorem in p dimensions) *Let $X_1, \dots, X_p$ be random variables with distribution functions $F_1, \dots, F_p$, respectively, and joint distribution function H. Then there exists a p-copula $C(u_1, \dots, u_p)$ such that*

$$H(x_1, \dots, x_p) = C(F_1(x_1), \dots, F_p(x_p)).$$

If $F_1, \dots, F_p$ are all continuous, C is unique. Otherwise, C is uniquely determined on the range $Ran F_1 \times \cdots \times Ran F_p$. When $\mathbf{x} = (X_1, \dots, X_p)'$ is a continuous random vector, its density $h(x_1, \dots, x_p)$ is expressed through this copula density $c(\cdot)$ and marginal densities $f_i(x_i)$

$$h(x_1, \dots, x_p) = c(F_1(x_1), \dots, F_p(x_p)) f_1(x_1) \dots f_p(x_p).$$

The complete proof of the theorem in p dimensions was first given in Moore and Spruill [40], later independently in Sklar [47].

Let $\mathbf{R}$ be a positive definite correlation matrix and $\Phi_{\mathbf{R}}$ the standardized multivariate normal distribution function with correlation matrix $\mathbf{R}$. Then the Gaussian copula has implicit definition (see Cherubini et al. [10, pp. 147–148], for example)

$$C_{\mathbf{R}}^{Ga}(\mathbf{u}) = \Phi_{\mathbf{R}} \left(\Phi^{-1}(u_1), \dots, \Phi^{-1}(u_p) \right).$$

The density of this Gaussian copula is given by

$$c_{\mathbf{R}}^{Ga}(u_1, \dots, u_p) = \frac{1}{|\mathbf{R}|^{1/2}} \exp\left(-\frac{1}{2} \mathbf{z}'(\mathbf{R}^{-1} - \mathbf{I}_p)\mathbf{z} \right),$$

where $\mathbf{z} = (\Phi^{-1}(u_1), \dots, \Phi^{-1}(u_p))'$ and $\mathbf{R}$ is the correlation matrix. In Fig. 2.6, the contour plot of a bivariate Gaussian copula density is given.

In the similar way, we can construct the density of a t copula. Let $t_{\mathbf{R},\nu}$ be the standardized multivariate t distribution with correlation matrix $\mathbf{R}$ and ν degrees of freedom. Then the t copula density is (Cherubini et al. [10, p. 148], for example)

$$c_{\mathbf{R},\nu}(u_1, \dots, u_p) = |\mathbf{R}|^{-\frac{1}{2}} \frac{\Gamma\left(\frac{\nu+p}{2}\right)}{\Gamma\left(\frac{\nu}{2}\right)} \left[\frac{\Gamma\left(\frac{\nu}{2}\right)}{\Gamma\left(\frac{\nu+1}{2}\right)} \right]^p \frac{\left(1 + \frac{1}{\nu}\mathbf{z}'\mathbf{R}^{-1}\mathbf{z}\right)^{-\frac{\nu+p}{2}}}{\prod_{j=1}^p \left(1 + \frac{z_j^2}{\nu}\right)^{-\frac{\nu+1}{2}}},$$

where $z_j = T_{1,\nu}^{-1}(u_j)$ and $T_{1,\nu}$ is the distribution function of the univariate standard t distribution.

The copulas built from elliptical multivariate distributions are symmetric (Gaussian and t copula, for instance). It seems natural to join skewed marginals into a multivariate distribution by a skewed copula density. Skew elliptical families offer a tool for that. We shall consider bivariate skew normal and skew t copulas.

Let us define first a skew normal copula.

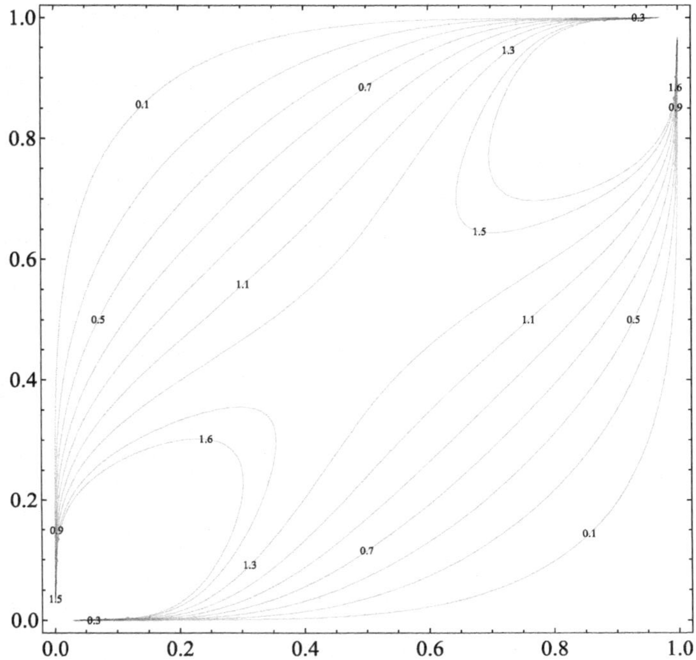

Fig. 2.6 Contour plot of the density of a Gaussian copula, $r_{12} = 0.7$

Definition 2.6 A two-dimensional copula $C_{2,SN}$ is called skew normal copula with parameters $\boldsymbol{\mu}$, $\mathbf{R}$ and $\boldsymbol{\alpha}$, if

$$C_{2,SN}(\mathbf{u}; \boldsymbol{\mu}, \mathbf{R}, \boldsymbol{\alpha}) = F_{2,SN}(F_1^{-1}(u_1; \mu_1, 1, \alpha_1^*), F_1^{-1}(u_2; \mu_2, 1, \alpha_2^*); \boldsymbol{\mu}, \mathbf{R}, \boldsymbol{\alpha}),$$

where $F_{2,SN}(\cdot)$ is the distribution function of the bivariate skew normal distribution (see Definition 2.4) and $F_1^{-1}(u_i; \mu_i, 1, \alpha_i^*)$, $i \in \{1, 2\}$, denotes the inverse of the distribution function of the univariate skew normal distribution $SN(\mu_i, 1, \alpha_i^*)$. The skewing parameters α_i^*, $i \in \{1, 2\}$, are defined by the equalities

$$\alpha_1^* = \frac{\alpha_1 + r_{12}\alpha_2}{\sqrt{1 + \alpha_2^2(1 - r_{12}^2)}}, \quad \alpha_2^* = \frac{\alpha_2 + r_{12}\alpha_1}{\sqrt{1 + \alpha_1^2(1 - r_{12}^2)}},$$

where r_{12} is the correlation coefficient.

We get the expressions of α_i^* in a special case from the formulae of the skewing parameters of marginal distributions (Azzalini and Capitanio [6, p. 130]).

The corresponding copula density is

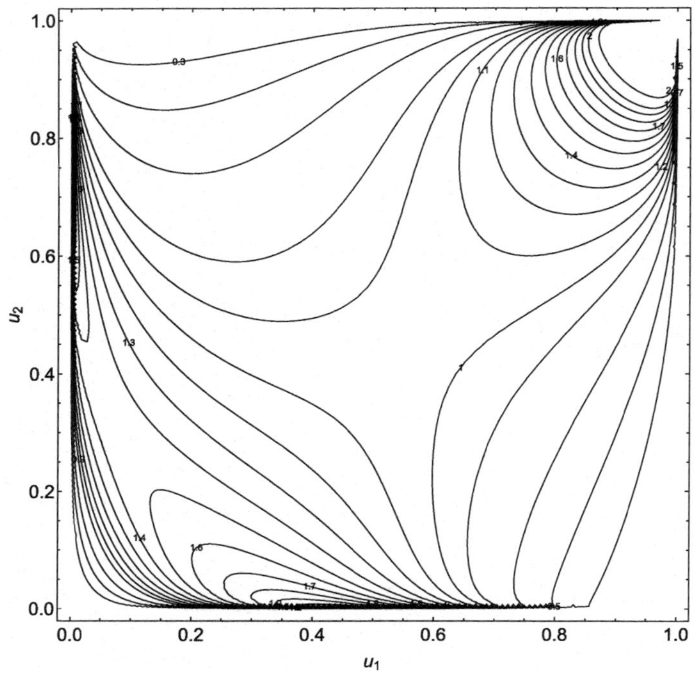

Fig. 2.7 Contour plot of the bivariate skew normal copula density, $r_{12} = 0.7$, $\boldsymbol{\alpha} = (5, 2)'$

$$c_{2,SN}(\mathbf{u}; \boldsymbol{\mu}, \mathbf{R}, \boldsymbol{\alpha}) = \frac{f_{2,SN}(F_1^{-1}(u_1; \mu_1, 1, \alpha_1^*), F_1^{-1}(u_2; \mu_2, 1, \alpha_2^*); \boldsymbol{\mu}, \mathbf{R}, \boldsymbol{\alpha})}{\prod_{i=1}^{2} f_{1,SN}(F_1^{-1}(u_i; \mu_i, 1, \alpha_i^*))},$$

where $f_{2,SN}(\cdot)$ is the density of the bivariate skew normal distribution, and functions $F_1^{-1}(u_i; \mu_i, 1, \alpha_i^*)$, $i \in \{1, 2\}$, are as in Definition 2.6. In Fig. 2.7, the contour plot of a skew normal copula density is given when $\boldsymbol{\mu} = \mathbf{0}$.

The skew $t_{2,\nu}$ copula is defined in a similar way (Kollo and Pettere [24]).

Definition 2.7 A bivariate copula $C_{2,\nu}$ is called skew $t_{2,\nu}$ copula with parameters $\boldsymbol{\mu}, \boldsymbol{\Sigma}, \boldsymbol{\alpha}$, if

$$C_{2,\nu}(u_1, u_2 : \boldsymbol{\mu}, \boldsymbol{\Sigma}, \boldsymbol{\alpha}) = T_{2,\nu}(T_{1,\nu}^{-1}(u_1; \mu_1, 1, \alpha_1^*), T_{1,\nu}^{-1}(u_2; \mu_2, 1, \alpha_2^*); \boldsymbol{\mu}, \boldsymbol{\Sigma}, \boldsymbol{\alpha}),$$

where $T_{1,\nu}^{-1}(u_i; \mu_i, 1, \alpha_i^*)$, $i \in \{1, 2\}$, denotes the inverse of the distribution function of the univariate skew $t_{1,\nu}$-distribution from the location family, $T_{2,\nu}$ is the distribution function of the bivariate skew $t_{2,\nu}$-distribution and the skewing parameters α_i^* are defined in Definition 6.

When $\boldsymbol{\mu} = \mathbf{0}$, in the arguments of the distribution function $T_{2,\nu}$, we have inverse functions of the distribution function $T_{1,\nu}(u_i, \alpha_i*)$, $i \in \{1, 2\}$, of the univariate standard skew t_ν distribution. In this case, the density of the skew t copula is of the form

$$c_{2,\nu}(\mathbf{u}, \mathbf{R}, \boldsymbol{\alpha}) = \frac{f_{2,\nu}[(T_{1,\nu}^{-1}(u_1, \alpha_1^*), T_{1,\nu}^{-1}(u_2, \alpha_2^*)), \mathbf{R}, \boldsymbol{\alpha}]}{\prod_{i=1}^{2} f_{1,\nu}[(T_{1,\nu}^{-1}(u_i, \alpha_i), \alpha_i^*]}, \tag{2.3}$$

where in the expression of the density $f_{p,\nu}$ in (2.2), we have taken $p = 2$, $\boldsymbol{\Sigma} = \mathbf{R}$ and $\boldsymbol{\mu} = \mathbf{0}$. The density (2.3) is the most often used version of the skew t copula density in data analysis.

Which copula to choose to join the marginals? In risk estimation, tail dependence is an important property.

Let $(X_1, X_2)'$ be a bivariate random vector with univariate marginal distribution functions $F_1(x)$ and $F_2(x)$. Denote

$$\lambda_U = \lim_{u \to 1} \lambda_U(u),$$

where

$$\lambda_U(u) = P(F_1(x) > u | F_2(x) > u).$$

The quantity λ_U is known as the *coefficient of upper tail dependence*. Similarly, the *lower tail dependence coefficient* λ_L is defined as

$$\lambda_L = \lim_{u \to 0} \lambda_L(u),$$

where

$$\lambda_L(u) = P(F_1(x) < u | F_2(x) < u).$$

For symmetric elliptical distributions $\lambda_L = \lambda_U = \lambda$, in the case of normal distribution $\lambda = 0$ while for the bivariate t distribution with ν degrees of freedom

$$\lambda = 2T_{1,\nu+1}\left(-\sqrt{\frac{(\nu+1)(1-r)}{1+r}}\right),$$

where $T_{1,\nu+1}(\cdot)$ denotes the distribution function of the standard t-variable with $\nu + 1$ degrees of freedom and r the correlation coefficient between the marginals (Demarta and McNeil [12]). Tail dependence of the skew t distribution is studied in Padoan [44]. From the expression of the density function $f_{2,\nu}(\mathbf{x}, \mathbf{R}, \boldsymbol{\alpha})$ of the skew t distribution, it follows that

$$f_{2,\nu}(\mathbf{x}, \mathbf{R}, \boldsymbol{\alpha}) = f_{2,\nu}(-\mathbf{x}, \mathbf{R}, -\boldsymbol{\alpha}),$$

therefore it is sufficient to study the upper tail dependence only. For the upper tail dependence coefficient of the skew t distribution, the following inequality holds (Kollo et al. [25]):

$$\lambda_U \geq \lambda \frac{T_{1,\nu+2}\left((\alpha_1+\alpha_2)\sqrt{\frac{(\nu+2)(1+r)}{2}}\right)}{T_{1,\nu+1}\left(\alpha_2^*\sqrt{\nu+1}\right)},$$

where λ is the value of the tail dependence coefficient for the corresponding bivariate t distributed random vector $(Y_1, Y_2)'$ and

$$\alpha_2^* = \frac{\alpha_2 + \alpha_1 r}{\sqrt{1+\alpha_1^2(1-r^2)}}.$$

When using multivariate copulas constructed from elliptical and skew elliptical distributions, a complication arises. In copula densities, the Pearson correlation matrix appears, but linear correlations are not invariant under the transformations by inverses of distribution functions. Fortunately, there is a simple relation between Kendall's τ and the linear correlation coefficient r for continuous elliptical distributions (Lindskog et al. [34]), which is

$$\tau(X_i, X_j) = \frac{2}{\pi} \arcsin r_{ij}.$$

This is the formula for the normal distribution and it holds for all continuous elliptical distributions.

2.7 Conclusion

Let us look back in time.

60 years ago, T. W. Anderson's classical book presented the main methods of multivariate analysis for the multivariate normal population.
40 years ago, elliptical distributions came into use.
30 years ago, multivariate Edgeworth type expansions enabled to transform multivariate normal distribution into non-symmetric data models.
20 years ago, skew elliptical distributions and copula models were introduced.

Today, we can join in a multivariate data model marginals with different distributions and certain dependence structures using multivariate copulas.

Acknowledgements I am most grateful to the Referees and Editors for their careful work with the manuscript. Their comments and suggestions improved considerably the presentation of the material and helped to avoid several misprints. The figures in the paper have been prepared by A. Selart, H. Visk and I. Allekand. I am extremely thankful to K. Filipiak for careful editorial work with the manuscript resulting in higher quality figures. The study was financially supported by the Estonian Research Foundation, grants IUT34-5 and PRG 1197.

References

1. Anderson, T.W.: Introduction to Multivariate Statistical Analysis. Wiley (1958)
2. Anderson, D.N.: A multivariate Linnik distribution. Stat. Probab. Lett. **14**, 333–336 (1992)
3. Anderson, T.W.: Introduction to Multivariate Statistical Analysis. Wiley (2003)
4. Azzalini, A.: A class of distributions which includes the normal ones. Scand. J. Stat. **12**, 171–178 (1985)
5. Azzalini, A., Capitanio, A.: Distributions generated by perturbation of symmetry with emphasis on a multivariate skew t-distribution. J. R. Stat. Soc. Ser. B. Stat. Methodol. **65**, 367–389 (2003)
6. Azzalini, A., Capitanio, A.: The Skew-Normal and Related Families. Cambridge University Press (2014)
7. Azzalini, A., Dalla Valle, A.: The multivariate skew normal distribution. Biometrika **83**, 715–726 (1996)
8. Cornish, E.A., Fisher, R.A.: Moments and cumulants in the specification of distributions. Rev. Int. Stat. Instit. **5**, 307–322 (1937)
9. Chen, L., Guo, S.: Copulas and Its Applications in Hydrology and Water Resources. Springer (2019)
10. Cherubini, U., Luciano, E., Vecchiato, W.: Copula Methods in Finance. Wiley (2004)
11. Cramér, H.: Mathematical Methods of Statistics. Princeton University Press (1974)
12. Demarta, S., McNeil, A.J.: The *t* copula and related copulas. Int. Stat. Rev. **73**, 111–129 (2005)
13. Fang, K.-T., Kotz, S., Ng, K.W.: Symmetric Multivariate and Related Distributions. Chapman & Hall (1990)
14. Fang, K.-T., Zhang, Y.-T.: Generalized Multivariate Analysis. Science Press, Springer-Verlag (1990)
15. Finney, D.J.: Some properties of a distribution specified by its cumulants. Technometrics **5**, 63–69 (1963)
16. Genton, M.G. (Ed.): Skew-Elliptical Distributions and Their Applications: A Journey Beyond Normality. Chapman & Hall/CRC (2004)
17. Gerber, H.U.: An Introduction to Mathematical Risk Theory. University of Pennsylvania, Huebner Foundation for Insurance Education (1979)
18. Hall, P.: Chi-squared approximations to the distribution of a sum of independent random variables. Ann. Probab. **11**, 1028–1036 (1983)
19. Isogai, T.: On a measure of multivariate skewness and a test for multivariate normality. Ann. Inst. Stat. Math. **34**, 531–541 (1982)
20. Jensen, D.R.: Multivariate distributions. In: Kotz, S., Johnson, N.L., Read, C.B. (eds.) Encyclopedia of Statistical Sciences, vol 6, pp. 43–55. Wiley (1985)
21. Kariya, T., Sinha, B.K.: Robustness of Statistical Tests. Academic Press (1989)
22. Kelker, D.: Distribution theory of spherical distributions and a location-scale parameter generalization. Sankhyā A **32**, 419–430 (1970)
23. Kollo, T.: Multivariate skewness and kurtosis measures with an application in ICA. J. Multivariate Anal. **99**, 2328–2338 (2008)
24. Kollo, T., Pettere, G.: Parameter estimation and application of the multivariate skew *t*-copula. In: Jaworski, P., Durante, F., Härdle, W.K., Rychlik, T. (eds.) Copula Theory and Its Applications, Chapter 15, pp. 289–298. Springer-Verlag (2010)
25. Kollo, T., Pettere, G., Valge, M.: Tail dependence of skew *t*-copulas. Comm. Stat. Simul. Comput. **46**, 1024–1034 (2017)
26. Kollo, T., Roos, A., von Rosen, D.: Approximation of the distribution of the location parameter in the Growth Curve model. Scand. J. Stat. **34**, 499–510 (2007)
27. Kollo, T., von Rosen, D.: Approximating by the Wishart distribution. Ann. Inst. Stat. Math. **47**, 767–783 (1995)
28. Kollo, T., von Rosen, D.: Unified approach to the approximation of multivariate densities. Scand. J. Stat. **25**, 93–109 (1998)
29. Kollo, T., von Rosen, D.: Advanced Multivariate Statistics with Matrices. Springer (2005)

30. Kollo, T., Srivastava, M.S.: Estimation and testing of parameters in multivariate Laplace distribution. Comm. Stat. Theory Methods **33**, 2363–2687 (2004)
31. Kotz, S., Kozubowski, T.J., Podgorski, K.: The Laplace Distribution, and Generalizations: A Revisit with Applications to Communications, Economics. Engineering and Finance, Birkhäuser (2001)
32. Kotz, S., Nadarajah, S.: Multivariate T-Distributions and Their Applications. Cambridge University Press (2004)
33. Käärik, M., Selart, A., Käärik, E.: On parametrization of multivariate skew-normal distribution. Comm. Stat. Theory Methods **44**, 1869–1885 (2015)
34. Lindskog, F., McNeil, A.J., Schmock, U.: Kendall's tau for elliptical distributions. In: Nakhaeizadeh, G., Rachev, S.T., Ridder T., Vollmer, K.H. (eds.) Credit Risk (Contributions to Economics), pp. 149–156. Physica-Verlag (2003)
35. MacRae, E.C.: Matrix derivatives with an application to an adaptive linear decision problem. Ann. Stat. **2**, 337–346 (1974)
36. Magnus, J.R., Neudecker, H.: Matrix Differential Calculus with Applications in Statistics and Econometrics, 2nd edn. Wiley (1999)
37. Malkovich, J.F., Afifi, A.A.: On tests for multivariate normality. J. Am. Stat. Assoc. **68**, 176–179 (1973)
38. Mardia, K.V.: Measures of multivariate skewness and kurtosis with applications. Biometrika **57**, 519–530 (1970)
39. Maxwell, J.C.: Illustration of the dynamic theory of gases: part I on the motions and collisions of perfectly elastic bodies. Phil. Mag. **19**, 19–32 (1860)
40. Moore, D.S., Spruill, M.C.: Unified large-sample theory of general chi-squared statistics for tests of fit. Ann. Stat. **3**, 599–616 (1975)
41. Móri, T.F., Rohatgi, V.K., Székely, G.J.: On multivariate skewness and kurtosis. Theory Probab. Appl. **38**, 547–551 (1993)
42. Muirhead, R.J.: Aspects of Multivariate Statistical Theory, 2nd edn. Wiley (2005)
43. Nelsen, R.B.: An Introduction to Copulas. Springer (1999)
44. Padoan, S.A.: Multivariate extreme models based on underlying skew-t and skew-normal distributions. J. Multivariate Anal. **102**, 977–991 (2011)
45. Rao, C.R.: Linear Statistical Inference and Its Applications. Wiley (1965)
46. Sklar, A.: Fonctions de répartition á n dimensions et Leurs Marges, vol. 8. Institut Statistique de l'Université de Paris, Paris, 229–231 (1959)
47. Sklar, A.: Random variables, distribution functions, and copulas—a personal look backward and forward. In: Rüschendorf, L., et al. (eds.) Distributions with Fixed Marginals and Related Topics, pp. 1–14. Institute of Mathematical Statistics, Hayward, CA (1996)
48. Song, K.-S.: Renyi information, loglikelihood and an intrinsic distribution measure. J. Stat. Plan. Inference **93**, 51–69 (2001)
49. Srivastava, M.S.: A measure of skewness and kurtosis and a graphical method for assessing multivariate normality. Stat. Probab. Lett. **2**, 263–267 (1984)
50. Traat, I.: Matrix calculus for multivariate distributions. Acta Comment. Univ. Tartu. Math. **733**, 64–84 (1986)
51. Visk, H.: Shifted multivariate Laplace distribution. Comm. Stat. Theory Methods **38**, 461–470 (2009)

Chapter 3
Multivariate Moments in Multivariate Analysis

Jolanta Pielaszkiewicz and Dietrich von Rosen

Abstract Moments for the normal, Wishart and beta-type distributions are presented. A number of relations involving the trace function are also considered due to their connection to spectral moments of random matrices. For a couple of specific relations, the technique of obtaining the results is described in detail. The majority of the results are presented for real-valued matrices, but complex-valued matrices are occasionally also treated.

3.1 Introduction

Moments play a fundamental role in statistics. When performing inference, a statistic is usually created, for example, an estimator or a test quantity. Ultimately, we would like to know the distribution of the statistic, which usually is not available. Therefore, over the years the focus has often been directed on deriving asymptotic distributions for the statistic, which only to some extent can replace the finite sample distribution.

The drawback with studying asymptotic distributions can to some extent be overcome through calculations of moments and cumulants. For example, the mean and dispersion of a statistic under consideration often indicate, when a comparison is made with data, whether or not the statistic is appropriate to use. Moreover, if moments up to an order of three can be derived, Edgeworth-type expansions can be obtained, meaning that an approximate finite sample distribution can be obtained (for details, see Chap. 2).

J. Pielaszkiewicz (✉) · D. von Rosen (✉)
Linköping University, Linköping, Sweden
e-mail: jolanta.pielaszkiewicz@liu.se

D. von Rosen
e-mail: Dietrich.von.Rosen@slu.se

J. Pielaszkiewicz
Stockholm University, Stockholm, Sweden

D. von Rosen
Swedish University of Agricultural Sciences, Uppsala, Sweden

K. Filipiak et al. (eds.), *Multivariate, Multilinear and Mixed Linear Models*, Contributions to Statistics, https://doi.org/10.1007/978-3-030-75494-5_3

The basic ideas in this article originate from univariate ideas where moments are obtained by differentiation of the moment generating functions or the characteristic functions. However, the multivariate setting leads to the consideration of different types of derivatives.

In this article, we present moments for the normal distribution, the Wishart distribution and different types of beta distributions. The idea is to show a collection of results and present different useful techniques for moment-related calculations. The presented techniques are mostly connected to matrix derivatives and the ordering of partial derivatives. Moreover, elements of the so-called free probability approach are presented due to their usefulness for deriving results concerning the spectral density.

3.2 Mathematical Background

If not explicitly stated otherwise, matrices and vectors are supposed to be real valued. However, occasionally complex matrices are discussed. A real-valued matrix of size $p \times q$ is denoted as $\mathbf{A} \in \mathbb{R}^{p\times q}$, whereas a complex-valued matrix of size $p \times q$ is denoted as $\mathbf{A} \in \mathbb{C}^{p\times q}$.

3.2.1 The vec-Operator and the Kronecker Product

In many places in this article, unit basis vectors are used. They are denoted by $\mathbf{d}_i$, $\mathbf{e}_j$, $\mathbf{f}_k$, $\mathbf{g}_l$, etc. For example, $\mathbf{d}_i \in \mathbb{R}^p$ means that $\mathbf{d}_i$ of size p equals 1 in the ith position and 0 elsewhere. Alternatively, it can be stated that $\mathbf{d}_i$ denotes the ith column of $\mathbf{I}_p$. Moreover, note that any matrix $\mathbf{A} \in \mathbb{R}^{p\times q}$ or $\mathbf{A} \in \mathbb{C}^{p\times q}$ can be presented as

$$\mathbf{A} = \sum_I a_{ij}\mathbf{d}_i\mathbf{e}_j',$$

where $\mathbf{d}_i \in \mathbb{R}^p$ and $\mathbf{e}_j \in \mathbb{R}^q$ are unit basis vectors, a_{ij} is the ijth-element of $\mathbf{A}$, $I = \{i, j; i \in \{1, \dots, p\}, j \in \{1, \dots, q\}\}$ and $'$ denotes the transpose.

Definition 3.1 The Kronecker product of the two matrices $\mathbf{A} \in \mathbb{C}^{p\times q}$ and $\mathbf{B} \in \mathbb{C}^{r\times s}$ is denoted as $\mathbf{A} \otimes \mathbf{B} \in \mathbb{C}^{pr\times qs}$ and equals $(a_{ij}\mathbf{B})$, which is expressed as follows:

$$(a_{ij}\mathbf{B}) = \begin{pmatrix} a_{11}\mathbf{B} & \dots & a_{1q}\mathbf{B} \\ \vdots & \ddots & \vdots \\ a_{p1}\mathbf{B} & \dots & a_{pq}\mathbf{B} \end{pmatrix}.$$

We also present an identity that can be seen as an equivalent definition of the Kronecker product and which appears to be useful for our derivations:

$$\mathbf{A} \otimes \mathbf{B} = \sum_{I} a_{ij} b_{kl} \mathbf{d}_i \mathbf{e}_j' \otimes \mathbf{f}_k \mathbf{g}_l',$$

where $\mathbf{d}_i \in \mathbb{R}^p$, $\mathbf{e}_j \in \mathbb{R}^q$, $\mathbf{f}_k \in \mathbb{R}^r$ and $\mathbf{g}_l \in \mathbb{R}^s$ are unit basis vectors and $I = \{i, j, k, l; i \in \{1, \ldots, p\}, j \in \{1, \ldots, q\}, k \in \{1, \ldots, r\}, l \in \{1, \ldots, s\}\}$. The above-given definition also holds for real-valued matrices. Kronecker product relations, as well as a formula for $r(\mathbf{A} \otimes \mathbf{B})$, where $\mathrm{r}(\mathbf{A})$ denotes the rank of $\mathbf{A}$, are presented in the next theorem and are often used in this article.

Theorem 3.1 *All the expressions below assume that the matrices are of proper sizes and either real- or complex-valued:*

(i) $(\mathbf{A} \otimes \mathbf{B})' = \mathbf{A}' \otimes \mathbf{B}'$;
(ii) $(\mathbf{A} + \mathbf{B}) \otimes (\mathbf{C} + \mathbf{D}) = \mathbf{A} \otimes \mathbf{C} + \mathbf{A} \otimes \mathbf{D} + \mathbf{B} \otimes \mathbf{C} + \mathbf{B} \otimes \mathbf{D}$;
(iii) $(\mathbf{A} \otimes \mathbf{B})(\mathbf{C} \otimes \mathbf{D}) = \mathbf{AB} \otimes \mathbf{CD}$;
(iv) *letting* $\mathbf{A}$ *and* $\mathbf{B}$ *be non-singular matrices,* $(\mathbf{A} \otimes \mathbf{B})^{-1} = \mathbf{A}^{-1} \otimes \mathbf{B}^{-1}$;
(v) $\mathrm{r}(\mathbf{A} \otimes \mathbf{B}) = \mathrm{r}(\mathbf{A})\mathrm{r}(\mathbf{B})$.

We also present a definition of the Kroneckerian power that will be used a few times in the presentation of results.

Definition 3.2 The Kroneckerian power of a matrix $\mathbf{A}$ is given by $\mathbf{A}^{\otimes k} = \underbrace{\mathbf{A} \otimes \cdots \otimes \mathbf{A}}_{k \text{ times}}$, where $k \in \{2, 3, \ldots\}$. If $k = 0$ then $\mathbf{A}^{\otimes 0} = \mathbf{I}$ or if $k = 1$ then $\mathbf{A}^{\otimes 1} = \mathbf{A}$.

Let the matrix $\mathbf{A} = (a_{ij})$ be real- or complex-valued. Then the *vec*-operator, which stacks the columns of a matrix underneath, starting with the first one, can be defined via

$$\mathrm{vec}\, \mathbf{A} = \sum_{i=1}^{p} \sum_{j=1}^{q} a_{ij} \mathbf{e}_j \otimes \mathbf{d}_i,$$

where $\mathbf{d}_i \in \mathbb{R}^p$ and $\mathbf{e}_j \in \mathbb{R}^q$ are unit basis vectors. A formal definition now follows.

Definition 3.3 The *vec*-operator is defined through

$$\mathrm{vec} : \mathbb{R}^{p \times q} \to \mathbb{R}^{pq}, \quad \mathrm{vec}\, (\mathbf{ab}') = \mathbf{b} \otimes \mathbf{a}, \quad \forall \mathbf{a} \in \mathbb{R}^p, \ \forall \mathbf{b} \in \mathbb{R}^q.$$

Definition 3.4 Letting $\mathbf{A}$ be a square matrix, then $\mathrm{Tr}\, (\mathbf{A})$ is the sum of its diagonal elements.

In the case of complex matrices, we use the notation $\mathbf{A}^*$ for the conjugate transpose, $\mathrm{vec}^* \mathbf{A}$ means $\mathrm{vec}\, \overline{\mathbf{A}'}$, $\overline{\mathbf{A}}$ is the conjugate of $\mathbf{A}$ and, as before, $'$ denotes the transpose. Almost trivial (but important) results are given in the next theorem.

Theorem 3.2 (i) *Let* $\mathbf{A} \in \mathbb{R}^{p \times q}$ *and* $\mathbf{B} \in \mathbb{R}^{p \times q}$. *Then,*

$$\mathrm{Tr}\, (\mathbf{A}'\mathbf{B}) = \mathrm{vec}'\mathbf{A} \mathrm{vec}\, \mathbf{B}.$$

(ii) *Let* $\mathbf{A} \in \mathbb{C}^{p\times q}$ *and* $\mathbf{B} \in \mathbb{C}^{p\times q}$ *be complex matrices. Then,*

$$\mathrm{Tr}\,(\mathbf{A}^*\mathbf{B}) = \mathrm{vec}^*\mathbf{A}\mathrm{vec}\,\mathbf{B} = \mathrm{vec}'\mathbf{B}\mathrm{vec}\,\overline{\mathbf{A}}.$$

(iii) *Let* $\mathbf{A} \in \mathbb{C}^{p\times p}$, $\mathbf{B} \in \mathbb{C}^{n\times n}$ *and* $\mathbf{T} \in \mathbb{C}^{p\times n}$. *Then,*

$$\mathrm{Tr}\,(\mathbf{ATBT}^*) = \mathrm{vec}^*\mathbf{T}(\mathbf{B}' \otimes \mathbf{A})\mathrm{vec}\,\mathbf{T}.$$

Proof Only statement (i) is established:

$$\mathrm{vec}'\mathbf{A}\mathrm{vec}\,\mathbf{B} = \sum_{i,j,k,l} a_{ij}b_{kl}(\mathbf{e}_j \otimes \mathbf{d}_i)'(\mathbf{e}_l \otimes \mathbf{d}_k) = \sum_{i,j} a_{ij}b_{ij} = \mathrm{Tr}\,(\mathbf{A}'\mathbf{B}).$$

□

Connected to the Kronecker product and the *vec*-operator is the commutation matrix. It is an orthogonal matrix which can be defined as follows.

Definition 3.5 Let $\mathbf{d}_i \in \mathbb{R}^p$ and $\mathbf{e}_j \in \mathbb{R}^q$ be unit basis vectors. The commutation matrix $\mathbf{K}_{p,q} : pq \times pq$ is defined by

$$\mathbf{K}_{p,q} = \sum_{i=1}^{p}\sum_{j=1}^{q}(\mathbf{d}_i\mathbf{e}_j') \otimes (\mathbf{e}_j\mathbf{d}_i').$$

Now some useful relations for the commutation matrix are presented. For proofs and additional relations, see Kollo and von Rosen [24, Sect. 1.3].

Theorem 3.3 *Let* $\mathbf{K}_{p,q}$ *be the commutation matrix. Then,*

(i) $\mathbf{K}_{p,q} = \mathbf{K}_{q,p}'$;
(ii) $\mathbf{K}_{p,q}\mathbf{K}_{q,p} = \mathbf{I}_{pq}$;
(iii) $\mathbf{K}_{p,1} = \mathbf{K}_{1,p} = \mathbf{I}_p$;
(iv) $\mathbf{K}_{p,q}\mathrm{vec}\,\mathbf{A} = \mathrm{vec}\,\mathbf{A}'$, $\mathbf{A} \in \mathbb{C}^{p\times q}$;
(v) *for* $\mathbf{A} \in \mathbb{R}^{p\times q}$, $\mathbf{B} \in \mathbb{R}^{q\times r}$, $\mathbf{C} \in \mathbb{R}^{r\times s}$ *and* $\mathbf{D} \in \mathbb{R}^{s\times p}$
$$\begin{aligned}\mathrm{Tr}\,(\mathbf{ABCD}) &= (\mathrm{vec}'\mathbf{C}' \otimes \mathrm{vec}'\mathbf{A})(\mathbf{I}_r \otimes \mathbf{K}_{s,q} \otimes \mathbf{I}_p)(\mathrm{vec}\,\mathbf{B} \otimes \mathrm{vec}\,\mathbf{D}')\\ &= (\mathrm{vec}'\mathbf{B} \otimes \mathrm{vec}'\mathbf{D})\mathbf{K}_{r,pqs}(\mathrm{vec}\,\mathbf{A} \otimes \mathrm{vec}\,\mathbf{C});\end{aligned}$$
(vi) *letting* $\mathbf{A} \in \mathbb{C}^{p\times q}$ *and* $\mathbf{B} \in \mathbb{C}^{r\times s}$, *then* $\mathbf{A} \otimes \mathbf{B} = \mathbf{K}_{p,r}(\mathbf{B} \otimes \mathbf{A})\mathbf{K}_{s,q}$.

Statement (iv) of the theorem is sometimes used as a definition of the commutation matrix.

3.2.2 Matrix Derivatives

In this section, four types of matrix derivatives are defined. One will be used when the matrices are unstructured, two are designed for handling symmetric matrices and one matrix derivative is suitable for dealing with complex matrices.

Definition 3.6 Let $\mathbf{Y} \in \mathbb{R}^{m\times n}$ be a function of $\mathbf{X} \in \mathbb{R}^{p\times q}$, which are both supposed to be real valued. The kth matrix derivative for unstructured matrices is, for $k \in \{1, 2, \ldots\}$, defined by

$$\frac{d^k\mathbf{Y}}{d\mathbf{X}^k} = \frac{d}{d\mathbf{X}}\frac{d^{k-1}\mathbf{Y}}{d\mathbf{X}^{k-1}},$$

and

$$\frac{d\mathbf{Y}}{d\mathbf{X}} = \frac{d\,\mathrm{vec}'\mathbf{Y}}{d\,\mathrm{vec}\,\mathbf{X}}, \quad pq \times mn, \qquad \frac{d^0\mathbf{Y}}{d\mathbf{X}^0} = \mathbf{Y},$$

where

$$\frac{d}{d\mathbf{X}} = \left(\frac{d}{dx_{11}}, \ldots, \frac{d}{dx_{p1}}, \frac{d}{dx_{12}}, \ldots, \frac{d}{dx_{p2}}, \ldots, \frac{d}{dx_{1q}}, \ldots, \frac{d}{dx_{pq}}\right)'.$$

Another way to write the first derivative in Definition 3.6 is to state that for $\mathbf{Y} \in \mathbb{R}^{m\times n}$ and $\mathbf{X} \in \mathbb{R}^{p\times q}$,

$$\frac{d\mathbf{Y}}{d\mathbf{X}} = \sum_I \frac{d\,y_{ij}}{d\,x_{rs}}(\mathbf{g}_s \otimes \mathbf{f}_r)(\mathbf{e}_j \otimes \mathbf{d}_i)', \tag{3.1}$$

where $I = \{i, j, r, s : 1 \le i \le m; 1 \le j \le n; 1 \le r \le q; 1 \le s \le p\}$ and $\mathbf{d}_i \in \mathbb{R}^m$, $\mathbf{e}_j \in \mathbb{R}^n$, $\mathbf{f}_r \in \mathbb{R}^q$ and $\mathbf{g}_s \in \mathbb{R}^p$ are unit basis vectors defined as in Sect. 3.2.1.

Now a matrix derivative designed for handling symmetric matrices is defined. This derivative gives us, for example, an adequate tool for deriving inverse Wishart moments, which are presented in Sect. 3.3.5.

Definition 3.7 Let $\mathbf{Y} \in \mathbb{R}^{m\times n}$ be a real-valued matrix which is a function of a real-valued symmetric matrix $\mathbf{X} \in \mathbb{R}^{q\times q}$. The kth matrix derivative, for $k \in \{1, 2, \ldots\}$, is defined, similarly to Definition 3.6, by

$$\frac{\widetilde{d}^k\mathbf{Y}}{d\mathbf{X}^k} = \frac{\widetilde{d}}{d\mathbf{X}}\frac{d^{k-1}\mathbf{Y}}{d\mathbf{X}^{k-1}},$$

and

$$\frac{\widetilde{d}\mathbf{Y}}{d\mathbf{X}} = \sum_I \frac{d\,y_{ij}}{d\,x_{rs}}(\mathbf{f}_s \otimes \mathbf{f}_r)(\mathbf{e}_j \otimes \mathbf{d}_i)'\varepsilon_{rs}, \quad q^2 \times mn,$$

where I is the same set as in (3.1) (with $q = p$) and

$$\varepsilon_{rs} = \begin{cases} 1, & r = s, \\ \frac{1}{2}, & r \neq s. \end{cases}$$

Throughout the article, $\dfrac{\widetilde{d}\,\mathbf{Y}}{d\,\mathbf{X}}$ will be used for representing this derivative.

From the above definitions, it follows that the matrix derivative is only an organization of partial derivatives. Accordingly, also other types of matrix derivatives also exist, e.g.,

$$\frac{d\,\mathbf{Y}}{d\,\mathbf{X}} = \frac{d}{d\,\mathrm{vec}\,'\mathbf{X}}\mathrm{vec}\,\mathbf{Y}, \qquad \text{or} \qquad \frac{d\,\mathbf{Y}}{d\,\mathbf{X}} = \mathbf{Y} \otimes \frac{d}{d\,\mathrm{vec}\,\mathbf{X}}.$$

The choice of derivative is governed by the aims of the applications. For the derivative in Definition 3.6, there is a nice expression for the chain rule, but other versions of a matrix derivative can, for example, yield more easily interpretable moment expressions when differentiating the moment generating function several times. After the next two theorems, two additional types of matrix derivatives are presented. One is designed for handling complex matrices and the other is useful when deriving moments for the Wishart distribution. For more relations and complete proofs, we refer to Kollo and von Rosen [24, Sect. 1.4].

Some useful relations for the derivatives given in Definitions 3.6 and 3.7 are now presented.

Theorem 3.4 *Let* $\dfrac{d\,\mathbf{Y}}{d\,\mathbf{X}}$ *be as in Definition 3.6 and suppose that all constant* $\mathbf{A}$ *and* $\mathbf{B}$ *matrices and all random* $\mathbf{X}$*,* $\mathbf{Y}$ *and* $\mathbf{Z}$ *matrices are of a proper size. Then,*

(i) *if* $\mathbf{X} \in \mathbb{R}^{p\times q}$, $\dfrac{d\,\mathbf{X}}{d\,\mathbf{X}} = \mathbf{I}_{pq}$;

(ii) *if* $z \in \mathbb{R}$ *and* $\mathbf{Y} \in \mathbb{R}^{r\times s}$ *are functions of* $\mathbf{X}$, $\dfrac{d\,\mathbf{Y}z}{d\,\mathbf{X}} = \dfrac{d\,\mathbf{Y}}{d\,\mathbf{X}}z + \dfrac{d\,z}{d\,\mathbf{X}}\mathrm{vec}\,'\mathbf{Y}$;

(iii) $\dfrac{d\,(\mathbf{A}'\mathrm{vec}\,\mathbf{X})}{d\,\mathbf{X}} = \mathbf{A}$;

(iv) *if* $\mathbf{Y}$ *and* $\mathbf{Z}$ *are functions of* $\mathbf{X}$, $\dfrac{d\,(\mathbf{Y}+\mathbf{Z})}{d\,\mathbf{X}} = \dfrac{d\,\mathbf{Y}}{d\,\mathbf{X}} + \dfrac{d\,\mathbf{Z}}{d\,\mathbf{X}}$;

(v) *if* $\mathbf{Y}$ *is a function of* $\mathbf{X}$ *and* $\mathbf{Z}$ *is a function of* $\mathbf{Y}$, $\dfrac{d\,\mathbf{Z}}{d\,\mathbf{X}} = \dfrac{d\,\mathbf{Y}}{d\,\mathbf{X}}\dfrac{d\,\mathbf{Z}}{d\,\mathbf{Y}}$ *(chain rule);*

(vi) $\dfrac{d\,(\mathbf{AXB})}{d\,\mathbf{X}} = \mathbf{B} \otimes \mathbf{A}'$;

(vii) *if* $\mathbf{Y}$ *is a function of* $\mathbf{X}$, $\dfrac{d\,(\mathbf{AYB})}{d\,\mathbf{X}} = \dfrac{d\,\mathbf{Y}}{d\,\mathbf{X}}(\mathbf{B} \otimes \mathbf{A}')$;

(viii) *if* $\mathbf{Y} \in \mathbb{R}^{r\times s}$ *and* $\mathbf{Z} \in \mathbb{R}^{s\times n}$ *are functions of* $\mathbf{X}$,

$$\frac{d\,(\mathbf{YZ})}{d\,\mathbf{X}} = \frac{d\,\mathbf{Y}}{d\,\mathbf{X}}(\mathbf{Z}\otimes\mathbf{I}_r) + \frac{d\,\mathbf{Z}}{d\,\mathbf{X}}(\mathbf{I}_n\otimes\mathbf{Y}');$$

(ix) *if* $\mathbf{X}^{-1}$ *exists,* $\dfrac{d\,\mathbf{X}^{-1}}{d\,\mathbf{X}} = -(\mathbf{X}^{-1}\otimes\mathbf{X}^{-1'});$

(x) *if* $\mathbf{Y} \in \mathbb{R}^{r\times s}$ *and* $\mathbf{Z} \in \mathbb{R}^{m\times n}$ *are functions of* $\mathbf{X}$

$$\frac{d\,(\mathbf{Y}\otimes\mathbf{Z})}{d\,\mathbf{X}} = \left(\frac{d\,\mathbf{Y}}{d\,\mathbf{X}}\otimes \mathrm{vec}\,'\mathbf{Z} + \mathrm{vec}\,'\mathbf{Y}\otimes\frac{d\,\mathbf{Z}}{d\,\mathbf{X}}\right)(\mathbf{I}_s\otimes\mathbf{K}_{r,n}\otimes\mathbf{I}_m);$$

(xi) $\dfrac{d\,\mathrm{Tr}\,(\mathbf{A}'\mathbf{X})}{d\,\mathbf{X}} = \mathrm{vec}\,\mathbf{A};$

(xii) $\dfrac{d\,\mathrm{Tr}\,(\mathbf{AXBX}')}{d\,\mathbf{X}} = \mathrm{vec}\,(\mathbf{A}'\mathbf{XB}') + \mathrm{vec}\,(\mathbf{AXB});$

(xiii) *if* $\mathbf{X}$ *is non-singular,* $\dfrac{d\,|\mathbf{X}|^s}{d\,\mathbf{X}} = s|\mathbf{X}|\mathrm{vec}\,(\mathbf{X}^{-1'}).$

Moreover, in the next theorem, those relations of Theorem 3.4 are presented which will change when $\mathbf{X}$ is symmetric and the derivative in Definition 3.7 is applied.

Theorem 3.5 *Let* $\mathbf{X} \in \mathbb{R}^{p\times p}$ *be symmetric and let* $\dfrac{\widetilde{d}\,\mathbf{Y}}{d\,\mathbf{X}}$ *be as in Definition 3.7. Then,*

(i) $\dfrac{\widetilde{d}\,\mathbf{X}}{d\,\mathbf{X}} = \frac{1}{2}(\mathbf{I}_{p^2} + \mathbf{K}_{p,p});$

(ii) $\dfrac{\widetilde{d}\,c\mathbf{X}}{d\,\mathbf{X}} = c\frac{1}{2}(\mathbf{I}_{p^2} + \mathbf{K}_{p,p});$

(iii) $\dfrac{\widetilde{d}\,(\mathbf{A}'\mathrm{vec}\,\mathbf{X})}{d\,\mathbf{X}} = \frac{1}{2}(\mathbf{I}_{p^2} + \mathbf{K}_{p,p})\mathbf{A};$

(iv) $\dfrac{\widetilde{d}\,(\mathbf{AXB})}{d\,\mathbf{X}} = \frac{1}{2}(\mathbf{I}_{p^2} + \mathbf{K}_{p,p})(\mathbf{B}\otimes\mathbf{A}');$

(v) *if* $\mathbf{X}^{-1}$ *exists,* $\dfrac{\widetilde{d}\,\mathbf{X}^{-1}}{d\,\mathbf{X}} = -\frac{1}{2}(\mathbf{I}_{p^2} + \mathbf{K}_{p,p})(\mathbf{X}^{-1}\otimes\mathbf{X}^{-1});$

(vi) $\dfrac{\widetilde{d}\,\mathrm{Tr}\,(\mathbf{A}'\mathbf{X})}{d\,\mathbf{X}} = \frac{1}{2}\mathrm{vec}\,(\mathbf{A}+\mathbf{A}');$

(vii) *if* $\mathbf{X}$ *is non-singular,* $\dfrac{\widetilde{d}\,|\mathbf{X}|^s}{d\,\mathbf{X}} = s|\mathbf{X}|\mathrm{vec}\,(\mathbf{X}^{-1});$

Note that in Theorem 3.5, the matrix $\frac{1}{2}(\mathbf{I}_{p^2} + \mathbf{K}_{p,p})$ is an idempotent matrix and hence a projector. Therefore, several expressions in Theorem 3.5 have a geometrical interpretation.

Another matrix derivative which is useful for handling symmetric matrices can be defined as follows.

Definition 3.8 Let $\mathbf{Y} \in \mathbb{R}^{m\times n}$ be a real-valued matrix which is a function of a real-valued symmetric matrix $\mathbf{X} \in \mathbb{R}^{q\times q}$. The kth matrix derivative, for $k \in \{1, 2, \ldots\}$, is defined, similarly to Definition 3.6, by

$$\frac{\tilde{\tilde{d}}^k\, \mathbf{Y}}{d\,\mathbf{X}^k} = \frac{\tilde{\tilde{d}}}{d\mathbf{X}} \frac{d^{k-1}\,\mathbf{Y}}{d\,\mathbf{X}^{k-1}},$$

and

$$\frac{\tilde{\tilde{d}}\,\mathbf{Y}}{d\,\mathbf{X}} = \sum_I \frac{d\,y_{ij}}{d\,x_{rs}} (\mathbf{d}_i\mathbf{e}_j' \otimes \mathbf{f}_r\mathbf{f}_s')\varepsilon_{rs},$$

where I and ε_{rs} are the same as in Definition 3.7. The notation $\frac{\tilde{\tilde{d}}\,\mathbf{Y}}{d\,\mathbf{X}}$ will be used for this derivative.

The next theorem comprises a number of relations which will be used when considering the Wishart matrix (see Definition 3.23).

Theorem 3.6 *Let* $\mathbf{X} \in \mathbb{R}^{p\times p}$ *be symmetric and* $\frac{\tilde{\tilde{d}}\,\mathbf{Y}}{d\,\mathbf{X}}$ *be given as in Definition 3.8. Then,*

(i) $\dfrac{\tilde{\tilde{d}}\,(\mathbf{Z}+\mathbf{Y})}{d\,\mathbf{X}} = \dfrac{\tilde{\tilde{d}}\,\mathbf{Z}}{d\,\mathbf{X}} + \dfrac{\tilde{\tilde{d}}\,\mathbf{Y}}{d\,\mathbf{X}}$;

(ii) $\dfrac{\tilde{\tilde{d}}\,(\mathbf{ZY})}{d\,\mathbf{X}} = \dfrac{\tilde{\tilde{d}}\,\mathbf{Z}}{d\,\mathbf{X}}(\mathbf{Y} \otimes \mathbf{I}_p) + (\mathbf{Z} \otimes \mathbf{I}_p)\dfrac{\tilde{\tilde{d}}\,\mathbf{Y}}{d\,\mathbf{X}}$;

(iii) $\dfrac{\tilde{\tilde{d}}\,\mathbf{Z}y}{d\,\mathbf{X}} = \dfrac{\tilde{\tilde{d}}\,\mathbf{Z}}{d\,\mathbf{X}}y + \mathbf{Z} \otimes \dfrac{\tilde{\tilde{d}}\,\mathbf{y}}{d\,\mathbf{X}}$;

(iv) $\dfrac{\tilde{\tilde{d}}\,(\mathbf{AYB})}{d\,\mathbf{X}} = (\mathbf{A} \otimes \mathbf{I}_p)\dfrac{\tilde{\tilde{d}}\,\mathbf{Y}}{d\,\mathbf{X}}(\mathbf{B} \otimes \mathbf{I}_p)$;

(v) *letting* $\mathbf{Y} \in \mathbb{R}^{q\times r}$ *and* $\mathbf{Z} \in \mathbb{R}^{s\times t}$, *then*

$$\frac{\tilde{\tilde{d}}\,(\mathbf{Y} \otimes \mathbf{Z})}{d\,\mathbf{X}} = \mathbf{Y} \otimes \frac{\tilde{\tilde{d}}\,\mathbf{Z}}{d\,\mathbf{X}} + (\mathbf{K}_{q,s} \otimes \mathbf{I}_p)\left(\mathbf{Z} \otimes \frac{\tilde{\tilde{d}}\,\mathbf{Y}}{d\,\mathbf{X}}\right)(\mathbf{K}_{t,r} \otimes \mathbf{I}_p),$$

$$\frac{\tilde{\tilde{d}}\,(\operatorname{vec}\mathbf{Y}\operatorname{vec}'\mathbf{Z})}{d\,\mathbf{X}} = \frac{\tilde{\tilde{d}}\operatorname{vec}\mathbf{Y}}{d\,\mathbf{X}}(\operatorname{vec}'\mathbf{Z} \otimes \mathbf{I}_p) + (\operatorname{vec}\mathbf{Y} \otimes \mathbf{I}_p)\frac{\tilde{\tilde{d}}\operatorname{vec}'\mathbf{Z}}{d\,\mathbf{X}};$$

(vi) *letting* $\mathbf{Y}$ *be non-singular, then* $\dfrac{\tilde{\tilde{d}}\,\mathbf{Y}^{-1}}{d\,\mathbf{X}} = -(\mathbf{Y}^{-1} \otimes \mathbf{I}_p)\dfrac{\tilde{\tilde{d}}\,\mathbf{Y}}{d\,\mathbf{X}}(\mathbf{Y}^{-1} \otimes \mathbf{I}_p)$;

(vii) $\dfrac{\tilde{\tilde{d}}\,\mathbf{X}}{d\,\mathbf{X}} = \dfrac{1}{2}(\operatorname{vec}\mathbf{I}_p\operatorname{vec}'\mathbf{I}_p + \mathbf{K}_{p,p})$;

(viii) $\dfrac{\widetilde{\widetilde{d}}\,\mathrm{Tr}\,(\mathbf{A}'\mathbf{X})}{d\,\mathbf{X}} = \dfrac{1}{2}(\mathbf{A}+\mathbf{A}')$;

(ix) $\dfrac{\widetilde{\widetilde{d}}\,|\mathbf{X}|^r}{d\,\mathbf{X}} = r\mathbf{X}^{-1}|\mathbf{X}|^r$;

(x) *letting* $\mathbf{X}$ *be non-singular, then*

$$\frac{\widetilde{\widetilde{d}}\,\mathrm{vec}\,\mathbf{X}^{-1}}{d\,\mathbf{X}} = -(\mathbf{I}_p \otimes \mathbf{X}^{-1} \otimes \mathbf{I}_p)\left(\mathbf{I}_p \otimes \frac{\widetilde{\widetilde{d}}\,\mathbf{X}}{d\,\mathbf{X}}\right)(\mathrm{vec}\,\mathbf{X}^{-1} \otimes \mathbf{I}_p).$$

It is also useful to define a matrix derivative for complex-valued matrices as follows.

Definition 3.9 Let $\mathbf{Y}$ and $\mathbf{X}$ be complex matrices. Then,

$$\frac{{}^{\mathbb{C}}d\mathbf{Y}}{d\,\mathbf{X}} = \frac{d\,\mathbf{Y}}{d\,\Re\mathbf{X}} + \imath\frac{d\,\mathbf{Y}}{d\,\Im\mathbf{X}},$$

where $\imath$ is the imaginary unit, $\Re$ and $\Im$ select the real and the imaginary parts, respectively, and $\frac{d\,\mathbf{Y}}{d\,\mathbf{X}}$ is given in Definition 3.6. Higher-order complex matrix derivatives are defined by

$$\frac{{}^{\mathbb{C}}d^k\,\mathbf{Y}}{d\,\mathbf{X}^k} = \frac{{}^{\mathbb{C}}d}{d\,\mathbf{X}}\overline{\frac{{}^{\mathbb{C}}d^{k-1}\,\mathbf{Y}}{d\,\mathbf{X}^{k-1}}}, \quad k \in \{2, 3, \ldots\},$$

where $\overline{\mathbf{A}}$ represents the conjugate of $\mathbf{A}$.

Now several relations involving the "complex derivative" are presented which all will be used in the article. Note that some relations are counterintuitive, which is linked to the non-differentiability of certain complex variables. Our definition should just be considered as an operator which is designed to solve certain statistical problems and is not connected to differentiability, in contrast to the case with the real matrix derivatives which were presented earlier.

Theorem 3.7 *Let* $\mathbf{T} \in \mathbb{C}^{p\times n}$. *Then,*

(i) $\dfrac{{}^{\mathbb{C}}d\overline{\mathbf{T}}}{d\,\mathbf{T}} = 2\mathbf{I}_{pn}, \dfrac{{}^{\mathbb{C}}d\mathbf{T}}{d\,\mathbf{T}} = \mathbf{0}$;

(ii) *if* $\mathbf{A} \in \mathbb{C}^{p\times n}$, $\dfrac{{}^{\mathbb{C}}d}{d\,\mathbf{T}}\mathrm{Tr}\,(\mathbf{T}^*\mathbf{A}) = 2\mathrm{vec}\,\mathbf{A}, \dfrac{{}^{\mathbb{C}}d}{d\,\mathbf{T}}\mathrm{Tr}\,(\mathbf{T}'\mathbf{A}) = \mathbf{0}$;

(iii) *if* $\mathbf{X} \in \mathbb{C}^{p\times n}$ *and* $y \in \mathbb{C}$ *are functions in* $\mathbf{T}$,

$$\frac{{}^{\mathbb{C}}d\mathbf{X}y}{d\,\mathbf{T}} = \frac{{}^{\mathbb{C}}d\mathbf{X}}{d\,\mathbf{T}}y + \frac{{}^{\mathbb{C}}dy}{d\,\mathbf{T}}\mathrm{vec}\,'\mathbf{X};$$

(iv) *if* $\mathbf{X} \in \mathbb{C}^{r\times s}$ *and* $\mathbf{Y} \in \mathbb{C}^{s\times n}$ *are functions in* $\mathbf{T}$,

$$\frac{{}^{\mathbb{C}}d(\mathbf{XY})}{d\,\mathbf{T}} = \frac{{}^{\mathbb{C}}d\mathbf{X}}{d\,\mathbf{T}}(\mathbf{Y}\otimes\mathbf{I}_r) + \frac{{}^{\mathbb{C}}d\mathbf{Y}}{d\,\mathbf{T}}(\mathbf{I}_n\otimes\mathbf{X}');$$

(v) *if* $\mathbf{X} \in \mathbb{C}^{r\times s}$ *and* $\mathbf{Y} \in \mathbb{C}^{m\times n}$ *are functions in* $\mathbf{T}$,
$$\frac{{}^{\mathbb{C}}d(\mathbf{X}\otimes\mathbf{Y})}{d\,\mathbf{T}} = \left(\frac{{}^{\mathbb{C}}d\mathbf{X}}{d\,\mathbf{T}} \otimes \mathrm{vec}'\mathbf{Y} + \mathrm{vec}'\mathbf{X} \otimes \frac{{}^{\mathbb{C}}d\mathbf{Y}}{d\,\mathbf{T}}\right)(\mathbf{I}_s \otimes \mathbf{K}_{r,n} \otimes \mathbf{I}_m);$$

(vi) *if* $\boldsymbol{\Sigma}$ *and* $\boldsymbol{\Psi}$ *are Hermitian and positive semi-definite,*
$$\frac{{}^{\mathbb{C}}d}{d\,\mathbf{T}}\mathrm{Tr}\,(\boldsymbol{\Sigma}\mathbf{T}\boldsymbol{\Psi}\mathbf{T}^*) = 2(\boldsymbol{\Psi}' \otimes \boldsymbol{\Sigma})\mathrm{vec}\,\mathbf{T};$$

(vii) *if* $\boldsymbol{\Sigma}$ *and* $\boldsymbol{\Psi}$ *are Hermitian and positive semi-definite,*
$$\frac{{}^{\mathbb{C}}d^2}{d\,\mathbf{T}^2}\mathrm{Tr}\,(\boldsymbol{\Sigma}\mathbf{T}\boldsymbol{\Psi}\mathbf{T}^*) = 4\boldsymbol{\Psi}' \otimes \boldsymbol{\Sigma}.$$

Proof Only statements (i), (ii), (vi), and (vii) will be proven. The relations in (iii), (iv), and (v) follow from Theorem 3.4, since these relations are only connected to the organization of the partial derivatives.

Concerning statements (i) and (ii), let $\mathbf{T} = \mathbf{T}_0 + \imath\mathbf{T}_1$, where $\mathbf{T}_0$ and $\mathbf{T}_1$ are real matrices. Then,

$$\frac{{}^{\mathbb{C}}d\overline{\mathbf{T}}}{d\,\mathbf{T}} = \frac{d\,(\mathbf{T}_0 - \imath\mathbf{T}_1)}{d\,\mathbf{T}_0} + \imath\frac{d\,(\mathbf{T}_0 - \imath\mathbf{T}_1)}{d\,\mathbf{T}_1} = \frac{d\,\mathbf{T}_0}{d\,\mathbf{T}_0} + \frac{d\,\mathbf{T}_1}{d\,\mathbf{T}_1} = 2\mathbf{I}_{pn},$$
$$\frac{{}^{\mathbb{C}}d\mathbf{T}}{d\,\mathbf{T}} = \frac{d\,(\mathbf{T}_0 + \imath\mathbf{T}_1)}{d\,\mathbf{T}_0} + \imath\frac{d\,(\mathbf{T}_0 + \imath\mathbf{T}_1)}{d\,\mathbf{T}_1} = \frac{d\,\mathbf{T}_0}{d\,\mathbf{T}_0} - \frac{d\,\mathbf{T}_1}{d\,\mathbf{T}_1} = \mathbf{0},$$

which establishes statement (i). Moreover, letting $\mathbf{A} = \mathbf{A}_0 + \imath\mathbf{A}_1$, where $\mathbf{A}_0$ and $\mathbf{A}_1$ are real matrices,

$$\begin{aligned}\frac{{}^{\mathbb{C}}d\mathrm{Tr}\,(\mathbf{T}^*\mathbf{A})}{d\,\mathbf{T}} &= \frac{d\mathrm{Tr}\,\left[\mathbf{T}_0'\mathbf{A}_0 + \mathbf{T}_1'\mathbf{A}_1 + \imath(\mathbf{T}_0'\mathbf{A}_1 - \mathbf{T}_1'\mathbf{A}_0)\right]}{d\,\mathbf{T}_0}\\ &+\imath\frac{d\mathrm{Tr}\,\left[\mathbf{T}_0'\mathbf{A}_0 + \mathbf{T}_1'\mathbf{A}_1 + \imath(\mathbf{T}_0'\mathbf{A}_1 - \mathbf{T}_1'\mathbf{A}_0)\right]}{d\,\mathbf{T}_1} = 2(\mathrm{vec}\,\mathbf{A}_0 + \imath\,\mathrm{vec}\,\mathbf{A}_1) = 2\mathrm{vec}\,\mathbf{A},\\ \frac{{}^{\mathbb{C}}d\mathrm{Tr}\,(\mathbf{T}'\mathbf{A})}{d\,\mathbf{T}} &= \frac{d\mathrm{Tr}\,\left[\mathbf{T}_0'\mathbf{A}_0 - \mathbf{T}_1'\mathbf{A}_1 + \imath(\mathbf{T}_0'\mathbf{A}_1 + \mathbf{T}_1'\mathbf{A}_0)\right]}{d\,\mathbf{T}_0}\\ &\quad+\imath\frac{d\mathrm{Tr}\,\left[\mathbf{T}_0'\mathbf{A}_0 - \mathbf{T}_1'\mathbf{A}_1 + \imath(\mathbf{T}_0'\mathbf{A}_1 + \mathbf{T}_1'\mathbf{A}_0)\right]}{d\,\mathbf{T}_1} = \mathbf{0}.\end{aligned}$$

Now statements (vi) and (vii) are proven. Firstly, it can be noted (see also Theorem 3.2 (iii)) that

$$\mathrm{Tr}\,(\boldsymbol{\Sigma}\mathbf{T}\boldsymbol{\Psi}\mathbf{T}^*) = \mathrm{vec}'\mathbf{T}(\boldsymbol{\Psi} \otimes \boldsymbol{\Sigma}')\mathrm{vec}\,\overline{\mathbf{T}}.$$

Using statement (iv) and thereafter statement (i) yields

$$\frac{{}^{\mathbb{C}}d}{d\mathbf{T}}\mathrm{Tr}\,(\boldsymbol{\Sigma}\mathbf{T}\boldsymbol{\Psi}\mathbf{T}^*) = \frac{{}^{\mathbb{C}}d\mathbf{T}}{d\mathbf{T}}(\boldsymbol{\Psi}\otimes\boldsymbol{\Sigma}')\mathrm{vec}\,\overline{\mathbf{T}} + \frac{{}^{\mathbb{C}}d(\boldsymbol{\Psi}\otimes\boldsymbol{\Sigma}')\mathrm{vec}\,\overline{\mathbf{T}}}{d\mathbf{T}}\mathrm{vec}\,\mathbf{T}$$
$$= \frac{{}^{\mathbb{C}}d\overline{\mathbf{T}}}{d\mathbf{T}}(\boldsymbol{\Psi}'\otimes\boldsymbol{\Sigma})\mathrm{vec}\,\mathbf{T} = 2(\boldsymbol{\Psi}'\otimes\boldsymbol{\Sigma})\mathrm{vec}\,\mathbf{T},$$

proving statement (vi).

According to Definition 3.9, where higher-order complex matrix derivatives are defined, and statement (vi),

$$\frac{{}^{\mathbb{C}}d^2}{d\mathbf{T}^2}\mathrm{Tr}\,(\boldsymbol{\Sigma}\mathbf{T}\boldsymbol{\Psi}\mathbf{T}^*) = \frac{{}^{\mathbb{C}}d}{d\mathbf{T}}2(\boldsymbol{\Psi}\otimes\overline{\boldsymbol{\Sigma}})\mathrm{vec}\,\overline{\mathbf{T}} = 2\frac{{}^{\mathbb{C}}d\overline{\mathbf{T}}}{d\mathbf{T}}(\boldsymbol{\Psi}'\otimes\boldsymbol{\Sigma}) = 4(\boldsymbol{\Psi}'\otimes\boldsymbol{\Sigma}).$$

Thus, statement (vii) has been derived. □

3.2.3 *Transforms*

Transforms have a prominent role in statistics. They are linked to distributions and the knowledge of distributions bears statistics. In this section, several transforms are presented.

3.2.3.1 The Characteristic Function and The Cumulant Generating Function

The Fourier transform is in statistics often termed a characteristic function or moment generating function. The Laplace transform is also often used as a moment generating function.

Definition 3.10 Let $\mathbf{X} \in \mathbb{R}^{p\times n}$ be an unstructured random matrix. Under the assumption that the integral in the next expression exists, the characteristic function, $\varphi_{\mathbf{X}}(\mathbf{T})$, $\mathbf{T} \in \mathbb{R}^{p\times n}$, is given by

$$\varphi_{\mathbf{X}}(\mathbf{T}) = \mathrm{E}\big(\exp\{\imath\,\mathrm{Tr}\,(\mathbf{T}'\mathbf{X})\}\big),$$

where $\imath$ is the imaginary unit and E (•) stands for the expectation, in this case with respect to **X**.

Definition 3.11 Let $\mathbf{X} \in \mathbb{R}^{p\times n}$ be an unstructured random matrix. Under the assumption that the characteristic function exists, the cumulant generating function, $\psi_{\mathbf{X}}(\mathbf{T})$, is given by $\psi_{\mathbf{X}}(\mathbf{T}) = \ln\varphi_{\mathbf{X}}(\mathbf{T})$, $\mathbf{T} \in \mathbb{R}^{p\times n}$.

In the above-given definitions, **X** was supposed to be unstructured. However, if there exist patterns in **X**, e.g., **X** being symmetric or being a correlation matrix, the definitions have to be modified so that there is a t_{ij}-element for each unique element in $\mathbf{X}$, e.g., each of the elements in the upper triangular part (including the diagonal) when a symmetric matrix is considered.

Supposing now that $\mathbf{X} \in \mathbb{C}^{p\times n}$, meaning that $\mathbf{X} = \mathbf{X}_0 + \imath\mathbf{X}_1$, where $\mathbf{X}_0$ and $\mathbf{X}_1$ are real-valued matrices of size $p \times n$, the following definition can be made.

Definition 3.12 Let $\mathbf{X} \in \mathbb{C}^{p\times n}$. Its characteristic function is given by

$$\varphi_{\mathbf{X}}(\mathbf{T}) = \mathrm{E}\big(\exp\{\imath\, \Re(\mathrm{Tr}\,(\mathbf{T}^{\star}\mathbf{X}))\}\big) = \mathrm{E}\big(\exp\{\imath\,\mathrm{Tr}\,(\mathbf{T}_0'\mathbf{X}_0 + \mathbf{T}_1'\mathbf{X}_1)\}\big),$$

where $\mathbf{T} \in \mathbb{C}^{p\times n}$, $\mathbf{T} = \mathbf{T}_0 + \imath\mathbf{T}_1$, $\Re(\bullet)$ selects the real part and $\mathbf{T}^*$ denotes the conjugate transpose.

3.2.3.2 The Stieltjes–Cauchy, the R- and the S-Transform

The Stieltjes–Cauchy transform is an alternative to the Fourier transform when deriving useful results, in particular in multivariate statistical analysis. For example, the Stieltjes–Cauchy transform can be used to handle the empirical spectral distribution function for symmetric matrices, i.e., to derive the spectral measure of random matrices of a large size. Thanks to its good algebraic properties, the Stieltjes–Cauchy transform is often used to simplify calculations. In the present article, this transform will appear in the results regarding the eigenvalue distribution of large Wishart matrices, see Theorem 3.20. Moreover, the eigenvalue distribution of a sum of large Wishart matrices is presented in Theorem 3.21, and corresponding results for large complex Wishart matrices are given in Theorem 3.22 using the Stieltjes–Cauchy transform.

In addition, in this section, the R-transform and S-transform are also presented. Both these transforms are related to the Stieltjes–Cauchy transform and have been found to be useful when considering asymptotic results for linear combinations and products of random matrices.

Definition 3.13 For the probability distribution function $F_{\mathbf{X}}$ on $\mathbb{R}$, the Stieltjes–Cauchy transform of $F_{\mathbf{X}}(x)$ is defined by

$$m_{\mathbf{X}}(z) = \int_{\mathbb{R}} \frac{1}{z-x} dF_{\mathbf{X}}(x),$$

where $z \in \{z : z \in \mathbb{C}, \Im(z) > 0\}$.

Note that the Stieltjes–Cauchy transform is also often referred to as the Cauchy transform. In some publications, $\widetilde{m}_{\mathbf{X}}(z) = \int_{\mathbb{R}} \frac{1}{x-z} dF_{\mathbf{X}}(x) = -m_{\mathbf{X}}(z)$ is also called the Stieltjes transform.

Definition 3.14 Let $\lambda_1 \le \cdots \le \lambda_n$ be all the eigenvalues of the square matrix $\mathbf{X}$. Then, the empirical spectral distribution (normalized spectral distribution) function $F_{\mathbf{X}}$ is given by

$$F_n^{\mathbf{X}}(u) = \frac{1}{n}\sum_{i=1}^{n} \mathbf{1}_{\{\lambda_i \le u\}},$$

where $\mathbf{1}_A$ is the indicator function of the set A.

The Stieltjes–Cauchy transform of the empirical spectral distribution F of the real symmetric matrix or, more generally, the Hermitian matrix $\mathbf{X}$, can be expressed via the sum of the diagonal elements of the matrix $(z\mathbf{I}_p - \mathbf{X})^{-1}$, namely

$$m_{\mathbf{X}}(z) = \int_{\mathbb{R}} \frac{1}{z-x} dF_p^{\mathbf{X}}(x) = \frac{1}{p}\sum_{i=1}^{p} \frac{1}{z - \lambda_i(\mathbf{X})} = \frac{1}{p}\operatorname{Tr}\left[(z\mathbf{I}_p - \mathbf{X})^{-1}\right], \quad (3.2)$$

where $\lambda_i(\mathbf{X})$ denotes the i-th eigenvalue of the matrix $\mathbf{X}$. Note that for the Hermitian matrix $\mathbf{X}$, all the eigenvalues are real valued.

For the well-defined Stieltjes–Cauchy transform, there exists a uniquely defined measure which can be obtained using the Stieltjes inversion formula presented in Theorem 3.9. Moreover, the convergence of the Stieltjes–Cauchy transform implies the convergence of the corresponding probability measures.

Lemma 3.1 *Let $\mathbf{X} \in \mathbb{C}^{p\times n}$ be a random matrix, while $\alpha \in \mathbb{R}\setminus\{0\}$ is a constant and $z \in \{z : z \in \mathbb{C}, \Im(z) > 0\}$. Then,*

$$\frac{n}{p} m_{\mathbf{X}^*\mathbf{X}}(z) = m_{\mathbf{X}\mathbf{X}^*}(z) - \frac{p-n}{pz}, \quad m_{\alpha\mathbf{X}}(\alpha z) = \frac{1}{\alpha} m_{\mathbf{X}}(z).$$

Another useful transform is the R-transform, which is defined via the Stieltjes–Cauchy transform as follows.

Definition 3.15 Let μ be a probability measure with compact support (i.e., for any open covering of the domain set, there exists a finite subcovering), while $m(z)$ is its Stieltjes–Cauchy transform. Then, the R-transform $R(z)$ is given by

$$R(z) = m^{-1}(z) - \frac{1}{z},$$

where $m^{-1}(\cdot)$ is the inverse with respect to composition, i.e., $m(m^{-1}(z)) = z$.

Theorem 3.8 *Let the distributions of the random matrices $\mathbf{X}$, $\mathbf{Y}$, $\mathbf{X}_n$, for all $n \in \mathbb{N}$, of size $p \times p$, have a compact support. The R-transform satisfies the following properties:*

(i) *(non–linearity) $R_{\alpha\mathbf{X}}(z) = \alpha R_{\mathbf{X}}(\alpha z)$ for every $\mathbf{X}$ and $\alpha \in \mathbb{C}$;*

(ii) *for any two free random matrices $\mathbf{X}$, $\mathbf{Y}$ (see Sect. 3.2.4 for definition of freeness),*

$$R_{\mathbf{X}+\mathbf{Y}}(z) = R_{\mathbf{X}}(z) + R_{\mathbf{Y}}(z);$$

(iii) *if* $\mathbf{X}$ *is a square matrix*

$$\lim_{n\to\infty} \frac{1}{p}\mathrm{E}\big(\mathrm{Tr}\,(\mathbf{X}_n^k)\big) = \frac{1}{p}\mathrm{E}\big(\mathrm{Tr}\,(\mathbf{X}^k)\big), \quad k \in \{1, 2, \ldots\},$$

and if there exists a neighborhood U *of 0 such that* $R_{\mathbf{X}}$ *and* $R_{\mathbf{X}_n}$ *are well-defined for* $n \in \mathbb{N}$*, then*

$$\lim_{n\to\infty} R_{\mathbf{X}_n}(y) = R_{\mathbf{X}}(y)$$

for all $y \in U$ *as a formal power series (convergence of* k_i *in Definition 3.15).*

Definition 3.16 The S-transform is defined through the R-transform via

$$S(z) = \frac{1}{R(zS(z))}.$$

3.2.3.3 An Inverse Transform

The Stieltjes inversion formula available in Akhiezer [1], for example, allows us to use the transform $m(z)$ to derive the measure μ.

Theorem 3.9 *For any open interval* $I = (a, b)$*, such that neither* a *nor* b *are atoms with respect to the probability measure* μ*, the Stieltjes inversion formula is given by*

$$\mu(I) = -\frac{1}{\pi} \lim_{y\to 0} \int_I \Im m(x + \imath y)dx.$$

Proof The proof is a detailed version of the proof given, e.g., in Couillet and Debbah [6]. We have

$$\begin{aligned}
-\frac{1}{\pi} \lim_{y\to 0} \int_I \Im m(x + \imath y)dx &= -\frac{1}{\pi} \lim_{y\to 0} \int_I \int_{\mathbb{R}} \Im \frac{1}{x + \imath y - t} d\mu(t)dx \\
&= \frac{1}{\pi} \lim_{y\to 0} \int_{\mathbb{R}} \int_a^b \frac{y}{(t-x)^2 + y^2} dx d\mu(t) \\
&= \frac{1}{\pi} \lim_{y\to 0} \int_{\mathbb{R}} \arctan\left(\frac{b-t}{y}\right) - \arctan\left(\frac{a-t}{y}\right) d\mu(t) \\
&= \frac{1}{\pi} \int_{\mathbb{R}} \lim_{y\to 0} \left[\arctan\left(\frac{b-t}{y}\right) - \arctan\left(\frac{a-t}{y}\right)\right] d\mu(t).
\end{aligned}$$

The possibility of interchanging the order of integration and taking the limit in the last equality follows by the bounded convergence theorem as $\mu(\mathbb{R}) = 1 < \infty$, and there exists a constant M such that

$$\arctan\left(\frac{b-t}{y}\right) - \arctan\left(\frac{a-t}{y}\right) < M, \qquad \forall y,\ \forall t,$$

meaning that for all y we have a uniformly bounded real-valued measurable function. Then, using the fact that $\lim_{y\to 0}\arctan(\frac{t}{y}) = \frac{\pi}{2}\mathrm{sgn}(t)$ for $t \in \mathbb{R}$, we obtain

$$\arctan\left(\frac{b-t}{y}\right) - \arctan\left(\frac{a-t}{y}\right) \xrightarrow{y\to 0} \begin{cases} 0, & \text{if } t < a \text{ or } t > b, \\ \pi, & \text{if } t \in (a, b), \end{cases}$$

which, using the dominated convergence theorem, completes the proof. □

More generally, the statement of Theorem 3.9 holds for any μ being a probability measure on $\mathbb{R}$, and for any $a < b$:

$$\mu((a, b)) + \frac{1}{2}\mu(\{a\}) + \frac{1}{2}\mu(\{b\}) = -\frac{1}{\pi}\lim_{y\to 0}\int_I \Im G(x + \imath y)dx.$$

3.2.4 *Definition of Moments*

There are several ways to define moments and cumulants. In this article, differentiation of the characteristic function for a "matrix distribution" is used to yield matrix moments and differentiation of the cumulant generating function is used to define "matrix cumulants". Moreover, there exist relatively deep mathematical approaches (algebraic/combinatorial approaches) for the treatment of cumulants and moments, e.g., see Speed and Silcock [39], Speed [38], and Rota and Shen [34], but these approaches will not be considered herein.

Definition 3.17 Let the characteristic function $\varphi_{\mathbf{X}}(\mathbf{T})$ be k times differentiable.

(i) The kth moment of an unstructured random matrix $\mathbf{X} \in \mathbb{R}^{p\times n}$, if it exists, equals

$$m_k(\mathbf{X}) = \frac{1}{i^k}\frac{d^k}{d\,\mathbf{T}^k}\varphi_{\mathbf{X}}(\mathbf{T})\bigg|_{\mathbf{T}=\mathbf{0}}, \qquad \mathbf{T} \in \mathbb{R}^{p\times n},$$

where the kth order matrix derivative is given in Definition 3.6.

(ii) The dispersion matrix $\mathrm{D}(\mathbf{X})$ is defined via

$$\mathrm{D}(\mathbf{X}) = m_2(\mathbf{X}) - m_1(\mathbf{X})m_1'(\mathbf{X}).$$

Using Definition 3.17 (i), one can obtain the next theorem.

Theorem 3.10 *Let* $\mathbf{X}$ *be an unstructured random matrix. If the moments up to order* k *exist, then for* $k \in \{1, 2, \ldots\}$,

$$m_k(\mathbf{X}) = \mathrm{E}\big(\mathrm{vec}\,\mathbf{X}(\mathrm{vec}'\mathbf{X})^{\otimes k-1}\big)\,.$$

Corollary 3.1 *Let* $\mathbf{x}$ *be an unstructured random matrix. If the moments up to order* k *exist, then for* $k \in \{1, 2, \ldots\}$,

$$m_k(\mathbf{x}) = \mathrm{E}\big(\mathbf{x}(\mathbf{x}')^{\otimes k-1}\big)\,.$$

A useful feature of the characteristic function is that it can be expanded via a Taylor series expansion, and then the expansion turns out to be an expansion based on the moments in Theorem 3.10.

Theorem 3.11 *Let* $\mathbf{X}$ *be an unstructured matrix. Then,*

$$\varphi_{\mathbf{X}}(\mathbf{T}) = 1 + \sum_{k=1}^{n} \frac{1}{k!}(\mathrm{vec}'\mathbf{T})^{\otimes k}\mathrm{vec}\,(m'_k(\mathbf{X})) + r_n,$$

where $m'_k(\mathbf{X})$ *is the transpose of* $m_k(\mathbf{X})$ *and* r_n *is a remainder term.*

Similarly to the definition of moments, the cumulants can be defined via differentiation of the cumulant generating function.

Definition 3.18 Let the cumulant generating function $\psi_{\mathbf{X}}(\mathbf{T})$ be k times differentiable. Then, the kth cumulant of an unstructured random matrix $\mathbf{X} \in \mathbb{R}^{p\times n}$ equals

$$c_k(\mathbf{X}) = \frac{1}{i^k}\frac{d^k}{d\,\mathbf{T}^k}\psi_{\mathbf{X}}(\mathbf{T})\bigg|_{\mathbf{T}=\mathbf{0}}, \qquad \mathbf{T} \in \mathbb{R}^{p\times n},$$

where the kth order matrix derivative is given in Definition 3.6.

The cumulant generating function can also be expanded with the help of a Taylor series expansion.

Theorem 3.12 *Let* $\mathbf{X}$ *be an unstructured matrix. Then,*

$$\psi_{\mathbf{X}}(\mathbf{T}) = \sum_{k=1}^{n} \frac{1}{k!}(\mathrm{vec}'\mathbf{T})^{\otimes k}\mathrm{vec}\,(c'_k(\mathbf{X})) + r_n,$$

where $c'_k(\mathbf{X})$ *is the transpose of* $c_k(\mathbf{X})$ *and* r_n *is a remainder term.*

The moments for complex variables can be defined in a similar way to that presented above. The characteristic function given by Definition 3.12 will be used and a suitable matrix derivative for differentiating the characteristic function is given by Definition 3.9.

Via Definitions 3.12 and 3.9, similarly to Theorem 3.10, the moments for complex random variables can be defined.

Definition 3.19 Let the characteristic function $\varphi_{\mathbf{X}}(\mathbf{T})$ be given in Definition 3.12.

(i) The kth moment of an unstructured random matrix $\mathbf{X} \in \mathbb{C}^{p\times n}$, if it exists, equals

$$m_k(\mathbf{X}) = \frac{1}{\imath^k}\frac{{}^{\mathbb{C}}d^k}{d\,\mathbf{T}^k}\varphi_{\mathbf{X}}(\mathbf{T})\bigg|_{\mathbf{T}=\mathbf{0}}, \qquad \mathbf{T} \in \mathbb{C}^{p\times n},$$

where the kth order matrix derivative is given in Definition 3.9.

(ii) The dispersion matrix $\mathrm{D}\,(\mathbf{X})$ for complex matrices is given by

$$\mathrm{D}\,(\mathbf{X}) = m_2(\mathbf{X}) - m_1(\mathbf{X})m_1^*(\mathbf{X}),$$

where $m_1^*(\mathbf{X})$ denotes the conjugate transpose of $m_1(\mathbf{X})$.

The idea of defining moments via complex "differentiation" of the characteristic function, in the case of the complex normal distribution, was first considered by Sultan and Tracy [43], but their technique differed somewhat from the technique utilized in this article.

Theorem 3.13 *Let* $\mathbf{X}$ *be an unstructured complex random matrix. Then,*

(i) $m_1(\mathbf{X}) = \mathrm{E}\,(\mathrm{vec}\,\mathbf{X})$;
(ii) $m_2(\mathbf{X}) = \mathrm{E}\,(\mathrm{vec}\,\mathbf{X}\mathrm{vec}^*\mathbf{X})$;
(iii) $m_3(\mathbf{X}) = \mathrm{E}\,(\mathrm{vec}\,\mathbf{X}(\mathrm{vec}'\mathbf{X} \otimes \mathrm{vec}^*\mathbf{X}))$;
(iv) $m_4(\mathbf{X}) = \mathrm{E}\,(\mathrm{vec}\,\mathbf{X}(\mathrm{vec}^*\mathbf{X} \otimes \mathrm{vec}'\mathbf{X} \otimes \mathrm{vec}^*\mathbf{X}))$.

Proof Let, with obvious notation, $\mathbf{X} = \mathbf{X}_0 + \imath\mathbf{X}_1$ and $\mathbf{T} = \mathbf{T}_0 + \imath\mathbf{T}_1$. Using Definition 3.19 (i) and applying $\frac{d^c\mathbf{Y}}{d\mathbf{X}}$ and $\varphi_{\mathbf{X}}(\mathbf{T})$ in Definitions 3.9 and3.12, respectively,

$$m_1(\mathbf{X}) = \frac{1}{\imath}\imath\,\mathrm{E}\big((\mathrm{vec}\,\mathbf{X}_0 + \imath\,\mathrm{vec}\,\mathbf{X}_1)\exp\{\imath\mathrm{Tr}\,(\mathbf{T}_0'\mathbf{X}_0 + \mathbf{T}_1'\mathbf{X}_1)\}\big)\bigg|_{\mathbf{T}=\mathbf{0}} = \mathrm{E}\,(\mathrm{vec}\,\mathbf{X}).$$

Again applying Definition 3.9 establishes

$$m_2(\mathbf{X}) = \frac{1}{\imath^2}\imath^2\,\mathrm{E}\big((\mathrm{vec}\,\mathbf{X}\mathrm{vec}^*\mathbf{X})\exp\{\imath\mathrm{Tr}\,(\mathbf{T}_0'\mathbf{X}_0 + \mathbf{T}_1'\mathbf{X}_1)\}\big)\bigg|_{\mathbf{T}=\mathbf{0}} = \mathrm{E}\,(\mathrm{vec}\,\mathbf{X}\mathrm{vec}^*\mathbf{X}).$$

The other two statements of the theorem are obtained by differentiating the characteristic function three and four times, respectively. □

From Definition 3.19 (i), it follows that moments of a higher order than four could have been presented, but they are rarely used.

Note that

$$\begin{aligned}\mathrm{vec}\,(m_2(\mathbf{X})) &= \mathrm{E}\,(\mathrm{vec}\,\overline{\mathbf{X}} \otimes \mathrm{vec}\,\mathbf{X}),\\ \mathrm{vec}^*(m_4(\mathbf{X})) &= \mathrm{E}\,(\mathrm{vec}'\mathbf{X} \otimes \mathrm{vec}^*\mathbf{X} \otimes \mathrm{vec}'\mathbf{X} \otimes \mathrm{vec}^*\mathbf{X}),\\ \mathrm{vec}\,(m_4(\mathbf{X})) &= \mathrm{E}\,(\mathrm{vec}\,\overline{\mathbf{X}} \otimes \mathrm{vec}\,\mathbf{X} \otimes \mathrm{vec}\,\overline{\mathbf{X}} \otimes \mathrm{vec}\,\mathbf{X}).\end{aligned}$$

Using the first and second relations above, the next theorem can be proved.

Theorem 3.14 *Let $\mathbf{X} \in \mathbb{C}^{p\times n}$ be an unstructured complex random matrix. Then,*

(i) $\mathrm{vec}\,\mathrm{E}\,(\mathrm{vec}\,(\mathbf{X}\mathbf{X}^*)) = (\mathrm{vec}'\mathbf{I}_n \otimes \mathbf{I}_{p^2})(\mathbf{I}_n \otimes \mathbf{K}_{n,p} \otimes \mathbf{I}_p)\mathrm{vec}\,(m_2(\mathbf{X}))$;

(ii) $\mathrm{vec}\,\mathrm{E}\,(\mathrm{vec}\,(\mathbf{X}\mathbf{X}^*)\mathrm{vec}^*(\mathbf{X}\mathbf{X}^*))$
$= \left[(\mathrm{vec}'\mathbf{I}_n \otimes \mathbf{I}_{p^2})(\mathbf{I}_n \otimes \mathbf{K}_{n,p} \otimes \mathbf{I}_p)\right]^{\otimes 2} (\mathbf{K}_{pn,pn} \otimes \mathbf{I}_{(pn)^2})\mathrm{vec}\,(m_4(\mathbf{X}))$.

This theorem is of use when deriving the dispersion for complex quadratic forms, for example, deriving the dispersion for the complex Wishart distribution, which will be defined later.

3.2.5 Moments of the Spectral Distribution Based on Freeness

In this section, the concept of freeness is introduced.

The free additive convolution of a measure μ (associated with a random matrix $\mathbf{X}$) and a measure ν (associated with a random matrix $\mathbf{Y}$), following Voiculescu [45], is defined as a measure (associated with the random matrix $\mathbf{X}+\mathbf{Y}$) which is such that statement (ii) of Theorem 3.8 holds.

We start by presenting a moment-cumulant relation for random variables and its free counterpart for square random matrices. We define free mixed cumulants, i.e., free cumulants of matrices that are not all identical, through mixed moments. Note that in the case of mixed moments and mixed cumulants of random variables, the moment-cumulant relation formula is given in the following form:

$$\mathrm{E}\,(X_1X_2 \times \cdots \times X_k) = \sum_{\pi\in P(k)} \prod_{p_j\in\pi} c_{|p_j|}(\mathbf{X}_{p_j}), \tag{3.3}$$

where $P(k)$ stands for the set of all the partitions of $\{1, 2, \ldots, k\}$, p_i are disjoint non-empty subsets of the partition π, such that $\bigcup_i p_i = \{1, 2, \ldots, k\}$, and $|\cdot|$ denotes cardinality. Moreover, vector $\mathbf{X}_{p_j} = (X_{i_1}, \ldots, X_{i_{|p_j|}})$, $i_t \in p_j$, $t = 1, 2, \ldots, |p_j|$ consists of variables X_i with indices from subset p_j and is trivially of length $|p_j|$. Note that the relation (3.3) can be used as a recursive definition of cumulants equivalent to Definition 3.18; see Mingo and Speicher [29, page 18] for comments on (3.3).

Let us consider random matrices $\mathbf{X}_i$ of size $p \times p$ and a functional

$$m_k(\mathbf{X}_1, \mathbf{X}_2, \ldots, \mathbf{X}_k) := \frac{1}{p}\mathrm{E}(\mathrm{Tr}\,(\mathbf{X}_1 \times \cdots \times \mathbf{X}_k))$$

that corresponds to the moment of a spectral distribution of $\mathbf{X}_1, \ldots, \mathbf{X}_k$, i.e., the mixed moment of spectral distribution. Note that due to the trace operator, the considered moments and later free cumulants are scalar objects.

A formula that defines recursively free mixed cumulants through mixed moments is derived from a relation that includes the sum over all the non-crossing partitions instead of all the partitions, in contrast to the statement above, namely

$$m_k(\mathbf{X}_1, \mathbf{X}_2, \ldots, \mathbf{X}_k) = \sum_{\pi \in NC(k)} \prod_{p_j \in \pi} k_{|p_j|}(\mathbf{X}_{i_1}, \ldots, \mathbf{X}_{i_{|p_j|}}), \tag{3.4}$$

with

$$k_1(\mathbf{X}_i) = m_1(\mathbf{X}_i), \qquad i \in \{1, 2, \ldots, n\},$$

where $NC(k)$ stands for the set of all the non-crossing partitions of $\{1, 2, \ldots, k\}$. Note that $k_t(\cdot)$ is a functional of t random matrices $\mathbf{X}_{i_r}$, where indices $i_r \in p_j$. One possible way to identify the class of non-crossing partitions $NC(k)$ from the set of all the possible partitions $P(k)$ of $\{1, 2, \ldots, k\}$ is to consider convex hulls of different blocks of the partition where all the vertices are placed on the circle so that they divide it in arcs of equal length. If the obtained convex hulls are disjoint, then the corresponding partition is non-crossing; see Fig. 3.1.

Following the relations (3.3) and (3.4), we conclude that

$$\begin{aligned} c_2(\mathbf{X}_{\{1,2\}}) &= c_2((X_1, X_2)) = \mathrm{E}\,(X_1X_2) - \mathrm{E}\,(X_1)\mathrm{E}\,(X_2), \\ k_2(\mathbf{X}_1, \mathbf{X}_2) &= m_2(\mathbf{X}_1, \mathbf{X}_2) - m_1(\mathbf{X}_1)m_1(\mathbf{X}_2), \end{aligned}$$

as all the partitions of the two-element set are non-crossing and it follows that

$$m_2(\mathbf{X}_1, \mathbf{X}_2) = k_2(\mathbf{X}_1, \mathbf{X}_2) + k_1(\mathbf{X}_1)k_1(\mathbf{X}_2) = k_2(\mathbf{X}_1, \mathbf{X}_2) + m_1(\mathbf{X}_1)m_1(\mathbf{X}_2).$$

However, the fourth cumulant and the fourth free cumulant differ from each other, since for $\mathbf{X} = (X, X, X, X)$

$$\begin{aligned} c_4(\mathbf{X}) &= \mathrm{E}\,(X^4) - 4\mathrm{E}\,(X)\mathrm{E}\,(X^3) - 3(\mathrm{E}\,(X^2))^2 + 12\mathrm{E}\,(X^2)(\mathrm{E}\,(X))^2 - 6(\mathrm{E}\,(X))^4, \\ k_4(\mathbf{X}) &= m_4(\mathbf{X}) - 4m_1(\mathbf{X})m_3(\mathbf{X}) - 2m_2^2(\mathbf{X}) + 10m_2(\mathbf{X})m_1^2(\mathbf{X}) - 5m_1^4(\mathbf{X}), \end{aligned}$$

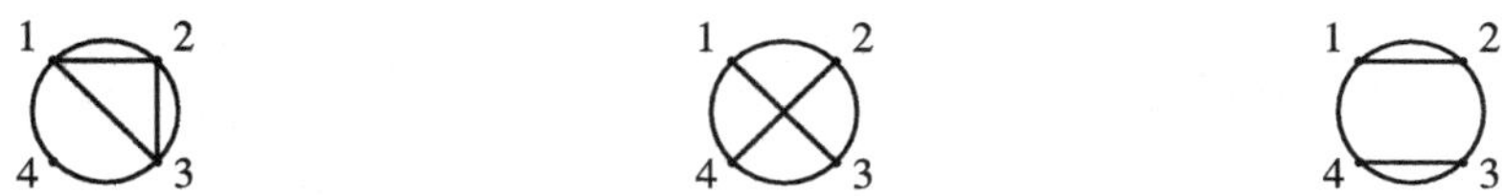

$\{\{1,2,3\},\{4\}\} \in NC(4)$ $\{\{1,3\},\{2,4\}\} \notin NC(4)$ $\{\{1,2\},\{3,4\}\} \in NC(4)$

Fig. 3.1 Examples of crossing and non-crossing partitions of the set $\{1, 2, 3, 4\}$

due to the existence of the crossing partition $\{\{1,3\},\{2,4\}\}$. See Speicher [40] for further details on the relation between moments and free cumulants.

Following Bożejko and Wysoczański [4], note that the R-transform introduced in Definition 3.15 is the generating function of the sequence of free cumulants $\{k_i\}_{i=1}^{\infty}$ given above. By series expansion of $R(z)$, we can state Theorem 3.15.

Theorem 3.15 *Let μ be a probability measure with compact support. Then, the R-transform can be expressed as a power series of the free cumulants $\{k_i\}$, $i \in \{1, 2, \ldots\}$, i.e.,*

$$R(z) = \sum_{i=0}^{\infty} k_{i+1} z^i.$$

The R-transform and the free cumulants $\{k_i\}$ essentially give us the same information about the underlying compactly supported probability measure.

Moreover, it can be proven that the vanishing of all the mixed cumulants is a sufficient and necessary condition for the freeness of the underlying variables, see Mingo and Speicher [29]. Note that for random matrices, freeness can only be achieved asymptotically. In the case of random matrices, asymptotic freeness can be defined as follows.

Definition 3.20 Let $\mathbf{X}$ and $\mathbf{Y}$ be random matrices of size $p \times p$. Then, $\mathbf{X}$ and $\mathbf{Y}$ are asymptotically free if all the mixed cumulants of the matrices $\mathbf{X}$ and $\mathbf{Y}$ vanish asymptotically for $p \to \infty$.

The S-transform is utilized for finding asymptotic results with regard to the products of free matrices. The free multiplicative convolution of measures μ (associated with a random matrix $\mathbf{X}$) and ν (associated with a random matrix $\mathbf{Y}$) is defined as a proper measure (associated with the random matrix $\mathbf{XY}$). In particular, for Hermitian matrices the following results hold.

Theorem 3.16 *For asymptotically free Hermitian matrices, the following statement holds:*

$$S_{\mu_{\mathbf{XY}}}(z) = S_{\mu_{\mathbf{X}}}(z) S_{\mu_{\mathbf{Y}}}(z).$$

In the above theorem, the matrices have to be self-adjoint. An additional condition is that at least one of the matrices must have a positive spectrum, which assures real eigenvalues of $\mathbf{XY}$, as they are the same as those of the Hermitian matrix $(\mathbf{X}^{1/2})'\mathbf{YX}^{1/2}$. Note that if the eigenvalues of a matrix are real, the S-transform contains enough information to recover the associated spectral measure. For further information on this topic, the reader is referred to Mingo and Speicher [29].

3.2.6 Common Multivariate Distributions

In this section, we start by defining the matrix normal distribution. A special case of this distribution will be used to define the Wishart distribution. Thereafter, with the help of two Wishart matrices, multivariate β-distributions will be defined. Moreover, with the help of the matrix normal distribution and the Wishart distribution, another β-type distribution will be defined.

3.2.6.1 The Matrix Normal Distribution

For any positive semi-definite matrix $\boldsymbol{\Sigma}$, there exists a square root $\boldsymbol{\Sigma}^{1/2}$, which for notational convenience will be supposed to be symmetric. There exist an infinite number of square roots, but our results do not depend on the specific choice of one of them.

Definition 3.21 A $p \times n$ real-valued matrix $\mathbf{X}$ has a matrix normal distribution when

$$\mathbf{X} = \mathbf{M} + \boldsymbol{\Sigma}^{1/2}\mathbf{E}\boldsymbol{\Psi}^{1/2},$$

where $\boldsymbol{\Sigma} = \boldsymbol{\Sigma}^{1/2}(\boldsymbol{\Sigma}^{1/2})'$, $\boldsymbol{\Sigma}^{1/2} : p \times r$, $\boldsymbol{\Psi} = \boldsymbol{\Psi}^{1/2}(\boldsymbol{\Psi}^{1/2})'$, $\boldsymbol{\Psi}^{1/2} : n \times s$, $\mathbf{M} : p \times n$ is non-random, and $\mathbf{E} = (e_{ij})$, with e_{ij} having an independent and identical standard normal distribution, i.e., $N(0, 1)$. When the $p \times n$ matrix $\mathbf{X}$ has a matrix normal distribution, it is denoted by $\mathbf{X} \sim N_{p,n}(\mathbf{M}, \boldsymbol{\Psi}, \boldsymbol{\Sigma})$.

An important special case is when $\boldsymbol{\Psi} = \mathbf{I}$, which corresponds to the case when the columns of $\mathbf{X}$ are independent. This particular case will be used when defining the Wishart distribution.

In the next theorem, a number of known results are listed. For proofs see, for example, Kollo and von Rosen [24, Sect. 2.2], where many more results are also given.

Theorem 3.17 *Let* $\mathbf{X} \sim N_{p,n}(\mathbf{M}, \boldsymbol{\Sigma}, \boldsymbol{\Psi})$.

(i) $\mathbf{AXB}' \sim N_{q,m}(\mathbf{AMB}', \mathbf{A\Sigma A}', \mathbf{B\Psi B}')$, *where* $\mathbf{A} \in \mathbb{R}^{q\times p}$ *and* $\mathbf{B} \in \mathbb{R}^{m\times n}$*;*

(ii) $\mathbf{AXK}$ *is independent of* $\mathbf{CXL}$ *for all constant matrices* $\mathbf{K}$ *and* $\mathbf{L}$ *of a proper size if* $\mathbf{A\Sigma C}' = \mathbf{0}$*;*

(iii) $\mathbf{KXB}'$ *is independent of* $\mathbf{LXD}'$ *for all constant matrices* $\mathbf{K}$ *and* $\mathbf{L}$ *of a proper size if* $\mathbf{B\Psi D}' = \mathbf{0}$*;*

(iv) *if* $\boldsymbol{\Sigma}$ *and* $\boldsymbol{\Psi}$ *are positive definite the density for* $\mathbf{X}$, $f_{\mathbf{X}}(\mathbf{X}_0)$, *equals*
$$f_{\mathbf{X}}(\mathbf{X}_0) = (2\pi)^{-\frac{1}{2}pn}|\boldsymbol{\Sigma}|^{-n/2}|\boldsymbol{\Psi}|^{-p/2}\exp\{-\tfrac{1}{2}\mathrm{Tr}\,[\boldsymbol{\Sigma}^{-1}(\mathbf{X}_0-\mathbf{M})\boldsymbol{\Psi}^{-1}(\mathbf{X}_0-\mathbf{M})']\};$$

(v) *the characteristic function for* $\mathbf{X}$ *equals*

$$\varphi_{\mathbf{X}}(\mathbf{T}) = \exp\{\iota\,\mathrm{Tr}\,(\mathbf{T}'\mathbf{M}) - \tfrac{1}{2}\mathrm{Tr}\,(\boldsymbol{\Sigma}\mathbf{T}\boldsymbol{\Psi}\mathbf{T}')\}, \qquad \mathbf{T} \in \mathbb{R}^{p\times n},$$

where ι *is the imaginary unit;*

(vi) *the cumulant generating function for* $\mathbf{X}$ *equals*

$$\psi_{\mathbf{X}}(\mathbf{T}) = \iota\,\mathrm{Tr}\,(\mathbf{T}'\mathbf{M}) - \tfrac{1}{2}\mathrm{Tr}\,(\boldsymbol{\Sigma}\mathbf{T}\boldsymbol{\Psi}\mathbf{T}'), \qquad \mathbf{T} \in \mathbb{R}^{p\times n}.$$

Now the complex version of the matrix normal distribution is defined.

Definition 3.22 The complex matrix $\mathbf{X} \in \mathbb{C}^{p\times n}$ follows a complex normal distribution if

$$\varphi_{\mathbf{X}}(\mathbf{T}) = \exp\{\imath\, \Re(\mathrm{Tr}\,(\mathbf{T}^*\mathbf{M})) - \tfrac{1}{4}\mathrm{Tr}\,(\boldsymbol{\Sigma}\mathbf{T}\boldsymbol{\Psi}\mathbf{T}^*)\}, \qquad \mathbf{T} \in \mathbb{C}^{p\times n},$$

where $\boldsymbol{\Sigma} = \boldsymbol{\Sigma}_0 + \imath\boldsymbol{\Sigma}_1 : p \times p$ and $\boldsymbol{\Psi} = \boldsymbol{\Psi}_0 + \imath\boldsymbol{\Psi}_1 : n \times n$ ($\boldsymbol{\Sigma}_0$ and $\boldsymbol{\Psi}_0$ are positive semi-definite, $\boldsymbol{\Sigma}_1$ and $\boldsymbol{\Psi}_1$ are skew-symmetric), and $\mathbf{M} \in \mathbb{C}^{p\times n}$. $\mathbf{X}$ having distribution given here will be denoted as $\mathbf{X} \sim CN_{p,n}(\mathbf{M}, \boldsymbol{\Sigma}, \boldsymbol{\Psi})$.

The complex matrix normal distribution is the same distribution as the vector (multivariate) complex normal distribution with the parameters $\mathrm{vec}\,\mathbf{M}$, $\boldsymbol{\Psi}' \otimes \boldsymbol{\Sigma}$. An early reference dealing with the complex normal distribution is Wooding [49] (see also Withers and Nadarajah [48], where many contributions are presented). A useful and clarifying interpretation of the real-valued matrix normal distribution is that it is a model for the distribution of data influenced by spatial-temporal effects. There is no equally clear interpretation of the complex matrix normal distribution. Moreover, put

$$\mathbf{t} = (\mathrm{vec}\,'\mathbf{T}_0 : \mathrm{vec}\,'\mathbf{T}_1)'$$

and it is worth noting that

$$\begin{aligned}\mathrm{Tr}\,(\boldsymbol{\Sigma}\mathbf{T}\boldsymbol{\Psi}\mathbf{T}^*) &= \mathrm{vec}\,^*\mathbf{T}(\boldsymbol{\Psi} \otimes \boldsymbol{\Sigma}')\mathrm{vec}\,\mathbf{T}\\ &= \mathbf{t}'\begin{pmatrix} \boldsymbol{\Psi}_0 \otimes \boldsymbol{\Sigma}_0 + \boldsymbol{\Psi}_1 \otimes \boldsymbol{\Sigma}_1 & -(\boldsymbol{\Psi}_0 \otimes \boldsymbol{\Sigma}_1 - \boldsymbol{\Psi}_1 \otimes \boldsymbol{\Sigma}_0) \\ \boldsymbol{\Psi}_0 \otimes \boldsymbol{\Sigma}_1 - \boldsymbol{\Psi}_1 \otimes \boldsymbol{\Sigma}_0 & \boldsymbol{\Psi}_0 \otimes \boldsymbol{\Sigma}_0 + \boldsymbol{\Psi}_1 \otimes \boldsymbol{\Sigma}_1 \end{pmatrix}\mathbf{t},\end{aligned}$$

showing the connections with the usual complex multivariate distribution. Moreover, if $\boldsymbol{\Psi} = \boldsymbol{\Psi}_0$, i.e., $\boldsymbol{\Psi}$ is real valued,

$$\mathrm{Tr}\,(\boldsymbol{\Sigma}\mathbf{T}\boldsymbol{\Psi}\mathbf{T}^*) = \mathrm{vec}\,^*\mathbf{T}(\boldsymbol{\Psi} \otimes \boldsymbol{\Sigma}')\mathrm{vec}\,\mathbf{T} = \mathbf{t}'\left(\boldsymbol{\Psi}_0 \otimes \begin{pmatrix} \boldsymbol{\Sigma}_0 & -\boldsymbol{\Sigma}_1 \\ \boldsymbol{\Sigma}_1 & \boldsymbol{\Sigma}_0 \end{pmatrix}\right)\mathbf{t}.$$

It is often assumed that $\boldsymbol{\Psi}_0 = \mathbf{I}$, which implies independent columns of the $\mathbf{X}$ matrix following a complex matrix normal distribution.

In Definition 3.22, in the exponent of the exponential function, $-\frac{1}{4}$ appears instead of $-\frac{1}{2}$, which appears for the real-valued matrix normal distribution (see Theorem 3.17 (v)). The explanation of this difference is that a complex matrix consists of two random matrices. Consider the special case of a variable Z which is supposed to follow a univariate standard complex normal distribution. The variable Z consists of two independently and equally distributed random variables, $\Re(Z)$ and $\Im(Z)$, and if one assumes that $\mathrm{D}\,(Z) = 1$, it follows that $\mathrm{D}\,(\Re Z) = \mathrm{D}\,(\Im Z) = \frac{1}{2}$ and, therefore, $\varphi_Z(t) = \exp(-\frac{1}{4}t^2)$, explaining why the factor $-\frac{1}{4}$ is used in Definition 3.22.

3.2.6.2 The Wishart Distribution

The Wishart distribution is in multivariate statistical analysis almost as important as the multivariate normal distribution.

Definition 3.23 The matrix $\mathbf{W} \in \mathbb{R}^{p\times p}$ has a Wishart distribution if and only if $\mathbf{W} = \mathbf{X}\mathbf{X}'$ for some matrix $\mathbf{X} \sim N_{p,n}(\mathbf{M}, \boldsymbol{\Sigma}, \mathbf{I}_n)$, where $\boldsymbol{\Sigma} : p \times p$ is supposed to be positive semi-definite. If the constant matrix $\mathbf{M} = \mathbf{0}$, we have a central Wishart distribution, which will be denoted as $W_p(\boldsymbol{\Sigma}, n)$, and if $\mathbf{M} \neq \mathbf{0}$, we have a non-central Wishart distribution, which will be denoted as $W_p(\boldsymbol{\Sigma}, n, \boldsymbol{\Delta})$, where $\boldsymbol{\Delta} = \boldsymbol{\Sigma}^{-1}\mathbf{M}\mathbf{M}'$ is of the size $p \times p$.

As for the matrix normal distribution, a few basic properties will now be presented. There are many more useful properties available (see any basic book on multivariate analysis and since Kollo and von Rosen, [24] has been mentioned before, the reader is referred to Sect. 2.3 of that book). The selection of properties is based on the need for them in this article.

Theorem 3.18 *The Wishart distribution satisfies the following statements:*

(i) *Let* $\mathbf{W}_1 \sim W_p(\boldsymbol{\Sigma}, m, \boldsymbol{\Delta}_1)$ *be independent of* $\mathbf{W}_2 \sim W_p(\boldsymbol{\Sigma}, n, \boldsymbol{\Delta}_2)$. *Then,* $\mathbf{W}_1 + \mathbf{W}_2 \sim W_p(\boldsymbol{\Sigma}, m+n, \boldsymbol{\Delta}_1 + \boldsymbol{\Delta}_2)$.

(ii) *If* $\mathbf{W} \sim W_p(\boldsymbol{\Sigma}, n, \boldsymbol{\Delta})$ *and* $\mathbf{A} \in \mathbb{R}^{p\times q}$, *then* $\mathbf{A}'\mathbf{W}\mathbf{A} \sim W_q(\mathbf{A}'\boldsymbol{\Sigma}\mathbf{A}, n, \mathbf{A}'\boldsymbol{\Delta}\mathbf{A})$.

(iii) *If* $\mathbf{W} \sim W_p(\boldsymbol{\Sigma}, n)$, $n > p$, *and* $\mathbf{A} \in \mathbb{R}^{p\times q}$, *then*

$$\mathbf{A}(\mathbf{A}'\mathbf{W}^{-1}\mathbf{A})^{-}\mathbf{A}' \sim W_p(\mathbf{A}(\mathbf{A}'\boldsymbol{\Sigma}^{-1}\mathbf{A})^{-}\mathbf{A}', n - p + \mathrm{r}(\mathbf{A}))$$

where "−" denotes an arbitrary generalized inverse.

(iv) *Let*

$$\mathbf{W} = \begin{pmatrix} \mathbf{W}_{11} & \mathbf{W}_{12} \\ \mathbf{W}_{21} & \mathbf{W}_{22} \end{pmatrix}, \ \boldsymbol{\Sigma} = \begin{pmatrix} \boldsymbol{\Sigma}_{11} & \boldsymbol{\Sigma}_{12} \\ \boldsymbol{\Sigma}_{21} & \boldsymbol{\Sigma}_{22} \end{pmatrix} \begin{pmatrix} r\times r & r \times (p-r) \\ (p-r)\times r & (p-r)\times(p-r) \end{pmatrix},$$

and put $\mathbf{W}_{1\cdot 2} = \mathbf{W}_{11} - \mathbf{W}_{12}\mathbf{W}_{22}^{-1}\mathbf{W}_{21}$, $\boldsymbol{\Sigma}_{1\cdot 2} = \boldsymbol{\Sigma}_{11} - \boldsymbol{\Sigma}_{12}\boldsymbol{\Sigma}_{22}^{-1}\boldsymbol{\Sigma}_{21}$. *Then,* $\mathbf{W}_{1\cdot 2} \sim W_r(\boldsymbol{\Sigma}_{1\cdot 2}, n - p + r)$.

(v) *Let* $\mathbf{W} \sim W_p(\boldsymbol{\Sigma}, n)$. *The characteristic function for* $\{W_{ij}, i \leq j\}$ *equals*

$$\varphi_{\mathbf{W}}(\mathbf{T}) = |\mathbf{I}_p - \imath\, \mathbf{K}(\mathbf{T})\boldsymbol{\Sigma}|^{-\frac{n}{2}},$$

where $\mathbf{K}(\mathbf{T}) = \mathbf{T} + \mathbf{T}_d$ *($\mathbf{T}_d$ denotes $\mathbf{T}$ when the off-diagonal elements of $\mathbf{T}$ have been replaced by* 0*).*

(vi) *The density for* $\mathbf{W} \sim W_p(\boldsymbol{\Sigma}, n)$, *if* $p < n$ *and* $\boldsymbol{\Sigma} > 0$, *is given by*

$$f_{\mathbf{W}}(\mathbf{W}) = \frac{1}{2^{\frac{pn}{2}}\Gamma_p(\frac{n}{2})|\boldsymbol{\Sigma}|^{\frac{n}{2}}}|\mathbf{W}|^{\frac{1}{2}(n-p-1)}\exp\{-\tfrac{1}{2}\mathrm{Tr}\,(\boldsymbol{\Sigma}^{-1}\mathbf{W})\}, \quad \mathbf{W} > 0.$$

(vii) *Let* $\mathbf{W} \sim W_p(\boldsymbol{\Sigma}, n)$, $p < n$, *and* $\boldsymbol{\Sigma} > 0$. *The density for* $\mathbf{V} = \mathbf{W}^{-1}$ *is given by*

$$f_{\mathbf{V}}(\mathbf{V}) = \frac{1}{2^{\frac{pn}{2}}\Gamma_p(\frac{n}{2})|\boldsymbol{\Sigma}|^{\frac{n}{2}}}|\mathbf{V}|^{-\frac{1}{2}(n+p+1)}\exp\{-\tfrac{1}{2}\mathrm{Tr}\,(\boldsymbol{\Sigma}^{-1}\mathbf{V}^{-1})\}, \quad \mathbf{V} > 0.$$

In statement (vii), the density for the inverse Wishart distribution is presented. It is of some interest to note that in statements (iii) and (iv) the expressions cannot be written as $\mathbf{ZZ}'$ for some $\mathbf{Z}$ following a matrix normal distribution, but we still have Wishart distributions. The explanation is that the expressions can be written as $\mathbf{ZPZ}'$ when $\mathbf{P}$ is an idempotent matrix which is distributed independently of $\mathbf{Z}$. This can happen if $\mathbf{Z}$ is generated by fewer random variables than the size of $\mathbf{Z}$.

Now the complex central Wishart distribution is defined.

Definition 3.24 The complex Wishart distribution is denoted $CW_p(\boldsymbol{\Sigma}, n)$. The matrix $\mathbf{W} \sim CW_p(\boldsymbol{\Sigma}, n)$ if $\mathbf{W} = \mathbf{XX}^*$, where $\mathbf{X} \sim CN_{p,n}(\mathbf{0}, \boldsymbol{\Sigma}, \mathbf{I}_n)$.

There exist other general definitions of a complex Wishart distribution (see Graczyk et al. [13]) and other general Wishart distributions than the complex one (see Andersson and Wojnar [2, 3] and Díaz-García and Gutiérrez-Jáimez [8]). References where many results connected to the complex Wishart matrix can be found are Goodman [12], Nagar and Gupta [31] and Withers and Nadarajah [48]. An early reference dealing with the complex Wishart distribution is Turin [44], while Srivastava [41] clearly derived the density for the complex Wishart distribution.

3.2.6.3 Spectral Distribution for Wishart Matrices

Let λ_i, $i \in \{1, 2, \ldots, p\}$, be the eigenvalues of a matrix $\mathbf{W}$ of size $p \times p$ and let $F_p^{\mathbf{W}}$ denote the empirical spectral distribution function as defined in Definition 3.14.

One of the most famous results for the spectral distribution of the Wishart matrix is the Marčenko–Pastur law, see Marčenko and Pastur [27]. These authors obtained asymptotic results for the sample covariance matrix with the degrees of freedom and the sample size simultaneously tending to infinity. That was accomplished in a similar fashion to the way in which the results for square matrices were obtained by Wigner [47].

Theorem 3.19 *[Marčenko–Pastur law] Let* $\mathbf{W} \sim W_p(\frac{\sigma^2}{n}\mathbf{I}_p, n)$. *Then, the empirical spectral distribution given in Definition 3.14 converges to the Marčenko–Pastur law given by the density* ($\frac{p}{n} \to c$):

$$\mu'(x) = \frac{\sqrt{[\sigma^2(1+\sqrt{c})^2 - x][x - \sigma^2(1-\sqrt{c})^2]}}{2\pi c\sigma^2 x}\mathbf{1}_{((1-\sqrt{c})^2\sigma^2, (1+\sqrt{c})^2\sigma^2)}(x)$$
$$+(1-c^{-1})\mathbf{1}_{[1,\infty)}(c)\delta_0(x),$$

where δ_0 *denotes the Dirac delta function.*

Note that if $c > 1$ in Theorem 3.19 then we have $p - n$ zero eigenvalues, and hence the asymptotic spectral distribution will have the point mass $\frac{p-n}{p} = 1 - \frac{1}{c}$ at 0.

In the special case when $c = 1$ we obtain the spectral density given by Wigner's semi-circle law, namely,

$$\mu'(x) = \frac{1}{2\pi x}\sqrt{4x - x^2},$$

which is also a β–density.

An extension covering the case of the Wishart matrix with a general parameter $\mathbf{\Sigma}$ was presented in Girko and von Rosen [10].

Theorem 3.20 (Girko and von Rosen [10]) *Let* $\mathbf{X} \sim N_{p,n}(\mathbf{0}, \mathbf{\Sigma}, \mathbf{\Psi})$, *where the eigenvalues of* $\mathbf{\Sigma}$ *and* $\mathbf{\Psi}$ *are bounded by some constant. Suppose that the Kolmogorov condition* $0 < c = \lim_{n\to\infty} \frac{p}{n} < \infty$ *holds and let* $F_p^{\mathbf{AA}'+\frac{1}{n}\mathbf{XX}'}(x)$ *be a normalized spectral distribution function of* $\mathbf{AA}' + \frac{1}{n}\mathbf{XX}'$, *where* $\mathbf{A}$ *is a non-random matrix. Then, for every* $x \geq 0$,

$$F_p^{\mathbf{AA}'+\frac{1}{n}\mathbf{XX}'}(x) - F_n(x) \xrightarrow{p} 0, \qquad n \to \infty,$$

where $\xrightarrow{p}$ *denotes convergence in probability and where for a large* n, $\{F_n(x)\}$ *are distribution functions satisfying*

$$\int_0^\infty \frac{dF_n(x)}{1+tx} = \frac{1}{p}\mathrm{Tr}\left[(\mathbf{I}_p + t\mathbf{AA}' + t\mathbf{\Sigma}a(t))^{-1}\right],$$

where for all $t > 0$, $a(t)$ *is a unique non-negative analytic function which exists and which satisfies the nonlinear equation*

$$a(t) = \frac{1}{n}\mathrm{Tr}\left\{\mathbf{\Psi}\left[\mathbf{I}_n + \frac{t}{n}\mathbf{\Psi}\mathrm{Tr}\left(\mathbf{\Sigma}(\mathbf{I}_p + t\mathbf{AA}' + t\mathbf{\Sigma}a(t))^{-1}\right)\right]^{-1}\right\}.$$

Note that $\int_0^\infty \frac{dF_n(x)}{1+tx}$ in Theorem 3.20 is essentially the Stieltjes–Cauchy transform $m(z)$, see (3.2), as $\int_0^\infty \frac{dF_n(x)}{1+tx} = -\frac{1}{z}m(-\frac{1}{z})$.

Theorem 3.21 (Girko and von Rosen [10]) *Consider*

$$\frac{1}{n_1}\mathbf{X}_1\mathbf{X}_1' + \frac{1}{n_2}\mathbf{X}_2\mathbf{X}_2',$$

where the matrices $\mathbf{X}_1$ *and* $\mathbf{X}_2$ *are independent and* $\mathbf{X}_i \sim N_{p,n}(\mathbf{0}, \mathbf{\Sigma}_i, \mathbf{\Psi}_i)$, $i \in \{1, 2\}$. *Let*

$$a(t) = \frac{1}{n_2}\text{Tr}\left[\boldsymbol{\Psi}_2\left(\mathbf{I}_{n_2} + \frac{t}{n_2}\boldsymbol{\Psi}_2 b(t)\right)^{-1}\right],$$

$$b(t) = \frac{1}{n_2}\text{Tr}\left[\boldsymbol{\Sigma}_2\left(\mathbf{I}_p + t\boldsymbol{\Sigma}_2 a(t) + t\boldsymbol{\Sigma}_1 c(t)\right)^{-1}\right],$$

$$c(t) = \frac{1}{n_1}\text{Tr}\left(\boldsymbol{\Psi}_1\left[\mathbf{I}_{n_1} + \frac{t}{n_1}\boldsymbol{\Psi}_1\text{Tr}\left(\boldsymbol{\Sigma}_1(\mathbf{I}_p + t\boldsymbol{\Sigma}_2 a(t) + t\boldsymbol{\Sigma}_1 b(t))^{-1}\right)\right]^{-1}\right),$$

$$d(t) = \frac{1}{n_1}\text{Tr}\left(\boldsymbol{\Psi}_1\left[\mathbf{I}_{n_1} + \frac{t}{n_1}\boldsymbol{\Psi}_1\text{Tr}\left(\boldsymbol{\Sigma}_1(\mathbf{I}_p + t\boldsymbol{\Sigma}_2 a(t) + t\boldsymbol{\Sigma}_1 d(t))^{-1}\right)\right]^{-1}\right).$$

Put $g(t) = \frac{1}{p}\text{Tr}\left[(\mathbf{I}_p + \frac{t}{n_1}\mathbf{X}_1\mathbf{X}_1' + \frac{t}{n_2}\mathbf{X}_2\mathbf{X}_2')^{-1}\right]$. *If*

$$0 < \lim_{n_1\to\infty}\frac{p}{n_1} < \infty \quad \textit{and} \quad 0 < \lim_{n_2\to\infty}\frac{p}{n_2} < \infty,$$

it follows that

$$g(t) \to \frac{1}{p}\left(\mathbf{I}_p + t\boldsymbol{\Sigma}_1 d(t) + t\boldsymbol{\Sigma}_2 a(t)\right)^{-1}, \qquad n_i \to \infty,\ i \in \{1, 2\}.$$

The previously stated theorems assume that the matrix **X** is real. A theorem regarding the empirical spectral distribution for the complex random matrix **X** can be given in the following form.

Theorem 3.22 (Silverstein and Bai [37]) *Let us consider* $\mathbf{X} = \mathbf{A} + \mathbf{Z}\mathbf{T}\mathbf{Z}^*$ *under the Kolmogorov condition* $0 < \lim_{n_1\to\infty}\frac{p}{n} = c < \infty$ *as* $p, n \to \infty$.

- $\mathbf{A}$ *is a Hermitian* $p \times p$ *matrix for which* $F^{\mathbf{A}}$ *converges vaguely to* ν *(i.e., without preservation of the total variation);* ν *is a possibly defective distribution function (a distribution function with discontinuities), i.e., there exists an everywhere dense subset* D *of* $\mathbb{R}$*, such that*

$$\forall_{a,b\in D,\, a<b}\ F^{\mathbf{A}}(a, b] \to \nu(a, b], \qquad n, p \to \infty.$$

- $\mathbf{Z}$, $\mathbf{T}$ *and* $\mathbf{A}$ *are independent, and* $\mathbf{Z}$ *and* $\mathbf{T}$ *are such that*
  - $\mathbf{Z} = (\frac{1}{\sqrt{p}}Z_{ij})_{p\times n}$, $Z_{ij} \in \mathbb{C}$, *i.i.d. with* $\mathrm{E}\,|Z_{11} - \mathrm{E}\,Z_{11}|^2 = 1$, *where* $|\cdot|$ *stands for the absolute value;*
  - $\mathbf{T} = diag(\Lambda_1, \Lambda_2, \ldots, \Lambda_n)$, *where* $\Lambda_i \in \mathbb{R}$ *and the empirical spectral distribution of the matrix* $\mathbf{T}$, *i.e.,* $\{\Lambda_1, \Lambda_2, \ldots, \Lambda_n\}$, *converges almost surely in distribution to a probability distribution function* H *as* $p, n \to \infty$.

Then $F^{\mathbf{X}}$, *the empirical distribution function of the eigenvalues of* $\mathbf{X}$, *converges almost surely, as* $p, n \to \infty$, *to a distribution function* F *whose Stieltjes–Cauchy transform equals* $m(z)$*, where* z *is a complex argument with a positive imaginary part* ($z \in \mathbb{C}^+$) *and satisfies the canonical equation*

$$m(z) = m_\nu\Big(z - \frac{1}{c}\int \frac{\tau dH(\tau)}{1-\tau m(z)}\Big),$$

where the integral is calculated with respect to the asymptotic distribution function H *of the matrix* $\mathbf{T}$ *mentioned above.*

3.2.6.4 Multivariate β-type Distributions

Definition 3.25 Let $\mathbf{W}_1 \sim W_p(\mathbf{I}_p, n)$, $p \le n$, $\mathbf{W}_2 \sim W_p(\mathbf{I}_p, m)$, $p \le m$, and $\mathbf{Y} \sim N_{p,m}(\mathbf{0}, \mathbf{I}_p, \mathbf{I}_m)$, $m < p$, be pairwise independently distributed. All the square roots which are presented below are supposed to be symmetric (it follows from the presentation that the choice of square roots will be immaterial).

(i) $\mathbf{F} = (\mathbf{W}_1 + \mathbf{W}_2)^{-1/2}\mathbf{W}_2(\mathbf{W}_1 + \mathbf{W}_2)^{-1/2}$. Then, $\mathbf{F}$ is said to follow a multivariate beta distribution of type I, which will be denoted $M\beta_I(p, m, n)$.
(ii) $\mathbf{Z} = \mathbf{W}_2^{-1/2}\mathbf{W}_1\mathbf{W}_2^{-1/2}$. Then, $\mathbf{Z}$ is said to follow a multivariate beta distribution of type II, which will be denoted $M\beta_{II}(p, m, n)$.
(iii) $\mathbf{G} = \mathbf{Y}'(\mathbf{W}_1 + \mathbf{Y}\mathbf{Y}')^{-1}\mathbf{Y}$.

In the literature, there also exist a multivariate beta distribution of type III and non-central versions of multivariate beta-type distributions of type I and II, but these will not be considered herein.

Theorem 3.23 *Let* $c(p, n) = \left(2^{\frac{pn}{2}}\Gamma_p(\frac{n}{2})\right)^{-1}$, *where* $\Gamma_p(\cdot)$ *is a multivariate gamma function (see, e.g., Muirhead [30]).*

(i) *The density for* $\mathbf{F}$ *given in Definition 3.25 (i) equals*

$$f_{\mathbf{F}}(\mathbf{F}) = \begin{cases} \frac{c(p,n)c(p,m)}{c(p,n+m)}|\mathbf{F}|^{\frac{1}{2}(m-p-1)}|\mathbf{I}_p - \mathbf{F}|^{\frac{1}{2}(n-p-1)}, & |\mathbf{I}_p - \mathbf{F}| > 0, |\mathbf{F}| > 0, \\ 0, & \textit{otherwise}. \end{cases}$$

(ii) *The density for* $\mathbf{Z}$ *given in Definition 3.25 (ii) equals*

$$f_{\mathbf{Z}}(\mathbf{Z}) = \begin{cases} \frac{c(p,n)c(p,m)}{c(p,n+m)}|\mathbf{Z}|^{\frac{1}{2}(n-p-1)}|\mathbf{I}_p + \mathbf{Z}|^{-\frac{1}{2}(n+m)}, & |\mathbf{Z}| > 0, \\ 0, & \textit{otherwise}. \end{cases}$$

(iii) *The density for* $\mathbf{G}$ *given in Definition 3.25 (iii) equals*

$$f_{\mathbf{G}}(\mathbf{G}) = \begin{cases} \frac{c(p,n)c(m,p)}{c(p,n+m)}|\mathbf{G}|^{\frac{1}{2}(p-m-1)}|\mathbf{I}_m - \mathbf{G}|^{\frac{1}{2}(n-p-1)}, & |\mathbf{I}_m - \mathbf{G}| > 0, |\mathbf{G}| > 0, \\ 0, & \textit{otherwise}. \end{cases}$$

Note that since $\mathbf{F}$ has compact support, its distribution is determined by its moments. The next theorem is proved in detail and, indeed, all the results of Theorem 3.23 could have been proven in a similar fashion.

Theorem 3.24 *Let* $\mathbf{W}_1 \sim W_p(\mathbf{I}_p, n)$, $p \leq n$, *and* $\mathbf{W}_2 \sim W_p(\mathbf{I}_p, m)$, $p \leq m$, *be independently distributed. The distributions for* $\mathbf{F} = (\mathbf{W}_1 + \mathbf{W}_2)^{-1/2}\mathbf{W}_2(\mathbf{W}_1 + \mathbf{W}_2)^{-1/2}$ *and* $\mathbf{F}_1 = \mathbf{W}_2^{1/2}(\mathbf{W}_1 + \mathbf{W}_2)^{-1}\mathbf{W}_2^{1/2}$ *are the same.*

Proof It will now be proven that the density for $\mathbf{F}_1$ equals that for $\mathbf{F}$ given in Theorem 3.23 (i). The joint density for $\mathbf{W}_1$ and $\mathbf{W}_2$ is given by

$$
\begin{aligned}
&f_{\mathbf{W}_1,\mathbf{W}_2}(\mathbf{W}_1, \mathbf{W}_2)\\
&\quad = c(p,n)c(p,m)|\mathbf{W}_1|^{\frac{1}{2}(n-p-1)}|\mathbf{W}_2|^{\frac{1}{2}(m-p-1)} \exp\{-\tfrac{1}{2}\mathrm{Tr}\,(\mathbf{W}_1 + \mathbf{W}_2)\}d\mathbf{W}_1 d\mathbf{W}_2.
\end{aligned}
$$

Put $\mathbf{W} = \mathbf{W}_1 + \mathbf{W}_2$ and the Jacobian of the transformation $(\mathbf{W}_1, \mathbf{W}_2) \to (\mathbf{W}, \mathbf{W}_2)$ equals 1. Then,

$$
\begin{aligned}
&f_{\mathbf{W},\mathbf{W}_2}(\mathbf{W}_1, \mathbf{W}_2)\\
&\quad = c(p,n)c(p,m)|\mathbf{W} - \mathbf{W}_2|^{\frac{1}{2}(n-p-1)}|\mathbf{W}_2|^{\frac{1}{2}(m-p-1)} \exp\{-\tfrac{1}{2}\mathrm{Tr}\,(\mathbf{W})\}d\mathbf{W}d\mathbf{W}_2\\
&\quad = c(p,n)c(p,m)|\mathbf{I}_p - \mathbf{W}_2^{1/2}\mathbf{W}^{-1}\mathbf{W}_2^{1/2}|^{\frac{1}{2}(n-p-1)}\\
&\qquad |\mathbf{W}|^{\frac{1}{2}(n-p-1)}|\mathbf{W}_2|^{\frac{1}{2}(m-p-1)} \exp\{-\tfrac{1}{2}\mathrm{Tr}\,(\mathbf{W})\}d\mathbf{W}d\mathbf{W}_2.
\end{aligned}
$$

Since from the definition of $\mathbf{F}_1$ it follows that $\mathbf{W} = \mathbf{W}_2^{1/2}\mathbf{F}_1^{-1}\mathbf{W}_2^{1/2}$, implying $\mathrm{Tr}\,(\mathbf{W}) = \mathrm{Tr}\,(\mathbf{F}_1^{-1}\mathbf{W}_2)$ and since the Jacobian of the transformation equals

$$
\begin{aligned}
&|J((\mathbf{W}_2^{1/2}\mathbf{F}_1^{-1}\mathbf{W}_2^{1/2}, \mathbf{W}_2) \to (\mathbf{F}_1, \mathbf{W}_2))| = |J(\mathbf{W}_2^{1/2}\mathbf{F}_1^{-1}\mathbf{W}_2^{1/2} \to \mathbf{F}_1)|\\
&\quad = |J(\mathbf{W}_2^{1/2}\mathbf{F}_1^{-1}\mathbf{W}_2^{1/2} \to \mathbf{F}_1^{-1})||J(\mathbf{F}_1^{-1} \to \mathbf{F}_1)| = |\mathbf{W}_2|^{\frac{1}{2}(p+1)}|\mathbf{F}_1|^{-(p+1)},
\end{aligned}
$$

where the last equality follows from Kollo and von Rosen [24, Ths. 1.4.13 (i) and 1.4.17 (ii)],

$$
\begin{aligned}
&f_{\mathbf{F}_1,\mathbf{W}_2}(\mathbf{F}_1, \mathbf{W}_2)\\
&\qquad = c(p,n)c(p,m)|\mathbf{I}_p - \mathbf{F}_1|^{\frac{1}{2}(n-p-1)}|\mathbf{F}_1|^{-\frac{1}{2}(n-p-1)}|\mathbf{W}_2|^{\frac{1}{2}(n+m-2(p+1))}\\
&\qquad\quad |\mathbf{W}_2|^{\frac{1}{2}(p+1)}|\mathbf{F}_1|^{-(p+1)} \exp\{-\tfrac{1}{2}\mathrm{Tr}\,(\mathbf{F}_1^{-1}\mathbf{W}_2)\}d\mathbf{F}_1 d\mathbf{W}_2.
\end{aligned}
$$

Now, using the Wishart density with $n + m$ degrees of freedom, $\mathbf{W}_2$ is integrated out from this expression and, therefore,

$$
f_{\mathbf{F}_1}(\mathbf{F}_1) = \frac{c(p,n)c(p,m)}{c(p,m+n)}|\mathbf{I}_p - \mathbf{F}_1|^{\frac{1}{2}(n-p-1)}|\mathbf{F}_1|^{\frac{1}{2}(m-p-1)}d\mathbf{F}_1,
$$

which is identical to the density for **F**. □

It can be noted that for **Z** given in Definition 3.25 (ii),

$$(\mathbf{I}_p + \mathbf{Z})^{-1} = \mathbf{W}_2^{1/2}(\mathbf{W}_1 + \mathbf{W}_2)^{-1}\mathbf{W}_2^{1/2};$$

the Jacobian of the transformation $\mathbf{Z} \to (\mathbf{I}_p + \mathbf{Z})^{-1}$ equals $|(\mathbf{I}_p + \mathbf{Z})|^{p+1}$ and the next theorem is established.

Theorem 3.25 *Let* **Z** *be given in Definition 3.25 (ii). Then,* $(\mathbf{I}_p + \mathbf{Z})^{-1}$ *has the same distribution as* $\mathbf{F}_1$ *in Theorem 3.24.*

When deriving the moments for **F** in Definition 3.25 (i), Theorems 3.24 and 3.25 will be utilized.

3.3 Moment Expressions

3.3.1 Matrix Normal Distribution

The moments for the matrix normal distribution will now be derived by differentiating the characteristic function, $\varphi_{\mathbf{X}}(\mathbf{T})$, in Theorem 3.17, using the matrix derivative given in Definition 3.6 and the rules for differentiating presented in Theorem 3.4. For details and references, see Kollo and von Rosen [24, Sect. 2.2]. A similar approach has been presented by Jammalamadaka et al. [22]. Let $\varphi^k_{\mathbf{X}}(\mathbf{T})$ denote $\dfrac{d^k\,\varphi_{\mathbf{X}}(\mathbf{T})}{d\,\mathbf{T}^k}$ with $\varphi^0_{\mathbf{X}}(\mathbf{T}) = \varphi_{\mathbf{X}}(\mathbf{T})$.

Theorem 3.26 *Let* $\mathbf{X} \sim N_{p,n}(\mathbf{M}, \mathbf{\Sigma}, \mathbf{\Psi})$ *with the characteristic function*

$$\varphi_{\mathbf{X}}(\mathbf{T}) = \exp\{\imath\,\mathrm{Tr}\,(\mathbf{T}'\mathbf{M}) - \tfrac{1}{2}\mathrm{Tr}\,(\mathbf{\Sigma}\mathbf{T}\mathbf{\Psi}\mathbf{T}')\}.$$

Then, $\varphi^1_{\mathbf{X}}(\mathbf{T}) = \mathbf{A}(\mathbf{T})\varphi_{\mathbf{X}}(\mathbf{T})$, *where*

$$\mathbf{A}(\mathbf{T}) = \frac{d\,\ln\varphi_{\mathbf{X}}(\mathbf{T})}{d\,\mathbf{T}} = \imath\,\mathrm{vec}\,\mathbf{M} - (\mathbf{\Psi}\otimes\mathbf{\Sigma})\mathrm{vec}\,\mathbf{T}.$$

Moreover, for $k > 1$

$$\begin{aligned}\varphi^k_{\mathbf{X}}(\mathbf{T}) &= \mathrm{vec}\,'\mathbf{A}(\mathbf{T})\otimes\varphi^{k-1}_{\mathbf{X}}(\mathbf{T}) + \mathbf{A}^1(\mathbf{T})\otimes\mathrm{vec}\,'\varphi^{k-2}_{\mathbf{X}}(\mathbf{T})\\ &\quad+\sum_{i=0}^{k-3}\left[\mathrm{vec}\,'\mathbf{A}^1(\mathbf{T})\otimes\varphi^{k-2}_{\mathbf{X}}(\mathbf{T})\right]\left(\mathbf{I}_{pn}\otimes\mathbf{K}_{pn,(pn)^{k-i-3}}\otimes\mathbf{I}_{(pn)^i}\right),\end{aligned}$$

where $\mathbf{A}^1(\mathbf{T}) = -\mathbf{\Psi}\otimes\mathbf{\Sigma}$ *is the first derivative of* $\mathbf{A}(\mathbf{T})$, *and for* $k = 2$ *the sum* $\sum_{i=0}^{k-3}$ *equals* **0**.

Proof Note that $\mathbf{A}(\mathbf{T})$ is a linear function in $\mathbf{T}$, implying that all the derivatives of a higher order than 1 equal $\mathbf{0}$. In order to catch the structure of the characteristic function, $\varphi_{\mathbf{X}}(\mathbf{T})$ will be differentiated four times and the general result will thereafter follow by applying an induction argument. Theorem 3.4 (viii), (xi), and (xii) yield

$$\varphi^1_{\mathbf{X}}(\mathbf{T}) = \mathbf{A}(\mathbf{T})\varphi_{\mathbf{X}}(\mathbf{T}).$$

Thereafter, differentiating $\varphi_{\mathbf{X}}(\mathbf{T})$ a second time and using Theorem 3.4 (viii),

$$\varphi^2_{\mathbf{X}}(\mathbf{T}) = \mathbf{A}^1(\mathbf{T})\varphi_{\mathbf{X}}(\mathbf{T}) + \mathbf{A}(\mathbf{T}) \otimes \varphi^1_{\mathbf{X}}(\mathbf{T}).$$

Among other differentiation rules, additionally applying Theorem 3.4 (x) to this expression leads to

$$\varphi^3_{\mathbf{X}}(\mathbf{T}) = \mathrm{vec}'\mathbf{A}^1(\mathbf{T}) \otimes \varphi^1_{\mathbf{X}}(\mathbf{T}) + \mathbf{A}^1(\mathbf{T}) \otimes \mathrm{vec}'\varphi^1_{\mathbf{X}}(\mathbf{T}) + \mathbf{A}(\mathbf{T}) \otimes \varphi^2_{\mathbf{X}}(\mathbf{T})$$

and differentiating once again implies that

$$\begin{aligned}\varphi^4_{\mathbf{X}}(\mathbf{T}) &= \mathbf{A}(\mathbf{T}) \otimes \varphi^3_{\mathbf{X}}(\mathbf{T}) + \mathbf{A}^1(\mathbf{T}) \otimes \mathrm{vec}'\varphi^2_{\mathbf{X}}(\mathbf{T}) \\ &+ \left[\mathrm{vec}'\mathbf{A}^1(\mathbf{T}) \otimes \varphi^2_{\mathbf{X}}(\mathbf{T})\right](\mathbf{I}_{pn} \otimes \mathbf{K}_{pn,pn}) + \mathrm{vec}'\mathbf{A}^1(\mathbf{T}) \otimes \varphi^2_{\mathbf{X}}(\mathbf{T}). \qquad (3.5)\end{aligned}$$

Hereafter, the proof is completed by induction and it is supposed that the statement of the theorem is true for $k-1$. Moreover, for $k=4$, the statement is identical to (3.5). Now $\varphi^{k-1}_{\mathbf{X}}(\mathbf{T})$ is differentiated and

$$\begin{aligned}\varphi^k_{\mathbf{X}}(\mathbf{T}) &= \tfrac{d}{d\mathbf{T}}\varphi^{k-1}_{\mathbf{X}}(\mathbf{T}) \\ &= \tfrac{d}{d\mathbf{T}}\left[\mathbf{A}(\mathbf{T}) \otimes \varphi^{k-2}_{\mathbf{X}}(\mathbf{T})\right] + \tfrac{d}{d\mathbf{T}}\left[\mathbf{A}^1(\mathbf{T}) \otimes \mathrm{vec}'\varphi^{k-3}_{\mathbf{X}}(\mathbf{T})\right] \\ &+ \tfrac{d}{d\mathbf{T}}\left[\sum_{i=0}^{k-4}\left(\mathrm{vec}'\mathbf{A}^1(\mathbf{T}) \otimes \varphi^{k-3}_{\mathbf{X}}(\mathbf{T})\right)\left(\mathbf{I}_{pn} \otimes \mathbf{K}_{pn,(pn)^{k-i-4}} \otimes \mathbf{I}_{(pn)^i}\right)\right].\end{aligned}$$

The three terms on the right-hand side of this equation are now evaluated separately and, in particular, Theorem 3.4 (x) is used:

$$\tfrac{d}{d\mathbf{T}}\left[\mathbf{A}(\mathbf{T}) \otimes \varphi^{k-2}_{\mathbf{X}}(\mathbf{T})\right] = \mathbf{A}(\mathbf{T}) \otimes \varphi^{k-1}_{\mathbf{X}}(\mathbf{T}) + \mathbf{A}^1(\mathbf{T}) \otimes \mathrm{vec}'\varphi^{k-2}_{\mathbf{X}}(\mathbf{T}), \qquad (3.6)$$

$$\tfrac{d}{d\mathbf{T}}\left[\mathbf{A}^1(\mathbf{T}) \otimes \mathrm{vec}'\varphi^{k-3}_{\mathbf{X}}(\mathbf{T})\right] = \left[\mathrm{vec}'\mathbf{A}^1(\mathbf{T}) \otimes \varphi^{k-2}_{\mathbf{X}}(\mathbf{T})\right]\left(\mathbf{I}_{pn} \otimes \mathbf{K}_{pn,(pn)^{k-3}}\right), \qquad (3.7)$$

$$\begin{aligned}&\tfrac{d}{d\mathbf{T}}\left\{\left[\mathrm{vec}'\mathbf{A}^1(\mathbf{T}) \otimes \varphi^{k-3}_{\mathbf{X}}(\mathbf{T})\right]\left(\mathbf{I}_{pn} \otimes \mathbf{K}_{pn,(pn)^{k-i-4}} \otimes \mathbf{I}_{(pn)^i}\right)\right\} \\ &\quad = \left[\mathrm{vec}'\mathbf{A}^1(\mathbf{T}) \otimes \varphi^{k-2}_{\mathbf{X}}(\mathbf{T})\right]\left(\mathbf{I}_{pn} \otimes \mathbf{K}_{pn,(pn)^{k-i-4}} \otimes \mathbf{I}_{(pn)^{i+1}}\right). \qquad (3.8)\end{aligned}$$

Summing (3.8) over i and adding (3.7) leads to

$$\sum_{i=0}^{k-4} \left[\mathrm{vec}'\mathbf{A}^1(\mathbf{T}) \otimes \varphi_{\mathbf{X}}^{k-2}(\mathbf{T})\right] \left(\mathbf{I}_{pn} \otimes \mathbf{K}_{pn,(pn)^{k-i-4}} \otimes \mathbf{I}_{(pn)^{i+1}}\right)$$
$$+\left[\mathrm{vec}'\mathbf{A}^1(\mathbf{T}) \otimes \varphi_{\mathbf{X}}^{k-2}(\mathbf{T})\right] \left(\mathbf{I}_{pn} \otimes \mathbf{K}_{pn,(pn)^{k-3}}\right).$$

In this relation the index is changed, i.e., $i \to i-1$, which implies

$$\sum_{i=1}^{k-3} \left[\mathrm{vec}'\mathbf{A}^1(\mathbf{T}) \otimes \varphi_{\mathbf{X}}^{k-2}(\mathbf{T})\right] \left(\mathbf{I}_{pn} \otimes \mathbf{K}_{pn,(pn)^{k-i-3}} \otimes \mathbf{I}_{(pn)^{i}}\right)$$
$$+\left[\mathrm{vec}'\mathbf{A}^1(\mathbf{T}) \otimes \varphi_{\mathbf{X}}^{k-2}(\mathbf{T})\right] \left(\mathbf{I}_{pn} \otimes \mathbf{K}_{pn,(pn)^{k-3}}\right)$$
$$= \sum_{i=0}^{k-3} \left[\mathrm{vec}'\mathbf{A}^1(\mathbf{T}) \otimes \varphi_{\mathbf{X}}^{k-2}(\mathbf{T})\right] \left(\mathbf{I}_{pn} \otimes \mathbf{K}_{pn,(pn)^{k-i-3}} \otimes \mathbf{I}_{(pn)^{i}}\right).$$

Thus, by adding (3.6) the statement of the lemma is established. □

Via Definition 3.17 and Theorem 3.26, a number of results can be obtained and some of them will be presented in the next theorem.

Theorem 3.27 *Let* $\mathbf{X} \sim N_{p,n}(\mathbf{M}, \boldsymbol{\Sigma}, \boldsymbol{\Psi})$. *Then,*

(i) $m_1(\mathbf{X}) = \mathrm{vec}\,\mathbf{M}$;
(ii) $m_2(\mathbf{X}) = \boldsymbol{\Psi} \otimes \boldsymbol{\Sigma} + \mathrm{vec}\,\mathbf{M}\mathrm{vec}'\mathbf{M}$;
(iii) $m_3(\mathbf{X}) = \mathrm{vec}\,\mathbf{M}(\mathrm{vec}'\mathbf{M})^{\otimes 2} + \mathrm{vec}'\mathbf{M} \otimes \boldsymbol{\Psi} \otimes \boldsymbol{\Sigma} + (\boldsymbol{\Psi} \otimes \boldsymbol{\Sigma}) \otimes \mathrm{vec}'\mathbf{M}$
$+\mathrm{vec}\,\mathbf{M}\mathrm{vec}'(\boldsymbol{\Psi} \otimes \boldsymbol{\Sigma})$;
(iv) $m_4(\mathbf{X})$
$= \mathrm{vec}\,\mathbf{M}(\mathrm{vec}'\mathbf{M})^{\otimes 3} + (\mathrm{vec}'\mathbf{M})^{\otimes 2} \otimes \boldsymbol{\Psi} \otimes \boldsymbol{\Sigma} + \mathrm{vec}'\mathbf{M} \otimes (\boldsymbol{\Psi} \otimes \boldsymbol{\Sigma}) \otimes \mathrm{vec}'\mathbf{M}$
$+\mathrm{vec}\,\mathbf{M}\mathrm{vec}'\mathbf{M} \otimes \mathrm{vec}'(\boldsymbol{\Psi} \otimes \boldsymbol{\Sigma}) + \boldsymbol{\Psi} \otimes \boldsymbol{\Sigma} \otimes \mathrm{vec}'(\boldsymbol{\Psi} \otimes \boldsymbol{\Sigma}) + \boldsymbol{\Psi} \otimes \boldsymbol{\Sigma} \otimes (\mathrm{vec}'\mathbf{M})^{\otimes 2}$
$+\left[\mathrm{vec}\,\mathbf{M}\mathrm{vec}'(\boldsymbol{\Psi} \otimes \boldsymbol{\Sigma}) \otimes \mathrm{vec}'\mathbf{M} + \mathrm{vec}'(\boldsymbol{\Psi} \otimes \boldsymbol{\Sigma}) \otimes \boldsymbol{\Psi} \otimes \boldsymbol{\Sigma}\right](\mathbf{I}_{(pn)^3} + \mathbf{I}_{pn} \otimes \mathbf{K}_{pn,pn})$;
(v) $m_k(\mathbf{X}) = \mathrm{vec}'\mathbf{M} \otimes m_{k-1}(\mathbf{X}) + \boldsymbol{\Psi} \otimes \boldsymbol{\Sigma} \otimes \mathrm{vec}'(m_{k-2}(\mathbf{X}))$
$+\sum_{i=0}^{k-3}\left[\mathrm{vec}'(\boldsymbol{\Psi} \otimes \boldsymbol{\Sigma}) \otimes m_{k-2}(\mathbf{X})\right]\left(\mathbf{I}_{pn} \otimes \mathbf{K}_{pn,(pn)^{k-i-3}} \otimes \mathbf{I}_{(pn)^i}\right), \quad k > 1.$

Proof These results follow from Theorem 3.26. Statements (i) to (iv) follow from the proof of the expressions for $\varphi_{\mathbf{X}}^k(\mathbf{T})$, $k \le 4$, by setting $\mathbf{T} = \mathbf{0}$, since $\mathbf{A}(\mathbf{0}) = \iota\,\mathrm{vec}'\mathbf{M}$ and $\mathbf{A}^1(\mathbf{0}) = -\boldsymbol{\Psi} \otimes \boldsymbol{\Sigma}$. The general case in statement (v) is verified by noting that

$$\varphi_{\mathbf{X}}^k(\mathbf{0}) = \iota^k m_k(\mathbf{X}), \qquad \mathbf{A}(\mathbf{0}) \otimes \varphi_{\mathbf{X}}^{k-1}(\mathbf{0}) = \iota^k \mathrm{vec}'\mathbf{M} \otimes m_{k-1}(\mathbf{X}),$$
$$\mathrm{vec}'\mathbf{A}^1(\mathbf{0}) \otimes \varphi_{\mathbf{X}}^{k-2}(\mathbf{0}) = \iota^k \mathrm{vec}'(\boldsymbol{\Psi} \otimes \boldsymbol{\Sigma}) \otimes m_{k-2}(\mathbf{X}).$$

□

Corollary 3.2 *Let* $\mathbf{x} \sim N_p(\mathbf{m}, \boldsymbol{\Sigma})$. *Then,*

(i) $m_1(\mathbf{x}) = \mathbf{m}$;
(ii) $m_2(\mathbf{x}) = \boldsymbol{\Sigma} + \mathbf{m}\mathbf{m}'$;
(iii) $m_3(\mathbf{x}) = \mathbf{m}(\mathbf{m}')^{\otimes 2} + \mathbf{m}' \otimes \boldsymbol{\Sigma} + \boldsymbol{\Sigma} \otimes \mathbf{m}' + \mathbf{m}\mathrm{vec}'\boldsymbol{\Sigma}$;

(iv) $m_4(\mathbf{x}) = \mathbf{m}(\mathbf{m}')^{\otimes 3} + (\mathbf{m}')^{\otimes 2} \otimes \boldsymbol{\Sigma} + \mathbf{m}' \otimes \boldsymbol{\Sigma} \otimes \mathbf{m}'$
$+\mathbf{m}\mathbf{m}' \otimes \mathrm{vec}'\boldsymbol{\Sigma} + \boldsymbol{\Sigma} \otimes \mathrm{vec}'\boldsymbol{\Sigma} + \boldsymbol{\Sigma} \otimes (\mathbf{m}')^{\otimes 2}$
$+ \left(\mathbf{m}\mathrm{vec}'\boldsymbol{\Sigma} \otimes \mathbf{m}' + \mathrm{vec}'\boldsymbol{\Sigma} \otimes \boldsymbol{\Sigma}\right)(\mathbf{I}_{p^3} + \mathbf{I}_p \otimes \mathbf{K}_{p,p})$;

(v) $$m_k(\mathbf{x}) = \mathbf{m}' \otimes m_{k-1}(\mathbf{x}) + \boldsymbol{\Sigma} \otimes \mathrm{vec}'(m_{k-2}(\mathbf{x})) + \sum_{i=0}^{k-3} \left[\mathrm{vec}'\boldsymbol{\Sigma} \otimes m_{k-2}(\mathbf{x})\right](\mathbf{I}_p \otimes \mathbf{K}_{p,p^{k-i-3}} \otimes \mathbf{I}_{p^i}), \quad k > 1.$$

Theorem 3.28 *Let* $\mathbf{Y} \sim N_{p,n}(\mathbf{0}, \boldsymbol{\Sigma}, \boldsymbol{\Psi})$. *Then, if k is odd,* $m_k(\mathbf{Y}) = \mathbf{0}$ *and*

(i) $m_2(\mathbf{Y}) = \boldsymbol{\Psi} \otimes \boldsymbol{\Sigma}$;

(ii) $$m_4(\mathbf{Y}) = \boldsymbol{\Psi} \otimes \boldsymbol{\Sigma} \otimes \mathrm{vec}'(\boldsymbol{\Psi} \otimes \boldsymbol{\Sigma}) + \left[\mathrm{vec}'(\boldsymbol{\Psi} \otimes \boldsymbol{\Sigma}) \otimes \boldsymbol{\Psi} \otimes \boldsymbol{\Sigma}\right](\mathbf{I}_{(pn)^3} + \mathbf{I}_{pn} \otimes \mathbf{K}_{pn,pn});$$

(iii) $$m_k(\mathbf{Y}) = \boldsymbol{\Psi} \otimes \boldsymbol{\Sigma} \otimes \mathrm{vec}'(m_{k-2}(\mathbf{Y})) + \sum_{i=0}^{k-3}\left[\mathrm{vec}'(\boldsymbol{\Psi} \otimes \boldsymbol{\Sigma}) \otimes m_{k-2}(\mathbf{Y})\right](\mathbf{I}_{pn} \otimes \mathbf{K}_{pn,(pn)^{k-i-3}} \otimes \mathbf{I}_{(pn)^i}), \; k \in \{2, 4, \ldots\}.$$

In statement (iii), $m_0(\mathbf{Y}) = 1$. It is also possible to express alternative moment relations, for example, by using the Kroneckerian power, which, in fact, is only a reorganization of the elements of $m_k(\mathbf{Y})$ and $m_k(\mathbf{X})$.

Theorem 3.29 *Let* $\mathbf{Y} \sim N_{p,n}(\mathbf{0}, \boldsymbol{\Sigma}, \boldsymbol{\Psi})$ *and* $\mathbf{X} \sim N_{p,n}(\mathbf{M}, \boldsymbol{\Sigma}, \boldsymbol{\Psi})$. *Then,*

(i) $\mathrm{E}\left(\mathbf{Y}^{\otimes 2}\right) = \mathrm{vec}\,\boldsymbol{\Sigma}\mathrm{vec}'\boldsymbol{\Psi}$;

(ii) $\mathrm{E}\left(\mathbf{X}^{\otimes 2}\right) = \mathrm{vec}\,\boldsymbol{\Sigma}\mathrm{vec}'\boldsymbol{\Psi} + \mathbf{M} \otimes \mathbf{M}$;

(iii) $$\mathrm{E}\left(\mathbf{X}^{\otimes 3}\right) = \mathrm{vec}\,\boldsymbol{\Sigma}\mathrm{vec}'\boldsymbol{\Psi} \otimes \mathbf{M} + (\mathbf{K}_{p,p} \otimes \mathbf{I}_p)(\mathbf{M} \otimes \mathrm{vec}\,\boldsymbol{\Sigma}\mathrm{vec}'\boldsymbol{\Psi})(\mathbf{K}_{n,n} \otimes \mathbf{I}_n) + \mathbf{M} \otimes \mathrm{vec}\,\boldsymbol{\Sigma}\mathrm{vec}'\boldsymbol{\Psi} - 3\mathbf{M}^{\otimes 3};$$

(iv) $$\mathrm{E}\left(\mathbf{Y}^{\otimes 4}\right) = (\mathrm{vec}\,\boldsymbol{\Sigma}\mathrm{vec}'\boldsymbol{\Psi})^{\otimes 2} + (\mathbf{I}_p \otimes \mathbf{K}_{p,p} \otimes \mathbf{I}_p)(\mathrm{vec}\,\boldsymbol{\Sigma}\mathrm{vec}'\boldsymbol{\Psi})^{\otimes 2}(\mathbf{I}_n \otimes \mathbf{K}_{n,n} \otimes \mathbf{I}_n) + (\mathbf{I}_p \otimes \mathbf{K}_{p^2,p})(\mathrm{vec}\,\boldsymbol{\Sigma}\mathrm{vec}'\boldsymbol{\Psi})^{\otimes 2}(\mathbf{I}_n \otimes \mathbf{K}_{n,n^2});$$

(v) $$\mathrm{E}\left(\mathbf{Y}^{\otimes k}\right) = \sum_{i=2}^{k}(\mathbf{I}_p \otimes \mathbf{K}_{p^{i-2},p} \otimes \mathbf{I}_{p^{k-i}})\left[\mathrm{vec}\,\boldsymbol{\Sigma}\mathrm{vec}'\boldsymbol{\Psi} \otimes \mathrm{E}\left(\mathbf{Y}^{\otimes k-2}\right)\right](\mathbf{I}_n \otimes \mathbf{K}_{n,n^{i-2}} \otimes \mathbf{I}_{n^{k-i}}), \quad k \in \{2, 4, \ldots\}.$$

Corollary 3.3 *Let* $\mathbf{y} \sim N_p(\mathbf{0}, \boldsymbol{\Sigma})$ *and* $\mathbf{x} \sim N_p(\mathbf{m}, \boldsymbol{\Sigma})$. *Then,*

(i) $\mathrm{E}\left(\mathbf{y}^{\otimes 2}\right) = \mathrm{vec}\,\boldsymbol{\Sigma}$;

(ii) $\mathrm{E}\left(\mathbf{x}^{\otimes 2}\right) = \mathrm{vec}\,\boldsymbol{\Sigma} + \mathbf{m} \otimes \mathbf{m}$;

(iii) $\mathrm{E}\left(\mathbf{x}^{\otimes 3}\right) = \mathrm{vec}\,\boldsymbol{\Sigma} \otimes \mathbf{m} + (\mathbf{K}_{p,p} \otimes \mathbf{I}_p)(\mathbf{m} \otimes \mathrm{vec}\,\boldsymbol{\Sigma}) + \mathbf{m} \otimes \mathrm{vec}\,\boldsymbol{\Sigma} - 3\mathbf{m}^{\otimes 3}$;

(iv) $\mathrm{E}\left(\mathbf{y}^{\otimes 4}\right) = \mathrm{vec}\,\boldsymbol{\Sigma}^{\otimes 2} + (\mathbf{I}_p \otimes \mathbf{K}_{p,p} \otimes \mathbf{I}_p)\mathrm{vec}\,\boldsymbol{\Sigma}^{\otimes 2} + (\mathbf{I}_p \otimes \mathbf{K}_{p^2,p})\mathrm{vec}\,\boldsymbol{\Sigma}^{\otimes 2}$;

(v) $$\mathrm{E}\left(\mathbf{y}^{\otimes k}\right) = \sum_{i=2}^{k}(\mathbf{I}_p \otimes \mathbf{K}_{p^{i-2},p} \otimes \mathbf{I}_{p^{k-i}})\left[\mathrm{vec}\,\boldsymbol{\Sigma} \otimes \mathrm{E}\left(\mathbf{y}^{\otimes k-2}\right)\right], \quad k \in \{2, 4, \ldots\}.$$

There exists a one-to-one relation between moments and cumulants, meaning that any moment can be expressed as a function of cumulants and vice versa. For the matrix normal distribution as for the corresponding univariate normal distribution, cumulants of a higher order than 2 equal 0. The proof follows from the fact that the cumulant generating function given in Theorem 3.26 (vi) consists of a linear function in $\mathbf{T}$, as well as a second term which is a quadratic function in $\mathbf{T}$.

Theorem 3.30 *Let* $\mathbf{X} \sim N_{p,n}(\mathbf{M}, \boldsymbol{\Sigma}, \boldsymbol{\Psi})$. *Then,*

(i) $c_1(\mathbf{X}) = \mathrm{vec}\,\mathbf{M}$;
(ii) $c_2(\mathbf{X}) = \boldsymbol{\Psi} \otimes \boldsymbol{\Sigma}$;
(iii) $c_k(\mathbf{X}) = \mathbf{0}, \quad k \geq 3$.

Using the matrix normal distribution and its moments derived in Theorem 3.26, moments of quadratic forms can be obtained.

Theorem 3.31 *Let* $\mathbf{X} \sim N_{p,n}(\mathbf{M}, \boldsymbol{\Sigma}, \boldsymbol{\Psi})$ *and let* $\mathbf{A}$ *and* $\mathbf{B}$ *be non-random matrices of a proper size. Then,*

(i) $\mathrm{E}\,(\mathbf{XAX}') = \mathrm{Tr}\,(\boldsymbol{\Psi}\mathbf{A})\boldsymbol{\Sigma} + \mathbf{MAM}'$;
(ii) $\mathrm{E}\,(\mathbf{XAX}' \otimes \mathbf{XBX}') = \mathrm{Tr}\,(\boldsymbol{\Psi}\mathbf{A})\mathrm{Tr}\,(\boldsymbol{\Psi}\mathbf{B})\boldsymbol{\Sigma} \otimes \boldsymbol{\Sigma} + \mathrm{Tr}\,(\boldsymbol{\Psi}\mathbf{A}\boldsymbol{\Psi}\mathbf{B}')\mathrm{vec}\,\boldsymbol{\Sigma}\mathrm{vec}\,'\boldsymbol{\Sigma}$
$+\mathrm{Tr}\,(\boldsymbol{\Psi}\mathbf{A}\boldsymbol{\Psi}\mathbf{B})\mathbf{K}_{p,p}(\boldsymbol{\Sigma} \otimes \boldsymbol{\Sigma}) + \mathrm{Tr}\,(\boldsymbol{\Psi}\mathbf{A})\boldsymbol{\Sigma} \otimes \mathbf{MBM}' + \mathrm{vec}\,(\mathbf{MB}\boldsymbol{\Psi}\mathbf{A}'\mathbf{M}')\mathrm{vec}\,'\boldsymbol{\Sigma}$
$+\mathbf{K}_{p,p}(\mathbf{MB}\boldsymbol{\Psi}\mathbf{AM}' \otimes \boldsymbol{\Sigma} + \boldsymbol{\Sigma} \otimes \mathbf{MA}\boldsymbol{\Psi}\mathbf{BM}') + \mathrm{vec}\,\boldsymbol{\Sigma}\mathrm{vec}\,'(\mathbf{MB}'\boldsymbol{\Psi}\mathbf{AM}')$
$+\mathrm{Tr}\,(\boldsymbol{\Psi}\mathbf{B})\mathbf{MAM}' \otimes \boldsymbol{\Sigma} + \mathbf{MAM}' \otimes \mathbf{MBM}'$.

Proof Statement (i) is obtained by considering $\mathrm{E}\,[\mathbf{X} \otimes \mathbf{X}]\mathrm{vec}\,\mathbf{A}$ and from Theorem 3.29 (ii), it follows that

$$\mathrm{E}(\mathrm{vec}\,(\mathbf{XAX}')) = \mathrm{Tr}\,(\boldsymbol{\Psi}\mathbf{A})\mathrm{vec}\,\boldsymbol{\Sigma} + \mathrm{vec}\,(\mathbf{MAM}'),$$

which is "equivalent" to statement (i).

For statement (ii), note that $\mathbf{X}$ and $\mathbf{Y} + \mathbf{M}$ have the same distribution, and that $\mathbf{Y} \sim N_{p,n}(\mathbf{0}, \boldsymbol{\Sigma}, \boldsymbol{\Psi})$. Then,

$$\begin{aligned}\mathrm{E}\,[(\mathbf{XAX}') \otimes (\mathbf{XBX}')] &= \mathrm{E}\,[(\mathbf{YAY}') \otimes (\mathbf{YBY}')] + (\mathbf{MAM}') \otimes (\mathbf{MBM}')\\ &+\mathrm{E}\,[(\mathbf{YAM}') \otimes (\mathbf{YBM}')] + \mathrm{E}\,[(\mathbf{YAM}') \otimes (\mathbf{MBY}')]\\ &+\mathrm{E}\,[(\mathbf{MAY}') \otimes (\mathbf{YBM}')] + \mathrm{E}\,[(\mathbf{MAY}') \otimes (\mathbf{MBY}')]\\ &+\mathrm{E}\,[(\mathbf{MAM}') \otimes (\mathbf{YBY}')] + \mathrm{E}\,[(\mathbf{YAY}') \otimes (\mathbf{MBM}')]. \qquad (3.9)\end{aligned}$$

Each term on the right-hand side of (3.9) can be considered separately, which with the help of Theorem 3.29 verifies statement (ii) of the theorem. □

We end this section by stating some facts for the complex normal distribution.

Theorem 3.32 *Let* $\mathbf{X} \in CN_{p,n}(\mathbf{M}, \boldsymbol{\Sigma}, \boldsymbol{\Psi})$ *follow the complex matrix normal distribution presented in Definition 3.22. Then,*

(i) $m_1(\mathbf{X}) = \mathrm{vec}\,\mathbf{M}$;
(ii) $m_2(\mathbf{X}) = \boldsymbol{\Psi}' \otimes \boldsymbol{\Sigma} + \mathrm{vec}\,\mathbf{M}\mathrm{vec}\,^*\mathbf{M}$;
(iii) $\mathrm{D}\,(\mathbf{X}) = \boldsymbol{\Psi}' \otimes \boldsymbol{\Sigma}$.

Proof The proof is based on differentiation of the characteristic function

$$\varphi_{\mathbf{X}}(\mathbf{T}) = \exp\{\imath\,\Re(\mathrm{Tr}\,(\mathbf{T}^*\mathbf{M})) - \tfrac{1}{4}\mathrm{Tr}\,(\boldsymbol{\Sigma}\mathbf{T}\boldsymbol{\Psi}\mathbf{T}^*)\}, \qquad \mathbf{T} \in \mathbb{C}^{p\times n},$$

twice, using the derivative given in Definition 3.9. Then,

$$\frac{{}^{\mathbb{C}}d}{d\mathbf{T}}\varphi_{\mathbf{X}}(\mathbf{T}) = \left[\imath \,\mathrm{vec}\,\mathbf{M} - \frac{1}{4}\frac{{}^{\mathbb{C}}d}{d\mathbf{T}}\mathrm{Tr}\,(\boldsymbol{\Sigma}\mathbf{T}\boldsymbol{\Psi}\mathbf{T}^{*})\right]\exp\{\imath\,\Re(\mathrm{Tr}\,(\mathbf{T}^{*}\mathbf{M})) - \tfrac{1}{4}\mathrm{Tr}\,(\boldsymbol{\Sigma}\mathbf{T}\boldsymbol{\Psi}\mathbf{T}^{*})\}.$$

When the somewhat special definition of the complex derivative has been used (referring to the application of the conjugate in Definition 3.9), differentiating a second time yields

$$\begin{aligned}&\frac{{}^{\mathbb{C}}d^2}{d\mathbf{T}^2}\varphi_{\mathbf{X}}(\mathbf{T})\\ &= \imath^2\left[\mathrm{vec}\,\mathbf{M}\mathrm{vec}\,{}^{*}\mathbf{M} + \frac{1}{4}\frac{{}^{\mathbb{C}}d^2}{d\mathbf{T}^2}\mathrm{Tr}\,(\boldsymbol{\Sigma}\mathbf{T}\boldsymbol{\Psi}\mathbf{T}^{*})\right]e^{\imath\,\Re(\mathrm{Tr}\,(\mathbf{T}^{*}\mathbf{M})) - \frac{1}{4}\mathrm{Tr}\,(\boldsymbol{\Sigma}\mathbf{T}\boldsymbol{\Psi}\mathbf{T}^{*})} + \mathbf{R}(\mathbf{T}),\end{aligned}$$

where the remainder term satisfies $\mathbf{R}(\mathbf{0}) = \mathbf{0}$. Thus, the theorem is established when the derivatives of the trace function are calculated (see Theorem 3.7 for the expressions), together with putting $\mathbf{T} = \mathbf{0}$. □

Theorem 3.33 *Let* $\mathbf{X} \in CN_{p,n}(\mathbf{0}, \boldsymbol{\Sigma}, \boldsymbol{\Psi})$. *Then,* $m_3(\mathbf{X}) = \mathbf{0}$ *and*

$$m_4(\mathbf{X}) = \mathrm{vec}\,'(\boldsymbol{\Psi}' \otimes \boldsymbol{\Sigma}) \otimes \boldsymbol{\Psi}' \otimes \boldsymbol{\Sigma} + (\boldsymbol{\Psi}' \otimes \boldsymbol{\Sigma}) \otimes \mathrm{vec}\,'(\boldsymbol{\Psi} \otimes \boldsymbol{\Sigma}').$$

Proof Now some hints are presented as to $m_4(\mathbf{X})$ is obtained. If we put $\mathbf{M} = \mathbf{0}$ in the proof of Theorem 3.32,

$$\begin{aligned}&\frac{{}^{\mathbb{C}}d^2}{d\mathbf{T}^2}\varphi_{\mathbf{X}}(\mathbf{T})\\ &= \left[-\tfrac{1}{4}\frac{{}^{\mathbb{C}}d^2}{d\mathbf{T}}\mathrm{Tr}\,(\boldsymbol{\Sigma}\mathbf{T}\boldsymbol{\Psi}\mathbf{T}^{*}) + \tfrac{1}{8}\frac{{}^{\mathbb{C}}d}{d\mathbf{T}}\mathrm{Tr}\,(\boldsymbol{\Sigma}\mathbf{T}\boldsymbol{\Psi}\mathbf{T}^{*})\mathrm{vec}\,{}^{*}\mathbf{T}(\boldsymbol{\Psi}' \otimes \boldsymbol{\Sigma})\right]\exp\{-\tfrac{1}{4}\mathrm{Tr}\,(\boldsymbol{\Sigma}\mathbf{T}\boldsymbol{\Psi}\mathbf{T}^{*})\}\\ &= \left[-(\boldsymbol{\Psi}' \otimes \boldsymbol{\Sigma}) + \tfrac{1}{4}(\boldsymbol{\Psi}' \otimes \boldsymbol{\Sigma})\mathrm{vec}\,\mathbf{T}\mathrm{vec}\,{}^{*}\mathbf{T}(\boldsymbol{\Psi}' \otimes \boldsymbol{\Sigma})\right]\exp\{-\tfrac{1}{4}\mathrm{Tr}\,(\boldsymbol{\Sigma}\mathbf{T}\boldsymbol{\Psi}\mathbf{T}^{*})\}, \qquad (3.10)\end{aligned}$$

where Theorem 3.7 has been used. To find the third order derivative, the conjugate of (3.10) is needed and then Theorem 3.7 is applied again:

$$\begin{aligned}\frac{{}^{\mathbb{C}}d^3}{d\mathbf{T}^3}\varphi_{\mathbf{X}}(\mathbf{T}) = \tfrac{1}{2}\big[&(\boldsymbol{\Psi}' \otimes \boldsymbol{\Sigma})\mathrm{vec}\,\mathbf{T}\mathrm{vec}\,'(\boldsymbol{\Psi}' \otimes \boldsymbol{\Sigma})\\ &+\mathrm{vec}\,'\mathbf{T}(\boldsymbol{\Psi} \otimes \boldsymbol{\Sigma}') \otimes (\boldsymbol{\Psi}' \otimes \boldsymbol{\Sigma})\big]\exp\{-\tfrac{1}{4}\mathrm{Tr}\,(\boldsymbol{\Sigma}\mathbf{T}\boldsymbol{\Psi}\mathbf{T}^{*})\} + \mathbf{R}_3(\mathbf{T}),\end{aligned} \qquad (3.11)$$

where $\mathbf{R}_3(\mathbf{T})$ is a remainder term which is not necessary to specify. It follows that $\mathbf{R}_3(\mathbf{0}) = \mathbf{0}$ and then

$$\left.\frac{{}^{\mathbb{C}}d^3}{d\mathbf{T}^3}\varphi_{\mathbf{X}}(\mathbf{T})\right|_{\mathbf{T}=\mathbf{0}} = \mathbf{0},$$

proving that $m_3(\mathbf{X}) = \mathbf{0}$. Finally, after taking the conjugate of (3.11) and then differentiating, we obtain

$$\frac{{}^{\mathbb{C}}d^4}{d\mathbf{T}^4}\varphi_{\mathbf{X}}(\mathbf{T})\bigg|_{\mathbf{T}=\mathbf{0}} = \mathrm{vec}'(\boldsymbol{\Psi}' \otimes \boldsymbol{\Sigma}) \otimes \boldsymbol{\Psi}' \otimes \boldsymbol{\Sigma} + (\boldsymbol{\Psi}' \otimes \boldsymbol{\Sigma}) \otimes \mathrm{vec}'(\boldsymbol{\Psi} \otimes \boldsymbol{\Sigma}').$$

□

Note that in the complex case the commutation matrix is not involved in the expression for $m_4(\mathbf{X})$ as it was for the real-valued normal distribution. From Theorems 3.14, 3.32 and 3.33, the next result for $\mathbf{XX}^*$ can be established.

Theorem 3.34 *Let* $\mathbf{X} \sim CN_{p,n}(\mathbf{0}, \boldsymbol{\Sigma}, \boldsymbol{\Psi})$. *Then,*

(i) $\mathrm{E}(\mathrm{vec}\,(\mathbf{XX}^*)) = \mathrm{Tr}\,(\boldsymbol{\Psi})\mathrm{vec}\,\boldsymbol{\Sigma}$;
(ii) $\mathrm{D}\,(\mathbf{XX}^*) = \mathrm{Tr}\,(\boldsymbol{\Psi}\boldsymbol{\Psi}')(\boldsymbol{\Sigma}' \otimes \boldsymbol{\Sigma})$.

3.3.2 Moments for Rotationally Invariant Symmetric Matrices

In many applications, there exist symmetric matrices $\mathbf{U}$ which are rotationally invariant, i.e., $\boldsymbol{\Gamma}\mathbf{U}\boldsymbol{\Gamma}'$ and $\mathbf{U}$ have the same distribution for all the orthogonal matrices $\boldsymbol{\Gamma}$. The next theorem will be verified in some detail (see also Cook and Forzani, [5] and related results in continuum mechanics to be found in Jog [23]).

Theorem 3.35 *Let* $\mathbf{U} \in \mathbb{R}^{p\times p}$ *be symmetric and rotationally invariant. Then,*

(i) $\mathrm{E}\,(\mathbf{U}) = c\mathbf{I}_p, \quad c = \mathrm{E}\,(U_{11})$;
(ii) $\mathrm{E}\,(\mathbf{U} \otimes \mathbf{U}) = c_1\mathbf{I}_{p^2} + c_2\mathrm{vec}\,\mathbf{I}_p\mathrm{vec}'\mathbf{I}_p + c_2\mathbf{K}_{p,p}, c_1 = \mathrm{E}\,(U_{11}U_{22})$, $c_2 = \mathrm{E}\,(U_{12}^2)$;
(iii) $\mathrm{D}\,(\mathbf{U}) = c_2\mathbf{I}_{p^2} + (c_1 - c^2)\mathrm{vec}\,\mathbf{I}_p\mathrm{vec}'\mathbf{I}_p + c_2\mathbf{K}_{p,p}$.

Proof Statement (i): Since $\mathrm{E}\,(\boldsymbol{\Gamma}\mathbf{U}\boldsymbol{\Gamma}') = \mathrm{E}\,(\mathbf{U})$ for all the orthogonal matrices $\boldsymbol{\Gamma}$, it follows that by using different choices of $\boldsymbol{\Gamma}$, $\mathrm{E}\,(\mathbf{U}) = c\mathbf{I}_p$ for some constant c which is determined by the relation $c = \mathbf{e}_1'\mathrm{E}\,(\mathbf{U})\mathbf{e}_1 = \mathrm{E}\,(U_{11})$.
Statement (ii): Consider

$$\mathrm{E}(\mathrm{vec}\,(\mathbf{U} \otimes \mathbf{U})) = \sum_{ijkl} \mathrm{E}\,(U_{ij}U_{kl})\mathbf{e}_j \otimes \mathbf{e}_l \otimes \mathbf{e}_i \otimes \mathbf{e}_k.$$

The pairs of indices (i, j) and (k, l) can be interchanged without affecting the moment expression, and interchanging either i and j or k and l will also not affect the moment expression. Thus, for some elements a_{ij}, where $a_{ij} = a_{ji}$, and constants d_1, d_2 and d_3,

$$\mathrm{E}(\mathrm{vec}\,(\mathbf{U}\otimes\mathbf{U})) = \sum_{ijkl}(d_1 a_{ij}a_{kl} + d_2 a_{ik}a_{jl} + d_2 a_{il}a_{jk} + d_3)\mathbf{e}_j \otimes \mathbf{e}_l \otimes \mathbf{e}_i \otimes \mathbf{e}_k,$$

which for a symmetric $\mathbf{A} = (a_{ij})$, $i, j \in \{1, \dots, p\}$, equals

$$\begin{aligned} &\mathrm{E}[\mathrm{vec}\,(\mathbf{U}\otimes\mathbf{U})] \\ &= d_1 \mathrm{vec}\,(\mathbf{A}\otimes\mathbf{A}) + d_2 \mathrm{vec}\,\mathbf{A} \otimes \mathrm{vec}\,\mathbf{A} + d_2 \mathrm{vec}\,(\mathbf{K}_{p,p}(\mathbf{A}\otimes\mathbf{A})) + d_3 \mathrm{vec}\,\mathbf{I}_{p^2}. \end{aligned}$$

Because $\mathbf{A} = \mathbf{\Gamma}\mathbf{D}\mathbf{\Gamma}'$ for an orthogonal matrix $\mathbf{\Gamma}$ and diagonal matrix $\mathbf{D}$

$$\begin{aligned} &\mathrm{vec}\,\mathrm{E}\,(\mathbf{U}\otimes\mathbf{U}) = (\mathbf{\Gamma}\otimes\mathbf{\Gamma}\otimes\mathbf{\Gamma}\otimes\mathbf{\Gamma})\mathrm{vec}\,\mathrm{E}\,(\mathbf{U}\otimes\mathbf{U}) \\ &= d_1 \mathrm{vec}\,(\mathbf{D}\otimes\mathbf{D}) + d_2 \mathrm{vec}\,\mathbf{D} \otimes \mathrm{vec}\,\mathbf{D} + d_2 \mathrm{vec}\,(\mathbf{K}_{p,p}(\mathbf{D}\otimes\mathbf{D})) + d_3 \mathrm{vec}\,\mathbf{I}_{p^2}. \end{aligned}$$

Moreover, by premultiplying with properly chosen orthogonal matrices, it can be proven that $\mathbf{D} = c\mathbf{I}_p$ for some constant c. Putting $c_1 = c^2 d_1 + d_3$ and $c_2 = c^2 d_2$, then

$$\begin{aligned} \mathrm{vec}\,\mathrm{E}\,(\mathbf{U}\otimes\mathbf{U}) &= (\mathbf{\Gamma}\otimes\mathbf{\Gamma}\otimes\mathbf{\Gamma}\otimes\mathbf{\Gamma})\mathrm{vec}\,\mathrm{E}\,(\mathbf{U}\otimes\mathbf{U}) \\ &= c_1 \mathrm{vec}\,\mathbf{I}_{p^2} + c_2 \mathrm{vec}\,\mathbf{I}_p \otimes \mathrm{vec}\,\mathbf{I}_p + c_2 \mathrm{vec}\,\mathbf{K}_{p,p}. \end{aligned}$$

It also follows that by premultiplying $\mathrm{vec}\,\mathrm{E}\,(\mathbf{U}\otimes\mathbf{U})$ by $\mathbf{e}_1'\otimes\mathbf{e}_1'\otimes\mathbf{e}_2'\otimes\mathbf{e}_2'$ and $\mathbf{e}_1'\otimes\mathbf{e}_2'\otimes\mathbf{e}_1'\otimes\mathbf{e}_2'$, that $\mathrm{E}\,(U_{11}U_{22}) = c_1$ and $\mathrm{E}\,(U_{12}^2) = c_2$.
Statement (iii): Note that instead of statement (ii), it is possible to write as follows:

$$\mathrm{E}\,(\mathrm{vec}\,\mathbf{U}\mathrm{vec}\,'\mathbf{U}) = c_2\mathbf{I}_{p^2} + c_1 \mathrm{vec}\,\mathbf{I}_p \mathrm{vec}\,'\mathbf{I}_p + c_2\mathbf{K}_{p,p}$$

and then the dispersion matrix $\mathrm{D}\,(\mathbf{U})$ is equal to $c_2\mathbf{I}_{p^2} + (c_1 - c^2)\mathrm{vec}\,\mathbf{I}_p\mathrm{vec}\,'\mathbf{I}_p + c_2\mathbf{K}_{p,p}$. □

Corollary 3.4 *Let $\mathbf{U} \in \mathbb{R}^{p\times p}$ be symmetric and rotationally invariant, and let c, c_1 and c_2 be as in Theorem 3.35 (i) and (ii). Then, c, c_1, and c_2 satisfy the following equations, from which the constants can be determined:*

$$\begin{aligned} \mathrm{Tr}\;[\mathrm{E}\,(\mathbf{U})] &= \mathrm{E}\,(\mathrm{Tr}\,\mathbf{U}) = cp, \\ \mathrm{Tr}\;[\mathrm{E}\,(\mathbf{U}\otimes\mathbf{U})] &= \mathrm{E}\,((\mathrm{Tr}\,\mathbf{U})^2) = c_1 p^2 + 2c_2 p, \\ \mathrm{vec}\,'\mathbf{I}_p\mathrm{E}\,(\mathbf{U}\otimes\mathbf{U})\mathrm{vec}\,\mathbf{I}_p &= \mathrm{E}\,(\mathrm{Tr}\,(\mathbf{U}^2)) = (c_1 + c_2)p + c_2 p^2 \end{aligned}$$

and

$$\begin{aligned} c &= \tfrac{1}{p}\mathrm{E}\,(\mathrm{Tr}\,\mathbf{U}), \\ c_1 &= \tfrac{p+1}{p(p+2)(p-1)}\mathrm{E}\,((\mathrm{Tr}\,\mathbf{U})^2) - \tfrac{2}{p^2(p+2)(p-1)}\mathrm{E}\,(\mathrm{Tr}\,(\mathbf{U}^2)), \\ c_2 &= \tfrac{1}{(p+2)(p-1)}\mathrm{E}\,(\mathrm{Tr}\,(\mathbf{U}^2)) - \tfrac{1}{p(p+2)(p-1)}\mathrm{E}\,((\mathrm{Tr}\,\mathbf{U})^2). \end{aligned}$$

3.3.3 Moments for the Wishart Distribution

In this section, only expressions related to the central Wishart distribution are presented. Since, if $\mathbf{W} \sim W_p(\boldsymbol{\Sigma}, n)$, the vectorised form of $\mathbf{W}$ can be expressed as $\mathrm{vec}\,\mathbf{W} = (\mathbf{X} \otimes \mathbf{X})\mathrm{vec}\,\mathbf{I}_n$, where $\mathbf{X} \sim N_{p,n}(\mathbf{0}, \boldsymbol{\Sigma}, \mathbf{I}_n)$, basic moments can be derived from Theorem 3.29. However, in the proof of the next theorem, we will mostly refer to Theorem 3.35.

Theorem 3.36 *Let* $\mathbf{W} \sim W_p(\boldsymbol{\Sigma}, n)$, $p < n$. *Then,*

(i) $\mathrm{E}\,(\mathbf{W}) = n\boldsymbol{\Sigma}$;

(ii) $\mathrm{E}\,(\mathbf{W} \otimes \mathbf{W}) = n^2\boldsymbol{\Sigma} \otimes \boldsymbol{\Sigma} + n\mathrm{vec}\,\boldsymbol{\Sigma}\mathrm{vec}'\boldsymbol{\Sigma} + n\mathbf{K}_{p,p}(\boldsymbol{\Sigma} \otimes \boldsymbol{\Sigma})$;

(iii) $\mathrm{D}\,(\mathbf{W}) = n(\mathbf{I}_{p^2} + \mathbf{K}_{p,p})(\boldsymbol{\Sigma} \otimes \boldsymbol{\Sigma})$;

(iv) $\mathrm{E}\,(\mathbf{W}^{\otimes k}) = n\mathrm{E}\,(\mathbf{W}^{\otimes (k-1)}) \otimes \boldsymbol{\Sigma} - 2\frac{\widetilde{d}}{d\,\boldsymbol{\Sigma}^{-1}}\mathrm{E}\,(\mathbf{W}^{\otimes (k-1)})$, $k > 2$,
where $\frac{\widetilde{d}}{d\mathbf{X}}$ *is defined in Definition 3.8;*

(v) $\mathrm{E}\,(\mathbf{W}^{\otimes 3}) = n\mathrm{E}\,(\mathbf{W}^{\otimes 2}) \otimes \boldsymbol{\Sigma} - 2\frac{\widetilde{d}}{d\,\boldsymbol{\Sigma}^{-1}}\mathrm{E}\,(\mathbf{W}^{\otimes 2})$,
where

$$\mathrm{E}\,(\mathbf{W}^{\otimes 2}) = n^2\boldsymbol{\Sigma} \otimes \boldsymbol{\Sigma} + n\mathrm{vec}\,\boldsymbol{\Sigma}\mathrm{vec}'\boldsymbol{\Sigma} + n\mathbf{K}_{p,p}(\boldsymbol{\Sigma} \otimes \boldsymbol{\Sigma}),$$

$$\frac{\widetilde{d}}{d\,\boldsymbol{\Sigma}^{-1}}\boldsymbol{\Sigma} \otimes \boldsymbol{\Sigma} = \boldsymbol{\Sigma} \otimes \frac{\widetilde{d}\,\boldsymbol{\Sigma}}{d\,\boldsymbol{\Sigma}^{-1}} + (\mathbf{K}_{p,p} \otimes \mathbf{I}_p)\left(\boldsymbol{\Sigma} \otimes \frac{\widetilde{d}\,\boldsymbol{\Sigma}}{d\,\boldsymbol{\Sigma}^{-1}}\right)(\mathbf{K}_{p,p} \otimes \mathbf{I}_p),$$

$$\frac{\widetilde{d}}{d\,\boldsymbol{\Sigma}^{-1}}\mathrm{vec}\,\boldsymbol{\Sigma}\mathrm{vec}'\boldsymbol{\Sigma} = \frac{\widetilde{d}\,\mathrm{vec}\,\boldsymbol{\Sigma}}{d\,\boldsymbol{\Sigma}^{-1}}(\mathrm{vec}'\boldsymbol{\Sigma} \otimes \mathbf{I}_p) + (\mathrm{vec}'\boldsymbol{\Sigma} \otimes \mathbf{I}_p)\frac{\widetilde{d}\,\mathrm{vec}\,\boldsymbol{\Sigma}}{d\,\boldsymbol{\Sigma}^{-1}},$$

$$\frac{\widetilde{d}}{d\,\boldsymbol{\Sigma}^{-1}}\mathbf{K}_{p,p}(\boldsymbol{\Sigma} \otimes \boldsymbol{\Sigma}) = (\mathbf{K}_{p,p} \otimes \mathbf{I}_p)\frac{\widetilde{d}}{d\,\boldsymbol{\Sigma}^{-1}}\boldsymbol{\Sigma} \otimes \boldsymbol{\Sigma},$$

$$\frac{\widetilde{d}\,\mathrm{vec}\,\boldsymbol{\Sigma}}{d\,\boldsymbol{\Sigma}^{-1}} = -(\mathbf{I}_p \otimes \boldsymbol{\Sigma} \otimes \mathbf{I}_p)\left(\mathbf{I}_p \otimes \frac{\widetilde{d}\,\boldsymbol{\Sigma}^{-1}}{d\,\boldsymbol{\Sigma}^{-1}}\right)(\mathrm{vec}\,\boldsymbol{\Sigma} \otimes \mathbf{I}_p),$$

$$\frac{\widetilde{d}\,\boldsymbol{\Sigma}}{d\,\boldsymbol{\Sigma}^{-1}} = -(\boldsymbol{\Sigma} \otimes \mathbf{I}_p)\frac{\widetilde{d}\,\boldsymbol{\Sigma}^{-1}}{d\,\boldsymbol{\Sigma}^{-1}}(\boldsymbol{\Sigma} \otimes \mathbf{I}_p), \qquad \frac{\widetilde{d}\,\boldsymbol{\Sigma}^{-1}}{d\,\boldsymbol{\Sigma}^{-1}} = \tfrac{1}{2}(\mathrm{vec}\,\boldsymbol{\Sigma}\mathrm{vec}'\boldsymbol{\Sigma} + \mathbf{K}_{p,p});$$

(vi) $\mathrm{E}\,(\mathbf{W}^2) = (n^2 + n)\boldsymbol{\Sigma}^2 + n\mathrm{Tr}\,(\boldsymbol{\Sigma})\boldsymbol{\Sigma}$;

(vii) $\mathrm{E}\,(\mathbf{W}^3) = \sum_{f,g=1}^{p}(\mathbf{I}_p \otimes \mathbf{e}'_f \otimes \mathbf{e}'_g)\mathrm{E}\,(\mathbf{W}^{\otimes 3})(\mathbf{e}_f \otimes \mathbf{e}_g \otimes \mathbf{I}_p)$,
where $\mathbf{e}_\bullet$ *is a unit basis vector of size* p.

Proof Note that $\mathrm{E}\,(\mathbf{W}) = \boldsymbol{\Sigma}^{1/2}\mathrm{E}\,(\mathbf{V})\boldsymbol{\Sigma}^{1/2}$ and $\mathrm{E}\,(\mathbf{W} \otimes \mathbf{W}) = (\boldsymbol{\Sigma}^{1/2} \otimes \boldsymbol{\Sigma}^{1/2})\mathrm{E}\,(\mathbf{V} \otimes \mathbf{V})(\boldsymbol{\Sigma}^{1/2} \otimes \boldsymbol{\Sigma}^{1/2})$, where $\mathbf{V} \sim W_p(\mathbf{I}_p, n)$. The matrix $\mathbf{V}$ is rotationally invariant and, therefore, according to Theorem 3.35, the following three constants have to be determined:

$$c = \mathrm{E}\,(V_{11}), \quad c_1 = \mathrm{E}\,(V_{11}V_{22}), \quad c_2 = \mathrm{E}\,(V_{12}^2).$$

Since $V_{11} \sim \chi^2(n)$, i.e., is chi-square distributed with n degrees of freedom, $c = n$. Moreover, V_{11} and V_{22} are independently distributed and, therefore, $c_1 = n^2$. Finally, it can be noted that $c_2 = n$ and statements (i)-(iii) have been established.

Turning to statement (iv), the Wishart density equals

$$f_{\mathbf{W}}(\mathbf{W}_0) = c|\boldsymbol{\Sigma}^{-1}|^{\frac{n}{2}}|\mathbf{W}_0|^{\frac{1}{2}(n-p-1)}\exp\{-\tfrac{1}{2}\mathrm{Tr}\,(\boldsymbol{\Sigma}^{-1}\mathbf{W}_0)\},$$

where c is a normalizing constant, and this expression is going to be differentiated. Differentiation of the density with respect to $\boldsymbol{\Sigma}^{-1}$ yields

$$\frac{\widetilde{\widetilde{d}}\, f_{\mathbf{W}}(\mathbf{W}_0)}{d\,\boldsymbol{\Sigma}^{-1}} = \frac{1}{2}(n\boldsymbol{\Sigma} - \mathbf{W}_0) f_{\mathbf{W}}(\mathbf{W}_0),$$

where Theorem 3.6 (viii) and (ix) have been utilized. Moreover, using Theorem 3.6 (iii),

$$\frac{\widetilde{\widetilde{d}}\,\mathrm{E}\,(\mathbf{W}^{\otimes(k-1)})}{d\,\boldsymbol{\Sigma}^{-1}} = \frac{n}{2}\mathrm{E}\,(\mathbf{W}^{\otimes(k-1)}) \otimes \boldsymbol{\Sigma} - \frac{1}{2}\mathrm{E}\,(\mathbf{W}^{\otimes k}),$$

which is identical to statement (iv).

Statement (v) follows directly from statement (iv) and, therefore, the derivative of $\mathrm{E}\,(\mathbf{W} \otimes \mathbf{W})$ is needed. Premultiplying statement (ii) by $\mathrm{vec}\,\mathbf{I}_p$ yields statement (vi). Some further calculations establish statement (vii). □

In Theorem 3.36 (iii), $n(\mathbf{I}_{p^2} + \mathbf{K}_{p,p}) = 2n\frac{1}{2}(\mathbf{I}_{p^2} + \mathbf{K}_{p,p})$ and $\frac{1}{2}(\mathbf{I}_{p^2} + \mathbf{K}_{p,p})$ is idempotent, i.e., it is a projection matrix.

Finally, in the next theorem, the expectation and dispersion for the complex Wishart distribution are given. The proof of the theorem is based on Theorem 3.34 with $\boldsymbol{\Psi} = \mathbf{I}_n$. It is of interest to compare the expression for the real-valued Wishart distribution in Theorem 3.36 (iii) with statement (ii) of the next theorem which indeed has a somewhat easier structure.

Theorem 3.37 *Let* $\mathbf{W} \sim CW_p(\boldsymbol{\Sigma}, n)$. *Then,*

(i) $\mathrm{E}(\mathbf{W}) = n\boldsymbol{\Sigma}$;
(ii) $\mathrm{D}(\mathbf{W}) = n\boldsymbol{\Sigma}' \otimes \boldsymbol{\Sigma}$.

3.3.4 Spectral Moments for Wishart Matrices

In this section, no theorems are presented. Instead alternative expressions for certain moment relations are given, so that the section becomes more like a discussion section. Any kind of moments for the empirical spectral distribution based on the Wishart matrix $\mathbf{W} \sim W_p(\boldsymbol{\Sigma}, n)$ can be expressed via

$$\mathrm{E}\left(\prod_{i=0}^{k} \mathrm{Tr}\left(\mathbf{W}^{m_i}\right)\right), \qquad m_i \geq 0, \quad i \in \{0, \ldots, k\}. \tag{3.12}$$

The case of $\mathrm{E}\left[\mathrm{Tr}\left(\mathbf{W}^k\right)\right]$ coincides with the moments of the empirical spectral distribution when $\mathbf{W}$ is used as an estimator of $n\boldsymbol{\Sigma}$. The moments of the Marčenko–Pastur distribution given in Theorem 3.19 can be derived when we consider expression (3.12) with $\boldsymbol{\Sigma} = \mathbf{I}_p$ under the Kolmogorov condition; i.e., for $\mathbf{W} \sim W_p(\frac{1}{n}\mathbf{I}_p, n)$, the moments of the spectral distribution are as follows:

$$\mathrm{E}\left(\frac{1}{p}\mathrm{Tr}\left(\mathbf{W}^k\right)\right) \to \sum_{i=0}^{k} \frac{c^i}{i+1}\binom{k-1}{i}\binom{k}{i}$$

as $n, p \to \infty$ with $\frac{p}{n} \to c \in (0, \infty)$.

We now present a selection of moment relations connected to expectation (3.12), for real as well as complex Wishart matrices.

Let $\mathbf{V} \sim W_p(\sigma^2\mathbf{I}_p, n, \boldsymbol{\Delta})$ be the non-central Wishart matrix as given in Definition 3.23. For specific choices of power k, exact formulas for (3.12) are given in a publication by Nel [32]. For any power $k \in \mathbb{N}$, it was proven that

$$\mathrm{E}\left((\mathrm{Tr}\,\mathbf{V})^k\right) = \sigma^{2k}\sum_{i=0}^{k}\binom{k}{i}2^{k-i}\left(\frac{np}{2}+i\right)^{k-i}(\mathrm{Tr}\,\boldsymbol{\Delta})^i.$$

Nel [32] also provided the first four moments for the central Wishart matrix with an arbitrary $\boldsymbol{\Sigma}$. Moreover, De Waal [46] presented

$$\mathrm{E}(\mathrm{Tr}\,\mathbf{V}) = n\mathrm{Tr}\,(\boldsymbol{\Sigma}) + \mathrm{Tr}\,(\boldsymbol{\Sigma}\boldsymbol{\Delta})$$

for the non-central $\mathbf{V}$, and if $\boldsymbol{\Sigma} = \sigma^2\mathbf{I}_p$, this simplifies to

$$\mathrm{E}\,(\mathrm{Tr}\,\mathbf{V}) = np\sigma^2 + \sigma^2\mathrm{Tr}\,\boldsymbol{\Delta} = \sigma^2\left[2\frac{np}{2} + \mathrm{Tr}\,\boldsymbol{\Delta}\right]$$

which agrees with Nel's result. These two authors differentiated a moment generating function to obtain the results. De Waal [46] also conjectured on the expectation of elementary symmetric functions of the roots of $\mathbf{V} \sim W_p(\boldsymbol{\Sigma}, n, \boldsymbol{\Delta})$, and his conjecture was later proven to hold by Saw [35] and Shah and Khatri [36].

For a non-central Wishart matrix $\mathbf{V} \sim W_p(\mathbf{I}_p, n, \boldsymbol{\Delta})$

$$\mathrm{E}\big(\mathrm{Tr}\,(\mathbf{V}^2)\big) = np(n+p+1) + 2(n+p+1)\mathrm{Tr}\,\boldsymbol{\Delta} + \mathrm{Tr}\,(\boldsymbol{\Delta}^2),$$

which is straightforward to derive. For more details see Gupta and Nadarajah [14].

A central Wishart matrix $\mathbf{W} \sim W_p(\boldsymbol{\Sigma}, n)$ is considered in Gupta and Nagar [15, Th. 3.3.23], where a general formula involving zonal polynomials is presented:

$$E\big((\mathrm{Tr}\,\mathbf{W})^k\big) = 2^k \sum_{\kappa} \left[\frac{n}{2}\right]_{\kappa} Z_{\kappa}(\boldsymbol{\Sigma}),$$

where $Z_\kappa(\cdot)$ stands for the zonal polynomial corresponding to κ, $\kappa = (k_1, \ldots, k_m)$ is a partition of k, such that $k_i \geq 0$ for all $i \in \{1, \ldots, m\}$, $m \in \{1, \ldots, k\}$ and $\sum_{i=1}^{m} k_i = k$, $[a]_\kappa = \prod_{i=1}^{m}[a - i + 1]_{k_i}$ with $[a]_k = (a + k)!/a!$ for $k \in \mathbb{N}_0$.

Moreover, since $\boldsymbol{\Sigma}^{-1/2}\mathbf{W}\boldsymbol{\Sigma}^{-1/2} \sim W_p(\mathbf{I}_p, n)$ and $\sum_\kappa [n]_\kappa Z_\kappa(\mathbf{I}_p) = [np]_k$ we have

$$E\Big(\big(\mathrm{Tr}\;(\boldsymbol{\Sigma}^{-1}\mathbf{W})\big)^k\Big) = 2^k \left[\frac{np}{2}\right]_k,$$

see, e.g., Subrahmaniam [42].

Following a theorem stated in Letac and Massam [25] (Th. 5 with a more general statement in Th. 1), an alternative formula equals

$$\begin{aligned} &E\left(\big(\mathrm{Tr}\;(\boldsymbol{\Sigma}^{-1}\mathbf{W})\big)^k\right) \\ &= \sum_{(i)\in I_k} \frac{k!}{i_1! \times \cdots \times i_k! 1^{i_1} \times \cdots \times k^{i_k}} \left(\frac{n}{2}\right)^{i_1+\ldots+i_k} \prod_{j=1}^{k} \left[\mathrm{Tr}\;(2\boldsymbol{\Sigma}^{-1}\boldsymbol{\Sigma})^j\right]^{i_j} \\ &= \sum_{(i)\in I_k} \frac{k!}{i_1! \times \cdots \times i_k! 1^{i_1} \times \cdots \times k^{i_k}} \left(\frac{n}{2}\right)^{i_1+\ldots+i_k} \prod_{j=1}^{k} \left(2^j p\right)^{i_j} \\ &= \sum_{(i)\in \mathcal{I}_k} \frac{k!}{i_1! \times \cdots \times i_k! 1^{i_1} \times \cdots \times k^{i_k}} (np)^{i_1+\cdots+i_k} 2^{k-\sum_{j=1}^{k} i_j}, \end{aligned} \tag{3.13}$$

where the set $\mathcal{I}_k$ consists of k-tuples $(i) = (i_1, \ldots, i_k)$, such that $i_1 + 2i_2 + \cdots + ki_k = k$ and i_j, $j \in \{1, \ldots, k\}$ are non-negative integers. An alternative version of the closed formula for the expectation of the power of the trace of the Wishart matrix $\mathbf{W}$ can be found in Mathai [28].

If $\mathrm{Tr}_j(\cdot)$ denotes the sum of all the principal minors of order j of the underlying matrix, then for the central Wishart matrix $\mathbf{W} \sim W_p(\boldsymbol{\Sigma}, n)$

$$E\,(\mathrm{Tr}_j \mathbf{W}) = n(n-1) \times \cdots \times (n - j + 1)\mathrm{Tr}_j \boldsymbol{\Sigma}.$$

For a central Wishart distribution with $\boldsymbol{\Sigma} = \mathbf{I}_p$, i.e., $\mathbf{W} \sim W_p(\mathbf{I}_p, n)$, a recursive formula for expectation (3.12) has been derived in Pielaszkiewicz et al. [33]. For all $k \in \mathbb{N}$ and all $m_0, m_1, \ldots, m_k$, such that $m_0 = 0$, $m_k \in \mathbb{N}$, $m_i \in \mathbb{N}_0$,

$$\begin{aligned}\mathrm{E}\left(\prod_{i=0}^{k}\mathrm{Tr}\,(\mathbf{W}^{m_i})\right) &= (n-p+m_k-1)\mathrm{E}\left(\mathrm{Tr}\,(\mathbf{W}^{m_k-1})\prod_{i=0}^{k-1}\mathrm{Tr}\,(\mathbf{W}^{m_i})\right)\\ &+2\sum_{i=0}^{k-1}m_i\mathrm{E}\left(\mathrm{Tr}\,(\mathbf{W}^{m_k+m_i-1})\prod_{\substack{j=0\\ j\neq i}}^{k-1}\mathrm{Tr}\,(\mathbf{W}^{m_j})\right)\\ &+\sum_{i=0}^{m_k-1}\mathrm{E}\left(\mathrm{Tr}\,(\mathbf{W}^{i})\mathrm{Tr}\,(\mathbf{W}^{m_k-1-i})\prod_{j=0}^{k-1}\mathrm{Tr}\,(\mathbf{W}^{m_j})\right). \quad (3.14)\end{aligned}$$

In the case of $m_1 = \ldots = m_k = 1$, the above formula simplifies to

$$\mathrm{E}\left((\mathrm{Tr}\,\mathbf{W})^k\right) = (np+2(k-1))\mathrm{E}\left((\mathrm{Tr}\,\mathbf{W})^{k-1}\right).$$

Based on formula (3.14), for $\mathbf{W} \sim W_p(\mathbf{\Sigma}, n)$,

$$\mathrm{E}\left(\mathrm{Tr}\left[(\mathbf{\Sigma}^{-1}\mathbf{W})^2\right]\right) = (n+p+1)np,$$

which (see, e.g., Fujikoshi et al. [9]) is a special case of

$$\mathrm{E}\big(\mathrm{Tr}\,\left(\mathbf{W}^2\right)\big) = (n+n^2)\mathrm{Tr}\,\left(\mathbf{\Sigma}^2\right) + n\,(\mathrm{Tr}\,\mathbf{\Sigma})^2.$$

An expression for

$$\mathrm{E}\left(\prod_{i=0}^{k}\mathrm{Tr}\,(\mathbf{W}\mathbf{H}_i)\right),$$

where $\mathbf{H}_1, \ldots, \mathbf{H}_k$ are arbitrary real symmetric matrices, can be found in Letac and Massam [26]. By choosing $\mathbf{H}_i$ appropriately, the moments of all the monomials are available. We refer to Di Nardo [7] for a generalization of the above expression of Letac, Massam [26] for a non-central Wishart matrix using a symbolic method called umbral calculations (e.g., see Rota and Shen [34]).

Let $\mathbf{W} \sim CW_p(\mathbf{I}_p, n)$ be the complex Wishart matrix defined as in Definition 3.24. The explicit formula for the kth spectral moment of the Wishart matrix $\mathbf{W}$ can be formulated (see Th. 2.5 in Hanlon et al. [20]) as follows:

$$\mathrm{E}\left(\mathrm{Tr}\,(\mathbf{W}^k)\right) = \frac{1}{k}\sum_{j=1}^{k}(-1)^{j-1}\frac{[n+k-j]_k[p+k-j]_k}{(k-j)!(j-1)!}, \qquad k \in \mathbb{N}. \quad (3.15)$$

A corresponding recursive result has been derived for all $k \in \mathbb{N}$

$$\begin{aligned}
\mathrm{E}\left(\mathrm{Tr}\,(\mathbf{W}^0)\right) &= p,\\
\mathrm{E}\left(\mathrm{Tr}\,(\mathbf{W}^1)\right) &= np,\\
\mathrm{E}\left(\mathrm{Tr}\,(\mathbf{W}^{k+1})\right) &= \frac{(2k+1)(n+p)}{k+2}\mathrm{E}\left(\mathrm{Tr}\,(\mathbf{W}^k)\right)\\
&\quad+\frac{(k-1)(k^2-(n-p)^2)}{k+2}\mathrm{E}\left(\mathrm{Tr}\,(\mathbf{W}^{k-1})\right); \qquad (3.16)
\end{aligned}$$

see Sect. 8 in Haagerup and Thorbjørnsen [17]. The recursive formula given above as (3.16) was inspired by the Harer–Zagier recursion formula. The original paper by Harer and Zagier [21] gives a recursive relation for even moments of the spectral distribution of a complex self-adjoint random matrix $\mathbf{Z} = (Z_{ij})$, formed of p^2 independent real variables which have a standard normal distribution and are such that Z_{ij} has mean 0 and variance 1, i.e., an expression for $\frac{1}{p}\mathrm{E}\left(\mathrm{Tr}(\mathbf{Z}^{2k})\right)$. The recursive formula was proven once again and extended to the interesting case of Wishart matrices in Haagerup and Thorbjørnsen [17]. Both the explicit result given in Haagerup and Thorbjørnsen [20] and the recursive result of Haagerup and Thorbjørnsen [17] derive $\mathrm{E}\left(\mathrm{Tr}(\mathbf{W}^k)\right)$, although equation (3.16) was said by [17] to be more efficient for generating moment tables than relation (3.15).

The spectral moments of a complex Wishart matrix were also presented in Letac and Massam [25] in a complex version of formula (3.13). The general version of that particular result had already been published in an earlier work by Graczyk et al. [13].

3.3.5 Inverse Wishart Moments

The moments for the inverted Wishart distribution are the same as the inverse moments for the Wishart distribution. In the next theorem, a few results are presented where the proofs are of particular interest.

Theorem 3.38 *Let* $\mathbf{W} \sim W_p(\mathbf{\Sigma}, n)$*,* $\mathbf{\Sigma} > 0$*, and*

$$c_2^{-1} = (n-p)(n-p-1)(n-p-3), \quad c_1 = (n-p-2)c_2, \quad c_3 = (n-p-1)c_2.$$

Then,

(i) $\mathrm{E}\,(\mathbf{W}^{-1}) = \frac{1}{n-p-1}\mathbf{\Sigma}^{-1}, \qquad n-p-1>0;$

(ii) *If* $n-p-3>0$,
$\mathrm{E}\,(\mathbf{W}^{-1}\otimes\mathbf{W}^{-1}) = c_1\mathbf{\Sigma}^{-1}\otimes\mathbf{\Sigma}^{-1} + c_2\mathrm{vec}\,\mathbf{\Sigma}^{-1}\mathrm{vec}'\mathbf{\Sigma}^{-1} + c_2\mathbf{K}_{p,p}(\mathbf{\Sigma}^{-1}\otimes\mathbf{\Sigma}^{-1}),$
$\mathrm{E}\,(\mathrm{vec}\,\mathbf{W}^{-1}\mathrm{vec}'\mathbf{W}^{-1}) = c_1\mathrm{vec}\,\mathbf{\Sigma}^{-1}\mathrm{vec}'\mathbf{\Sigma}^{-1} + c_2(\mathbf{I}_{p^2}+\mathbf{K}_{p,p})(\mathbf{\Sigma}^{-1}\otimes\mathbf{\Sigma}^{-1});$

(iii) $\mathrm{E}\,(\mathrm{vec}\,\mathbf{W}^{-1}\mathrm{vec}'\mathbf{W}) = \frac{n}{n-p-1}\mathrm{vec}\,\mathbf{\Sigma}^{-1}\mathrm{vec}'\mathbf{\Sigma} - \frac{1}{n-p-1}(\mathbf{I}_{p^2}+\mathbf{K}_{p,p}),$
$n-p-1>0;$

(iv) $\mathrm{E}\,(\mathbf{W}^{-1}\mathbf{W}^{-1}) = c_3\mathbf{\Sigma}^{-1}\mathbf{\Sigma}^{-1} + c_2\mathrm{Tr}\,(\mathbf{\Sigma}^{-1})\mathbf{\Sigma}^{-1}, \qquad n-p-3>0;$

(v) $\mathrm{E}\,(\mathrm{Tr}\,(\mathbf{W}^{-1})\mathbf{W}) = \frac{1}{n-p-1}(n\mathrm{Tr}\,(\mathbf{\Sigma}^{-1})\mathbf{\Sigma} - 2\mathbf{I}_p), \qquad n-p-1>0;$

(vi) $\mathrm{E}\,(\mathrm{Tr}\,(\mathbf{W}^{-1})\mathbf{W}^{-1}) = 2c_1\boldsymbol{\Sigma}^{-1}\boldsymbol{\Sigma}^{-1} + c_2\mathrm{Tr}\,(\boldsymbol{\Sigma}^{-1})\boldsymbol{\Sigma}^{-1}, \quad n-p-3>0.$

Proof Statements (i) and (ii) can be verified via Theorem 3.35. However, now a somewhat different proof is presented which is based on partial integration (a multivariate version). The matrix derivative $\frac{\widetilde{d}\mathbf{Y}}{d\mathbf{X}}$ in Definition 3.7 will be utilized and to prove statement (i) the following can be noted (a similar technique and similar ideas were presented by Haff [18, 19]):

$$\int_{\mathbf{W}>0} \frac{\widetilde{d}\,\mathbf{f_W}(\mathbf{W})}{d\,\mathbf{W}} d\,\mathbf{W} = \mathbf{0}, \tag{3.17}$$

where the Wishart density for $\mathbf{W} \sim W_p(\boldsymbol{\Sigma}, n)$ is given by

$$f_{\mathbf{W}}(\mathbf{W}) = c|\boldsymbol{\Sigma}|^{-\frac{1}{2}n}|\mathbf{W}|^{\frac{1}{2}(n-p-1)}\exp\{-\tfrac{1}{2}\mathrm{Tr}\,(\boldsymbol{\Sigma}^{-1}\mathbf{W})\},$$

where c denotes the normalizing constant. Moreover, $\int_{\mathbf{W}>0}$ means an ordinary multiple integral with integration performed over the subset of $\mathbb{R}^{\frac{1}{2}p(p+1)}$, with $\mathbf{W}$ being positive definite and $d\,\mathbf{W}$ denoting the Lebesque measure $\prod_{k\leq l} d\,W_{kl}$ in $\mathbb{R}^{\frac{1}{2}p(p+1)}$. Establishing relation (3.17) is relatively easy (see Kollo and von Rosen [24, pp. 258–259]), since, when $|\mathbf{W}| \to 0$, the Wishart density converges also to 0. Now using the differentiation rules presented in Theorem 3.5 yield

$$\tfrac{1}{2}\int_{\mathbf{W}>0} \left[(n-p-1)\mathrm{vec}\,\mathbf{W}^{-1} - \mathrm{vec}\,\boldsymbol{\Sigma}^{-1}\right] f_{\mathbf{W}}(\mathbf{W})d\,\mathbf{W} = \mathbf{0},$$

which implies $(n-p-1)\mathrm{E}\,(\mathbf{W}^{-1}) = \boldsymbol{\Sigma}^{-1}$.

To verify statement (ii), consider

$$\int_{\mathbf{W}>0} \frac{\widetilde{d}}{d\,\mathbf{W}}(\mathbf{W}^{-1} f_{\mathbf{W}}(\mathbf{W}))d\,\mathbf{W} = \mathbf{0},$$

which gives the equation

$$-\mathrm{E}\,(\mathbf{W}^{-1}\otimes\mathbf{W}^{-1}) - \mathbf{K}_{p,p}\mathrm{E}\,(\mathbf{W}^{-1}\otimes\mathbf{W}^{-1}) + (n-p-1)\mathrm{E}\,(\mathrm{vec}\,\mathbf{W}^{-1}\mathrm{vec}\,'\mathbf{W}^{-1}) = \frac{1}{n-p-1}\mathrm{vec}\,\boldsymbol{\Sigma}^{-1}\mathrm{vec}\,'\boldsymbol{\Sigma}^{-1}.$$

From here one can obtain two more equivalent relations. For example, one relation is obtained by premultiplying by $\mathbf{K}_{p,p}$ and for all the technical details, see Kollo and von Rosen [24, p. 260]. Put

$$\mathbf{T} = \left(\mathrm{E}\,(\mathbf{W}^{-1} \otimes \mathbf{W}^{-1}) : \mathbf{K}_{p,p}\mathrm{E}\,(\mathbf{W}^{-1} \otimes \mathbf{W}^{-1}) : \mathrm{E}\,(\mathrm{vec}\,\mathbf{W}^{-1}\mathrm{vec}'\mathbf{W}^{-1})\right),$$
$$\mathbf{M} = (\mathrm{vec}\,\boldsymbol{\Sigma}^{-1}\mathrm{vec}'\boldsymbol{\Sigma}^{-1} : \mathbf{K}_{p,p}(\boldsymbol{\Sigma}^{-1} \otimes \boldsymbol{\Sigma}^{-1}) : \boldsymbol{\Sigma}^{-1} \otimes \boldsymbol{\Sigma}^{-1}),$$
$$\mathbf{Q} = \begin{pmatrix} -\mathbf{I}_{p^2} & -\mathbf{I}_{p^2} & (n-p-1)\mathbf{I}_{p^2} \\ -\mathbf{I}_{p^2} & (n-p-1)\mathbf{I}_{p^2} & -\mathbf{I}_{p^2} \\ (n-p-1)\mathbf{I}_{p^2} & -\mathbf{I}_{p^2} & -\mathbf{I}_{p^2} \end{pmatrix}$$

and we can write

$$\mathbf{QT} = \tfrac{1}{n-p-1}\mathbf{M}.$$

This can be considered as an equation in $\mathbf{T}$ with a solution $\mathbf{T} = (n-p-1)^{-1}\mathbf{Q}^{-1}\mathbf{M}$, where

$$\mathbf{Q}^{-1} = \tfrac{1}{(n-p)(n-p-3)} \begin{pmatrix} \mathbf{I}_{p^2} & \mathbf{I}_{p^2} & (n-p-2)\mathbf{I}_{p^2} \\ \mathbf{I}_{p^2} & (n-p-2)\mathbf{I}_{p^2} & \mathbf{I}_{p^2} \\ (n-p-2)\mathbf{I}_{p^2} & \mathbf{I}_{p^2} & \mathbf{I}_{p^2} \end{pmatrix}.$$

Thus, $\mathbf{T}$ is found as a function of $\boldsymbol{\Sigma}$ and, in particular, $\mathrm{E}\,(\mathbf{W}^{-1} \otimes \mathbf{W}^{-1})$ and $\mathrm{E}\,(\mathrm{vec}\,\mathbf{W}^{-1}\mathrm{vec}'\mathbf{W}^{-1})$ in statement (ii) are obtained.

To prove statement (iii), both sides of the equation in statement (i) have to be differentiated with respect to $\boldsymbol{\Sigma}^{-1}$, which yields

$$\mathrm{E}\,(n\mathrm{vec}\,\mathbf{W}^{-1}\mathrm{vec}'\boldsymbol{\Sigma} - \mathrm{vec}\,\mathbf{W}^{-1}\mathrm{vec}'\mathbf{W}) = \tfrac{1}{n-p-1}(\mathbf{I}_{p^2} + \mathbf{K}_{p,p})$$

and which is identical to the statement of the theorem.

Statement (iv) follows from statement (ii) by noting that

$$\mathrm{vec}\,\mathrm{E}\,(\mathbf{W}^{-1}\mathbf{W}^{-1}) = \mathrm{E}\big(W^{-1} \otimes \mathbf{W}^{-1})\mathrm{vec}\,\mathbf{I}_{p^2}\big).$$

Statement (v) is obtained by premultiplying statement (iii) by $\mathrm{vec}'\mathbf{I}$ and the relation in statement (vi) can be established by using

$$\sum_{m=1}^{p}(\mathbf{e}_m' \otimes \mathbf{I}_p)\mathrm{E}\,(\mathbf{W}^{-1} \otimes \mathbf{W}^{-1})(\mathbf{e}_m \otimes \mathbf{I}_p) = \mathrm{E}\big(\mathrm{Tr}\,(\mathbf{W}^{-1})\mathbf{W}^{-1}\big).$$

□

If it is assumed that $\mathbf{W}$ is singular, which in our case means that in $\mathbf{W} \sim W_p(\boldsymbol{\Sigma}, n)$, $p > n$ (a singular $\boldsymbol{\Sigma}$ leads to another type of singular Wishart distribution), then under the assumption that $\boldsymbol{\Sigma} = \mathbf{I}_p$, moment results for the Moore–Penrose inverse $\mathbf{W}^{+}$ exist.

Theorem 3.39 *Let* $\mathbf{W} \sim W_p(\mathbf{I}_p, n)$. *Then,*

(i) *if* $p > n + 1$, $\mathrm{E}(\mathbf{W}^+) = \frac{n}{p(p-n-1)}\mathbf{I}_p$;

(ii) *if* $p > n + 3$, $\mathrm{E}(\mathrm{vec}\,\mathbf{W}^+\mathrm{vec}'\mathbf{W}^+) = c_1(\mathbf{I}_{p^2} + \mathbf{K}_{p,p}) + c_2\mathrm{vec}\mathbf{I}_p\mathrm{vec}\mathbf{I}'_p$, *where*

$$c_1 = \frac{n(p(p-1)-n(p-n-2)-2)}{p(p-1)(p+2)(p-n)(p-n-1)(p-n-3)}, \quad c_2 = \frac{n(4+n(p+1)(p-n-2))}{p(p-1)(p+2)(p-n)(p-n-1)(p-n-3)}.$$

Proof Since the distribution for $\mathbf{W}$ is rotationally invariant, one way of proving statement (i) is to refer to Theorem 3.35 (i) and it follows that $\mathrm{E}(\mathbf{W}^+) = c\mathbf{I}_p$ for some positive constant c. The constant can be determined by taking the trace, i.e., $\mathrm{E}(\mathrm{Tr}\,(\mathbf{W}^+)) = cp$, and using the fact that $\mathbf{W}^+ = \mathbf{X}(\mathbf{X}'\mathbf{X})^{-1}(\mathbf{X}'\mathbf{X})^{-1}\mathbf{X}'$, where $\mathbf{X} \sim N_{p,n}(\mathbf{0}, \mathbf{I}_p, \mathbf{I}_n)$, yields

$$\mathrm{Tr}\,(\mathbf{W}^+) = \mathrm{Tr}\,(\mathbf{X}(\mathbf{X}'\mathbf{X})^{-1}(\mathbf{X}'\mathbf{X})^{-1}\mathbf{X}') = \mathrm{Tr}\,((\mathbf{X}'\mathbf{X})^{-1}),$$

where $\mathbf{X}'\mathbf{X} \sim W_n(\mathbf{I}_n, p)$. Then, Theorem 3.38 (i) determines c and establishes statement (i).

For the proof of the second statement, the reader is referred to Cook and Forzani [5], but statement (ii) can also be verified by using Theorem 3.35 (ii). □

3.3.6 Moments for Multivariate β-type Distributions

In Definition 3.25, three different β-type distributions were presented. For each of them the mean and dispersion will now be derived.

Theorem 3.40 *Let* $\mathbf{Z} \sim M\beta_{II}(p, m, n)$ *be given in Definition 3.25 (ii) and*

$$c_2^{-1} = (n-p)(n-p-1)(n-p-3), \qquad c_1 = (n-p-2)c_2,$$

$$d_1 = \frac{n^2(m-p-2)+2n}{(m-p)(m-p-1)(m-p-3)}, \; d_2 = \frac{n(m-p-1)+n^2}{(m-p)(m-p-1)(m-p-3)},$$

$$e_1 = m(c_1m + 2c_2), \; e_2 = m(c_2m + c_1 + c_2).$$

Then,

(i) $\mathrm{E}(\mathbf{Z}) = \frac{n}{m-p-1}\mathbf{I}_p, \quad m-p-1 > 0;$
(ii) $\mathrm{D}(\mathbf{Z}) = d_2(\mathbf{I}_{p^2} + \mathbf{K}_{p,p}) + (d_1 - \frac{n^2}{(m-p-1)^2})\mathrm{vec}\,\mathbf{I}_p\mathrm{vec}'\mathbf{I}_p, \quad m-p-3 > 0;$
(iii) $\mathrm{E}(\mathbf{Z} \otimes \mathbf{Z}) = d_1\mathbf{I}_{p^2} + d_2\mathrm{vec}\,\mathbf{I}_p\mathrm{vec}'\mathbf{I}_p + d_2\mathbf{K}_{p,p}, \quad m-p-3 > 0;$
(iv) $\mathrm{E}(\mathbf{Z}^{-1}) = \frac{m}{n-p-1};$
(v) $\mathrm{E}(\mathbf{Z}^{-1} \otimes \mathbf{Z}^{-1}) = e_1\mathbf{I}_{p^2} + e_2\mathrm{vec}\,\mathbf{I}_p\mathrm{vec}'\mathbf{I}_p + e_2\mathbf{K}_{p,p};$
(vi) $\mathrm{E}(\mathrm{vec}\,\mathbf{Z}^{-1}\mathrm{vec}'\mathbf{Z}^{-1}) = e_2(\mathbf{I}_{p^2} + \mathbf{K}_{p,p}) + e_1\mathrm{vec}\,\mathbf{I}_p\mathrm{vec}'\mathbf{I}_p.$

Proof By definition, $\mathbf{Z} = \mathbf{W}_2^{-1/2}\mathbf{W}_1\mathbf{W}_2^{-1/2}$, where $\mathbf{W}_1 \sim W_p(\mathbf{I}_p, n)$ and $\mathbf{W}_2 \sim W_p(\mathbf{I}_p, m)$ are independently distributed. Thus, Theorem 3.36 (i) and Theorem 3.38 (i) establish statement (i).

Concerning statement (ii), it can be noted that

$$\begin{aligned}\mathrm{E}\,(\mathrm{vec}\,\mathbf{Z}\mathrm{vec}\,'\mathbf{Z}) &= \mathrm{E}\,\left(\left(\mathbf{W}_2^{-1/2}\otimes\mathbf{W}_2^{-1/2}\right)\mathrm{vec}\,\mathbf{W}_1\mathrm{vec}\,'\mathbf{W}_1\left(\mathbf{W}_2^{-1/2}\otimes\mathbf{W}_2^{-1/2}\right)\right)\\ &= n\left(\mathbf{I}_{p^2}+\mathbf{K}_{p,p}\right)\mathrm{E}\,\left(\mathbf{W}_2^{-1}\otimes\mathbf{W}_2^{-1}\right)+n^2\mathrm{E}\,\left(\mathrm{vec}\,\mathbf{W}_2^{-1}\mathrm{vec}\,'\mathbf{W}_2^{-1}\right)\end{aligned}$$

and the expression in statement (ii) follows by using Theorem 3.38 (ii) and subtracting $\mathrm{E}\,(\mathrm{vec}\,\mathbf{Z})\mathrm{E}\,(\mathrm{vec}\,'\mathbf{Z})$.

By identifying the coefficients, statement (iii) is obtained via Theorem 3.35 (ii) and (iii). The other statements concerning the inverse $\mathbf{Z}^{-1}$ can be proven in a similar fashion. □

The moments $\mathbf{F}$ in Definition 3.25 (i) are more complicated to derive than the moments for $\mathbf{Z}$ and, in fact, these moments will be derived via knowledge of $(\mathbf{I}_p + \mathbf{Z})^{-1}$.

Theorem 3.41 *Let $\mathbf{F} \sim M\beta_I(p, m, n)$ be given in Definition 3.25 (i) and explicit expressions of coefficients f_1 and f_2 can be found using the equations*

$$\begin{aligned}f_1 + f_2(n+m+1) &= e_2\tfrac{(n-p-1)(n-p-2)}{n+m} - e_1\tfrac{n-p-1}{n+m},\\ f_1 + f_2 &= e_1\tfrac{(n-p-1)^2}{n+m} - 2e_2\tfrac{n-p-1}{n+m},\end{aligned}$$

where e_1 and e_2 are defined in Theorem 3.40. Then,

(i) $\mathrm{E}\,(\mathbf{F}) = \frac{m}{n+m}\mathbf{I}_p,$

(ii) $\mathrm{D}\,(\mathbf{F}) = f_2(\mathbf{I}_{p^2} + \mathbf{K}_{p,p}) + (f_1 - \frac{m^2}{(n+m)^2})\mathrm{vec}\,\mathbf{I}_p\mathrm{vec}\,'\mathbf{I}_p, \qquad m-p-3>0,$

(iii) $\mathrm{E}\,(\mathbf{F}\otimes\mathbf{F}) = f_1\mathbf{I}_p + f_2\mathrm{vec}\,\mathbf{I}_p\mathrm{vec}\,'\mathbf{I}_p + f_2\mathbf{K}_{p,p}, \qquad m-p-3>0.$

Proof According to Theorems 3.24 and 3.25, finding the expectation and dispersion for $\mathbf{F}$ is the same as finding the expectation and dispersion for $(\mathbf{I}_p + \mathbf{Z})^{-1}$. The moments for $(\mathbf{I}_p + \mathbf{Z})^{-1}$ are easier to derive than the moments for $\mathbf{F}$.

The same technique as that applied when deriving the inverse Wishart moments will now be applied. It follows that

$$\mathbf{0} = \int \frac{\widetilde{d}}{d\mathbf{Z}}c|\mathbf{Z}|^{\frac{1}{2}(n-p-1)}|\mathbf{I}_p + \mathbf{Z}|^{-\frac{1}{2}(n+m)}d\,\mathbf{Z},$$

where c is a normalizing constant and the derivative is given in Definition 3.7. By differentiation, using Theorem 3.5 (vii), we obtain

$$\mathbf{0} = \tfrac{1}{2}(n-p-1)\mathrm{E}\,(\mathrm{vec}\,\mathbf{Z}^{-1}) - \tfrac{1}{2}(n+m)\mathrm{E}\,[\mathrm{vec}\,(\mathbf{I}_p + \mathbf{Z})^{-1}],$$

which is identical to

$$\mathrm{E}\,(\mathrm{vec}\,(\mathbf{I}_p + \mathbf{Z})^{-1}) = \frac{n-p-1}{n+m}\mathrm{E}\,(\mathrm{vec}\,\mathbf{Z}^{-1}).$$

However, based on the definition of $\mathbf{Z}$,

$$\mathrm{E}\,(\mathbf{Z}^{-1}) = \mathrm{E}\,\left(\mathbf{W}_2^{1/2}\mathbf{W}_1^{-1}\mathbf{W}_2^{1/2}\right) = \tfrac{m}{n-p-1}\mathbf{I}_p,$$

implying statement (i).

Turning to statement (ii), consider

$$\mathbf{0} = \int \frac{\widetilde{d}}{d\,\mathbf{Z}}\left(\mathbf{Z}^{-1} f_{\mathbf{Z}}(\mathbf{Z})\right) d\,\mathbf{Z}, \qquad f_{\mathbf{Z}}(\mathbf{Z}) = c|\mathbf{Z}|^{\frac{1}{2}(n-p-1)}|\mathbf{I}_p + \mathbf{Z}|^{-\frac{1}{2}(n+m)}d\,\mathbf{Z},$$

which is identical to (see Theorem 3.4 (ii))

$$\mathbf{0} = \int \frac{\widetilde{d}\,\mathbf{Z}^{-1}}{d\,\mathbf{Z}} f_{\mathbf{Z}}(\mathbf{Z}) d\,\mathbf{Z} + \int \frac{\widetilde{d}\, f_{\mathbf{Z}}(\mathbf{Z})}{d\,\mathbf{Z}}\mathrm{vec}\,'\mathbf{Z}^{-1} d\,\mathbf{Z},$$

where (see Theorem 3.5)

$$\begin{aligned}
\frac{\widetilde{d}\,\mathbf{Z}^{-1}}{d\,\mathbf{Z}} &= -\tfrac{1}{2}(\mathbf{I}_{p^2} + \mathbf{K}_{p,p})(\mathbf{Z}^{-1} \otimes \mathbf{Z}^{-1}),\\
\frac{\widetilde{d}\, f_{\mathbf{Z}}(\mathbf{Z})}{d\,\mathbf{Z}} &= \tfrac{1}{2}(n-p-1)\mathrm{vec}\,\mathbf{Z}^{-1} f_{\mathbf{Z}}(\mathbf{Z}) - \tfrac{1}{2}\mathrm{vec}\,(\mathbf{I}_p + \mathbf{Z})^{-1} f_{\mathbf{Z}}(\mathbf{Z}).
\end{aligned}$$

Thus,

$$\begin{aligned}
\mathbf{0} = &-\tfrac{1}{2}(\mathbf{I}_{p^2} + \mathbf{K}_{p,p})\mathrm{E}\,\left(\mathbf{Z}^{-1} \otimes \mathbf{Z}^{-1}\right) + \tfrac{1}{2}(n-p-1)\mathrm{E}\,\left(\mathrm{vec}\,\mathbf{Z}^{-1}\mathrm{vec}\,'\mathbf{Z}^{-1}\right)\\
&-\tfrac{1}{2}(n+m)\mathrm{E}\,\left[\mathrm{vec}\,(\mathbf{I}_p + \mathbf{Z})^{-1}\mathrm{vec}\,'\mathbf{Z}^{-1}\right]. \qquad (3.18)
\end{aligned}$$

Equivalently, it is possible to start with

$$\mathbf{0} = \int \frac{\widetilde{d}}{d\,\mathbf{Z}}\left[(\mathbf{I}_p + \mathbf{Z})^{-1} f_{\mathbf{Z}}(\mathbf{Z})\right] d\,\mathbf{Z},$$

and obtain

$$\begin{aligned}
\mathbf{0} = &-\tfrac{1}{2}(\mathbf{I}_{p^2} + \mathbf{K}_{p,p})\mathrm{E}\,\left[(\mathbf{I}_p + \mathbf{Z})^{-1} \otimes (\mathbf{I}_p + \mathbf{Z})^{-1}\right]\\
&+\tfrac{1}{2}(n-p-1)\mathrm{E}\,\left[\mathrm{vec}\,\mathbf{Z}^{-1}\mathrm{vec}\,'(\mathbf{I}_p + \mathbf{Z})^{-1}\right]\\
&-\tfrac{1}{2}(n+m)\mathrm{E}\,\left[\mathrm{vec}\,(\mathbf{I}_p + \mathbf{Z})^{-1}\mathrm{vec}\,'(\mathbf{I}_p + \mathbf{Z})^{-1}\right]. \qquad (3.19)
\end{aligned}$$

By transposing (3.19), for example, and noting that

$$\mathbf{K}_{p,p}\mathrm{E}\left((\mathbf{I}_p+\mathbf{Z})^{-1}\otimes(\mathbf{I}_p+\mathbf{Z})^{-1}\right)=\mathrm{E}\left((\mathbf{I}_p+\mathbf{Z})^{-1}\otimes(\mathbf{I}_p+\mathbf{Z})^{-1}\right)\mathbf{K}_{p,p},$$

it follows that

$$\mathrm{E}\left(\mathrm{vec}\,(\mathbf{I}_p+\mathbf{Z})^{-1}\mathrm{vec}\,'\mathbf{Z}^{-1}\right)=\mathrm{E}\left(\mathrm{vec}\,\mathbf{Z}^{-1}\mathrm{vec}\,'(\mathbf{I}_p+\mathbf{Z})^{-1}\right).$$

Thus, (3.18) and (3.19) yield

$$\begin{aligned}(n+m)&\left[\mathrm{vec}\,(\mathbf{I}_p+\mathbf{Z})^{-1}\mathrm{vec}\,'(\mathbf{I}_p+\mathbf{Z})^{-1}\right]+(\mathbf{I}_{p^2}+\mathbf{K}_{p,p})\mathrm{E}\left((\mathbf{I}_p+\mathbf{Z})^{-1}\otimes(\mathbf{I}_p+\mathbf{Z})^{-1}\right)\\&=-\tfrac{n-p-1}{n+m}(\mathbf{I}_{p^2}+\mathbf{K}_{p,p})\mathrm{E}\left(\mathbf{Z}^{-1}\otimes\mathbf{Z}^{-1}\right)+\tfrac{(n-p-1)^2}{n+m}\mathrm{E}\left(\mathrm{vec}\,\mathbf{Z}^{-1}\mathrm{vec}\,'\mathbf{Z}^{-1}\right).\end{aligned}$$

From Theorem 3.40 (v) and (vi), we have

$$\begin{aligned}(\mathbf{I}_{p^2}+\mathbf{K}_{p,p})\mathrm{E}\,(\mathbf{Z}^{-1}\otimes\mathbf{Z}^{-1})&=(e_1+e_2)(\mathbf{I}_{p^2}+\mathbf{K}_{p,p})+2e_2\mathrm{vec}\,\mathbf{I}_p\mathrm{vec}\,'\mathbf{I}_p,\\ \mathrm{E}\,(\mathrm{vec}\,\mathbf{Z}^{-1}\mathrm{vec}\,'\mathbf{Z}^{-1})&=e_2(\mathbf{I}_{p^2}+\mathbf{K}_{p,p})+e_1\mathrm{vec}\,\mathbf{I}_p\mathrm{vec}\,'\mathbf{I}_p,\end{aligned}$$

where e_1 and e_2 are presented in Theorem 3.40. Moreover, again using Theorem 3.35

$$\begin{aligned}\mathrm{E}\left((\mathbf{I}_p+\mathbf{Z})^{-1}\otimes(\mathbf{I}_p+\mathbf{Z})^{-1}\right)&=f_1\mathbf{I}_{p^2}+f_2\mathrm{vec}\,\mathbf{I}_p\mathrm{vec}\,'\mathbf{I}_p+f_2\mathbf{K}_{p,p},\\ \mathrm{E}\left(\mathrm{vec}\,(\mathbf{I}_p+\mathbf{Z})^{-1}\mathrm{vec}\,'(\mathbf{I}_p+\mathbf{Z})^{-1}\right)&=f_1\mathrm{vec}\,\mathbf{I}_p\mathrm{vec}\,'\mathbf{I}_p+f_2(\mathbf{I}_{p^2}+\mathbf{K}_{p,p}),\end{aligned}$$

where f_1 and f_2 are constants which are to be determined. Since $\mathrm{vec}\,\mathbf{I}_p\mathrm{vec}\,'\mathbf{I}_p$ and $(\mathbf{I}_{p^2}+\mathbf{K}_{p,p})$ act as basis vectors in certain spaces, f_1 and f_2 can be found via the identification of coefficients in relation (3.19). The coefficients in front of $(\mathbf{I}_{p^2}+\mathbf{K}_{p,p})$ satisfy

$$(n+m)f_2+f_1+f_2=-\tfrac{n-p-1}{n+m}(e_1+e_2)+\tfrac{(n-p-1)^2}{n+m}e_2$$

and the coefficients in front of $\mathrm{vec}\,\mathbf{I}_p\mathrm{vec}\,'\mathbf{I}_p$ equal

$$f_1+f_2=-\tfrac{n-p-1}{n+m}2e_2+\tfrac{(n-p-1)^2}{n+m}e_1.$$

Some minor calculations give the equations of the theorem. □

Theorem 3.42 *Let* **G** *and* **F** *be given in Definition 3.25 (i) and (iii), respectively. Moreover, let f_1 and f_2 be given in Theorem 3.41. Then,*

(i) $\mathrm{E}(\mathrm{Tr}\,\mathbf{G})=p-\mathrm{E}\,(\mathrm{Tr}\,\mathbf{F})=p\frac{n}{m+n}$;
(ii) $\mathrm{E}\big((\mathrm{Tr}\,\mathbf{G})^2\big)=p^2-2p\mathrm{E}\,(\mathrm{Tr}\,\mathbf{F})+\mathrm{E}\,\left((\mathrm{Tr}\,\mathbf{F})^2\right)=p^2(1+f_1)+2p(f_2-\frac{m}{n+m})$;
(iii) $\mathrm{E}\big(\mathrm{Tr}\,(\mathbf{G}^2)\big)=\mathrm{E}\big(\mathrm{Tr}\,(\mathbf{I}_p-\mathbf{F})(\mathbf{I}_p-\mathbf{F})\big)=p(1+f_1+f_2-\frac{2m}{m+n})+p^2f_2$.

Proof The distribution of **G** is based on $\mathbf{G}=\mathbf{Y}'(\mathbf{W}_1+\mathbf{Y}\mathbf{Y}')^{-1}\mathbf{Y}$, where $\mathbf{Y}\sim N_{p,m}(\mathbf{0},\mathbf{I}_p,\mathbf{I}_m)$, $p>m$, and $\mathbf{W}_1\sim W_p(\mathbf{I}_p,n)$, $p<n$. Moreover, let $\mathbf{W}=\mathbf{W}_1+\mathbf{Y}\mathbf{Y}'\sim W_p(\mathbf{I}_p,m+n)$. Then,

$$\mathrm{Tr}\,\mathbf{G} = \mathrm{Tr}\,(\mathbf{W}^{-1}\mathbf{Y}\mathbf{Y}') = \mathrm{Tr}\,\left[\mathbf{W}^{-1}(\mathbf{W}-\mathbf{W}_1)\right] = p - \mathrm{Tr}\,(\mathbf{W}^{-1/2}\mathbf{W}_1\mathbf{W}^{-1/2}).$$

Thus, $\mathrm{E}\,(\mathrm{Tr}\,\mathbf{G}) = p - \mathrm{E}\,(\mathrm{Tr}\,\mathbf{F})$, where $\mathbf{F} \sim M\beta_I(p, m, n)$ (see Definition 3.25 (i)), and statement (i) follows from Theorem 3.41 (i).

After some manipulations one obtains the following equations:

$$\begin{aligned} \mathrm{E}\big((\mathrm{Tr}\,\mathbf{G})^2\big) &= p^2 - 2p\mathrm{E}\,(\mathrm{Tr}\,\mathbf{F}) + \mathrm{E}\,\left[(\mathrm{Tr}\,\mathbf{F})^2\right], \\ \mathrm{E}\big(\mathrm{Tr}\,(\mathbf{G}^2)\big) &= \mathrm{E}\,\left\{\mathrm{Tr}\,\left[(\mathbf{I}_p - \mathbf{F})(\mathbf{I}_p - \mathbf{F})\right]\right\}. \end{aligned}$$

The expressions in statements (ii) and (iii) follow now from Theorem 3.41 (iii), using $\mathrm{Tr}\,(\mathbf{F}\otimes\mathbf{F}) = (\mathrm{Tr}\,\mathbf{F})^2$ and $\mathrm{vec}\,'\mathbf{I}_p(\mathbf{F}\otimes\mathbf{F})\mathrm{vec}\,\mathbf{I} = \mathrm{Tr}\,(\mathbf{F}\mathbf{F})$. □

Theorem 3.43 *Let* $\mathbf{G}$ *be given in Definition 3.25 (iii), and let* $\mathrm{E}\,\left((\mathrm{Tr}\,\mathbf{G})^2\right)$ *and* $\mathrm{E}\,\left[\mathrm{Tr}\,(\mathbf{G}^2)\right]$ *be expressed as in Theorem 3.42. Put*

$$\begin{aligned} g_1 &= \tfrac{p+1}{p(p+2)(p-1)}\mathrm{E}\big((\mathrm{Tr}\,\mathbf{G})^2\big) - \tfrac{2}{p^2(p+2)(p-1)}\mathrm{E}\,\left(\mathrm{Tr}\,(\mathbf{G}^2)\right), \\ g_2 &= \tfrac{1}{(p+2)(p-1)}\mathrm{E}\big(\mathrm{Tr}\,(\mathbf{G}^2)\big) - \tfrac{1}{p(p+2)(p-1)}\mathrm{E}\,\left((\mathrm{Tr}\,\mathbf{G})^2\right). \end{aligned}$$

Then,

(i) $\mathrm{E}\,(\mathbf{G}) = \frac{n}{n+m}\mathbf{I}_p$;

(ii) $\mathrm{D}\,(\mathbf{G}) = g_2(\mathbf{I}_{p^2} + \mathbf{K}_{p,p}) + \left(g_1 - \frac{n^2}{(n+m)^2}\right)\mathrm{vec}\,\mathbf{I}_p\mathrm{vec}\,'\mathbf{I}_p, \qquad m - p - 3 > 0$;

(iii) $\mathrm{E}\,(\mathbf{G}\otimes\mathbf{G}) = g_1\mathbf{I}_p + g_2\mathrm{vec}\,\mathbf{I}_p\mathrm{vec}\,'\mathbf{I}_p + g_2\mathbf{K}_{p,p}, \qquad m - p - 3 > 0.$

Proof To prove statement (i), it follows due to invariance that $\mathrm{E}\,(\mathbf{G}) = c\mathbf{I}$ and $c = \frac{1}{p}\mathrm{E}\,(\mathrm{Tr}\,\mathbf{G})$, where $\mathrm{E}\,(\mathrm{Tr}\,\mathbf{G})$ is presented in Theorem 3.42. Statements (ii) and (iii) are both verified via Corollary 3.4. The condition $m - p - 3 > 0$ is needed for the calculation of f_1 and f_2 in Theorem 3.42. □

Acknowledgements The support provided for Dietrich von Rosen by the Swedish Research Council (2017-03003) is gratefully acknowledged.

References

1. Akhiezer, N.I.: The Classical Moment Problem. Oliver & Boyd, Edinburgh and London (1965)
2. Andersson, S.A., Wojnar, G.G.: The Wishart distributions on homogeneous cones. Acta Comment. Univ. Tartu. Math. **8**, 3–62 (2004)
3. Andersson, S.A., Wojnar, G.G.: Wishart distributions on homogeneous cones. J. Theor. Probab. **17**, 781–818 (2004)
4. Bożejko, M., Wysoczański, J.: New examples of convolutions and non-commutative central limit theorems. Banach Center Publ. **43**, 95–103 (1998)
5. Cook, R.D., Forzani, L.: On the mean and variance of the generalized inverse of a singular Wishart matrix. Electron. J. Stat. **5**, 146–158 (2011)

6. Couillet, R., Debbah, M.: Random Matrix Methods for Wireless Communications. Cambridge University Press, Cambridge (2011)
7. Di Nardo, E.: On a symbolic representation of non-central Wishart random matrices with applications. J. Multivariate Anal. **125**, 121–135 (2014)
8. Díaz-García, J.A., Gutiérrez-Jáimez, R.: Matricvariate and matrix multivariate T distributions and associated distributions. Metrika **75**, 963–976 (2012)
9. Fujikoshi, Y., Ulyanov, V.V., Shimizu, R.: Multivariate Statistics: High-Dimensional and Large-Sample Approximations. John Wiley & Sons, Hoboken, NJ (2011)
10. Girko, V.L., von Rosen, D.: Asymptotics for the normalized spectral function of matrix quadratic form. Random Oper. Stoch. Equ. **2**, 153–161 (1994)
11. Glueck, D.H., Muller, K.E.: On the trace of a Wishart. Comm. Stat. Theory Methods **27**, 2137–2141 (1998)
12. Goodman, N.R.: Statistical analysis based on a certain multivariate complex Gaussian distribution: an introduction. Ann. Math. Stat. **34**, 152–177 (1963)
13. Graczyk, P., Letac, G., Massam, H.: The complex Wishart distribution and the symmetric group. Ann. Stat. **31**, 287–309 (2003)
14. Gupta, A.K., Nadarajah, S.: Handbook of Beta Distribution and Its Applications. In: Statistics: a Series of Textbooks and Monographs. Marcel Dekker, New York (2004)
15. Gupta, A.K., Nagar, D.K.: Matrix Variate Distributions. Monographs and Surveys in Pure and Applied Mathematics, 104. Chapman & Hall/CRC, Boca Raton, Fl (2000)
16. Haagerup, U.: On Voiculescu's R– and S–transforms for free non–commuting random variables. In: Voiculescu, D.V. (ed.) Free Probability Theory pp. 127–148. American Mathematical Society (1997)
17. Haagerup, U., Thorbjørnsen, S.: Random matrices with complex Gaussian entries. Expo. Math. **21**, 293–337 (2003)
18. Haff, L.R.: Further identities for the Wishart distribution with applications in regression. Canad. J. Stat. **9**, 215–224 (1981)
19. Haff, L.R.: Identities for the inverse Wishart distribution with computational results in linear and quadratic discrimination. Sankhyā B **44**, 245–258 (1982)
20. Hanlon, P.J., Stanley, R.P., Stembridge, J.: Some combinatorial aspects of the spectra of normally distributed random matrices. Contemp. Math. **138**, 151–174 (1992)
21. Harer, J.L., Zagier, D.: The Euler characteristic of the moduli space of curves. Invent. Math. **85**, 457–485 (1986)
22. Jammalamadaka, S.R., Rao, T.S., Terdik, G.: Higher order cumulants of random vectors and applications to statistical inference and time series. Sankhyā A **68**, 326–356 (2006)
23. Jog, C.S.: A concise proof of the representation theorem for fourth-order isotropic tensors. J. Elast. **85**, 119–124 (2006)
24. Kollo, T., von Rosen, D.: Advanced Multivariate Statistics with Matrices. Springer, Dordrecht (2005)
25. Letac, G., Massam, H.: All invariant moments of the Wishart distribution. Scand. J. Stat. **31**, 295–318 (2004)
26. Letac, G., Massam, H.: The noncentral Wishart as an exponential family, and its moments. J. Multivariate Anal. **99**, 1393–1417 (2008)
27. Marčenko, V.A., Pastur, L.A.: Distribution of eigenvalues in certain sets of random matrices. Matematicheskii Sbornik (N.S.) **72**, 507–536 (1967)
28. Mathai, A.M.: Moments of the trace of a noncentral Wishart matrix. Comm. Stat. Theory Methods **9**, 795–801 (1980)
29. Mingo, J.A., Speicher, R.: Free Probability and Random Matrices. Springer, New York (2017)
30. Muirhead, R.J.: Aspects of Multivariate Statistical Theory. John Wiley & Sons, New York (1982)
31. Nagar, D.K., Gupta, A.K.: Expectations of functions of complex Wishart matrix. Acta Appl. Math. **113**, 265–288 (2011)
32. Nel, D.G.: The h-th moment of the trace of a noncentral Wishart matrix. S Afr. Stat. J. **5**, 41–52 (1971)

33. Pielaszkiewicz, J.M., von Rosen, D., Singull, M.: On $E\big[\prod_{i=0}^{k} tr\{W^{m_i}\}\big]$, where $W \sim \mathcal{W}_p(I, n)$. Comm. Stat. Theory Methods **46**, 2990–3005 (2017)
34. Rota, G.-C., Shen, J.: On the combinatorics of cumulants. In memory of Gian-Carlo Rota. J. Combin. Theory Ser. A **91**, 283–304 (2000)
35. Saw, J.G.: Expectation of elementary symmetric functions of a Wishart matrix. Ann. Stat. **1**, 580–582 (1973)
36. Shah, B.K., Khatri, C.G.: Proof of conjectures about the expected values of the elementary symmetric functions of a noncentral Wishart matrix. Ann. Stat. **2**, 833–836 (1974)
37. Silverstein, J.W., Bai, Z.D.: On the empirical distribution of eigenvalues of a class of large dimensional random matrices. J. Multivariate Anal. **54**, 175–192 (1995)
38. Speed, T.P.: Invariant moments and cumulants. In: Ray-Chaudhuri, D. (ed.) Coding Theory and Design Theory, Part II. IMA Vol. Math. Appl. vol. 21, pp. 319–335. Springer, New York (1990)
39. Speed, T.P., Silcock, H.L.: Cumulants and partition lattices. VI. Variances and covariances of mean squares. J. Aust. Math. Soc. **44**, 362–388 (1988)
40. Speicher, R.: Multiplicative functions on the lattice of non-crossing partitions and free convolution. Math. Ann. **298**, 611–628 (1994)
41. Srivastava, M.S.: On the complex Wishart distribution. Ann. Math. Stat. **36**, 313–315 (1965)
42. Subrahmaniam, K.: Recent trends in multivariate normal distribution theory: on the zonal polynomials and other functions of matrix argument. Sankhyā A **38**(3), 221–258 (1976)
43. Sultan, S.A., Tracy, D.S.: Moments of the complex multivariate normal distribution. Special issue honoring Calyampudi Radhakrishna Rao. Linear Algebra Appl. **237/238**, 191–204 (1996)
44. Turin, G.L.: The characteristic function of Hermitian quadratic forms in complex normal variables. Biometrika **47**, 199–201 (1960)
45. Voiculescu, D.: Addition of certain non-commuting random variables. J. Funct. Anal. **66**(3), 323–346 (1986)
46. De Waal, D.J.: On the expected values of the elementary symmetric functions of a noncentral Wishart matrix. Ann. Math. Stat. **43**, 344–347 (1972)
47. Wigner, E.P.: On the distribution of the roots of certain symmetric matrices. Ann. Math. **67**, 325–327 (1958)
48. Withers, C.S., Nadarajah, S.: Moments and cumulants for the complex Wishart. J. Multivariate Anal. **112**, 242–247 (2012)
49. Wooding, R.A.: The multivariate distribution of complex normal variables. Biometrika **43**, 212–215 (1956)

Chapter 4
Regularized Estimation of Covariance Structure Through Quadratic Loss Function

Defei Zhang, Xiangzhao Cui, Chun Li, Jine Zhao, Li Zeng, and Jianxin Pan

Abstract Estimation of high-dimensional covariance structure is an interesting topic in statistics. Motivated by the work of Lin et al. [9], in this paper, the quadratic loss function is proposed to measure the discrepancy between a real covariance matrix and its candidate covariance matrix, where the latter has a regular structure. A commonly encountered candidate structures including MA(1), compound symmetry, AR(1), and banded Toeplitz matrix are considered. Regularization is made by selecting the optimal structure from a potential class of candidate covariance structures through minimizing the discrepancy, i.e., the quadratic loss function, between the given matrix and the candidate covariance class. Analytical or numerical solutions to the optimization problems are obtained and simulation studies are also conducted, showing that the proposed approach provides a reliable method to regularize covariance structures. It is applied to analyze real data problems for illustration of the use of the proposed method.

4.1 Introduction

Structured covariance matrices were studied by many researchers in the statistical literature, for example, Ye and Pan [16], Lin and Jovanović [8], Ning et al. [11], and so on. In the analysis of covariance structure, one may want to find a regularized covariance matrix to estimate the unknown but true covariance matrix, since the true covariance matrix which has a regularized structure cannot be estimated directly by the use of the sample covariance matrix due to random noises, especially when the dimension of the covariance matrix is very high. Recently, Lin et al. [9] proposed a novel method to regularize the covariance structure by using the entropy loss function. Cui et al. [1] considered the similar regularization problem for covariance structure but using the Frobenius-norm discrepancy. The candidate covariance

D. Zhang · X. Cui · C. Li · J. Zhao · L. Zeng
Department of Mathematics, Honghe University, Mengzi, China

J. Pan (✉)
Department of Mathematics, The University of Manchester, Manchester, UK
e-mail: Jianxin.Pan@manchester.ac.uk

K. Filipiak et al. (eds.), *Multivariate, Multilinear and Mixed Linear Models*, Contributions to Statistics, https://doi.org/10.1007/978-3-030-75494-5_4

structures they considered include the order-1 moving average structure, compound symmetry, AR(1), and banded Toeplitz structure.

Motivated by their work, we propose to regularize the covariance structure through quadratic loss function, which was studied by Olkin and Selliah [12] and Haff [4] before. Specifically, we consider the quadratic loss function below

$$L(\boldsymbol{\Sigma}, \mathbf{B}) = \mathrm{Tr}\left(\boldsymbol{\Sigma}^{-1}\mathbf{B} - \mathbf{I}_m\right)^2, \tag{4.1}$$

which measures the distance between $\boldsymbol{\Sigma}$ and $\mathbf{B}$, where $\boldsymbol{\Sigma}$ and $\mathbf{B}$ are both $m \times m$ matrices. The quadratic loss function above is one of the commonly used criteria when measuring the difference between matrices and have good mathematical properties and statistical interpretations, see Olkin and Selliah [12], Haff [4], and James and Stein [6]. For example, it is invariant under the scale transformation $\boldsymbol{\Sigma} \to \mathbf{A}\boldsymbol{\Sigma}\mathbf{A}'$ and $\mathbf{B} \to \mathbf{A}\mathbf{B}\mathbf{A}'$ for any nonsingular matrix $\mathbf{A}$ (see, e.g., Muirhead [10]), where $\mathbf{A}'$ is the the transpose of the matrix $\mathbf{A}$. Obviously, $L(\boldsymbol{\Sigma}, \mathbf{B})$ is always nonnegative and is zero when $\boldsymbol{\Sigma} = \mathbf{B}$. We define the distance between a positive definite covariance matrix $\boldsymbol{\Sigma}$ and the matrix set $\mathbb{B}$ by

$$L(\boldsymbol{\Sigma}, \mathbb{B}) = \min_{\mathbf{B} \in \mathbb{B}} L(\boldsymbol{\Sigma}, \mathbf{B}),$$

where $L(\boldsymbol{\Sigma}, \mathbf{B})$ is the quadratic loss function defined in (4.1), and $\mathbb{B}$ is a covariance matrix set with regular structure.

In this paper, we consider the following four candidate covariance structures, that is

$$\mathbb{B} = \{\text{ MA(1), Compound symmetry, AR(1), banded Toeplitz matrices}\},$$

where the corresponding structures are given by, respectively

$$\mathbf{B}(c, \sigma) = \sigma^2 \begin{pmatrix} 1 & c & 0 & \cdots & 0 \\ c & 1 & c & \ddots & \vdots \\ 0 & \ddots & \ddots & \ddots & 0 \\ \vdots & \ddots & \ddots & 1 & c \\ 0 & \cdots & 0 & c & 1 \end{pmatrix}_{m \times m} \tag{4.2}$$

for MA(1), where $\sigma > 0$ and $-1/(2\cos(\pi/(m+1))) < c < 1/(2\cos(\pi/(m+1)))$ to ensure the matrix in (4.2) is positive definite

$$\mathbf{B}(c,\sigma)=\sigma^2\begin{pmatrix} 1 & c & c & \cdots & c \\ c & 1 & c & \ddots & \vdots \\ c & \ddots & \ddots & \ddots & c \\ \vdots & \ddots & \ddots & 1 & c \\ c & \cdots & c & c & 1 \end{pmatrix}_{m\times m} \tag{4.3}$$

for compound symmetry, where $\sigma > 0$ and $-1/(m-1) < c < 1$ to guarantee the matrix in (4.3) is positive definite

$$\mathbf{B}(c,\sigma)=\sigma^2\begin{pmatrix} 1 & c & c^2 & \cdots & c^{m-1} \\ c & 1 & c & \cdots & c^{m-2} \\ c^2 & c & 1 & \cdots & c^{m-3} \\ \vdots & \ddots & \ddots & \ddots & \vdots \\ c^{m-1} & c^{m-2} & \cdots & c & 1 \end{pmatrix}_{m\times m} \tag{4.4}$$

for AR(1), where $\sigma > 0$ and $-1 < c < 1$ to ensure the matrix in (4.4) is positive definite

$$\mathbf{B}(c,\sigma)=\sigma^2\begin{pmatrix} 1 & c_1 & \cdots & c_p & 0 & \cdots & 0 \\ c_1 & 1 & c_1 & \ddots & \ddots & \ddots & \vdots \\ \vdots & \ddots & \ddots & \ddots & \ddots & \ddots & 0 \\ c_p & \ddots & \ddots & \ddots & \ddots & \ddots & c_p \\ 0 & \ddots & \ddots & \ddots & \ddots & \ddots & \vdots \\ \vdots & \ddots & \ddots & \ddots & \ddots & 1 & c_1 \\ 0 & \cdots & 0 & c_p & \cdots & c_1 & 1 \end{pmatrix}_{m\times m}, \tag{4.5}$$

for a banded Toeplitz matrix, where $\sigma > 0$, $c_1, c_2, \ldots, c_p$ are in general nonzero constants. Note that the conditions for the constraint of positive definiteness of the matrix in (4.5) are not analytically available when $p \geq 2$. Filipiak et al. [3] showed that the constants $c_1, c_2, \ldots, c_p$ $(p \geq 2)$ must be inside the asymptotic cone formed by nonnegative definite banded Toeplitz matrices, in order to ensure that $\mathbf{B}(c,\sigma)$ in (4.4) is positive definite.

For the unknown population covariance matrix $\mathbf{\Sigma}$, the aim of this paper is to find a positive definite matrix $\mathbf{B}$ that minimizes the discrepancy $L(\mathbf{\Sigma}, \mathbf{B})$ over the set of matrices $\mathbb{B}$, implying the resulting positive definite matrix $\mathbf{B}$ which has a certain structure is the best covariance estimator of $\mathbf{\Sigma}$ among the candidate set $\mathbb{B}$. Obviously, the larger the class of the set $\mathbb{B}$, the closer the resulting estimator to the true covariance $\mathbf{\Sigma}$. In this paper, although we only consider the four candidates in $\mathbb{B}$, other structures can also be investigated in a similar manner.

The rest of this paper is organized as follows. In Sect. 4.2, we discuss the four typical covariance candidates under the quadratic loss function and obtain analytical or numerical results for each case. Simulation studies and real data analysis are considered in Sect. 4.3. Some further discussions are presented in Sect. 4.4.

4.2 Main Results

Denote $\mathbf{1}_m = (1, 1, \ldots, 1)'$, an m-dimensional vector of ones, $\mathbf{J}_m = \mathbf{1}_m\mathbf{1}_m'$, $\mathbf{I}_m$ is m-order unit matrix, and $\mathbf{T}_1$ is an $m \times m$ symmetric matrix with the first superdiagonal and subdiagonal equal to 1 and all others elements equal to 0.

MA(1) case. For the MA(1) matrix, the quadratic loss function is

$$\begin{aligned} L(c, \sigma) &= \mathrm{Tr}\left(\boldsymbol{\Sigma}^{-1}\sigma^2(\mathbf{I}_m + c\mathbf{T}_1) - \mathbf{I}_m\right)^2 \\ &= \sigma^4\,\mathrm{Tr}(\boldsymbol{\Sigma}^{-2}) + 2c\sigma^4\,\mathrm{Tr}(\boldsymbol{\Sigma}^{-2}\mathbf{T}_1) + c^2\sigma^4\,\mathrm{Tr}(\boldsymbol{\Sigma}^{-2}\mathbf{T}_1^2) \\ &\quad -2\sigma^2\,\mathrm{Tr}(\boldsymbol{\Sigma}^{-1}) - 2c\sigma^2\,\mathrm{Tr}(\boldsymbol{\Sigma}^{-1}\mathbf{T}_1) + m. \end{aligned}$$

Let $x = \mathrm{Tr}(\boldsymbol{\Sigma}^{-2}\mathbf{T}_1)$, $y = \mathrm{Tr}(\boldsymbol{\Sigma}^{-2}\mathbf{T}_1^2)$, and $z = \mathrm{Tr}(\boldsymbol{\Sigma}^{-1}\mathbf{T}_1)$, the first-order partial derivative for $L(c, \sigma)$ is given by

$$\nabla L(c, \sigma) := \begin{pmatrix} \frac{\partial L}{\partial c} \\ \frac{\partial L}{\partial \sigma} \end{pmatrix} = \begin{pmatrix} 2\sigma^4 x + 2c\sigma^4 y - 2\sigma^2 z \\ 4\sigma^3\,\mathrm{Tr}(\boldsymbol{\Sigma}^{-2}) + 8c\sigma^3 x + 4\sigma^3 c^2 y - 4\sigma\,\mathrm{Tr}(\boldsymbol{\Sigma}^{-1}) - 4c\sigma z \end{pmatrix},$$

so that the stationary points (c, σ) of $L(c, \sigma)$ must satisfy

$$\begin{cases} \sigma^2 = \dfrac{y\,\mathrm{Tr}(\boldsymbol{\Sigma}^{-1}) - xz}{y\,\mathrm{Tr}(\boldsymbol{\Sigma}^{-2}) - x^2}, \\ c = \dfrac{z\,\mathrm{Tr}(\boldsymbol{\Sigma}^{-2}) - x\,\mathrm{Tr}(\boldsymbol{\Sigma}^{-1})}{y\,\mathrm{Tr}(\boldsymbol{\Sigma}^{-1}) - xz}. \end{cases} \tag{4.6}$$

Furthermore, the Hessian matrix is provided by

$$\nabla^2 L(c, \sigma) = \begin{pmatrix} 2\sigma^4 y & 8\sigma^3 x + 8c\sigma^3 y - 4\sigma z \\ 8\sigma^3 x + 8c\sigma^3 y - 4\sigma z & 12\sigma^2\,\mathrm{Tr}(\boldsymbol{\Sigma}^{-2}) + 24c\sigma^2 x + 12\sigma^2 c^2 y - 4\,\mathrm{Tr}(\boldsymbol{\Sigma}^{-1}) - 4cz \end{pmatrix}.$$

Since $(\nabla^2 L(c, \sigma))_{11} = 2\sigma^4 y > 0$, and $\det(\nabla^2 L(c, \sigma)) = 16\sigma^6(y\,\mathrm{Tr}(\boldsymbol{\Sigma}^{-2}) - x^2) > 0$ (see the proof in the Appendix A1) then the minimum of $L(c, \sigma)$ is attained at (c, σ) defined by (4.6). We summarize the results above in the following theorem.

Theorem 4.1 *For any positive definite covariance matrix* $\boldsymbol{\Sigma}$*, there exists an unique tridiagonal matrix* $\boldsymbol{B}(c, \sigma)$ *in the form (4.2) that minimizes the quadratic loss function* $L(c, \sigma) := L(\boldsymbol{\Sigma}, \boldsymbol{B}(c, \sigma))$*, defined in (4.1), over* $\sigma \in (0, +\infty)$ *and* $c \in (-\infty, +\infty)$*. Furthermore, the minimum is attained at* (c, σ) *which are provided by*

$$\begin{cases} \sigma^2 = \dfrac{\mathrm{Tr}(\boldsymbol{\Sigma}^{-1})\,\mathrm{Tr}(\boldsymbol{\Sigma}^{-2}\mathbf{T}_1^2) - \mathrm{Tr}(\boldsymbol{\Sigma}^{-2}\mathbf{T}_1)\,\mathrm{Tr}(\boldsymbol{\Sigma}^{-1}\mathbf{T}_1)}{\mathrm{Tr}(\boldsymbol{\Sigma}^{-2})\,\mathrm{Tr}(\boldsymbol{\Sigma}^{-2}\mathbf{T}_1^2) - (\mathrm{Tr}(\boldsymbol{\Sigma}^{-2}\mathbf{T}_1))^2}, \\ c \;= \dfrac{\mathrm{Tr}(\boldsymbol{\Sigma}^{-2})\,\mathrm{Tr}(\boldsymbol{\Sigma}^{-1}\mathbf{T}_1) - \mathrm{Tr}(\boldsymbol{\Sigma}^{-1})\,\mathrm{Tr}(\boldsymbol{\Sigma}^{-2}\mathbf{T}_1)}{\mathrm{Tr}(\boldsymbol{\Sigma}^{-2}\mathbf{T}_1^2)\,\mathrm{Tr}(\boldsymbol{\Sigma}^{-1}) - \mathrm{Tr}(\boldsymbol{\Sigma}^{-2}\mathbf{T}_1)\,\mathrm{Tr}(\boldsymbol{\Sigma}^{-1}\mathbf{T}_1)}. \end{cases} \tag{4.7}$$

It is noted that the solution c in (4.7) may be any value in $(-\infty, +\infty)$. If $c \in (-1/(2\cos(\pi/(m+1))),\ 1/(2\cos(\pi/(m+1))))$, the resulting tridiagonal matrix $\mathbf{B}(c,\sigma)$ is positive definite. If $c \notin (-1/(2\cos(\pi/(m+1))),\ 1/(2\cos(\pi/(m+1))))$, the matrix $\mathbf{B}(c,\sigma)$ is no longer positive definite and, in this case, it is not a good approximation to the positive definite covariance matrix $\boldsymbol{\Sigma}$.

Compound symmetry case. For the compound symmetry structure, the quadratic loss function is

$$\begin{aligned} L(c,\sigma) &= \mathrm{Tr}\left(\boldsymbol{\Sigma}^{-1}\sigma^2(\mathbf{I}_m + c(\mathbf{J}_m - \mathbf{I}_m)) - \mathbf{I}_m\right)^2 \\ &= \sigma^4\,\mathrm{Tr}\left(\boldsymbol{\Sigma}^{-2}\mathbf{I}_m + 2c\boldsymbol{\Sigma}^{-2}(\mathbf{J}_m - \mathbf{I}_m) + c^2\boldsymbol{\Sigma}^{-2}(\mathbf{J}_m - \mathbf{I}_m)^2\right) \\ &\quad -2\sigma^2\,\mathrm{Tr}\left(\boldsymbol{\Sigma}^{-1}\mathbf{I}_m + c\boldsymbol{\Sigma}^{-1}(\mathbf{J}_m - \mathbf{I}_m)\right) + m \\ &= \sigma^4\,\mathrm{Tr}(\boldsymbol{\Sigma}^{-2}) + 2c\sigma^4\,\mathrm{Tr}(\boldsymbol{\Sigma}^{-2}(\mathbf{J}_m - \mathbf{I}_m)) + c^2\sigma^4\,\mathrm{Tr}(\boldsymbol{\Sigma}^{-2}(\mathbf{J}_m - \mathbf{I}_m)^2) \\ &\quad -2\sigma^2\,\mathrm{Tr}(\boldsymbol{\Sigma}^{-1}) - 2c\sigma^2\,\mathrm{Tr}\left(\boldsymbol{\Sigma}^{-1}(\mathbf{J}_m - \mathbf{I}_m)\right) + m. \end{aligned}$$

Denote $u = \mathrm{Tr}(\boldsymbol{\Sigma}^{-1}(\mathbf{J}_m - \mathbf{I}_m))$, $v = \mathrm{Tr}(\boldsymbol{\Sigma}^{-2}(\mathbf{J}_m - \mathbf{I}_m)^2)$, and $w = \mathrm{Tr}(\boldsymbol{\Sigma}^{-2}(\mathbf{J}_m - \mathbf{I}_m))$. Then the first-order partial derivative for $L(c,\sigma)$ is given by

$$\begin{aligned} \nabla L(c,\sigma) &:= \begin{pmatrix} \frac{\partial L}{\partial \sigma} \\ \frac{\partial L}{\partial c} \end{pmatrix} \\ &= \begin{pmatrix} 4\sigma^3\,\mathrm{Tr}(\boldsymbol{\Sigma}^{-2}) + 8c\sigma^3 w + 4\sigma^3 c^2 v - 4\sigma\,\mathrm{Tr}(\boldsymbol{\Sigma}^{-1}) - 4c\sigma u \\ 2\sigma^4 w + 2c\sigma^4 v - 2\sigma^2 u \end{pmatrix}, \end{aligned}$$

so that the stationary points (c,σ) of $L(c,\sigma)$ must be

$$\begin{cases} \sigma^2 = \dfrac{v\,\mathrm{Tr}(\boldsymbol{\Sigma}^{-1}) - wu}{v\,\mathrm{Tr}(\boldsymbol{\Sigma}^{-2}) - w^2}, \\ c \;= \dfrac{u\,\mathrm{Tr}(\boldsymbol{\Sigma}^{-2}) - w\,\mathrm{Tr}(\boldsymbol{\Sigma}^{-1})}{v\,\mathrm{Tr}(\boldsymbol{\Sigma}^{-1}) - wu}. \end{cases} \tag{4.8}$$

Furthermore, the Hessian matrix is of the form

$$\nabla^2 L(c,\sigma) = \begin{pmatrix} 2\sigma^4 v & 8\sigma^3 w + 8c\sigma^3 v - 4\sigma u \\ 8\sigma^3 w + 8c\sigma^3 v - 4\sigma u & 12\sigma^2\,\mathrm{Tr}(\boldsymbol{\Sigma}^{-2}) + 24c\sigma^2 w + 12\sigma^2 c^2 v - 4\,\mathrm{Tr}(\boldsymbol{\Sigma}^{-1}) - 4cu \end{pmatrix}.$$

Since $(\nabla^2 L(c,\sigma))_{11} = 2\sigma^4 v > 0$, and $\det(\nabla^2 L(c,\sigma)) = 16\sigma^6(v\,\mathrm{Tr}(\boldsymbol{\Sigma}^{-2}) - w^2) > 0$ (see the proof in the Appendix A2) then the minimum of $L(c,\sigma)$ is attained at (c,σ) defined by (4.8). We summarize the above results in the following theorem.

Theorem 4.2 *Given a positive definite covariance matrix* $\boldsymbol{\Sigma} \in \mathbb{R}^{m\times m}$, *denote* $L(c, \sigma) := L(\boldsymbol{\Sigma}, \boldsymbol{B}(c, \sigma))$, *where* $\boldsymbol{B}(c, \sigma) \in \mathbb{R}^{m\times m}$ *is a matrix of compound symmetry in the form of (4.3). The global minimum of* $L(c, \sigma)$ *over* $\sigma > 0$ *and* $c \in (-\infty, +\infty)$ *must be attained at the pair* (c, σ) *defined by (4.8).*

Note that the solution c in (4.8) may be any value in $(-\infty, +\infty)$. If $c \in (-1/(m-1), 1)$, the resulting compound symmetry matrix $\mathbf{B}(c, \sigma)$ is positive definite. If $c \notin (-1/(m-1), 1)$, the matrix $\mathbf{B}(c, \sigma)$ is not positive definite and, in this case, it should not be considered as a good approximation to the positive definite covariance matrix $\boldsymbol{\Sigma}$.

AR(1) case. For the AR(1) structure, the quadratic loss function is given by

$$
\begin{aligned}
L(c, \sigma) &= \operatorname{Tr}\left(\boldsymbol{\Sigma}^{-1}\sigma^2 \sum_{i=0}^{m-1} c^i \mathbf{T}_i - \mathbf{I}_m\right)^2 \\
&= \sigma^4\left(\sum_{i=0}^{m-1} c^{2i} \operatorname{Tr}((\boldsymbol{\Sigma}^{-1}\mathbf{T}_i)^2) + 2\sum_{i=0}^{m-2} c^{2i+1} \operatorname{Tr}(\boldsymbol{\Sigma}^{-1}\mathbf{T}_i\boldsymbol{\Sigma}^{-1}\mathbf{T}_{i+1})\right) \\
&\quad -2\sigma^2\left(\sum_{i=0}^{m-1} c^i \operatorname{Tr}(\boldsymbol{\Sigma}^{-1}\mathbf{T}_i)\right) + m,
\end{aligned}
$$

where $\mathbf{T}_0 = \mathbf{I}_m$ and $\mathbf{T}_i, i \in \{1 \ldots, m\}$, are $m \times m$ symmetric matrices with ith superdiagonal and subdiagonal equal to 1 and zeros elsewhere. Then the first-order partial derivative for $L(c, \sigma)$ is

$$
\nabla L(c, \sigma) := \begin{pmatrix} \frac{\partial L}{\partial \sigma} \\ \frac{\partial L}{\partial c} \end{pmatrix} = \begin{pmatrix} \nabla L_1(c, \sigma) \\ \nabla L_2(c, \sigma) \end{pmatrix},
$$

where

$$
\begin{aligned}
\nabla L_1(c, \sigma) &= 4\sigma^3\left[\sum_{i=0}^{m-1} c^{2i} \operatorname{Tr}((\boldsymbol{\Sigma}^{-1}\mathbf{T}_i)^2) + 2\sum_{i=0}^{m-2} c^{2i+1} \operatorname{Tr}(\boldsymbol{\Sigma}^{-1}\mathbf{T}_i\boldsymbol{\Sigma}^{-1}\mathbf{T}_{i+1})\right] \\
&\quad -4\sigma \sum_{i=0}^{m-1} c^i \operatorname{Tr}(\boldsymbol{\Sigma}^{-1}\mathbf{T}_i), \\
\nabla L_2(c, \sigma) &= \sigma^4\left[\sum_{i=0}^{m-1} 2ic^{2i-1} \operatorname{Tr}((\boldsymbol{\Sigma}^{-1}\mathbf{T}_i)^2) + 2\sum_{i=0}^{m-2} (2i+1)c^{2i} \operatorname{Tr}(\boldsymbol{\Sigma}^{-1}\mathbf{T}_i\boldsymbol{\Sigma}^{-1}\mathbf{T}_{i+1})\right] \\
&\quad -2\sigma^2 \textstyle\sum_{i=0}^{m-1} ic^{i-1} \operatorname{Tr}(\boldsymbol{\Sigma}^{-1}\mathbf{T}_i).
\end{aligned}
$$

Therefore the stationary points (c, σ) of $L(c, \sigma)$ must satisfy

$$\begin{cases} \sigma^2 = \dfrac{\sum_{i=0}^{m-1} c^i \operatorname{Tr}(\boldsymbol{\Sigma}^{-1}\mathbf{T}_i)}{\sum_{i=0}^{m-1} c^{2i} \operatorname{Tr}((\boldsymbol{\Sigma}^{-1}\mathbf{T}_i)^2)+2\sum_{i=0}^{m-2} c^{2i+1} \operatorname{Tr}(\boldsymbol{\Sigma}^{-1}\mathbf{T}_i\boldsymbol{\Sigma}^{-1}\mathbf{T}_{i+1})}, \\ \dfrac{2\sum_{i=0}^{m-1} ic^{i-1} \operatorname{Tr}(\boldsymbol{\Sigma}^{-1}\mathbf{T}_i)}{\sum_{i=0}^{m-1} c^i \operatorname{Tr}(\boldsymbol{\Sigma}^{-1}\mathbf{T}_i)} = \dfrac{\sum_{i=0}^{m-1} 2ic^{2i-1} \operatorname{Tr}((\boldsymbol{\Sigma}^{-1}\mathbf{T}_i)^2)+2\sum_{i=0}^{m-2} (2i+1)c^{2i} \operatorname{Tr}(\boldsymbol{\Sigma}^{-1}\mathbf{T}_i\boldsymbol{\Sigma}^{-1}\mathbf{T}_{i+1})}{\sum_{i=0}^{m-1} c^{2i} \operatorname{Tr}((\boldsymbol{\Sigma}^{-1}\mathbf{T}_i)^2)+2\sum_{i=0}^{m-2} c^{2i+1} \operatorname{Tr}(\boldsymbol{\Sigma}^{-1}\mathbf{T}_i\boldsymbol{\Sigma}^{-1}\mathbf{T}_{i+1})}. \end{cases} \tag{4.9}$$

Let $a_i := \operatorname{Tr}(\boldsymbol{\Sigma}^{-1}\mathbf{T}_i)$, $b_i := \operatorname{Tr}((\boldsymbol{\Sigma}^{-1}\mathbf{T}_i)^2)$, $d_i := \operatorname{Tr}(\boldsymbol{\Sigma}^{-1}\mathbf{T}_i\boldsymbol{\Sigma}^{-1}\mathbf{T}_{i+1})$ and set

$$\begin{aligned} h(c) := &\sum_{i=0}^{m-1} ia_ic^i \left(\sum_{i=0}^{m-1} b_ic^{2i} + 2\sum_{i=0}^{m-2} d_ic^{2i+1}\right) \\ &- \sum_{i=0}^{m-1} a_ic^i \left(\sum_{i=0}^{m-1} ib_ic^{2i} + \sum_{i=0}^{m-2} (2i+1)d_ic^{2i+1}\right) = 0. \end{aligned} \tag{4.10}$$

We now verify that $h(c)$ has at least one root for some $\boldsymbol{\Sigma}$ in $(-1, 1)$ by numerical solutions. Some typical simulation results are presented in Fig. 4.1. The four plots in Fig. 4.1 corresponds to four covariance structures for $\boldsymbol{\Sigma}$, respectively, MA structure ($c = 0.2$ and $\sigma^2 = 4$) on the top-left panel, Compound Symmetry (CS) structure ($c = 0.5$ and $\sigma^2 = 2$) on the top-right panel, AR(1) structure ($c = 0.2$ and $\sigma^2 = 2$) on the bottom-left panel, and Toeplitz structure ($c \sim \mathrm{U}(0,1)$, $\sigma^2 = 2$ and $p = m - 1$) on the bottom-right panel, where $m = 100$. From Fig. 4.1, it is clear that $h(c)$ has at least one root for the selected structures of $\boldsymbol{\Sigma}$ in the range of $(-1, 1)$. We, therefore, conclude that the system (4.9) has at least a solution pair (c, σ). In computational mathematics, there are many numerical algorithms that can be used to calculate the roots of a polynomial like $h(c)$ in (4.10), see Ellis and Watson [2] for example.

Furthermore, the Hessian matrix is provided by

$$\nabla^2 L(c,\sigma) = \begin{pmatrix} \nabla_{11} & \nabla_{12} \\ \nabla_{21} & \nabla_{22} \end{pmatrix},$$

where

$$\begin{aligned} \nabla_{11} &:= 12\sigma^2 \left(\sum_{i=0}^{m-1} b_ic^{2i} + 2\sum_{i=0}^{m-2} d_ic^{2i+1}\right) - 4\left(\sum_{i=0}^{m-1} a_ic^i\right), \\ \nabla_{12} &= \nabla_{21} := 4\sigma^3 \left(\sum_{i=0}^{m-1} 2ib_ic^{2i-1} + 2\sum_{i=0}^{m-2} (2i+1)d_ic^{2i}\right) - 4\sigma\left(\sum_{i=0}^{m-1} ia_ic^{i-1}\right), \\ \nabla_{22} &:= \sigma^4 \left(\sum_{i=0}^{m-1} 2i(2i-1)b_ic^{2i-2} + 2\sum_{i=0}^{m-2} (2i+1)(2i)d_ic^{2i}\right) \\ &\quad -2\sigma^2\left(\sum_{i=0}^{m-1} i(i-1)c^{i-2}a_i\right). \end{aligned}$$

Similarly, we can judge $\det(\nabla^2 L(c,\sigma)) > 0$ for $\boldsymbol{\Sigma}$ by numerical simulations. All these numerical simulations suggest that the solutions in (4.10) are the minima of

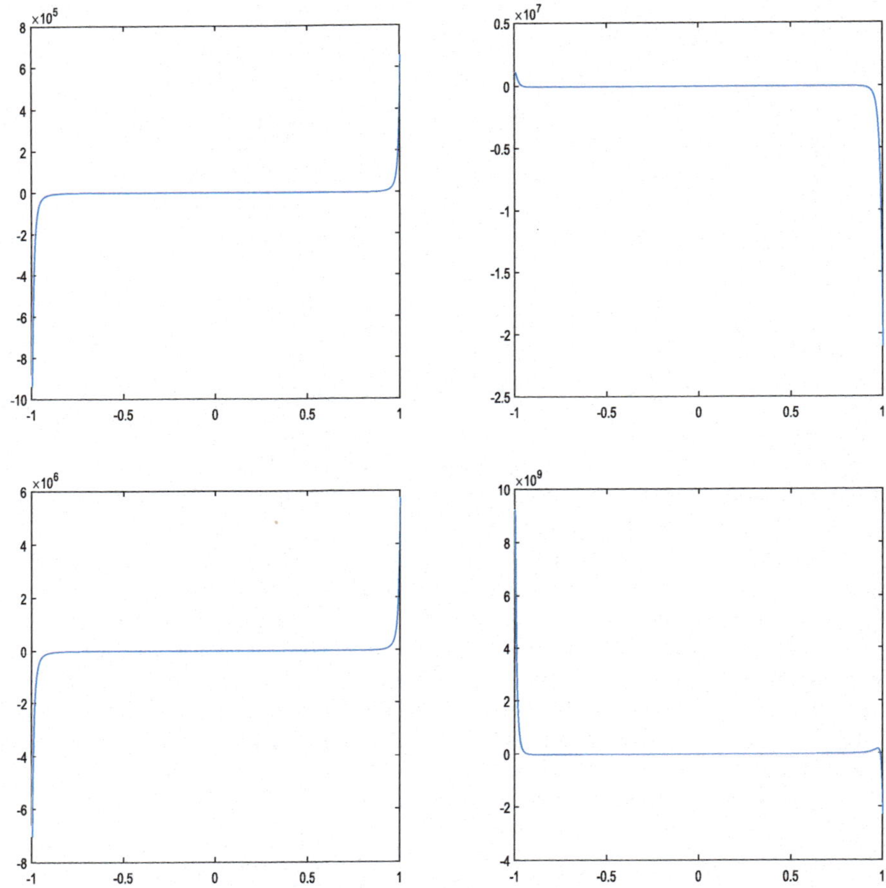

Fig. 4.1 The figures of $h(c)$ in $(-1, 1)$

$L(c, \sigma)$ when $\mathbf{B}(c, \sigma)$ is of an AR(1) structure. We summarize below the analysis and results above as a conjecture, though it has not been proved mathematically.

Conjecture. Given a positive definite covariance matrix $\mathbf{\Sigma} \in \mathbb{R}^{m \times m}$, denote $L(c, \sigma) := L(\mathbf{\Sigma}, \mathbf{B}(c, \sigma))$ where $\mathbf{B}(c, \sigma) \in \mathbb{R}^{m \times m}$ is a covariance matrix with the AR(1) structure in the form of (4.4). Then the local minima of $L(c, \sigma)$ are attained at the pair (c, σ) that satisfies (4.9).

Note that the solution c in (4.9) may be any value in $(-\infty, +\infty)$. If $c \in (-1, 1)$, the resulting AR(1) matrix $\mathbf{B}(c, \sigma)$ is positive definite. If $c \notin (-1, 1)$, the matrix $\mathbf{B}(c, \sigma)$ is not positive definite and in this case it cannot be viewed as a good approximation to the positive definite covariance matrix $\mathbf{\Sigma}$.

Banded Toeplitz case. The banded Toeplitz matrix has the following form:

$$\mathbf{B}(\mathbf{C},\sigma)=\begin{pmatrix} 1 & c_1 & \cdots & c_p & \cdots & 0 \\ c_1 & 1 & c_1 & \ddots & \ddots & \vdots \\ \vdots & \ddots & \ddots & \ddots & \ddots & c_p \\ c_p & \ddots & \ddots & \ddots & \ddots & \vdots \\ \vdots & \ddots & \ddots & \ddots & 1 & c_1 \\ 0 & \cdots & c_p & \cdots & c_1 & 1 \end{pmatrix}\sigma^2=(\mathbf{I}_m+\mathbf{TC})\sigma^2, \tag{4.11}$$

where $\mathbf{C}=(c_1\mathbf{I}_m, c_2\mathbf{I}_m, \ldots, c_p\mathbf{I}_m)'$ and $\mathbf{T}=(\mathbf{T}_1,\ldots,\mathbf{T}_p)$ are $pm\times m$ and $m\times pm$ matrices, respectively, and $\mathbf{T}_i$, $i\in\{1,\ldots,p\}$, are $m\times m$ symmetric matrices with the ith superdiagonal and subdiagonal equal to 1 and zeros elsewhere. We first analyze the banded Toeplitz matrix case by a general method. Note the quadratic loss function becomes

$$\begin{aligned} L(\mathbf{\Sigma},\mathbf{B}) &= \operatorname{Tr}\left(\mathbf{\Sigma}^{-1}(\mathbf{I}+\mathbf{TC})\sigma^2-\mathbf{I}_m)^2\right) \\ &= \sigma^4\operatorname{Tr}(\mathbf{\Sigma}^{-2})+2\sigma^4\operatorname{Tr}(\mathbf{\Sigma}^{-2}\mathbf{TC})+\sigma^4\operatorname{Tr}(\mathbf{\Sigma}^{-2}(\mathbf{TC})^2) \\ &\quad -2\sigma^2\operatorname{Tr}(\mathbf{\Sigma}^{-1})-2\sigma^2\operatorname{Tr}(\mathbf{\Sigma}^{-1}\mathbf{TC})+m. \end{aligned}$$

Let $\mathbf{c}=(c_1,c_2,\ldots,c_p)'\in\mathbb{R}^p$. The first-order partial derivative for $L(\mathbf{c},\sigma)$ is given by

$$\begin{aligned} \nabla L(\mathbf{c},\sigma) &:= \left(\frac{\partial L}{\partial\sigma},\frac{\partial L}{\partial c_1},\ldots,\frac{\partial L}{\partial c_p}\right)' \\ &= \begin{pmatrix} 4\sigma^3\operatorname{Tr}\left[\mathbf{\Sigma}^{-2}(\mathbf{I}_m+\mathbf{TC})^2\right]-4\sigma\operatorname{Tr}\left[\mathbf{\Sigma}^{-1}(\mathbf{I}_m+\mathbf{TC})\right] \\ 2\sigma^4\operatorname{Tr}(\mathbf{\Sigma}^{-2}\mathbf{T}_1)+2\sigma^4\operatorname{Tr}(\mathbf{\Sigma}^{-2}\mathbf{TCT}_1)-2\sigma^2\operatorname{Tr}(\mathbf{\Sigma}^{-1}\mathbf{T}_1) \\ \vdots \\ 2\sigma^4\operatorname{Tr}(\mathbf{\Sigma}^{-2}\mathbf{T}_p)+2\sigma^4\operatorname{Tr}(\mathbf{\Sigma}^{-2}\mathbf{TCT}_p)-2\sigma^2\operatorname{Tr}(\mathbf{\Sigma}^{-1}\mathbf{T}_p) \end{pmatrix} \end{aligned}$$

so that the stationary points $(\mathbf{c},\sigma)$ of $L(\mathbf{c},\sigma)$ must satisfy

$$\begin{cases} \sigma^2 = \dfrac{\operatorname{Tr}\left[\mathbf{\Sigma}^{-1}(\mathbf{I}_m+\mathbf{TC})\right]}{\operatorname{Tr}\left[\mathbf{\Sigma}^{-2}(\mathbf{I}_m+\mathbf{TC})^2\right]}, \\ \operatorname{Tr}(\mathbf{\Sigma}^{-2}\mathbf{TCT}_i)=\sigma^{-2}\operatorname{Tr}(\mathbf{\Sigma}^{-1}\mathbf{T}_i)-\sigma^2\operatorname{Tr}(\mathbf{\Sigma}^{-2}\mathbf{T}_i), \quad i\in\{1,\ldots,p\}. \end{cases} \tag{4.12}$$

We now need to verify that there are $p+1$ solutions $(c_1,c_2,\ldots,c_p,\sigma^2)\in\mathbb{R}^{p+1}$ for the system (4.12). In fact, we notice that

$$\begin{aligned} &\operatorname{Tr}(\mathbf{\Sigma}^{-2}\mathbf{TCT}_i)\cdot\operatorname{Tr}\left[\mathbf{\Sigma}^{-1}(\mathbf{I}_m+\mathbf{TC})\right]\cdot\operatorname{Tr}\left[\mathbf{\Sigma}^{-2}(\mathbf{I}_m+\mathbf{TC})^2\right] \\ &\quad=\operatorname{Tr}(\mathbf{\Sigma}^{-1}\mathbf{T}_i)\cdot\operatorname{Tr}\left[\mathbf{\Sigma}^{-2}(\mathbf{I}_m+\mathbf{TC})^2)\right]^2-\operatorname{Tr}(\mathbf{\Sigma}^{-2}\mathbf{T}_i)\cdot\operatorname{Tr}\left[\mathbf{\Sigma}^{-1}(\mathbf{I}_m+\mathbf{TC})\right]^2. \end{aligned}$$

Now, let $a_1 = \mathrm{Tr}(\boldsymbol{\Sigma}^{-1})$, $a_2 = \mathrm{Tr}(\boldsymbol{\Sigma}^{-2})$, $a_{3i} = \mathrm{Tr}(\boldsymbol{\Sigma}^{-1}\mathbf{T}_i)$, $a_{4i} = \mathrm{Tr}(\boldsymbol{\Sigma}^{-2}\mathbf{T}_i)$, and

$$\begin{aligned} x_1(\mathbf{c}) &= \mathrm{Tr}(\boldsymbol{\Sigma}^{-1}\mathbf{TC}), \quad x_2(\mathbf{c}) = \mathrm{Tr}(\boldsymbol{\Sigma}^{-2}\mathbf{TC}), \\ x_{3i}(\mathbf{c}) &= \mathrm{Tr}(\boldsymbol{\Sigma}^{-2}\mathbf{TCT}_i), \; x_4(\mathbf{c}) = \mathrm{Tr}(\boldsymbol{\Sigma}^{-2}(\mathbf{TC})^2), \end{aligned}$$

then

$$\begin{aligned} h_i(\mathbf{c}) := \; & x_{3i}(a_1a_2 + 2a_1x_2 + a_1x_4 + a_2x_2 + 2x_1x_2 + x_1x_4) + a_{4i}(a_1^2 + 2a_1x_1 + x_1^2) \\ & - a_{3i}(a_2^2 + 4x_2^2 + x_4^2 + 4a_2x_2 + 2a_2x_4 + 4x_2x_4) \\ = \; & 0, \quad i \in \{1, \ldots, p\}, \end{aligned} \tag{4.13}$$

so that the system (4.12) becomes p equations as shown in (4.13). Indeed, the equations (4.13) at least have a solution. In order to avoid computational complexity, we propose to use the Newton method as the one in Lin et al. [9] to solve the equations. We now change the notation to denote $x_0 = \sigma^2$ and $x_i = \sigma^2 c_i$, $i \in \{1, \ldots, p\}$. The matrix $\mathbf{B}$ in (4.11) can be rewritten as

$$\mathbf{B}(\mathbf{x}) = \sum_{i=0}^{p} x_i \mathbf{T}_i,$$

where $\mathbf{x} = (x_0, x_1, \ldots, x_p)' \in \mathbb{R}^{p+1}$, $\mathbf{T}_0 = \mathbf{I}_m$ and $\mathbf{T}_i$ $(1 \le i \le p)$ are the same as above. We define $\Omega \subset \mathbb{R}^{p+1}$ by

$$\Omega := \left\{ \mathbf{x} \in \mathbb{R}^{p+1} : \mathbf{B}(\mathbf{x}) = \sum_{i=0}^{p} x_i \mathbf{T}_i \text{ is positive definite} \right\} \tag{4.14}$$

and $f(\mathbf{x}) : \mathbb{R}^{p+1} \mapsto \mathbb{R}$,

$$f(\mathbf{x}) := L(\boldsymbol{\Sigma}, \mathbf{B}(\mathbf{x})) = \mathrm{Tr}\left(\boldsymbol{\Sigma}^{-1}\mathbf{B}(\mathbf{x}) - \mathbf{I}_m\right)^2. \tag{4.15}$$

Since Ω is isomorphic to the set of all positive definite Toeplitz matrices, the problem of minimizing $f(\mathbf{B})$ over positive definite matrix $\mathbf{B}$ with structure (4.11) is equivalent to

$$\begin{gathered} \min f(\mathbf{x}) \text{ in } (4.15) \\ \text{subject to } \mathbf{x} \in \Omega \text{ in } (4.14). \end{gathered} \tag{4.16}$$

Since $f(\mathbf{B}) := L(\boldsymbol{\Sigma}, \mathbf{B})$ is a strictly convex function of $\mathbf{B}$ and $\mathbf{B}(\mathbf{x}) = \sum_{i=0}^{p} x_i \mathbf{T}_i$ is an affine map of $\mathbf{x}$, by the fact that composition with an affine mapping preserves convexity $f(\mathbf{x}) := f(\mathbf{B}(\mathbf{x}))$ is strictly convex in $\mathbf{x}$. On the other hand, the set of all positive definite Toeplitz matrices is a convex set and so does Ω. Therefore, problem (4.16) is a convex optimization problem and hence has an unique minimizer.

Since $\nabla_{x_i}\mathbf{B} = \mathbf{T}_i$, by applying the chain rule we have the gradient of f

$$\nabla_{x_i} f = 2\,\mathrm{Tr}(\mathbf{T}_i(\boldsymbol{\Sigma}^{-1}\mathbf{B} - \mathbf{I}_m)\boldsymbol{\Sigma}^{-1}), \quad i \in \{0, \ldots, p\},$$

and the Hessian $\mathbf{H} = (h_{ij}) \in \mathbb{R}^{(p+1)\times(p+1)}$ of f

$$h_{ij} = \nabla^2_{x_i x_j} f = 2\,\mathrm{Tr}(\mathbf{T}_i \boldsymbol{\Sigma}^{-1} \mathbf{T}_j \boldsymbol{\Sigma}^{-1}), \quad i, j \in \{0, \ldots, p\}. \tag{4.17}$$

And since $f(\mathbf{x})$ is strictly convex in Ω, of which the Newton's method generally works very well, the Newton's method with backtracking line search is applied to the problem (4.16). Lin et al. [9] gave the algorithm to solve the problem (4.16). For the sake of completeness, we summarize the algorithm as follows: First, we choose an initial point $\mathbf{x}^{(0)}$ such that $\mathbf{B}(\mathbf{x}^{(0)})$ is positive definite. To ensure that the iterates remain in Ω, in the backtracking line search to choose the step size t, we first multiply the initial guess $t = 1$ by a constant $\beta \in (0, 1)$ until $\mathbf{B}(\mathbf{x} + t\Delta\mathbf{x}_{nt})$ is positive definite and then continue backtracking until a sufficient decrease condition is satisfied.

Algorithm 4.1 [Newton's method for solving problem (4.16)]
Given a starting point $\mathbf{x} \in \Omega$ in (4.14) and tolerance ϵ, repeat

1. Compute the Newton step and decrement: evaluate the gradient $\mathbf{g}$ and Hessian $\mathbf{H}$ (4.17) at $\mathbf{x}$; $\Delta\mathbf{x}_{nt} := -\mathbf{H}^{-1}\mathbf{g}$; $\lambda^2 := \mathbf{g}'\mathbf{H}\mathbf{g}$.
2. Stopping criterion: quit if $\lambda^2/2 \leq \epsilon$.
3. Backtracking line search: given parameters $\alpha \in (0, 0.5)$ and $\beta \in (0, 1)$,
   $t := 1$; while $\mathbf{x} + t\Delta\mathbf{x}_{nt} \notin \Omega$ in (4.14), $t := \beta t$;
   while $f(\mathbf{x} + t\Delta\mathbf{x}_{nt}) > f(\mathbf{x}) + \alpha t \mathbf{g}'\Delta\mathbf{x}_{nt}$
   $t := \beta t$.
4. Update: $\mathbf{x} = \mathbf{x} + t\Delta\mathbf{x}_{nt}$.

4.3 Numerical Experiments

4.3.1 *Simulation Studies*

Let m be the dimension of the covariance matrices we test. We first generate an $m \times n$ data matrix $\mathbf{R}$ with columns randomly drawn from a multivariate normal distribution $\mathcal{N}(\boldsymbol{\mu}, \boldsymbol{\Sigma})$ with the mean vector $\boldsymbol{\mu} = \sigma^2 \mathbf{1}_m \in \mathbb{R}^m$ and the covariance matrix $\boldsymbol{\Sigma}$. We then compute the sample covariance matrix $\mathbf{A}$ with the generated data $\mathbf{R}$. We test with the true covariance matrix $\boldsymbol{\Sigma}$ of various dimensions m, being either unstructured or having structures as discussed in the previous sections, where for each structure we consider several different values for σ^2 and $\mathbf{c}$. The sample size is chosen as $n = 1000$. We summarize the experimental results in Tables 4.1, 4.2, and 4.3, where the covariance matrix size is $m = 100$, and in Tables 4.4, 4.5, and 4.6, where $m = 200$. We choose $\mathbf{c} \in \{0.2, 0.5, 0.75\}$ and $\sigma^2 \in \{2, 4, 8\}$ for $\boldsymbol{\Sigma}$ having MA(1), CS, and AR(1) structures. For a general Toeplitz matrix, we use the above σ^2 but the correlation coefficients $\mathbf{c}$ is randomly chosen from the uniform distribution $U(0, 1)$ such that the resulting $\boldsymbol{\Sigma}$ is positive definite, where $\dim(\mathbf{c}) = p = m/2$. For each combination of the selected values of σ^2 and c, we repeat 1000 runs across

Table 4.1 Simulation results with $m = 100$, $\sigma^2 = 2$

$c = 0.20$		**B**							
		MA(1)		CS		AR(1)		Toep	
$\mathbf{\Sigma}$	$L_{\mathbf{\Sigma},\mathbf{A}}$	$L_{\mathbf{A},\mathbf{B}}$	$L_{\mathbf{\Sigma},\mathbf{B}}$	$L_{\mathbf{A},\mathbf{B}}$	$L_{\mathbf{\Sigma},\mathbf{B}}$	$L_{\mathbf{A},\mathbf{B}}$	$L_{\mathbf{\Sigma},\mathbf{B}}$	$L_{\mathbf{A},\mathbf{B}}$	$L_{\mathbf{\Sigma},\mathbf{B}}$
MA(1)	10.10	10.81	0.00	16.70	8.11	241.23	122.14	10.37	3.01
CS	10.10	10.86	0.92	10.51	0.00	72.92	31.94	10.67	4.18
AR(1)	10.11	10.42	0.29	15.51	6.65	18.63	0.02	10.43	3.16
$c = 0.50$		**B**							
		MA(1)		CS		AR(1)		Toep	
$\mathbf{\Sigma}$	$L_{\mathbf{\Sigma},\mathbf{A}}$	$L_{\mathbf{A},\mathbf{B}}$	$L_{\mathbf{\Sigma},\mathbf{B}}$	$L_{\mathbf{A},\mathbf{B}}$	$L_{\mathbf{\Sigma},\mathbf{B}}$	$L_{\mathbf{A},\mathbf{B}}$	$L_{\mathbf{\Sigma},\mathbf{B}}$	$L_{\mathbf{A},\mathbf{B}}$	$L_{\mathbf{\Sigma},\mathbf{B}}$
MA(1)	10.10	77.31	0.00	97.19	96.56	310.21	250.34	10.30	3.29
CS	10.11	10.90	0.98	10.78	0.00	631.22	461.14	11.46	2.39
AR(1)	10.10	16.49	7.55	29.69	23.54	19.71	0.04	10.64	2.61
$c = 0.75$		**B**							
		MA(1)		CS		AR(1)		Toep	
$\mathbf{\Sigma}$	$L_{\mathbf{\Sigma},\mathbf{A}}$	$L_{\mathbf{A},\mathbf{B}}$	$L_{\mathbf{\Sigma},\mathbf{B}}$	$L_{\mathbf{A},\mathbf{B}}$	$L_{\mathbf{\Sigma},\mathbf{B}}$	$L_{\mathbf{A},\mathbf{B}}$	$L_{\mathbf{\Sigma},\mathbf{B}}$	$L_{\mathbf{A},\mathbf{B}}$	$L_{\mathbf{\Sigma},\mathbf{B}}$
CS	10.10	10.90	0.99	10.86	0.00	575.02	432.13	11.32	5.24
AR(1)	10.11	24.36	17.09	36.26	30.84	15.91	0.07	10.91	1.71
c not assigned		**B**							
		MA(1)		CS		AR(1)		Toep	
$\mathbf{\Sigma}$	$L_{\mathbf{\Sigma},\mathbf{A}}$	$L_{\mathbf{A},\mathbf{B}}$	$L_{\mathbf{\Sigma},\mathbf{B}}$	$L_{\mathbf{A},\mathbf{B}}$	$L_{\mathbf{\Sigma},\mathbf{B}}$	$L_{\mathbf{A},\mathbf{B}}$	$L_{\mathbf{\Sigma},\mathbf{B}}$	$L_{\mathbf{A},\mathbf{B}}$	$L_{\mathbf{\Sigma},\mathbf{B}}$
UnStr	10.11	18.38	10.11	18.36	10.10	367.73	206.52	19.89	10.60
Toep	10.10	175.32	124.51	28.74	9.91	17.56	0.91	14.93	0.00

all the simulation studies and then report the averaged values of the loss function in Tables 4.1, 4.2, 4.3, 4.4, 4.5, and 4.6. The notation and abbreviation for the results reported in the tables are summarized:

- $\mathbf{\Sigma}$: true covariance matrix.
- $\mathbf{A}$: sample covariance matrix.
- $\mathbf{B}$: the computed covariance matrix that has a certain structure and minimizes the quadratic loss function $L(\mathbf{A}, \mathbf{B})$ in (4.1).
- $L_{\mathbf{\Sigma},\mathbf{A}}$, $L_{\mathbf{A},\mathbf{B}}$ and $L_{\mathbf{\Sigma},\mathbf{B}}$: the quadratic loss function $L(\mathbf{\Sigma}, \mathbf{A})$, $L(\mathbf{A}, \mathbf{B})$, and $L(\mathbf{\Sigma}, \mathbf{B})$, respectively.

Table 4.2 Simulation results with $m = 100,\ \sigma^2 = 4$

$c = 0.20$		**B**							
		MA(1)		CS		AR(1)		Toep	
$\boldsymbol{\Sigma}$	$L_{\boldsymbol{\Sigma},\mathbf{A}}$	$L_{\mathbf{A},\mathbf{B}}$	$L_{\boldsymbol{\Sigma},\mathbf{B}}$	$L_{\mathbf{A},\mathbf{B}}$	$L_{\boldsymbol{\Sigma},\mathbf{B}}$	$L_{\mathbf{A},\mathbf{B}}$	$L_{\boldsymbol{\Sigma},\mathbf{B}}$	$L_{\mathbf{A},\mathbf{B}}$	$L_{\boldsymbol{\Sigma},\mathbf{B}}$
MA(1)	10.11	10.83	0.00	16.72	8.11	525.31	443.62	10.29	3.19
CS	10.10	10.86	0.92	10.51	0.00	120.12	99.01	10.66	5.15
AR(1)	10.10	10.41	0.29	15.53	6.65	17.33	0.06	10.34	3.29
$c = 0.50$		**B**							
		MA(1)		CS		AR(1)		Toep	
$\boldsymbol{\Sigma}$	$L_{\boldsymbol{\Sigma},\mathbf{A}}$	$L_{\mathbf{A},\mathbf{B}}$	$L_{\boldsymbol{\Sigma},\mathbf{B}}$	$L_{\mathbf{A},\mathbf{B}}$	$L_{\boldsymbol{\Sigma},\mathbf{B}}$	$L_{\mathbf{A},\mathbf{B}}$	$L_{\boldsymbol{\Sigma},\mathbf{B}}$	$L_{\mathbf{A},\mathbf{B}}$	$L_{\boldsymbol{\Sigma},\mathbf{B}}$
MA(1)	10.10	77.5	0.00	97.20	96.56	421.03	341.11	11.73	3.26
CS	10.09	10.90	0.98	10.78	0.00	32.93	12.94	10.80	5.27
AR(1)	10.10	16.50	7.55	29.67	23.54	16.72	0.06	10.72	2.10
$c = 0.75$		**B**							
		MA(1)		CS		AR(1)		Toep	
$\boldsymbol{\Sigma}$	$L_{\boldsymbol{\Sigma},\mathbf{A}}$	$L_{\mathbf{A},\mathbf{B}}$	$L_{\boldsymbol{\Sigma},\mathbf{B}}$	$L_{\mathbf{A},\mathbf{B}}$	$L_{\boldsymbol{\Sigma},\mathbf{B}}$	$L_{\mathbf{A},\mathbf{B}}$	$L_{\boldsymbol{\Sigma},\mathbf{B}}$	$L_{\mathbf{A},\mathbf{B}}$	$L_{\boldsymbol{\Sigma},\mathbf{B}}$
CS	10.10	10.90	0.99	10.87	0.00	256.61	134.12	11.48	2.42
AR(1)	10.09	24.35	17.09	36.26	30.84	16.12	0.07	10.91	1.65
c not assigned		**B**							
		MA(1)		CS		AR(1)		Toep	
$\boldsymbol{\Sigma}$	$L_{\boldsymbol{\Sigma},\mathbf{A}}$	$L_{\mathbf{A},\mathbf{B}}$	$L_{\boldsymbol{\Sigma},\mathbf{B}}$	$L_{\mathbf{A},\mathbf{B}}$	$L_{\boldsymbol{\Sigma},\mathbf{B}}$	$L_{\mathbf{A},\mathbf{B}}$	$L_{\boldsymbol{\Sigma},\mathbf{B}}$	$L_{\mathbf{A},\mathbf{B}}$	$L_{\boldsymbol{\Sigma},\mathbf{B}}$
UnStr	10.11	18.39	10.11	18.33	10.08	210.23	141.25	19.89	10.60
Toep	10.02	228.31	171.82	173.74	126.96	36.22	18.28	15.13	0.00

For simplicity, "Toep" represents a covariance matrix with Toeplitz structure across all tables in this paper.

Note that the covariance matrix with MA(1) structure is not positive definite when $c = 0.75$, so that in Tables 4.1, 4.2, 4.3, 4.4, 4.5, and 4.6 such cases are not presented. From Tables 4.1, 4.2, 4.3, 4.4, 4.5, and 4.6, we note that when $\boldsymbol{\Sigma}$ is structured, the regularized estimator $\mathbf{B}$ that has the same structure as $\boldsymbol{\Sigma}$ is much better than the sample covariance matrix $\mathbf{A}$ in terms of the quadratic loss function, namely $L(\boldsymbol{\Sigma}, \mathbf{B}) < L(\boldsymbol{\Sigma}, \mathbf{A})$ for a regularized covariance matrix. In addition, if the matrix $\mathbf{B}$ has the same structure as $\boldsymbol{\Sigma}$ or the Toeplitz, then the corresponding discrepancies $L(\mathbf{A}, \mathbf{B})$ are smaller than any other structures. Note the Toeplitz candidate is always the smallest. It should not be surprising for the matrix $\mathbf{B}$ with Toeplitz structure to win out because all MA(1), CS, and AR(1) are indeed special Toeplitz structures. We also point out that with the bandwidth p of the general Toeplitz ranging from 1 to $m - 1$, the smallest minimum is always obtained as $p = m - 1$.

Table 4.3 Simulation results with $m = 100,\ \sigma^2 = 8$

$c = 0.20$		**B**							
		MA(1)		CS		AR(1)		Toep	
Σ	$L_{\Sigma,A}$	$L_{A,B}$	$L_{\Sigma,B}$	$L_{A,B}$	$L_{\Sigma,B}$	$L_{A,B}$	$L_{\Sigma,B}$	$L_{A,B}$	$L_{\Sigma,B}$
MA(1)	10.10	10.81	0.00	16.72	8.11	502.32	481.55	9.91	7.34
CS	10.11	10.86	0.92	10.51	0.00	50.83	48.61	71.46	68.90
AR(1)	10.10	10.40	0.29	15.52	6.65	15.92	0.02	74.90	72.57
$c = 0.50$		**B**							
		MA(1)		CS		AR(1)		Toep	
Σ	$L_{\Sigma,A}$	$L_{A,B}$	$L_{\Sigma,B}$	$L_{A,B}$	$L_{\Sigma,B}$	$L_{A,B}$	$L_{\Sigma,B}$	$L_{A,B}$	$L_{\Sigma,B}$
MA(1)	10.10	77.42	0.00	97.20	96.56	142.05	111.31	12.52	4.73
CS	10.10	10.91	0.98	10.78	0.00	37.15	35.73	10.71	4.77
AR(1)	10.11	16.49	7.55	29.69	23.54	13.12	0.02	64.18	61.48
$c = 0.75$		**B**							
		MA(1)		CS		AR(1)		Toep	
Σ	$L_{\Sigma,A}$	$L_{A,B}$	$L_{\Sigma,B}$	$L_{A,B}$	$L_{\Sigma,B}$	$L_{A,B}$	$L_{\Sigma,B}$	$L_{A,B}$	$L_{\Sigma,B}$
CS	10.11	10.91	0.99	10.87	0.00	75.54	48.17	10.79	5.25
AR(1)	10.11	24.37	17.09	36.26	30.84	14.82	0.09	40.14	39.76
c not assigned		**B**							
		MA(1)		CS		AR(1)		Toep	
Σ	$L_{\Sigma,A}$	$L_{A,B}$	$L_{\Sigma,B}$	$L_{A,B}$	$L_{\Sigma,B}$	$L_{A,B}$	$L_{\Sigma,B}$	$L_{A,B}$	$L_{\Sigma,B}$
UnStr	10.09	18.37	10.11	18.34	10.10	122.15	85.34	19.88	10.58
Toep	10.31	77.36	45.12	86.89	66.64	75.43	43.02	15.23	0.00

4.3.2 *Real Data Analysis*

4.3.2.1 Kenward's Cattle Data Analysis

We did experiments with Kenward's cattle data (Kenward [7]). Our experiments were carried out with the cattle data in a similar way as in Sect. 4.2 and the results are reported in Table 4.7. We also present under the column "Time" the time (in seconds) used to find the optimal matrix **B** for each structure, which is recorded by using the commands "tic" and "toc" in MATLAB. The data analysis was done using a desktop computer with CPU being Inter(R) Xeon(R) Platinum 8160 2.10GHz with 256 GB memory.

Since we do not know the true covariance matrix $\mathbf{\Sigma}$ from the real cattle data, the discrepancies $L(\mathbf{\Sigma}, \mathbf{A})$ and $L(\mathbf{\Sigma}, \mathbf{B})$ are not available in Table 4.7. Instead, the discrepancy $L_{\mathbf{A},\mathbf{B}}$ is used to identify the most likely covariance structure among the possible candidate structures, MA(1), CS, AR(1), and general Toeplitz.

From Table 4.7, it is clear that the underlying covariance structures are very likely to be Toeplitz for both groups, among the four possible candidate structures, as their

Table 4.4 Simulation results with $m = 200,\ \sigma^2 = 2$

$c = 0.20$		**B**							
		MA(1)		CS		AR(1)		Toep	
$\boldsymbol{\Sigma}$	$L_{\Sigma,\mathbf{A}}$	$L_{\mathbf{A},\mathbf{B}}$	$L_{\Sigma,\mathbf{B}}$	$L_{\mathbf{A},\mathbf{B}}$	$L_{\Sigma,\mathbf{B}}$	$L_{\mathbf{A},\mathbf{B}}$	$L_{\Sigma,\mathbf{B}}$	$L_{\mathbf{A},\mathbf{B}}$	$L_{\Sigma,\mathbf{B}}$
MA(1)	40.10	42.51	0.00	50.96	16.46	893.15	248.62	38.93	14.62
CS	40.23	40.88	0.96	40.80	0.00	453.14	170.51	39.62	0.81
AR(1)	40.23	41.04	0.60	49.05	13.54	103.56	0.05	43.76	29.13
$c = 0.50$		**B**							
		MA(1)		CS		AR(1)		Toep	
$\boldsymbol{\Sigma}$	$L_{\Sigma,\mathbf{A}}$	$L_{\mathbf{A},\mathbf{B}}$	$L_{\Sigma,\mathbf{B}}$	$L_{\mathbf{A},\mathbf{B}}$	$L_{\Sigma,\mathbf{B}}$	$L_{\mathbf{A},\mathbf{B}}$	$L_{\Sigma,\mathbf{B}}$	$L_{\mathbf{A},\mathbf{B}}$	$L_{\Sigma,\mathbf{B}}$
MA(1)	40.20	186.41	0.00	197.40	166.56	432.05	271.23	45.52	29.16
CS	40.21	40.87	0.99	40.91	0.00	371.93	208.34	44.88	14.03
AR(1)	40.21	51.42	15.15	72.47	47.77	83.42	0.03	46.18	29.69
$c = 0.75$		**B**							
		MA(1)		CS		AR(1)		Toep	
$\boldsymbol{\Sigma}$	$L_{\Sigma,\mathbf{A}}$	$L_{\mathbf{A},\mathbf{B}}$	$L_{\Sigma,\mathbf{B}}$	$L_{\mathbf{A},\mathbf{B}}$	$L_{\Sigma,\mathbf{B}}$	$L_{\mathbf{A},\mathbf{B}}$	$L_{\Sigma,\mathbf{B}}$	$L_{\mathbf{A},\mathbf{B}}$	$L_{\Sigma,\mathbf{B}}$
CS	40.23	40.93	0.99	40.90	0.00	604.19	462.72	43.28	34.45
AR(1)	40.23	63.91	30.01	83.24	62.37	90.13	0.08	49.28	33.59
c not assigned		**B**							
		MA(1)		CS		AR(1)		Toep	
$\boldsymbol{\Sigma}$	$L_{\Sigma,\mathbf{A}}$	$L_{\mathbf{A},\mathbf{B}}$	$L_{\Sigma,\mathbf{B}}$	$L_{\mathbf{A},\mathbf{B}}$	$L_{\Sigma,\mathbf{B}}$	$L_{\mathbf{A},\mathbf{B}}$	$L_{\Sigma,\mathbf{B}}$	$L_{\mathbf{A},\mathbf{B}}$	$L_{\Sigma,\mathbf{B}}$
UnStr	40.24	67.08	40.25	67.04	40.25	221.05	102.13	65.71	38.23
Toep	41.21	114.35	11.12	310.25	190.57	148.74	29.82	92.15	0.00

discrepancy $L_{\mathbf{A},\mathbf{B}}$ has smaller values than others. We can claim that Group 1 also tends to have an AR(1) covariance structure, due to the values $L_{\mathbf{A},\mathbf{B}}$ of the AR(1) and the Toeplitz are smaller than MA(1) and CS. Group 2 tends to have a CS covariance structure. This agrees with the finding by Pourahmadi [14] and Pan and Mackenzie [13].

4.3.2.2 Dental Data Analysis

We also did experiments for the dental data (Potthoff and Roy [15]). Dental measurements were made on 11 girls and 16 boys at ages 8, 10, 12, and 14 years. Each measurement is the distance, in millimeters, from the center of the pituitary to the pterygomaxillary fissure.

From Table 4.8, we find that the underlying covariance structures are very likely to be Toeplitz for the girl group, and also can claim that girl group tends to have an AR(1) covariance structure, due to the fact that the values $L_{\mathbf{A},\mathbf{B}}$ of the AR(1) and the Toeplitz are smaller than MA(1) and CS. The boy group tends to have a CS or Toeplitz covariance structure.

Table 4.5 Simulation results with $m = 200,\ \sigma^2 = 4$

$c = 0.20$		**B**							
		MA(1)		CS		AR(1)		Toep	
$\mathbf{\Sigma}$	$L_{\mathbf{\Sigma},\mathbf{A}}$	$L_{\mathbf{A},\mathbf{B}}$	$L_{\mathbf{\Sigma},\mathbf{B}}$	$L_{\mathbf{A},\mathbf{B}}$	$L_{\mathbf{\Sigma},\mathbf{B}}$	$L_{\mathbf{A},\mathbf{B}}$	$L_{\mathbf{\Sigma},\mathbf{B}}$	$L_{\mathbf{A},\mathbf{B}}$	$L_{\mathbf{\Sigma},\mathbf{B}}$
MA(1)	40.21	42.52	0.00	50.97	16.46	412.65	219.61	32.73	11.52
CS	40.25	40.90	0.96	40.78	0.00	143. 16	90.13	40.98	31.23
AR(1)	40.25	41.07	0.60	49.04	13.54	99.11	0.06	42.34	31.42
$c = 0.50$		**B**							
		MA(1)		CS		AR(1)		Toep	
$\mathbf{\Sigma}$	$L_{\mathbf{\Sigma},\mathbf{A}}$	$L_{\mathbf{A},\mathbf{B}}$	$L_{\mathbf{\Sigma},\mathbf{B}}$	$L_{\mathbf{A},\mathbf{B}}$	$L_{\mathbf{\Sigma},\mathbf{B}}$	$L_{\mathbf{A},\mathbf{B}}$	$L_{\mathbf{\Sigma},\mathbf{B}}$	$L_{\mathbf{A},\mathbf{B}}$	$L_{\mathbf{\Sigma},\mathbf{B}}$
MA(1)	40.21	186.53	0.00	197.40	196.56	421.34	290.25	47.32	28.11
CS	40.22	40.87	0.99	40.86	0.00	360.75	166.71	40.68	30.17
AR(1)	40.22	51.41	15.15	72.46	47.77	96.13	0.08	79.31	76.95
$c = 0.75$		**B**							
		MA(1)		CS		AR(1)		Toep	
$\mathbf{\Sigma}$	$L_{\mathbf{\Sigma},\mathbf{A}}$	$L_{\mathbf{A},\mathbf{B}}$	$L_{\mathbf{\Sigma},\mathbf{B}}$	$L_{\mathbf{A},\mathbf{B}}$	$L_{\mathbf{\Sigma},\mathbf{B}}$	$L_{\mathbf{A},\mathbf{B}}$	$L_{\mathbf{\Sigma},\mathbf{B}}$	$L_{\mathbf{A},\mathbf{B}}$	$L_{\mathbf{\Sigma},\mathbf{B}}$
CS	40.22	40.89	0.99	40.91	0.00	422.41	270.24	44.86	14.21
AR(1)	40.22	63.89	34.01	83.23	62.37	95.96	0.13	63.55	56.25
c not assigned		**B**							
		MA(1)		CS		AR(1)		Toep	
$\mathbf{\Sigma}$	$L_{\mathbf{\Sigma},\mathbf{A}}$	$L_{\mathbf{A},\mathbf{B}}$	$L_{\mathbf{\Sigma},\mathbf{B}}$	$L_{\mathbf{A},\mathbf{B}}$	$L_{\mathbf{\Sigma},\mathbf{B}}$	$L_{\mathbf{A},\mathbf{B}}$	$L_{\mathbf{\Sigma},\mathbf{B}}$	$L_{\mathbf{A},\mathbf{B}}$	$L_{\mathbf{\Sigma},\mathbf{B}}$
UnStr	40.21	67.05	40.23	67.04	40.21	633.21	442.18	64.23	38.62
Toep	40.52	365.41	167.25	640.23	450.24	209.62	68.21	91.02	0.00

4.4 Conclusions

Motivated by the work of Lin et al. [9], we estimate the underlying covariance structure by minimizing the quadratic loss function between a given/candidate covariance matrix and the unknown population covariance matrix that may have has certain regularized structures. Their method used the so-called entropy loss function which involves calculating the eigenvalues of matrices and its computations may be too intensive especially for high-dimensional covariance matrices.

In this paper, we work on an alternative but very commonly used criterion, the quadratic loss function, which measures certain distances between two matrices. The advantages of the proposed method here are the analytical or numerical solutions to the parameter estimators under the covariance structures MA(1), CS, AR(1), and Toeplitz, so that their computation is easy to implement. The simulation results and real data analysis show the agreement with Lin et al. [9]'s results. In contrast to their method, it is convenient to consider more candidate structures for the population covariance matrix by using the quadratic loss function. We are currently studying more candidate covariance structures including ARMA(1,1) structure, linearly struc-

Table 4.6 Simulation results with $m = 200,\ \sigma^2 = 8$

$c = 0.20$		**B**							
		MA(1)		CS		AR(1)		Toep	
$\boldsymbol{\Sigma}$	$L_{\boldsymbol{\Sigma},\mathbf{A}}$	$L_{\mathbf{A},\mathbf{B}}$	$L_{\boldsymbol{\Sigma},\mathbf{B}}$	$L_{\mathbf{A},\mathbf{B}}$	$L_{\boldsymbol{\Sigma},\mathbf{B}}$	$L_{\mathbf{A},\mathbf{B}}$	$L_{\boldsymbol{\Sigma},\mathbf{B}}$	$L_{\mathbf{A},\mathbf{B}}$	$L_{\boldsymbol{\Sigma},\mathbf{B}}$
MA(1)	40.21	42.51	0.00	50.94	16.46	872.36	608.42	149.52	139.55
CS	40.25	40.89	0.96	40.80	0.00	91.62	79.21	142.66	131.69
AR(1)	40.25	41.06	0.60	49.03	13.54	89.14	0.05	149.78	139.83
$c = 0.50$		**B**							
		MA(1)		CS		AR(1)		Toep	
$\boldsymbol{\Sigma}$	$L_{\boldsymbol{\Sigma},\mathbf{A}}$	$L_{\mathbf{A},\mathbf{B}}$	$L_{\boldsymbol{\Sigma},\mathbf{B}}$	$L_{\mathbf{A},\mathbf{B}}$	$L_{\boldsymbol{\Sigma},\mathbf{B}}$	$L_{\mathbf{A},\mathbf{B}}$	$L_{\boldsymbol{\Sigma},\mathbf{B}}$	$L_{\mathbf{A},\mathbf{B}}$	$L_{\boldsymbol{\Sigma},\mathbf{B}}$
MA(1)	40.21	186.53	0.00	197.40	196.56	612.35	442.63	104.05	100.13
CS	40.25	40.90	0.99	40.88	0.00	112.13	88.36	40.98	30.75
AR(1)	40.25	51.44	15.15	72.49	47.77	91.85	0.05	128.24	117.21
$c = 0.75$		**B**							
		MA(1)		CS		AR(1)		Toep	
$\boldsymbol{\Sigma}$	$L_{\boldsymbol{\Sigma},\mathbf{A}}$	$L_{\mathbf{A},\mathbf{B}}$	$L_{\boldsymbol{\Sigma},\mathbf{B}}$	$L_{\mathbf{A},\mathbf{B}}$	$L_{\boldsymbol{\Sigma},\mathbf{B}}$	$L_{\mathbf{A},\mathbf{B}}$	$L_{\boldsymbol{\Sigma},\mathbf{B}}$	$L_{\mathbf{A},\mathbf{B}}$	$L_{\boldsymbol{\Sigma},\mathbf{B}}$
CS	40.23	40.90	0.99	40.86	0.00	363.24	166.76	40.91	30.52
AR(1)	40.23	63.88	34.01	83.25	62.37	86.93	0.13	83.68	79.97
c not assigned		**B**							
		MA(1)		CS		AR(1)		Toep	
$\boldsymbol{\Sigma}$	$L_{\boldsymbol{\Sigma},\mathbf{A}}$	$L_{\mathbf{A},\mathbf{B}}$	$L_{\boldsymbol{\Sigma},\mathbf{B}}$	$L_{\mathbf{A},\mathbf{B}}$	$L_{\boldsymbol{\Sigma},\mathbf{B}}$	$L_{\mathbf{A},\mathbf{B}}$	$L_{\boldsymbol{\Sigma},\mathbf{B}}$	$L_{\mathbf{A},\mathbf{B}}$	$L_{\boldsymbol{\Sigma},\mathbf{B}}$
UnStr	40.24	67.06	40.25	65.71	40.25	421.03	212.35	60.12	37.51
Toep	40.95	178.62	41.46	471.36	382.06	191.11	58.62	88.64	0.00

Table 4.7 Results of experiments on Kenward's cattle data

	MA(1)		CS		AR(1)		Toep	
	$L_{\mathbf{A},\mathbf{B}}$	Time	$L_{\mathbf{A},\mathbf{B}}$	Time	$L_{\mathbf{A},\mathbf{B}}$	Time	$L_{\mathbf{A},\mathbf{B}}$	Time
Group 1	5.766	0.046	5.866	0.074	5.653	0.016	4.748	3.077
Group 2	4.969	0.054	4.905	0.041	5.103	0.018	4.359	1.842

Table 4.8 Results of experiments on Dental data

	MA(1)		CS		AR(1)		Toep	
	$L_{\mathbf{A},\mathbf{B}}$	Time	$L_{\mathbf{A},\mathbf{B}}$	Time	$L_{\mathbf{A},\mathbf{B}}$	Time	$L_{\mathbf{A},\mathbf{B}}$	Time
Girl group	1.344	0.032	1.322	0.025	1.286	0.016	1.038	0.813
Boy group	1.501	0.033	0.894	0.030	1.708	0.010	1.307	2.753

tured covariance, factor analytic, Hankel structure, and others. For such complicated covariance structures, however, challenging work is inevitably due to computational complexity in multivariate optimization problem, where the number of variables increases with the dimension m of the covariance matrix. It inevitably needs certain robust and valid algorithms to handle such issues.

As commented by the Editor and reviewer, the solutions of c provided in Theorems 4.1 and 4.2 and Conjecture may fall outside of the interval of constraints, which makes the resulting matrix $\mathbf{B}(c, \sigma)$ not positive definite. In this case, the matrix $\mathbf{B}(c, \sigma)$ cannot be considered as a good approximation to the matrix $\mathbf{\Sigma}$. However, it is possible to use certain techniques to modify $\mathbf{B}(c, \sigma)$, such that the modified matrix becomes positive definite. For example, the matrix $\mathbf{\Sigma}$ can be projected onto a cone of positive definite structured matrices, so that the projected matrix becomes positive definite (see, e.g., Filipiak et al. [3]). An alternative is to calibrate the matrix $\mathbf{B}(c, \sigma)$ by replacing its nonpositive eigenvalues with certain small positive values that are determined by a data-driven method, see Huang et al. [5] for the details.

Acknowledgements We would like to thank the Editor and one anonymous reviewer for their helpful comments and suggestions, which leads to substantial improvements to the paper. This work is partially supported by the Natural Science Foundations of China (11761028), the Reserve Talents Foundation of Yunnan Province (No.2015HB061), and the Reserve Talents Foundations of Honghe University (2014HB0204).

Appendix

We need the following lemma to judge the determinant sign of a matrix.

Lemma 4.1 *If* $\mathbf{A}$ *and* $\mathbf{B}$ *are positive semidefinite matrices of the same order, then*

$$0 \leq \mathrm{Tr}(\mathbf{AB}) \leq \mathrm{Tr}(\mathbf{A}) \cdot \mathrm{Tr}(\mathbf{B}),$$

and

$$\mathrm{Tr}(\mathbf{AB}) \leq (\mathrm{Tr}(\mathbf{A}^2))^{\frac{1}{2}} \cdot (\mathrm{Tr}(\mathbf{B}^2))^{\frac{1}{2}}.$$

A1. The proof of $\det(\nabla^2 L(c, \sigma)) > 0$ for MA(1) case.

We first note the first order partial derivative for $L(c, \sigma)$ is

$$\begin{aligned}\nabla L(c, \sigma) &:= \begin{pmatrix} \frac{\partial L}{\partial c} \\ \frac{\partial L}{\partial \sigma} \end{pmatrix} \\ &= \begin{pmatrix} 2\sigma^4 x + 2c\sigma^4 y - 2\sigma^2 z \\ 4\sigma^3 \mathrm{Tr}(\mathbf{\Sigma}^{-2}) + 8c\sigma^3 x + 4\sigma^3 c^2 y - 4\sigma \mathrm{Tr}(\mathbf{\Sigma}^{-1}) - 4c\sigma z \end{pmatrix}.\end{aligned}$$

Then the Hessian matrix is

$$\nabla^2 L(c,\sigma) = \begin{pmatrix} 2\sigma^4 y & 8\sigma^3 x + 8c\sigma^3 y - 4\sigma z \\ 8\sigma^3 x + 8c\sigma^3 y - 4\sigma z & 12\sigma^2 \mathrm{Tr}(\mathbf{\Sigma}^{-2}) + 24c\sigma^2 x + 12\sigma^2 c^2 y - 4\,\mathrm{Tr}(\mathbf{\Sigma}^{-1}) - 4cz \end{pmatrix}$$
$$= \begin{pmatrix} 2\sigma^4 y & 4\sigma^3 x + 4c\sigma^3 y \\ 4\sigma^3 x + 4c\sigma^3 y & 8\sigma^2 \mathrm{Tr}(\mathbf{\Sigma}^{-2}) + 16c\sigma^2 x + 8\sigma^2 c^2 y \end{pmatrix},$$

where $x = \mathrm{Tr}(\mathbf{\Sigma}^{-2}\mathbf{T}_1)$, $y = \mathrm{Tr}(\mathbf{\Sigma}^{-2}\mathbf{T}_1^2)$, $z = \mathrm{Tr}(\mathbf{\Sigma}^{-1}\mathbf{T}_1)$. Thus

$$\begin{aligned} \det(\nabla^2 L(c,\sigma)) &= 16\sigma^6(y\,\mathrm{Tr}(\mathbf{\Sigma}^{-2}) - x^2) \\ &= 16\sigma^6(\mathrm{Tr}(\mathbf{\Sigma}^{-2}\mathbf{T}_1^2)\cdot \mathrm{Tr}(\mathbf{\Sigma}^{-2}) - (\mathrm{Tr}(\mathbf{\Sigma}^{-2}\mathbf{T}_1))^2). \end{aligned}$$

According to Lemma 4.1, we have $\det(\nabla^2 L(c,\sigma)) > 0$.

A2. The proof of $\det(\nabla^2 L(c,\sigma)) > 0$ for CS case.
Since

$$\nabla^2 L(c,\sigma) = \begin{pmatrix} 2\sigma^4 v & 8\sigma^3 w + 8c\sigma^3 v - 4\sigma u \\ 8\sigma^3 w + 8c\sigma^3 v - 4\sigma u & 12\sigma^2 \mathrm{Tr}(\mathbf{\Sigma}^{-2}) + 24c\sigma^2 w + 12\sigma^2 c^2 v - 4\,\mathrm{Tr}(\mathbf{\Sigma}^{-1}) - 4cu \end{pmatrix}$$
$$= \begin{pmatrix} 2\sigma^4 v & 4\sigma^3 w + 4c\sigma^3 v \\ 4\sigma^3 w + 4c\sigma^3 v & 8\sigma^2 \mathrm{Tr}(\mathbf{\Sigma}^{-2}) + 16c\sigma^2 w + 8\sigma^2 c^2 v \end{pmatrix},$$

where $u = \mathrm{Tr}(\mathbf{\Sigma}^{-1}(\mathbf{J}_m - \mathbf{I}_m))$, $v = \mathrm{Tr}(\mathbf{\Sigma}^{-2}(\mathbf{J}_m - \mathbf{I}_m)^2)$, $w = \mathrm{Tr}(\mathbf{\Sigma}^{-2}(\mathbf{J}_m - \mathbf{I}_m))$. Then, following Lemma 4.1, we have

$$\begin{aligned} \det(\nabla^2 L(c,\sigma)) &= 16\sigma^6(v\,\mathrm{Tr}(\mathbf{\Sigma}^{-2}) - w^2) \\ &= 16\sigma^6\left[\mathrm{Tr}(\mathbf{\Sigma}^{-2}(\mathbf{J}_m - \mathbf{I}_m)^2)\cdot \mathrm{Tr}(\mathbf{\Sigma}^{-2}) - (\mathrm{Tr}(\mathbf{\Sigma}^{-2}(\mathbf{J}_m - \mathbf{I}_m)))^2\right] > 0. \end{aligned}$$

References

1. Cui, X., Li, C., Zhao, J., Zeng, L., Zhang, D., Pan, J.: Covariance structure regularization via Frobenius-norm discrepancy. Linear Algebra Appl. **510**, 124–145 (2016)
2. Ellis, G.H., Watson, L.T.: A parallel algorithm for simple roots for polynomials. Comput. Math. Appl. **10**, 107–121 (1984)
3. Filipiak, K., Markiewicz, A., Mieldzioc, A., Sawikowska, A.: On projection of a positive definite matrix on a cone of nonnegative definite Toeplitz matrices. Electron. J. Linear Algebra **33**, 74–82 (2018)
4. Haff, L.R.: Empirical Bayes estimation of the multivariate normal covariance matrix. Ann. Stat. **8**(3), 586–597 (1980)
5. Huang, C., Farewell, D., Pan, J.: A calibration method for non-positive definite covariance matrix in multivariate data analysis. J. Multivar. Anal. **157**, 45–52 (2017)

6. James, W., Stein, C.: Estimation with quadratic loss. In: Neyman, J. (ed.) Proceedings of the Fourth Berkeley Symposium on Mathematical Statistics and Probability, vol. 1, pp. 361–379. University of California Press, Berkeley (1961)
7. Kenward, M.: A method for comparing profiles of repeated measurements. Appl. Stat. **36**, 296–308 (1987)
8. Lin, F., Jovanović, M.R.: Least-squares approximation of structured covariances. IEEE Trans. Autom. Control **54**(7), 1643–1648 (2009)
9. Lin, L., Higham, N.J., Pan, J.: Covariance structure regularization via entropy loss function. Comput. Stat. Data Anal. **72**(4), 315–327 (2014)
10. Muirhead, R.J.: Aspects of Multivariate Statistical Theory. Wiley, New York (1982)
11. Ning, L., Jiang, X, Georgiou, T.: Geometric methods for structured covariance estimation. In: American Control Conference, pp. 1877–1882. IEEE (2012)
12. Olkin, I., Selliah, J.B.: Estimating covariance matrix in a multivariate normal distribution. In: Gupta, S.S., Moore, D.S (eds.) Statistical Decision Theory and Related Topics, vol. II, pp. 313–326. Academic Press, New York (1977)
13. Pan, J., Mackenzie, G.: On modelling mean-covariance structures in longitudinal studies. Biometrika **90**(1), 239–244 (2003)
14. Pourahmadi, M.: Joint mean-covariance models with applications to longitudinal data: unconstrained parameterisation. Biometrika **86**(3), 677–690 (1999)
15. Potthoff, R.F., Roy, S.N.: A generalized multivariate analysis of variance model useful especially for growth curve problems. Biometrika **51**, 313–326 (1964)
16. Ye, H., Pan, J.: Modelling of covariance structures in generalised estimating equations for longitudinal data. Biometrika **93**(4), 927–941 (2006)

Chapter 5
Separable Covariance Structure Identification for Doubly Multivariate Data

Katarzyna Filipiak, Daniel Klein, and Monika Mokrzycka

Abstract The aim of this paper is to present two methods for the identification of separable covariance structures with both components unstructured, or with one component additionally structured as compound symmetry or first-order autoregression, for doubly multivariate data. As measures of discrepancy between an unstructured covariance matrix and the structured one, the Frobenius norm and the entropy loss function are used. The minimum of each discrepancy function is presented, and then simulation studies are performed to verify whether the considered discrepancy functions recognize the true covariance structure properly. An interpretation of the presented approach using a real data example is also given. This paper is mainly an overview of the papers by van Loan and Pitsianis [18], Filipiak and Klein [8], Filipiak et al. [10], Filipiak et al. [12].

5.1 Introduction

The eagerness of experimenters to study as many variables as possible, in many locations, depths, etc., or the behavior of certain processes as often as possible, results in large, multi-level multivariate datasets. Such experiments occur in almost all fields of science, including biology, genetics, agricultural science, medicine and biomedicine, environmental science, and engineering. In most cases, the data are correlated, and identification of the relations between characteristics, locations, depths, time points, etc., is of interest. This makes it possible to extend knowledge about the variables behavior, but also provides an opportunity to analyze data with the use of more

K. Filipiak
Institute of Mathematics, Poznań University of Technology, Poznań, Poland
e-mail: katarzyna.filipiak@put.poznan.pl

D. Klein
Faculty of Science, Institute of Mathematics, P. J. Šafárik University, Košice, Slovakia
e-mail: daniel.klein@upjs.sk

M. Mokrzycka (✉)
Institute of Plant Genetics, Polish Academy of Sciences, Poznań, Poland
e-mail: mmok@igr.poznan.pl

K. Filipiak et al. (eds.), *Multivariate, Multilinear and Mixed Linear Models*, Contributions to Statistics, https://doi.org/10.1007/978-3-030-75494-5_5

precise statistical models. Identification of the covariance structure may reduce significantly the number of estimated parameters, reducing the degrees of freedom used up in the estimation of the covariance parameters, and leaving them for the estimation of parameters of interest, and thus improving the quality of statistical inference. Moreover, reduction of the number of parameters enables statistical inference when the sample size is too small compared with the number of variables; that is, when the estimator of an unstructured covariance matrix is singular.

In this paper, we consider doubly multivariate data, where variables are collected at two levels; for example, several characteristics are measured at various locations or time points. In such a situation, the most intuitive covariance structure is a separable structure, in which the time or location points are correlated "independently" of characteristics and characteristics are correlated "independently" of time or location points; see, e.g., Dutilleul [6], Lu and Zimmerman [19], Roy and Khattree [23, 24], Srivastava et al. [25], Filipiak and Klein [7]. Nevertheless, for large datasets, there is usually only a little a priori knowledge about some reasonable structure of these covariance components, but the number of parameters to estimate can still be reduced by further assuming some structure within the components, e.g., compound symmetry (CS), first-order autoregression (AR(1)), Toeplitz, banded Toeplitz, moving average, etc., which may follow from the nature of the data.

The main goal of this paper is to propose methods for the identification of the dependence structure using algebraic techniques based on the respectively derived discrepancy measure between a given matrix and a set of potential dependence structures. Under a simple multivariate model, Cui et al. [3] and Lin et al. [17] used the Frobenius norm as well as the entropy loss function to identify the covariance structure. In these papers CS, AR(1), Toeplitz, and banded Toeplitz are proposed as possible structures. We focus on the identification of a covariance matrix in doubly multivariate data over the set of separable matrices with both components unstructured (UN), or one component structured as CS or AR(1). When the Frobenius norm is considered as the discrepancy function, Filipiak and Klein [8] presented the best approximations among the set of considered structures, while Filipiak et al. [10] and Filipiak et al. [12] used the entropy loss function as a measure of discrepancy. On the basis of theoretical results, it is verified, whether the relevant discrepancy function recognizes the true covariance structure properly. In this paper, we collect these results together and compare them.

It should be noted, that aside from algebraic methods of covariance structure identification, graphical methods, such as neural networks or heatmaps (Gilson et al. [13]) or the graphical lasso algorithm (Devijver and Gallopin [5]), are also available and are often used by researchers.

The paper is organized as follows. In Sect. 5.2, we present the models, give necessary notation, and define the discrepancy functions and the partial and block trace operators. In Sect. 5.3, we give formulas to obtain the best approximation of the covariance matrix in the set of separable matrices, or in the set of separable matrices with one component as a CS or AR(1) correlation structure, with the use of the Frobenius norm and the entropy loss function as discrepancy measures. In Sect. 5.4, we present the results of simulation studies carried out to verify the usefulness of the considered discrepancy functions. In Sect. 5.5, we apply the presented methods to a real data example. Finally, we summarize the results.

5.2 Models and Discrepancy Functions

Let us consider an experiment regarding n independent objects, for which we measure q characteristics at p time or location points. The data collected in this way are called doubly multivariate. Denote by $\mathbf{Y}_i$ a $q \times p$ observation matrix for the i-th object, $i \in \{1, \ldots, n\}$, with $\mathrm{E}(\mathbf{Y}_i) = \mathbf{M}$. Assuming normality, we can write

$$\mathbf{y}_i = \operatorname{vec} \mathbf{Y}_i \sim N_{qp}(\operatorname{vec} \mathbf{M}, \boldsymbol{\Omega})$$

or

$$\mathbf{Y} = \left(\mathbf{y}_1, \ldots, \mathbf{y}_n\right)' \sim N_{n,qp}\left(\mathbf{1}_n \operatorname{vec}' \mathbf{M}, \mathbf{I}_n, \boldsymbol{\Omega}\right), \tag{5.1}$$

where $\operatorname{vec}(\cdot)$ is the operator stacking the columns of a matrix one below another forming a vector, $\boldsymbol{\Omega} \in \mathbb{R}^{qp \times qp}$ is an unknown positive definite (p.d.) covariance matrix, $\mathbf{1}_n$ is an n-dimensional vector of ones, and $\mathbf{I}_n$ is the identity matrix of order n.

If there are no assumptions about the structure of $\boldsymbol{\Omega}$ and if $n \leqslant qp$, then the sample covariance matrix (or maximum likelihood estimator of $\boldsymbol{\Omega}$) is singular. Very often, researchers have some knowledge about the possible covariance structure, which follows from the nature of the data. For example, for doubly multivariate data the most intuitive assumption would be a separable structure, which means that the time or location points are correlated "independently" of characteristics and characteristics are correlated "independently" of time or location points. For two such sources of variability, the set of separable structures can be presented as a set of Kronecker products of two covariance matrices, that is

$$\mathscr{S} = \{\boldsymbol{\Psi} \otimes \boldsymbol{\Sigma} \in \mathbb{R}^{qp \times qp} \text{ p.d.} : \quad \boldsymbol{\Psi} \in \mathbb{R}^{p \times p} \text{ p.d.}, \ \boldsymbol{\Sigma} \in \mathbb{R}^{q \times q} \text{ p.d.}\}.$$

The matrix $\boldsymbol{\Psi}$ is usually understood as the covariance matrix for time or location points and it is the same for each characteristic, and $\boldsymbol{\Sigma}$ is usually understood as the covariance matrix for the characteristics and is the same for each time or location point. Note, that the representation of a matrix by a Kronecker product of two matrices is not unique, because for an arbitrary $c > 0$, $\boldsymbol{\Psi} \otimes \boldsymbol{\Sigma} = c\boldsymbol{\Psi} \otimes \frac{1}{c}\boldsymbol{\Sigma}$. To avoid problems with the identification of Kronecker product components Srivastava et al. [25] proposed to fix one of the diagonal entries of one of the two component matrices at an arbitrary positive value, e.g., one. Throughout this paper, we set the first diagonal element of $\boldsymbol{\Psi}$ as 1. Moreover, the underlying structure can be further specified by assuming a particular structure of one of the Kronecker product components, say $\boldsymbol{\Psi}$. The most common structures for repeated measurements are CS and AR(1). Therefore, we consider two subsets of $\mathscr{S}$ of the following form:

$$\mathscr{S}_{\rm CS} = \{\boldsymbol{\Psi}_{\rm CS} \otimes \boldsymbol{\Sigma} \in \mathbb{R}^{qp\times qp} \text{ p.d.}: \quad \boldsymbol{\Psi}_{\rm CS} \in \mathbb{R}^{p\times p} \text{ p.d.}, \ \boldsymbol{\Sigma} \in \mathbb{R}^{q\times q} \text{ p.d.}\},$$
$$\mathscr{S}_{\rm AR} = \{\boldsymbol{\Psi}_{\rm AR} \otimes \boldsymbol{\Sigma} \in \mathbb{R}^{qp\times qp} \text{ p.d.}: \quad \boldsymbol{\Psi}_{\rm AR} \in \mathbb{R}^{p\times p} \text{ p.d.}, \ \boldsymbol{\Sigma} \in \mathbb{R}^{q\times q} \text{ p.d.}\}.$$

Recall that $\boldsymbol{\Psi}_{\rm CS}$ is a compound symmetry correlation matrix, which can be written as

$$\boldsymbol{\Psi}_{\rm CS} = (1-\rho)\mathbf{I}_p + \rho\mathbf{1}_p\mathbf{1}_p',$$

and $\boldsymbol{\Psi}_{\rm AR}$ is a first-order autoregression correlation matrix of the form

$$\boldsymbol{\Psi}_{\rm AR} = \mathbf{I}_p + \sum_{i=1}^{p-1} \rho^i \left(\mathbf{C}^i + \mathbf{C}^{i\prime}\right),$$

with $\mathbf{C} = (c_{ij})$, $c_{ij} = 1$ for $j - i = 1$ and 0 otherwise, $i, j \in \{1, \ldots, p\}$. To ensure the positive definiteness of these matrices, the correlation coefficient ρ must belong to the interval $\left(-\frac{1}{p-1}, \ 1\right)$ or $(-1, \ 1)$ for $\boldsymbol{\Psi}_{\rm CS}$ or $\boldsymbol{\Psi}_{\rm AR}$, respectively.

It should be noted that a more precise structure of the covariance makes it possible to weaken the sample size requirement. To ensure positive definiteness of the unstructured covariance matrix estimator, the number of columns of $\mathbf{Y}$ is required to be smaller than the number of rows of $\mathbf{Y}$. Similarly, in a separable structure, $\boldsymbol{\Psi}$ and $\boldsymbol{\Sigma}$ are the covariance matrices of columns and rows of each $\mathbf{Y}_i$, and hence, to obtain p.d. estimators of the separability components, the sample size must be greater than p and greater than q, giving $n > \max\{p, q\}$. In the case of $\boldsymbol{\Psi}$, structured as CS or AR(1) correlation matrices, $\boldsymbol{\Psi}_{\rm CS}$ and $\boldsymbol{\Psi}_{\rm AR}$ are p.d. as long as the correlation coefficients belong to the relevant intervals, which do not depend on the sample size, while the estimator of $\boldsymbol{\Sigma}$ is p.d. if $n > q$.

Observe that the structure of the first component of separability can easily be switched to the second component; it is enough to change the order of Kronecker product components with respect to the rule

$$\boldsymbol{\Psi} \otimes \boldsymbol{\Sigma} = \mathbf{K}_{q,p}(\boldsymbol{\Sigma} \otimes \boldsymbol{\Psi})\mathbf{K}_{p,q}$$

with $\mathbf{K}_{p,q}$ being the commutation matrix; see, e.g., Magnus and Neudecker [20], Kollo and von Rosen [16].

Besides the nature of the experiment, the question is how to choose the most relevant structure. Thus, our goal is to find a matrix from the set $\mathscr{S}$, $\mathscr{S}_{\rm CS}$ or $\mathscr{S}_{\rm AR}$ for which the discrepancy with respect to the given unstructured matrix $\boldsymbol{\Omega}$ is the smallest. Obviously, since $\mathscr{S}_{\rm CS}, \mathscr{S}_{\rm AR} \subset \mathscr{S}$, the smallest discrepancy is always obtained by the structure from $\mathscr{S}$. Nevertheless, the structure from $\mathscr{S}_{\rm CS}$ or $\mathscr{S}_{\rm AR}$ can also be interesting to consider, as it gives more freedom in performing statistical inference. It should be noted that none of the sets $\mathscr{S}$, $\mathscr{S}_{\rm CS}$ or $\mathscr{S}_{\rm AR}$ are convex. A graphical interpretation of our goal can be seen in Fig. 5.1. The process of choosing the most relevant structure is called regularization by some authors, e.g., Lin et al. [17], Cui et al. [3]. However, note that in the literature regularization is often related to certain improvements of the estimators properties (see, e.g., Bickel and Li [1],

Hastie et al. [14]). Therefore, to avoid possible misunderstanding, the process described in this paper will be called *covariance structure identification*.

To shorten the notation, by $\mathbf{\Gamma}$, we denote the appropriate symmetric p.d. structure, that is, $\mathbf{\Psi} \otimes \mathbf{\Sigma}$, $\mathbf{\Psi}_{\text{CS}} \otimes \mathbf{\Sigma}$ or $\mathbf{\Psi}_{\text{AR}} \otimes \mathbf{\Sigma}$. As a measure of discrepancy, we use

– the Frobenius norm

$$f_F(\mathbf{\Omega}; \mathbf{\Gamma}) = ||\mathbf{\Omega} - \mathbf{\Gamma}||_F^2 = \text{Tr}\left[(\mathbf{\Omega} - \mathbf{\Gamma})(\mathbf{\Omega} - \mathbf{\Gamma})'\right],$$

– the entropy loss function

$$f_E(\mathbf{\Omega}; \mathbf{\Gamma}) = \text{Tr}\left(\mathbf{\Omega}^{-1}\mathbf{\Gamma}\right) - \ln\left|\mathbf{\Omega}^{-1}\mathbf{\Gamma}\right| - qp.$$

The entropy loss function (Dey and Srinivasan [4], James and Stein [15], Lin et al. [17]) is also known as a Kullback–Leibler divergence between two multivariate distributions which differ in covariance matrices; see, e.g., Pan and Fang [22]. Obviously, $\mathbf{\Omega}$ and $\mathbf{\Gamma}$, as covariance matrices, must be p.d.

For convenience, by $\mathscr{G}$ we denote the appropriate set of structures, i.e., $\mathscr{S}$, $\mathscr{S}_{\text{CS}}$ or $\mathscr{S}_{\text{AR}}$, and we denote the respective discrepancy function by $L_\Omega^{(k)}(\mathbf{\Gamma})$, $k \in \{F, E\}$, for a given $\mathbf{\Omega}$. Our main goal can now be expressed as the determination of the minimum of the discrepancy function $L_\Omega^{(k)}(\mathbf{\Gamma})$ over the set $\mathscr{G}$, that is

$$\zeta^{(k)} = \min_{\mathbf{\Gamma} \in \mathscr{G}} L_\Omega^{(k)}(\mathbf{\Gamma}), \qquad k \in \{F, E\}.$$

Throughout the paper, we use the properties of the partial trace and block trace operators of block square matrices, defined by Filipiak et al. [11] for rectangular block matrices.

Definition 5.1 For an arbitrary matrix $\mathbf{A} = (\mathbf{A}_{ij}) \in \mathbb{R}^{qp \times qp}$ with blocks $\mathbf{A}_{ij} \in \mathbb{R}^{q \times q}$

(i) the partial trace operator, $\text{PTr}_q : \mathbb{R}^{qp \times qp} \to \mathbb{R}^{p \times p}$, is the matrix of the traces of blocks $\mathbf{A}_{ij}$, that is

$$\text{PTr}_q\mathbf{A} = (\text{Tr}\,\mathbf{A}_{ij}), \quad i, j \in \{1, \ldots, p\},$$

(ii) the block trace operator, $\text{BTr}_q : \mathbb{R}^{qp \times qp} \to \mathbb{R}^{q \times q}$, is the sum of all diagonal blocks $\mathbf{A}_{ii}$, that is

$$\text{BTr}_q\mathbf{A} = \sum_{i=1}^{p} \mathbf{A}_{ii}.$$

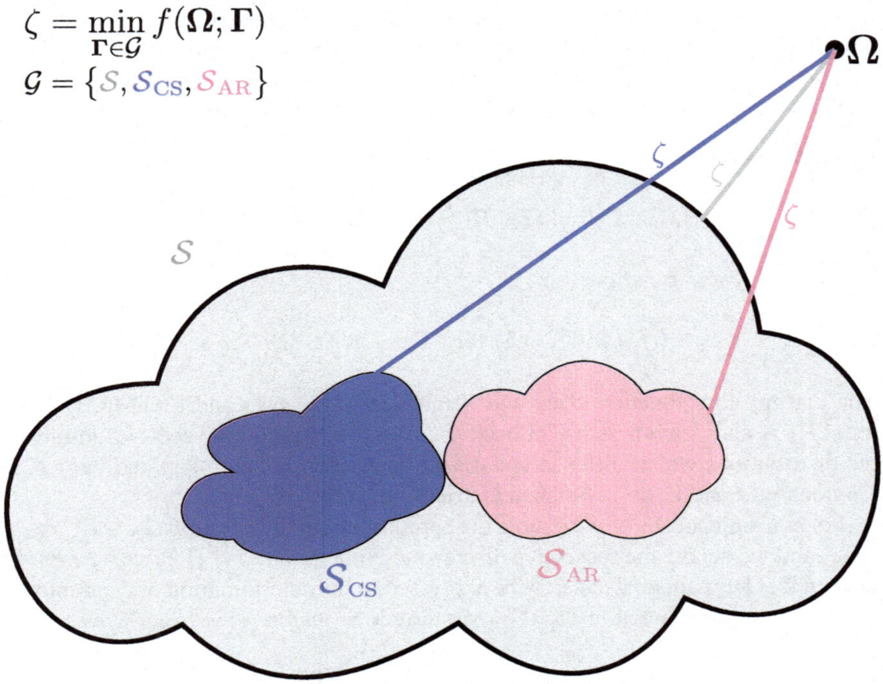

Fig. 5.1 Graphical interpretation of covariance structure identification

5.3 Minimum of Discrepancy Functions

In this section, formulas are given for the best approximations of a given matrix in the sense of minimizing the considered discrepancy functions.

5.3.1 Approximation via the Frobenius Norm

Separable structure approximation with the use of the Frobenius norm, and properties of the approximation, were discussed in van Loan and Pitsianis [18].

For a block matrix, $\mathbf{\Omega} \in \mathbb{R}^{qp \times qp}$, let us consider the following transformation

$$R(\mathbf{\Omega}) = \left(\text{vec}\,\mathbf{\Omega}_{11}, \quad \text{vec}\,\mathbf{\Omega}_{21}, \quad \ldots \quad, \text{vec}\,\mathbf{\Omega}_{p1}, \quad \ldots \quad, \text{vec}\,\mathbf{\Omega}_{pp}\right)' \in \mathbb{R}^{p^2 \times q^2},$$

and its singular value decomposition

$$R(\mathbf{\Omega}) = \mathbf{U}\mathbf{\Delta}\mathbf{V}',$$

with $\boldsymbol{\Delta} = \text{diag}(\delta_1, \cdots, \delta_r), \delta_1 \geqslant \cdots \geqslant \delta_r$ being the singular values of $R(\boldsymbol{\Omega})$, r being the rank of $\boldsymbol{\Omega}$ and with $\mathbf{U}, \mathbf{V}$ being matrices of left and right singular vectors, respectively.

The following theorem gives the best approximation of $L_{\Omega}^{(F)}(\boldsymbol{\Psi} \otimes \boldsymbol{\Sigma})$.

Theorem 5.1 (van Loan and Pitsianis [18]) *For a given symmetric p.d. matrix $\boldsymbol{\Omega}$, there exists a unique symmetric p.d. matrix $\boldsymbol{\Psi} \otimes \boldsymbol{\Sigma}$ that minimizes the Frobenius norm over the set $\mathscr{S}$, and the minimum is attained at*

$$\text{vec}\,\boldsymbol{\Psi} = \sqrt{\delta_1}\boldsymbol{u}_1 \quad \textit{and} \quad \text{vec}\,\boldsymbol{\Sigma} = \sqrt{\delta_1}\boldsymbol{v}_1,$$

where δ_1 is the largest singular value of $R(\boldsymbol{\Omega})$ and $\boldsymbol{u}_1$ and $\boldsymbol{v}_1$ are respective singular vectors ($\boldsymbol{u}_1$ — first column of U and $\mathbf{v}_1$ — first column of V).

Approximation of the separable structure $\boldsymbol{\Psi}_{\text{CS}} \otimes \boldsymbol{\Sigma}$ using the Frobenius norm was discussed in Filipiak and Klein [8]. The solution is based on the spectral decomposition $\boldsymbol{\Psi}_{\text{CS}} = \mathbf{UGU}'$, where $\mathbf{G} = \text{diag}(1 + (p-1)\rho, 1-\rho, \dots, 1-\rho)$ is the diagonal matrix and $\mathbf{U}$ is an orthogonal matrix which does not depend on ρ and whose first column is proportional to $\mathbf{1}_p$.

Theorem 5.2 (Filipiak and Klein [8]) *For a given symmetric p.d. matrix $\boldsymbol{\Omega}$, there exists a unique symmetric p.d. matrix $\boldsymbol{\Psi}_{\text{CS}} \otimes \boldsymbol{\Sigma}$ that minimizes the Frobenius norm over the set $\mathscr{S}_{\text{CS}}$, and this minimum is attained at ρ, $\boldsymbol{\Sigma}$ satisfying the following system of equations:*

$$\rho = \frac{a - (p-1)c + \sqrt{[(p-1)c - a]^2 + 4(p-1)b^2}}{2(p-1)b},$$
$$\boldsymbol{\Sigma} = \frac{1}{1 + (p-1)\rho^2}\left[\rho\widetilde{\boldsymbol{\Omega}}_{11} + (1-\rho)\,\mathrm{H}\right]$$

with

$$a = \mathrm{Tr}\left[\left(\widetilde{\boldsymbol{\Omega}}_{11} - \mathrm{H}\right)\left(\widetilde{\boldsymbol{\Omega}}_{11} - \mathrm{H}\right)'\right], \quad b = \mathrm{Tr}\left[\left(\widetilde{\boldsymbol{\Omega}}_{11} - \mathrm{H}\right)\mathrm{H}'\right], \quad c = \mathrm{Tr}\left[\mathrm{HH}'\right],$$
$$\widetilde{\boldsymbol{\Omega}} = \left(\mathrm{U}' \otimes \mathrm{I_q}\right)\boldsymbol{\Omega}\left(\mathrm{U} \otimes \mathrm{I_q}\right) = \left(\widetilde{\boldsymbol{\Omega}}_{ij}\right)_{1 \leqslant i,j \leqslant p} \quad \forall_{i,j} : \widetilde{\boldsymbol{\Omega}}_{i,j} \in \mathbb{R}^{q \times q},$$
$$\mathrm{H} = \tfrac{1}{\mathrm{p}}\mathrm{BTr_q}\widetilde{\boldsymbol{\Omega}}.$$

Approximation of the separable structure $\boldsymbol{\Psi}_{\text{AR}} \otimes \boldsymbol{\Sigma}$ using the Frobenius norm was discussed in Filipiak and Klein [8].

Theorem 5.3 (Filipiak and Klein [8]) *For a given symmetric p.d. matrix $\boldsymbol{\Omega}$, there exists a symmetric p.d. matrix $\boldsymbol{\Psi}_{\text{AR}} \otimes \boldsymbol{\Sigma}$ that minimizes the Frobenius norm over the set $\mathscr{S}_{\text{AR}}$, and this minimum is attained at ρ, $\boldsymbol{\Sigma}$ satisfying the following system of equations:*

$$\begin{aligned} &\mathrm{Tr}\left(\boldsymbol{\Psi}_{\mathrm{AR}}^2\right)\mathrm{Tr}\left[(\mathrm{F}\otimes\boldsymbol{\Sigma}_1')\boldsymbol{\Omega}\right]-\mathrm{Tr}\left(\mathrm{F}\boldsymbol{\Psi}_{\mathrm{AR}}\right)\mathrm{Tr}\left(\boldsymbol{\Sigma}_1\boldsymbol{\Sigma}_1'\right)=0,\\ &\boldsymbol{\Sigma}=\left[\mathrm{Tr}\left(\boldsymbol{\Psi}_{\mathrm{AR}}^2\right)\right]^{-1}\boldsymbol{\Sigma}_1 \end{aligned}$$

with

$$\boldsymbol{\Sigma}_1=\mathrm{BTr}_q\left[\left(\boldsymbol{\Psi}_{\mathrm{AR}}\otimes\mathrm{I_q}\right)\boldsymbol{\Omega}\right],\quad \boldsymbol{F}=\sum_{i=1}^{p-1}i\rho^{i-1}\left(\mathrm{C^i}+\mathrm{C'^i}\right).$$

The properties of the approximations given in the above theorems, such as nonnegativity, symmetry, and positive definiteness, were studied by van Loan and Pitsianis [18] and by Filipiak and Klein [8]. Moreover, Filipiak and Klein [8] compared the approximations from Theorem 5.2 and 5.3 with maximum likelihood estimators of structured covariances with respect to bias and variability. They showed that, despite the fact that the approximation gives a slightly less accurate estimator of the true covariance matrix than the maximum likelihood estimator, in the case of $\boldsymbol{\Psi}_{\mathrm{CS}}\otimes\boldsymbol{\Sigma}$, the Frobenius norm approximations of ρ and $\boldsymbol{\Sigma}$ are given in an explicit form, while the determination of a maximum likelihood estimator requires the solution of a polynomial of order higher than 2 or the application of an iterative procedure (more about the methods of maximum likelihood estimation can be found in Filipiak et al. [10]).

5.3.2 Approximation via the Entropy Loss Function

Approximation of the separable structure using the entropy loss function was discussed in Filipiak et al. [12]. The following theorem gives the best approximation of $L_{\Omega}^{(E)}(\boldsymbol{\Psi}\otimes\boldsymbol{\Sigma})$.

Theorem 5.4 (Filipiak et al. [12]) *For a given symmetric p.d. matrix* $\boldsymbol{\Omega}$*, there exists a symmetric p.d. matrix* $\boldsymbol{\Psi}\otimes\boldsymbol{\Sigma}$ *that minimizes the entropy loss function over the set of separable structures with unstructured components,* $\mathscr{S}$*, and this minimum is attained at* $\boldsymbol{\Psi}$, $\boldsymbol{\Sigma}$ *satisfying the following system of equations:*

$$\begin{aligned} \boldsymbol{\Psi}^{-1}&=\tfrac{1}{q}\mathrm{PTr}_q\left[\left(\mathrm{I_p}\otimes\boldsymbol{\Sigma}\right)\boldsymbol{\Omega}^{-1}\right],\\ \boldsymbol{\Sigma}^{-1}&=\tfrac{1}{p}\mathrm{BTr}_q\left[\left(\boldsymbol{\Psi}\otimes\mathrm{I_q}\right)\boldsymbol{\Omega}^{-1}\right]. \end{aligned}$$

Approximation of the separable structure $\boldsymbol{\Psi}_{\mathrm{CS}}\otimes\boldsymbol{\Sigma}$ using the entropy loss function was discussed in Filipiak et al. [10] and Filipiak et al. [12]. The following theorem gives the best approximation of $L_{\Omega}^{(E)}(\boldsymbol{\Psi}_{\mathrm{CS}}\otimes\boldsymbol{\Sigma})$.

Theorem 5.5 (Filipiak et al. [10]) *For a given symmetric p.d. matrix* $\boldsymbol{\Omega}$*, there exists a unique symmetric p.d. matrix* $\boldsymbol{\Psi}_{\mathrm{CS}}\otimes\boldsymbol{\Sigma}$ *that minimizes the entropy loss function over the set* $\mathscr{S}_{\mathrm{CS}}$*, and this minimum is attained at* ρ, $\boldsymbol{\Sigma}$ *satisfying the following system of equations:*

$$\rho = \frac{-(p-2)\alpha - pq(p-1) + \sqrt{[(p-2)\alpha + pq(p-1)]^2 + 4(p-1)\alpha^2}}{-2(p-1)\alpha},$$
$$p\boldsymbol{\Sigma}^{-1} = \mathrm{BTr}_q\left[\left(\boldsymbol{\Psi}_{\mathrm{CS}} \otimes \mathrm{I_q}\right)\boldsymbol{\Omega}^{-1}\right]$$

with $\alpha = \mathrm{Tr}\left\{\left[\left(\mathbf{1}_p\mathbf{1}_p' - \mathrm{I_p}\right) \otimes \boldsymbol{\Sigma}\right]\boldsymbol{\Omega}^{-1}\right\}.$

Filipiak et al. [10] proposed and compared algorithms (two iterative methods and one direct) for determination of the best approximation over the set $\mathscr{S}_{\mathrm{CS}}$, and compared statistical properties of the obtained solution and the maximum likelihood estimator of the covariance structure. They showed that for a separable structure $\boldsymbol{\Psi}_{\mathrm{CS}} \otimes \boldsymbol{\Sigma}$ the maximum likelihood estimators of ρ and $\boldsymbol{\Sigma}$ have less bias and have smaller mean square error and loss than the entropy loss estimators. These results indicate that further study on possible improvements to entropy loss estimators are possible, and this will be the subject of future research.

Approximation of the separable structure $\boldsymbol{\Psi}_{\mathrm{AR}} \otimes \boldsymbol{\Sigma}$ using the entropy loss function was discussed in Filipiak et al. [12]. The following theorem gives the best approximation of $L_{\Omega}^{(E)}(\boldsymbol{\Psi}_{\mathrm{AR}} \otimes \boldsymbol{\Sigma})$.

Theorem 5.6 (Filipiak et al. [12]) *For a given symmetric p.d. matrix* $\boldsymbol{\Omega}$*, there exists a symmetric p.d. matrix* $\boldsymbol{\Psi}_{\mathrm{AR}} \otimes \boldsymbol{\Sigma}$ *that minimizes the entropy loss function over the set* $\mathscr{S}_{\mathrm{AR}}$*, and this minimum is attained at* $\rho, \boldsymbol{\Sigma}$ *satisfying the following system of equations:*

$$(1-\rho^2)\cdot \mathrm{Tr}\left[(\mathrm{F} \otimes \boldsymbol{\Sigma})\boldsymbol{\Omega}^{-1}\right] + 2\rho q(p-1) = 0,$$
$$\boldsymbol{\Sigma}^{-1} = \tfrac{1}{p}\cdot \mathrm{BTr}_q\left[(\boldsymbol{\Psi}_{\mathrm{AR}} \otimes \mathrm{I_q})\boldsymbol{\Omega}^{-1}\right]$$

with $\mathrm{F} = \sum_{\mathrm{i}=1}^{\mathrm{p}-1} \mathrm{i}\rho^{\mathrm{i}-1}(\mathrm{C^i} + \mathrm{C^{i\prime}}).$

Note that the approximations given in Theorems 5.4, 5.5 and 5.6 are not given in explicit form. The systems of equations can be solved numerically using an iterative method, the so-called flip-flop algorithm (as the equation for $\boldsymbol{\Psi}/\rho$ depends on $\boldsymbol{\Sigma}$ and the equation for $\boldsymbol{\Sigma}$ depends on $\boldsymbol{\Psi}/\rho$). It should also be observed that the first equation in Theorem 5.5 is quadratic, while the corresponding equation in Theorem 5.6 is a polynomial of order p. Thus, determination of the solution in the $\mathscr{S}$ and $\mathscr{S}_{\mathrm{AR}}$ cases is more time-consuming than in the case of $\mathscr{S}_{\mathrm{CS}}$.

5.4 Simulation Studies

In this section, we apply the presented methods to verify whether the respective discrepancy function detects properly the true covariance structure. Due to the fact that $\mathscr{S}_{\mathrm{CS}}$ and $\mathscr{S}_{\mathrm{AR}}$ are subsets of $\mathscr{S}$, the discrepancy with respect to a separable structure set $\mathscr{S}$ is always the smallest and it is reasonable to consider only the sets

$\mathcal{S}_{\text{CS}}$ and $\mathcal{S}_{\text{AR}}$. Thus, we generate the data from a matrix normal distribution with $n = 100$, and $(p, q) \in \{(3, 3), (10, 3), (15, 3), (30, 3), (3, 5), (10, 5), (15, 5)\}$ and with respective covariance matrix $\mathbf{\Omega}$, that is $\mathbf{Y} \sim N_{n,qp}(\mathbf{0}, \mathbf{I}_n, \mathbf{\Omega})$, where $\mathbf{\Omega} = \mathbf{\Psi}_{\text{CS}} \otimes \mathbf{\Sigma}$ or $\mathbf{\Omega} = \mathbf{\Psi}_{\text{AR}} \otimes \mathbf{\Sigma}$ with various values of ρ. Since it is known from Filipiak et al. [12] that

- $\zeta_{\text{UN}}^{(E)}$ does not depend on the choice of $\mathbf{\Psi}$ and $\mathbf{\Sigma}$ in $\mathbf{\Psi} \otimes \mathbf{\Sigma}$ (in particular $\mathbf{\Psi}$ can be CS or AR(1)),
- $\zeta_{\text{CS}}^{(E)}$ does not depend on the choice of ρ and $\mathbf{\Sigma}$ in $\mathbf{\Psi}_{\text{CS}} \otimes \mathbf{\Sigma}$, and depends on the choice of ρ in $\mathbf{\Psi}_{\text{AR}} \otimes \mathbf{\Sigma}$,
- $\zeta_{\text{AR}}^{(E)}$ depends on the choice of ρ and does not depend on $\mathbf{\Sigma}$ in $\mathbf{\Psi}_{\text{CS}} \otimes \mathbf{\Sigma}$ or $\mathbf{\Psi}_{\text{AR}} \otimes \mathbf{\Sigma}$,

in all simulations we assume $\mathbf{\Sigma} = \mathbf{I}_q$.

We verify whether the Frobenius norm and the entropy loss function recognize properly the true $\mathbf{\Omega}$. Clearly, having the data without any knowledge of the true covariance structure, we usually estimate the covariance matrix by the sample covariance or maximum likelihood estimator. In this section, we use the maximum likelihood estimator of $\mathbf{\Omega}$ under model (5.1), which for arbitrary mean vector vec $\mathbf{M}$ is of the form

$$\mathbf{S} = \frac{1}{n}\mathbf{Y}'\left(\mathbf{I}_n - \frac{1}{n}\mathbf{1}_n\mathbf{1}_n'\right)\mathbf{Y}.$$

Obviously, $\mathbf{S}$ is a random matrix without any structure, even if the true $\mathbf{\Omega}$ is structured. Observe that for $n = 100$, $p = 30$, and $q = 5$, the matrix $\mathbf{S}$ would be singular, as $n < qp$, and the assumption of the invertibility of $\mathbf{S}$ required in the entropy loss function is not satisfied. Thus, we do not consider this case. We show that the Frobenius norm and the entropy loss function can find the underlying structure regardless of this unstructured estimator.

Note that $\zeta^{(F)}$ and $\zeta^{(E)}$ are not comparable. Moreover, Filipiak and Klein [8] used an adjusted version of the discrepancy based on the Frobenius norm, that is, they considered $\kappa^{(F)} = \zeta^{(F)}/||\mathbf{S}||_F$, to obtain the discrepancy in the interval [0, 1). Therefore, in this section, we use an adjusted version of the discrepancy also for the entropy loss function, namely $\kappa^{(E)} = 1 - 1/[1 + \log(\zeta^{(E)} + 1)]$. Note that in the simulation studies on the relevance of structure detection via the entropy loss function described by Filipiak et al. [12], the unadjusted discrepancy $\zeta^{(E)}$ was used directly, while an adjustment of the form $1 - 1/(1 + \zeta^{(E)})$ was considered in the power studies.

To achieve the aim of this section for a separable structure with one component structured and $k \in \{F, E\}$, we proceed as follows:

1. generate the data from $N_{n,qp}(\mathbf{0}, \mathbf{I}_n, \mathbf{\Omega})$;
2. compute $\mathbf{S}$;
3. compute discrepancies;

$$\zeta_{\mathrm{UN}}^{(k)} = \min_{\boldsymbol{\Psi},\boldsymbol{\Sigma}} L_S^{(k)}(\boldsymbol{\Psi} \otimes \boldsymbol{\Sigma}),$$
$$\zeta_{\mathrm{CS}}^{(k)} = \min_{\rho,\boldsymbol{\Sigma}} L_S^{(k)}(\boldsymbol{\Psi}_{\mathrm{CS}} \otimes \boldsymbol{\Sigma}),\ \zeta_{\mathrm{AR}}^{(k)} = \min_{\rho,\boldsymbol{\Sigma}} L_S^{(k)}(\boldsymbol{\Psi}_{\mathrm{AR}} \otimes \boldsymbol{\Sigma});$$

4. compute adjusted discrepancies $\kappa_{\mathrm{UN}}^{(k)}$, $\kappa_{\mathrm{CS}}^{(k)}$ and $\kappa_{\mathrm{AR}}^{(k)}$;
5. repeat 1000 times steps 1–4;
6. average $\kappa_{\mathrm{UN}}^{(k)}$, $\kappa_{\mathrm{CS}}^{(k)}$ and $\kappa_{\mathrm{AR}}^{(k)}$;
7. verify if

$$\overline{\kappa}_{\mathrm{UN}}^{(k)} \leqslant \overline{\kappa}_{\mathrm{CS}}^{(k)},\ \overline{\kappa}_{\mathrm{UN}}^{(k)} \leqslant \overline{\kappa}_{\mathrm{AR}}^{(k)},$$

$$\boldsymbol{\Omega} \in \mathscr{S}_{\mathrm{CS}} \quad \Longrightarrow \quad \overline{\kappa}_{\mathrm{CS}}^{(k)} \leqslant \overline{\kappa}_{\mathrm{AR}}^{(k)},$$
$$\boldsymbol{\Omega} \in \mathscr{S}_{\mathrm{AR}} \quad \Longrightarrow \quad \overline{\kappa}_{\mathrm{AR}}^{(k)} \leqslant \overline{\kappa}_{\mathrm{CS}}^{(k)}.$$

Recall that due to Theorems 5.1, 5.2 and 5.5 the values $\kappa_{\mathrm{UN}}^{(F)}$, $\kappa_{\mathrm{CS}}^{(F)}$, and $\kappa_{\mathrm{CS}}^{(E)}$ are determined uniquely for each simulated dataset. In the remaining cases, the uniqueness of the best approximation is not proven; however, in all simulated datasets, there was exactly one minimum.

In Tables 5.1, 5.2, 5.3, and 5.4, the averaged adjusted discrepancies with respect to the considered discrepancy functions are given. As expected, for both the Frobenius norm and the entropy loss function the averaged adjusted discrepancy $\overline{\kappa}_{\mathrm{UN}}^{(k)}$ is the smallest (among $\overline{\kappa}_{\mathrm{UN}}^{(k)}$, $\overline{\kappa}_{\mathrm{CS}}^{(k)}$ and $\overline{\kappa}_{\mathrm{AR}}^{(k)}$, as the set $\mathscr{S}$ contains $\mathscr{S}_{\mathrm{CS}}$ and $\mathscr{S}_{\mathrm{AR}}$). Moreover, if $\rho = 0$ then $\mathscr{S}_{\mathrm{CS}} = \mathscr{S}_{\mathrm{AR}}$, and hence $\overline{\kappa}_{\mathrm{CS}}^{(k)}$ and $\overline{\kappa}_{\mathrm{AR}}^{(k)}$ are equal. Slight differences in the discrepancies in Tables 5.1, 5.2, 5.3, and 5.4 result from the numerical procedure.

The most important conclusion from all of the tables is that both discrepancy functions detect the true structure properly (point 7 of the algorithm outline is satisfied).

Observe that $\overline{\kappa}_{\mathrm{UN}}^{(E)}$ does not change with ρ in all of the tables, and $\overline{\kappa}_{\mathrm{CS}}^{(E)}$ is constant for every ρ in Tables 5.1 and 5.2. Moreover, $\overline{\kappa}_{\mathrm{UN}}^{(E)}$ depends only on the experiment parameters (n, q, p), thus, the respective results in Tables 5.1 and 5.3 are the same, and similarly, respective $\overline{\kappa}_{\mathrm{UN}}^{(E)}$ are the same in Tables 5.2 and 5.4. These observations confirm that the respective indices do not depend on ρ. Note that the results for $\overline{\kappa}_{\mathrm{AR}}^{(E)}$ given in the last columns of Tables 5.3 and 5.4 seem to be equal among the respective sets of parameters; however, in fact, they are not. Recall, that the original discrepancies, presented in Filipiak et al. [12], were adjusted to obtain values in the interval $[0, 1)$, which provides a false impression of the independence of $\overline{\kappa}_{\mathrm{AR}}^{(E)}$ of ρ. Note that the Frobenius norm does not have this interesting property. On the other hand, the process of identification of the covariance structure with the use of the Frobenius norm is faster, as the algorithms are not so time-consuming.

Table 5.1 The averaged adjusted discrepancy in the case of $\boldsymbol{\Omega} = \boldsymbol{\Psi}_{\mathrm{CS}} \otimes \mathbf{I}_3$ with $n = 100$

		Frobenius norm			Entropy loss function		
p	ρ	$\overline{\kappa}_{\mathrm{UN}}^{(F)}$	$\overline{\kappa}_{\mathrm{CS}}^{(F)}$	$\overline{\kappa}_{\mathrm{AR}}^{(F)}$	$\overline{\kappa}_{\mathrm{UN}}^{(E)}$	$\overline{\kappa}_{\mathrm{CS}}^{(E)}$	$\overline{\kappa}_{\mathrm{AR}}^{(E)}$
3	−0.1	0.2568	0.2735	0.2862	0.1196	0.1299	0.1382
3	0	0.2608	0.2768	0.2767	0.1196	0.1299	0.1299
3	0.1	0.2568	0.2717	0.2802	0.1196	0.1299	0.1356
3	0.5	0.1762	0.1862	0.2235	0.1196	0.1299	0.1984
3	0.9	0.0618	0.0657	0.0701	0.1196	0.1299	0.2516
10	−0.1	0.4403	0.4694	0.5227	0.4403	0.4556	0.5445
10	0	0.4552	0.4846	0.4846	0.4403	0.4556	0.4556
10	0.1	0.4386	0.4661	0.5165	0.4403	0.4556	0.4649
10	0.5	0.2341	0.2487	0.3291	0.4403	0.4556	0.5086
10	0.9	0.0732	0.0779	0.0829	0.4403	0.4556	0.5508
15	0	0.5276	0.5615	0.5615	0.5356	0.5483	0.5483
15	0.1	0.5024	0.5340	0.5997	0.5356	0.5483	0.5532
15	0.5	0.2446	0.2601	0.3263	0.5356	0.5483	0.5714
15	0.9	0.0748	0.0797	0.0834	0.5356	0.5483	0.5907
30	0	0.6510	0.6916	0.6916	0.6728	0.6867	0.6867
30	0.1	0.6045	0.6418	0.7373	0.6728	0.6867	0.6873
30	0.5	0.2552	0.2714	0.3120	0.6728	0.6867	0.6886
30	0.9	0.0761	0.0811	0.0831	0.6728	0.6867	0.6903

5.5 Real Data Example

In this section, we use a medical dataset to illustrate the methods for choosing a covariance structure using the Frobenius norm and entropy loss function. McKiernan et al. [21] collected multivariate longitudinal data, where body weight, fat mass, and estimated upper leg lean muscle mass were measured over 12 years in 12 adult male rhesus monkeys from 16 to 22 years of age (after the 6th, 9th, and 12th year of research), to analyze early stage sarcopenia in aging rhesus monkeys. Clearly, in this dataset $q = 3$ characteristics were measured at $p = 3$ time points for $n = 12$ individuals. The data, originally presented by McKiernan et al. [21, Table 1] and reproduced in Filipiak et al. [9, Table 1], are described in detail in McKiernan et al. [21].

In this paper, we verify whether the structure chosen by the respective discrepancy function and the structure identified by the likelihood ratio test at a 0.05 significance level is the same. Thus, in Table 5.5, we present the relevant discrepancies as well as the p-values determined by empirical null distributions, i.e., simulated distributions of the test statistics (for more details see Filipiak et al. [9]). Besides all three characteristics we also study all pairs of variables, as these are used to analyze particular cases of Sarcopenia; see, e.g., Filipiak et al. [9].

Table 5.2 The averaged adjusted discrepancy in the case of $\boldsymbol{\Omega} = \boldsymbol{\Psi}_{\text{CS}} \otimes \mathbf{I}_5$ with $n = 100$

		Frobenius norm			Entropy loss function		
p	ρ	$\overline{\kappa}^{(F)}_{\text{UN}}$	$\overline{\kappa}^{(F)}_{\text{CS}}$	$\overline{\kappa}^{(F)}_{\text{AR}}$	$\overline{\kappa}^{(E)}_{\text{UN}}$	$\overline{\kappa}^{(E)}_{\text{CS}}$	$\overline{\kappa}^{(E)}_{\text{AR}}$
3	−0.1	0.3334	0.3410	0.3511	0.2459	0.2509	0.2577
3	0	0.3383	0.3456	0.3457	0.2459	0.2509	0.2510
3	0.1	0.3332	0.3401	0.3472	0.2459	0.2509	0.2555
3	0.5	0.2309	0.2357	0.2657	0.2459	0.2509	0.3057
3	0.9	0.0822	0.0841	0.0876	0.2459	0.2509	0.3483
10	−0.1	0.5500	0.5626	0.6026	0.5650	0.5692	0.6078
10	0	0.5666	0.5793	0.5792	0.5650	0.5692	0.5692
10	0.1	0.5483	0.5602	0.5985	0.5650	0.5692	0.5725
10	0.5	0.3047	0.3114	0.3735	0.5650	0.5692	0.5903
10	0.9	0.0973	0.0995	0.1032	0.5650	0.5692	0.6116
15	0	0.6430	0.6572	0.6572	0.6429	0.6469	0.6469
15	0.1	0.6168	0.6301	0.6788	0.6429	0.6469	0.6481
15	0.5	0.3177	0.3247	0.3754	0.6429	0.6469	0.6532
15	0.9	0.0992	0.1015	0.1043	0.6429	0.6469	0.6597

The discrepancies based on the Frobenius norm are presented in Filipiak and Klein [8], while the discrepancies obtained via the entropy loss function are presented in Filipiak et al. [12].

As expected, for all sets of variables the discrepancy $\kappa^{(k)}_{\text{UN}}$, $k \in \{F, E\}$, is the smallest. This observation confirms the choice of a separable structure made by testing, at a 0.05 significance level, that the hypothesis of separability (versus an unstructured covariance matrix) is not rejected.

Let us now compare the discrepancies and p-values for the sets of variables under the separable structure. Note that the Frobenius norm discrepancy, $\kappa^{(F)}_{\text{UN}}$, does not interact with the p-value: $\kappa^{(F)}_{\text{UN}}$ for the set of all three variables (0.2535) is relatively close to its counterpart for the pair 2, 3 (0.2407), while the p-values are much different from each other (p-value between 0.05 and 0.1 for the set of all three variables and between 0.45 and 0.5 for the pair 2, 3). Similarly, $\kappa^{(F)}_{\text{UN}}$ for the pair 1, 2 (0.0519) is relatively close to its counterpart for the pair 1, 3 (0.0877), while the p-values are not (p-value between 0.85 and 0.9 for the pair 1, 2 and between 0.1 and 0.15 for the pair 1, 3). For the entropy loss discrepancy, $\kappa^{(E)}_{\text{UN}}$, it can be observed that increasing discrepancy implies decreasing p-value, that is if the discrepancies are ordered in ascending order, 0.3891 for the pair 1, 2, 0.4971 for the pair 2, 3, 0.5335 for the pair 1, 3, and 0.7052 for all three variables, the p-values will appear decreasing: p-value between 0.85 and 0.9 for the pair 1, 2, between 0.45 and 0.5 for the pair 2, 3, between 0.1 and 0.15 for the pair 1, 3, and between 0.05 and 0.1 for all three variables. Moreover, in the case of the Frobenius norm, similar values of discrepancy do not correspond to similar p-values: the discrepancy 0.2535 for all three variables

Table 5.3 The averaged adjusted discrepancy in the case of $\mathbf{\Omega} = \mathbf{\Psi}_{\text{AR}} \otimes \mathbf{I}_3$ with $n = 100$

		Frobenius norm			Entropy loss function		
p	ρ	$\overline{\kappa}^{(F)}_{\text{UN}}$	$\overline{\kappa}^{(F)}_{\text{CS}}$	$\overline{\kappa}^{(F)}_{\text{AR}}$	$\overline{\kappa}^{(E)}_{\text{UN}}$	$\overline{\kappa}^{(E)}_{\text{CS}}$	$\overline{\kappa}^{(E)}_{\text{AR}}$
3	−0.9	0.0721	0.7215	0.0766	0.1196	0.4518	0.1299
3	−0.5	0.1959	0.4683	0.2070	0.1196	0.2568	0.1299
3	0	0.2608	0.2768	0.2767	0.1196	0.1299	0.1299
3	0.5	0.1957	0.2485	0.2066	0.1196	0.1927	0.1299
3	0.9	0.0716	0.0847	0.0760	0.1196	0.2508	0.1298
10	−0.9	0.1494	0.9003	0.1586	0.4403	0.5851	0.4556
10	−0.5	0.3720	0.6668	0.3961	0.4403	0.5232	0.4556
10	0	0.4552	0.4846	0.4846	0.4403	0.4556	0.4556
10	0.5	0.3727	0.5591	0.3967	0.4403	0.5119	0.4556
10	0.9	0.1502	0.2459	0.1591	0.4403	0.5536	0.4556
15	−0.9	0.1897	0.9213	0.2011	0.5356	0.6220	0.5483
15	−0.5	0.4400	0.7085	0.4685	0.5356	0.5840	0.5483
15	0	0.5276	0.5615	0.5615	0.5356	0.5483	0.5483
15	0.5	0.4402	0.6397	0.4689	0.5356	0.5795	0.5483
15	0.9	0.1894	0.3392	0.2009	0.5356	0.6070	0.5483
30	−0.9	0.2771	0.9401	0.2941	0.6728	0.6990	0.6867
30	−0.5	0.5631	0.7763	0.5987	0.6728	0.6915	0.6867
30	0	0.6510	0.6916	0.6916	0.6728	0.6867	0.6867
30	0.5	0.5632	0.7483	0.5988	0.6728	0.6912	0.6867
30	0.9	0.2764	0.5375	0.2934	0.6728	0.6970	0.6867

is relatively close to its counterpart for the pair 2, 3 (0.2407), while the p-values are much different.

Let us concentrate now on the $\mathbf{\Psi}_{\text{CS}} \otimes \mathbf{\Sigma}$ and $\mathbf{\Psi}_{\text{AR}} \otimes \mathbf{\Sigma}$ structures. Observe that for all three variables, the Frobenius norm discrepancy indicates separable structures with the first component as AR(1) as more relevant, while the entropy loss function chooses separability with the first component as CS. Nevertheless, the difference between $\kappa^{(E)}_{\text{CS}}$ and $\kappa^{(E)}_{\text{AR}}$ is very small and may result from the numerical algorithm used for the determination of discrepancy. Moreover, at a 0.05 significance level, both null hypotheses (separability with one component structured) are rejected.

When the pairs of variables are considered, usually $\kappa^{(k)}_{\text{AR}}$ is smaller than $\kappa^{(k)}_{\text{CS}}$, $k \in \{F, E\}$, which may indicate separability with the first component as AR(1) as more relevant. Observe, however, that when the first two characteristics are considered, at a 0.05 significance level both hypotheses of separability with the first component structured are not rejected, while for the remaining pairs the decision is the opposite. This may result from the fact that for the first two characteristics the discrepancies are much lower than in the case of the two remaining pairs. Moreover, for the two last pairs of characteristics, the discrepancies are close to their counterparts in the

Table 5.4 The averaged adjusted discrepancy in the case of $\mathbf{\Omega} = \mathbf{\Psi}_{\text{AR}} \otimes \mathbf{I}_5$ with $n = 100$

		Frobenius norm			Entropy loss function		
p	ρ	$\overline{\kappa}_{\text{UN}}^{(F)}$	$\overline{\kappa}_{\text{CS}}^{(F)}$	$\overline{\kappa}_{\text{AR}}^{(F)}$	$\overline{\kappa}_{\text{UN}}^{(E)}$	$\overline{\kappa}_{\text{CS}}^{(E)}$	$\overline{\kappa}_{\text{AR}}^{(E)}$
3	−0.9	0.0954	0.7240	0.0975	0.2459	0.5094	0.2510
3	−0.5	0.2555	0.4951	0.2607	0.2459	0.3525	0.2510
3	0	0.3383	0.3456	0.3457	0.2459	0.2509	0.2510
3	0.5	0.2560	0.2937	0.2611	0.2459	0.3008	0.2509
3	0.9	0.0954	0.1041	0.0975	0.2459	0.3471	0.2509
10	−0.9	0.1970	0.9030	0.2013	0.5650	0.6321	0.5692
10	−0.5	0.4727	0.7051	0.4835	0.5650	0.5970	0.5692
10	0	0.5666	0.5793	0.5792	0.5650	0.5692	0.5692
10	0.5	0.4737	0.6129	0.4842	0.5650	0.5916	0.5692
10	0.9	0.1983	0.2726	0.2023	0.5650	0.6131	0.5692
15	−0.9	0.2480	0.9242	0.2532	0.6429	0.6729	0.6469
15	−0.5	0.5497	0.7514	0.5620	0.6429	0.6573	0.6469
15	0	0.6430	0.6572	0.6572	0.6429	0.6469	0.6469
15	0.5	0.5503	0.6952	0.5625	0.6429	0.6558	0.6469
15	0.9	0.2490	0.3676	0.2543	0.6429	0.6661	0.6469

Table 5.5 Frobenius norm and entropy loss discrepancies as well as p-values based on empirical null distribution of the likelihood ratio test statistic for three covariance structures: $\mathbf{\Psi} \otimes \mathbf{\Sigma}$, $\mathbf{\Psi}_{\text{CS}} \otimes \mathbf{\Sigma}$ and $\mathbf{\Psi}_{\text{AR}} \otimes \mathbf{\Sigma}$

Variables		$\mathbf{\Psi} \otimes \mathbf{\Sigma}$	$\mathbf{\Psi}_{\text{CS}} \otimes \mathbf{\Sigma}$	$\mathbf{\Psi}_{\text{AR}} \otimes \mathbf{\Sigma}$
1, 2, 3	$\kappa^{(F)}$	0.2535	0.5954	0.5745
	$\kappa^{(E)}$	0.7052	0.7254	0.7260
	p-value	(0.05,0.10)	< 0.01	(0.01,0.05)
1, 2	$\kappa^{(F)}$	0.0519	0.3612	0.3522
	$\kappa^{(E)}$	0.3891	0.5443	0.5218
	p-value	(0.85,0.90)	(0.20,0.25)	(0.60,0.65)
2, 3	$\kappa^{(F)}$	0.2407	0.6012	0.5798
	$\kappa^{(E)}$	0.4971	0.6189	0.6176
	p-value	(0.45,0.50)	< 0.01	(0.01,0.05)
1, 3	$\kappa^{(F)}$	0.0877	0.6621	0.6310
	$\kappa^{(E)}$	0.5335	0.6317	0.6269
	p-value	(0.10,0.15)	< 0.01	(0.01,0.05)

comparison of all three variables, and similar observations are made for the respective p-values. Finally, in the case of the first two characteristics, the discrepancies are lower, and simultaneously the respective p-values are higher, than in the remaining cases.

All of these observations imply that both of the discrepancies may be useful in covariance structure identification, although further studies on this topic are needed.

5.6 Conclusions

In this paper, the best approximations of a separable covariance structure using the Frobenius norm and the entropy loss function as measures of discrepancy have been presented. The most important conclusion from the simulation results is that both functions recognize the true covariance structure properly, as $\overline{\kappa}_{\mathrm{UN}}^{(k)}$ is smaller than $\overline{\kappa}_{\mathrm{CS}}^{(k)}$ and $\overline{\kappa}_{\mathrm{AR}}^{(k)}$ for $k \in \{F, E\}$, and $\overline{\kappa}_{\mathrm{CS}}^{(k)} \leqslant \overline{\kappa}_{\mathrm{AR}}^{(k)}$ if the true $\mathbf{\Omega}$ is equal to $\mathbf{\Psi}_{\mathrm{CS}} \otimes \mathbf{I}_q$, while $\overline{\kappa}_{\mathrm{AR}}^{(k)} \leqslant \overline{\kappa}_{\mathrm{CS}}^{(k)}$ if the true $\mathbf{\Omega}$ is equal to $\mathbf{\Psi}_{\mathrm{AR}} \otimes \mathbf{I}_q$.

The entropy loss function has three main properties which may recommend it (in comparison with the Frobenius norm) for use by researchers. First of all, it represents the divergence between two distributions that differ with respect to the covariance matrices, while the Frobenius norm offers only an algebraic interpretation of the distance between two matrices. Moreover, entropy loss discrepancy for a separable structure is independent of the form of the true covariance structure, which can be seen in the equalities between $\overline{\kappa}_{\mathrm{UN}}^{(E)}$ in Tables 5.1, 5.2, 5.3, and 5.4. Finally, there is a relation between both discrepancies and p-values (in most cases increasing discrepancy implies decreasing p-value); however, only in the case of entropy loss do similar values of discrepancy correspond to similar p-values. Nevertheless, it must be noted that the identification of the structure via the entropy loss function is much more time-consuming than in the case of the Frobenius norm. Moreover, the Frobenius norm discrepancy can be used even for so-called high-dimensional datasets, in which the sample size is too small to estimate the covariance.

Finally, in this paper, we assume the covariance structure to be nonsingular; however, this requirement may be weakened. Recently, Chen et al. [2] proposed a method of covariance matrix identification via the entropy loss function under the multivariate model with a modified singular estimator of the unstructured covariance matrix. Its generalization to doubly multivariate observations will be the topic of our future research.

Acknowledgements The authors thank Stefan Banach International Mathematical Center, Institute of Mathematics of the Polish Academy of Sciences, Warsaw, for providing the opportunity and support for this paper. This research is partially supported by Scientific Activities No. 04/43/SBAD/0115 (Katarzyna Filipiak) and by the Slovak Research and Development Agency under contract no. APVV-17-0568 and VEGA MŠ SR 1/0311/18 (Daniel Klein).

References

1. Bickel, P.J., Li, B.: Regularization in statistics. Test **15**, 271–344 (2006)
2. Chen, C., Zhou, J., Pan, J.: Correlation structure regularization via entropy loss function for high-dimension and low-sample-size data. Commun. Stat. Simul. Comput. (2019). https://doi.org/10.1080/03610918.2019.1571607
3. Cui, X., Li, C., Zhao, J., Zeng, L., Zhang, D., Pan, J.: Covariance structure regularization via Frobenius norm discrepancy. Linear Algebra Appl. **510**, 124–145 (2016)
4. Dey, D.K., Srinivasan, C.: Estimation of a covariance matrix under Stein's loss. Ann. Stat. **13**, 1581–1591 (1985)
5. Devijver, E., Gallopin, M.: Block-diagonal covariance selection for high-dimensional Gaussian graphical models. J. Am. Stat. Assoc. **113**, 306–314 (2018)
6. Dutilleul, P.: The MLE algorithm for the matrix normal distribution. J. Stat. Comput. Simul. **64**, 105–123 (1999)
7. Filipiak, K., Klein, D.: Estimation of parameters under a generalized growth curve model. J. Multivar. Anal. **158**, 73–86 (2017)
8. Filipiak, K., Klein, D.: Approximation with a Kronecker product structure with one component as compound symmetry or autoregression. Linear Algebra Appl. **559**, 11–33 (2018)
9. Filipiak, K., Klein, D., Roy, A.: A comparison of likelihood ratio tests and Rao's score test for three separable covariance matrix structures. Biom. J. **59**, 192–215 (2017)
10. Filipiak, K., Klein, D., Mokrzycka, M.: Estimators comparison of separable covariance structure with one component as compound symmetry matrix. Electron. J. Linear Algebra **33**, 83–98 (2018)
11. Filipiak, K., Klein, D., Vojtková, E.: The properties of partial trace and block trace operators of partitioned matrix. Electron. J. Linear Algebra **33**, 2–15 (2018)
12. Filipiak, K., Klein, D., Markiewicz, A., Mokrzycka, M.: Approximation with a Kronecker product structure with one component as compound symmetry or autoregression via entropy loss function. Linear Algebra Appl. **610**, 625–646 (2021)
13. Gilson, M., Dahmen, D., Moreno-Bote, R., Insabato, A., Helias, M.: The covariance perceptron: a new framework for classification and processing of time series in recurrent neural networks. bioRxiv (2019). https://doi.org/10.1101/562546
14. Hastie, T., Tibshirani, R., Friedman, J.: The Elements of Statistical Learning: Data Mining, Inference, and Prediction. Springer, New York (2009)
15. James, W., Stein, C.: Estimation with quadratic loss. In: Neyman, J. (ed.) Proceedings of the Fourth Berkeley Symposium. Mathematical Statistics and Probability, vol. 1., pp. 361–379. The Statistical Laboratory, University of California, 30 June–30 July 1960. University of California Press (1961)
16. Kollo, T., von Rosen, D.: Advanced Multivariate Statistics with Matrices. Springer, Dordrecht (2005)
17. Lin, L., Higham, N.J., Pan, J.: Covariance structure regularization via entropy loss function. Comput. Stat. Data Anal. **72**, 315–327 (2014)
18. van Loan, C.F., Pitsianis, N.: Approximation with Kronecker products. In: De Moor, B.L.R., Moonen, M.S., Golub, G.H. (eds.) Linear Algebra for Large Scale and Real-Time Applications, pp. 293–314. Kluwer Publications, Dordrecht, The Netherlands (1992)
19. Lu, N., Zimmerman, D.: The likelihood ratio test for a separable covariance matrix. Stat. Probab. Lett. **73**, 449–457 (2005)
20. Magnus, J., Neudecker, H.: Symmetry, 0–1 matrices and Jacobians, a review. Econ. Theory **2**, 157–190 (1986)
21. McKiernan, S.H., Colman, R.J., Lopez, M., Beasley, T.M., Weindruch, R., Aiken, J.M.: Longitudinal analysis of early stage Sarcopenia in aging rhesus monkeys. Exp. Gerontol. **44**, 170–176 (2009)
22. Pan, J., Fang, K.: Growth Curve Models and Statistical Diagnostics. Springer, New York (2002)

23. Roy, A., Khattree, R.: Testing the hypothesis of a Kronecker product covariance matrix in multivariate repeated measures data. In: Proceedings of the 30th Annual SAS Users Group International Conference (SUGI 30), Philadelphia (2005)
24. Roy, A., Khattree, R.: Classification of multivariate repeated measures data with temporal autocorrelation. J. Appl. Stat. Sci. **15**, 283–294 (2007)
25. Srivastava, M., von Rosen, T., von Rosen, D.: Models with a Kronecker product covariance structure: estimation and testing. Math. Methods Stat. **17**, 357–370 (2008)

Chapter 6
Estimation and Testing of the Covariance Structure of Doubly Multivariate Data

Katarzyna Filipiak and Daniel Klein

Abstract The covariance matrix of doubly multivariate data often has a separable structure, that is, it can be presented as the Kronecker product of two positive definite matrices. In particular, one of the separability components can be further specified, for example, as compound symmetry or autoregression of order one. Another suitable structure for doubly multivariate data is a block compound symmetry structure. In this paper, two testing procedures for such covariance structures, namely the likelihood ratio and Rao score tests, will be discussed. Using simulation studies, it will be shown that the Rao score test outperforms the likelihood ratio test in a number of contexts, mainly for small and moderate sample size. Both of the testing methods will then be illustrated by two real data examples.

6.1 Introduction

Repeated measures data on more than one variable arise in almost all fields of science including agricultural, biological, biomedical, medical, environmental, and engineering research. In such doubly multivariate experiments, the observations are made on more than one response variable, say q, and each response variable is measured repeatedly over p time or location points for each of n individuals. It is of great importance for any statistical analysis to develop efficient techniques for accounting for variation among the measurements of such datasets.

For doubly multivariate data, there is usually no knowledge about the relations between observations, and thus the assumption of an unstructured (UN) covariance matrix may be robust. However, as there are $qp(qp + 1)/2$ free parameters in a UN structure, a sample size n of at least $qp + 1$ is required for estimation purposes; otherwise the problem of overparameterization will arise. Moreover, for small sam-

K. Filipiak
Institute of Mathematics, Poznań University of Technology, Poznań, Poland
e-mail: katarzyna.filipiak@put.poznan.pl

D. Klein (✉)
Faculty of Science, Institute of Mathematics, P. J. Šafárik University, Košice, Slovakia
e-mail: daniel.klein@upjs.sk

K. Filipiak et al. (eds.), *Multivariate, Multilinear and Mixed Linear Models*, Contributions to Statistics, https://doi.org/10.1007/978-3-030-75494-5_6

ples a UN structure can result in rather weak inference, in the sense that too many degrees of freedom are used up in estimating the covariance parameters, leaving too few for the parameters of interest; see Crowder and Hand [1, p. 60]. A common way to overcome this problem is to assume some structure.

The problem of interest in this paper is to test four kinds of special structures which are very natural for doubly multivariate data. Namely, we consider (1) a separable structure with unstructured components, (2) a separable structure with one component structured as compound symmetry (CS), (3) a separable structure with one component structured as autoregression of order one (AR(1)), and (4) a block compound symmetry (BCS) structure. Two testing procedures — likelihood ratio test (LRT), commonly used in the literature, and the Rao score test (RST) — will be presented and compared.

The paper is organized as follows. In Sect. 6.2, the model for doubly multivariate data is presented. In Sect. 6.3, the structures under consideration are presented and their maximum likelihood estimators (MLEs) are determined. Hypotheses about the special structures and two testing procedures are given in Sect. 6.4. In Sect. 6.5, some statistical properties of both tests are shown with the use of algebraic methods, and then the tests are compared via simulations. In Sect. 6.6, the tests are applied to two real data examples. Finally, some conclusions are given in Sect. 6.7.

Most of the presented results follow largely from Filipiak et al. [6, 7] and Roy et al. [25].

6.2 Model for Doubly Multivariate Data

Let $\boldsymbol{Y}_i, i \in \{1, \dots, n\}$, be independent and identically distributed $q \times p$ observation matrices, each containing measurements of q characteristics at p time or location points on the ith individual. Such doubly multivariate observations may be presented in a three-dimensional tensor form. Observe, however, that in such a case there are three possible directions of arrangement of the matrices $\boldsymbol{Y}_i$. In this paper, we consider the tensor of observations $\mathcal{Y} \in \mathbb{R}^{n,q,p}$ created from the n-dimensional sample of $q \times p$ matrices $\boldsymbol{Y}_i$, written in horizontal position one below another (see Fig. 6.1, left), and the position of each element is identified by its three indices; see Fig. 6.1, right.

Following Kolda and Bader [12], in tensor analysis the matrices $\boldsymbol{Y}_i$ are called qp-mode *slices* of $\mathcal{Y}$, and can also be written as $\boldsymbol{Y}_{i\bullet\bullet}$. To distinguish possible modes, observe, that $\boldsymbol{Y}_{\bullet j\bullet}$ and $\boldsymbol{Y}_{\bullet\bullet k}$ are called, respectively, np-mode and nq-mode slices.

We note that, following Kolda and Bader [12], the one-dimensional arrays (i.e., vectors) obtained by fixing every index but one are called *fibers*, and can be expressed as $\boldsymbol{y}_{\bullet jk}$, $\boldsymbol{y}_{i\bullet k}$ and $\boldsymbol{y}_{ij\bullet}$ for, respectively, n-mode, q-mode, and p-mode fibers.

We recall that the vec operator for a $q \times p$ matrix arranges the columns one below the other to form a qp-dimensional vector. The extension of vectorization to the tensor may be formulated in several ways. Kolda and Bader [12] used the following definition:

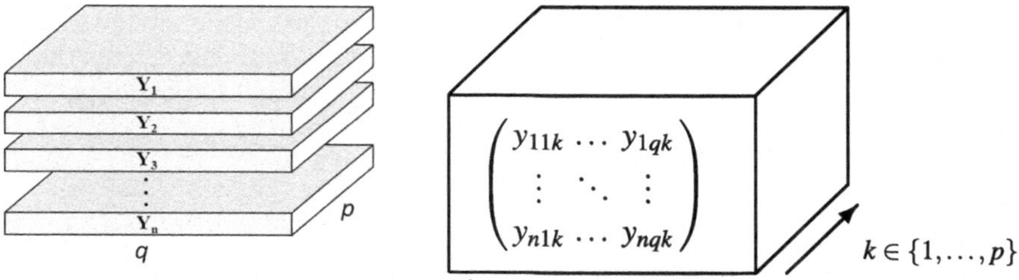

Fig. 6.1 Visualization of a three-dimensional tensor $\mathscr{Y} = (y_{ijk})$ as qp-mode slices $\boldsymbol{Y}_{i\cdot\cdot} = \boldsymbol{Y}_i$ (left) and as a cube of elements y_{ijk} (right)

$$\operatorname{vec} \mathscr{Y} = \sum_{i=1}^{n} \sum_{j=1}^{q} \sum_{k=1}^{p} y_{ijk}\, \boldsymbol{e}_{k,p} \otimes \boldsymbol{e}_{j,q} \otimes \boldsymbol{e}_{i,n},$$

where $\boldsymbol{e}_{t,s}$ is the tth column of the identity matrix $\boldsymbol{I}_s$. Throughout the paper, we will use the simplified notation $\operatorname{vec} \mathscr{Y} = \boldsymbol{y}$.

The natural way of working with a doubly multivariate data tensor $\mathscr{Y}$ is *matricization* (also called *unfolding* or *flattening*; see, e.g., Kolda and Bader [12]), that is, transformation of the tensor $\mathscr{Y}$ into matrix form. For a three-dimensional tensor, there are three possibilities for performing matricization (for more details see, e.g., Filipiak and Klein [4, Def. 2]), and the specific permutation of elements is not important as long as it is consistent across related calculations. Since we are used to working with multivariate models in which the observation vectors for all individuals are arranged as rows one below another, it is natural to vectorize and transpose each $\boldsymbol{Y}_i$ and to write the result as rows one below another forming an $n \times qp$ matrix $\boldsymbol{Y}$. Such a process is equivalent to matricization of $\mathscr{Y}$ with respect to nq-mode slices, $\boldsymbol{Y}_{\bullet\bullet k}$, given side by side, and algebraically can be represented as

$$\boldsymbol{Y} = \sum_{i=1}^{n} \sum_{j=1}^{q} \sum_{k=1}^{p} y_{ijk}\, \boldsymbol{e}_{i,n} (\boldsymbol{e}'_{k,p} \otimes \boldsymbol{e}'_{j,q}).$$

Let us assume that

$$\mathscr{Y} \sim N_{n,p,q} (\mathscr{M}, \boldsymbol{I}_n, \boldsymbol{\Sigma}, \boldsymbol{\Psi}), \tag{6.1}$$

where $\mathscr{M} \in \mathbb{R}^{n,q,p}$ is the general expectation, $\boldsymbol{\Sigma}$ is a $q \times q$ covariance matrix of characteristics (the same for each time or location point), and $\boldsymbol{\Psi}$ is a $p \times p$ covariance matrix of time or location points (the same for each characteristic). Obviously, both $\boldsymbol{\Sigma}$ and $\boldsymbol{\Psi}$ are symmetric, positive definite (p.d.) matrices.

Throughout the paper, we assume the same mean $\boldsymbol{M}$ for each qp-mode slice of $\mathrm{E}(\mathscr{Y}) = \mathscr{M}$, i.e., $\boldsymbol{M}_{i\bullet\bullet} = \boldsymbol{M}$ for all $i \in \{1, \ldots, n\}$. It is easy to see that after matricization $\mathrm{E}(\boldsymbol{Y}) = \mathbf{1}_n \operatorname{vec}' \boldsymbol{M}$, with $\mathbf{1}_n$ being an n-dimensional vector of ones. Observe that the dispersion of observations in (6.1) is separated for rows and columns;

however, even if this seems very natural, it need not be the case. That is, the covariance matrix can be any $qp \times qp$ symmetric p.d. matrix, say $\boldsymbol{\Omega}$. Using the matricized version of $\mathscr{Y}$, we can write this more general model (without the assumption on separability) as

$$\boldsymbol{Y} \sim N_{n,qp}\left(\mathbf{1}_n\boldsymbol{\mu}', \boldsymbol{I}_n, \boldsymbol{\Omega}\right).$$

with $\boldsymbol{\mu} = \operatorname{vec} \boldsymbol{M}$. Observe, moreover, that the vectorization of $\mathscr{Y}$ is equivalent to $\operatorname{vec} \boldsymbol{Y}$, since $\operatorname{vec}(\mathbf{1}_n\boldsymbol{\mu}') = \boldsymbol{\mu} \otimes \mathbf{1}_n = (\boldsymbol{I}_{qp} \otimes \mathbf{1}_n)\boldsymbol{\mu}$. Therefore, the general model can also be written as

$$\boldsymbol{y} \sim N_{nqp}\left((\boldsymbol{I}_{qp} \otimes \mathbf{1}_n)\boldsymbol{\mu}, \boldsymbol{\Omega} \otimes \boldsymbol{I}_n\right). \tag{6.2}$$

Similarly, the vectorized version of model (6.1), in which the separable structure of covariance follows from two different sources of variability, can be rewritten as

$$\boldsymbol{y} \sim N_{nqp}\left((\boldsymbol{I}_{qp} \otimes \mathbf{1}_n)\boldsymbol{\mu}, \boldsymbol{\Psi} \otimes \boldsymbol{\Sigma} \otimes \boldsymbol{I}_n\right). \tag{6.3}$$

For clarity, the vectors of unknown parameters under models (6.2) and (6.3) can be written, respectively, as

$$\boldsymbol{\theta}_{\Omega} = \left(\boldsymbol{\mu}', \operatorname{vech}' \boldsymbol{\Omega}\right)', \quad \boldsymbol{\theta}_{\text{sep}} = \left(\boldsymbol{\mu}', \operatorname{vech}' \boldsymbol{\Psi}, \operatorname{vech}' \boldsymbol{\Sigma}\right)', \tag{6.4}$$

where the vech operator puts the columns of a symmetric matrix underneath each other, starting with the first, by eliminating all the supradiagonal elements. The number of unknown parameters to be estimated in $\boldsymbol{\Omega}$ is $qp(qp+1)/2$, which increases very rapidly with an increase in either the number of characteristics q or the number of time or location points p. Moreover, when the sample size n is close to qp or less than qp, the sample covariance matrix or the MLE of $\boldsymbol{\Omega}$ is ill-conditioned or singular, respectively. To avoid this problem, researchers usually rely on structured covariance matrices, which depend on a smaller set of unknown parameters. The problem, though, is knowing what the true structure is. To detect the relevant covariance structure, some regularization techniques — graphical, e.g., neural networks or mapping (Gilson et al. [11]) or a graphical lasso algorithm (Devijver and Gallopin [2]), as well as algebraic (see, e.g., Filipiak and Klein [5], Filipiak et al. [10]) — can be used. On the other hand, in many experiments the underlying structure follows from the experimenter's a priori knowledge about the variables behavior which derives from the nature of the data, for example. Nevertheless, in both cases, the structure (regularized or a priori identified) should then be tested. Thus, the aim of this paper is to compare two testing procedures, the LRT and the RST, for testing one of the following structures: separability (that is, $\boldsymbol{\Omega} = \boldsymbol{\Psi} \otimes \boldsymbol{\Sigma}$), separability with one component additionally structured as CS or AR(1), or BCS; see Sect. 6.3.1 for the details.

6.3 Covariance Structures and Maximum Likelihood Estimators

In this section, we present the covariance structures under consideration and we state formulas for their maximum likelihood estimators. Relevant formulas follow from Filipiak et al. [6, 7] and Roy et al. [25]. It should be noted that, despite the lack of explicit formulas for separable structures (including one component additionally structured), the systems of equations to obtain MLEs are presented in elegant matrix forms. In the earlier literature, corresponding results usually consist of several sums of particular elements of covariance matrices; see, e.g., Dutilleul [3] or Lu and Zimmerman [15] for a separable structure with both components unstructured, Roy and Khattree [21] for a separable structure with one component as CS, Roy and Khattree [22] for a separable structure with one component as AR(1), and Leiva [14] for the BCS structure. Moreover, Roy et al. [24] considered the best unbiased estimators of unknown components of the BCS structure, which are also presented with the use of several sums. Note that the MLEs of components of the BCS structure determined in Roy et al. [25] have explicit matrix form, and it can be shown that they are proportional to the optimal estimators given by Roy et al. [24].

6.3.1 Covariance Structures

6.3.1.1 Separable Structure

We say that a covariance structure is *separable* if it can be expressed as the Kronecker product of two components, that is, $\boldsymbol{\Psi} \otimes \boldsymbol{\Sigma}$. Recall that $\boldsymbol{\Psi}$ represents the covariance matrix of p repeated measurements across time or location on a given characteristic, and it is assumed to be the same for all characteristics, while $\boldsymbol{\Sigma}$ represents the covariance matrix of q characteristics at any given time or location point, and it is assumed to be the same for all time or location points. Thus, both of these matrices must be p.d. The representation of a matrix as a Kronecker product of two matrices is not unique, because for an arbitrary $c > 0$, $\boldsymbol{\Psi} \otimes \boldsymbol{\Sigma} = c\boldsymbol{\Psi} \otimes \frac{1}{c}\boldsymbol{\Sigma}$. This identifiability problem has implications for estimation, and to circumvent this problem Srivastava et al. [28] proposed to fix one of the diagonal entries of one of the two component matrices as an arbitrary positive value, e.g., one. Throughout this work, we set the first diagonal element of $\boldsymbol{\Psi}$ as 1. Hence, the number of unknown parameters to be estimated in the separable structure is $p(p+1)/2 + q(q+1)/2 - 1$. Since in the following we consider further structures on one of the components of separability, say $\boldsymbol{\Psi}$, in the unstructured case it will be additionally subscripted, that is, $\boldsymbol{\Psi}_{\mathrm{UN}}$.

6.3.1.2 Separable Structure with One Component Structured as CS or AR(1)

Without loss of generality we assume that $\boldsymbol{\Psi}$ can be further structured, while $\boldsymbol{\Sigma}$ remains unstructured. Such an assumption seems to be reasonable, as the correlation between repeated measurements often remains the same (CS structure) or decreases in time (AR(1) structure). Nevertheless, if one wants to assume the structure on the covariance of characteristics, it is enough to change the order of the Kronecker product components with respect to the rule

$$\boldsymbol{\Psi} \otimes \boldsymbol{\Sigma} = \boldsymbol{K}_{q,p}(\boldsymbol{\Sigma} \otimes \boldsymbol{\Psi})\boldsymbol{K}_{p,q}$$

with $\boldsymbol{K}_{p,q}$ being the commutation matrix; see, e.g., Magnus and Neudecker [16], Kollo and von Rosen [13].

To satisfy the assumption that the first diagonal entry of $\boldsymbol{\Psi}$ is equal to one, the matrices $\boldsymbol{\Psi}_{\text{CS}}$ and $\boldsymbol{\Psi}_{\text{AR}}$ will be taken as correlation matrices, that is

$$\boldsymbol{\Psi}_{\text{CS}} = (1-\varrho)\boldsymbol{I}_p + \varrho \boldsymbol{J}_p = \begin{pmatrix} 1 & \varrho & \varrho & \cdots & \varrho \\ \varrho & 1 & \varrho & \cdots & \varrho \\ \varrho & \varrho & 1 & \cdots & \varrho \\ \vdots & \vdots & \vdots & \ddots & \vdots \\ \varrho & \varrho & \varrho & \cdots & 1 \end{pmatrix}, \tag{6.5}$$

$$\boldsymbol{\Psi}_{\text{AR}} = \boldsymbol{I}_p + \sum_{i=1}^{p-1} \varrho^i (\boldsymbol{C}_p^i + \boldsymbol{C}_p'^i) = \begin{pmatrix} 1 & \varrho & \varrho^2 & \cdots & \varrho^{p-1} \\ \varrho & 1 & \varrho & \cdots & \varrho^{p-2} \\ \varrho^2 & \varrho & 1 & \cdots & \varrho^{p-3} \\ \vdots & \vdots & \vdots & \ddots & \vdots \\ \varrho^{p-1} & \varrho^{p-2} & \varrho^{p-3} & \cdots & 1 \end{pmatrix}, \tag{6.6}$$

where $\boldsymbol{J}_p = \mathbf{1}_p\mathbf{1}_p'$ and $\boldsymbol{C}_p$ is a $p \times p$ circulant matrix with ones on the supradiagonal and zeros elsewhere. In both cases there is only one unknown correlation parameter; therefore, the number of unknown parameters in both structures, $\boldsymbol{\Psi}_{\text{CS}} \otimes \boldsymbol{\Sigma}$ and $\boldsymbol{\Psi}_{\text{AR}} \otimes \boldsymbol{\Sigma}$, is $q(q+1)/2+1$. It is worth noting that $\boldsymbol{\Psi}_{\text{CS}}$ and $\boldsymbol{\Psi}_{\text{AR}}$ are p.d. iff $-1/(p-1) < \varrho < 1$ and $|\varrho| < 1$, respectively.

For clarity, the vectors of unknown parameters under the considered models can be written, respectively, as

$$\boldsymbol{\theta}_{\text{CS}} = \left(\boldsymbol{\mu}', \varrho, \text{vech}'\,\boldsymbol{\Sigma}\right)', \quad \boldsymbol{\theta}_{\text{AR}} = \left(\boldsymbol{\mu}', \varrho, \text{vech}'\,\boldsymbol{\Sigma}\right)'. \tag{6.7}$$

6.3.1.3 Block Compound Symmetry Structure

Another covariance structure considered by researchers is defined as

$$\boldsymbol{\Gamma} = \boldsymbol{I}_p \otimes (\boldsymbol{\Gamma}_0 - \boldsymbol{\Gamma}_1) + \boldsymbol{J}_p \otimes \boldsymbol{\Gamma}_1 = \begin{pmatrix} \boldsymbol{\Gamma}_0 & \boldsymbol{\Gamma}_1 & \cdots & \boldsymbol{\Gamma}_1 \\ \boldsymbol{\Gamma}_1 & \boldsymbol{\Gamma}_0 & \cdots & \boldsymbol{\Gamma}_1 \\ \vdots & & \ddots & \vdots \\ \boldsymbol{\Gamma}_1 & \boldsymbol{\Gamma}_1 & \cdots & \boldsymbol{\Gamma}_0 \end{pmatrix},$$

where the $q \times q$ matrices $\boldsymbol{\Gamma}_0$ and $\boldsymbol{\Gamma}_1$ represent covariance matrices of q response variables, respectively, at any given time or location point, and between any two time or location points. To ensure the positive definiteness of $\boldsymbol{\Gamma}$, the positive definiteness of $\boldsymbol{\Gamma}_0$ must be assumed, while $\boldsymbol{\Gamma}_1$ should be symmetric and such that $\boldsymbol{\Gamma}$ is p.d.

The BCS structure is also called *exchangeable*, as the columns of the data matrix $\boldsymbol{Y}_i$ can be exchanged without changing the covariance matrix $\boldsymbol{\Gamma}$. To verify positive definiteness and to determine the maximum likelihood estimators of $\boldsymbol{\Gamma}$, it is convenient to work with another parameterization, that is,

$$\boldsymbol{\Gamma} = \boldsymbol{Q}_p \otimes \boldsymbol{\Delta}_1 + \boldsymbol{P}_p \otimes \boldsymbol{\Delta}_2,$$

where $\boldsymbol{\Delta}_1 = \boldsymbol{\Gamma}_0 - \boldsymbol{\Gamma}_1, \boldsymbol{\Delta}_2 = \boldsymbol{\Gamma}_0 + (p-1)\boldsymbol{\Gamma}_1$, matrix $\boldsymbol{P}_p = \frac{1}{p}\boldsymbol{J}_p$ denotes the orthogonal projector onto the column space of vector of ones, $\mathscr{C}(\mathbf{1}_p)$, and $\boldsymbol{Q}_p = \boldsymbol{I}_p - \boldsymbol{P}_p$. To ensure positive definiteness, it is enough now to assume the positive definiteness of $\boldsymbol{\Delta}_1$ and $\boldsymbol{\Delta}_2$. Note that the separable structure $\boldsymbol{\Psi}_{\text{CS}} \otimes \boldsymbol{\Sigma}$ is a special case of BCS, with the additional parameter space restriction $\boldsymbol{\Gamma}_1 = \varrho\boldsymbol{\Gamma}_0$.

For clarity, the vectors of unknown parameters under the considered model can be written as

$$\boldsymbol{\theta}_{\text{BCS}} = \left(\boldsymbol{\mu}', \text{vech}'\,\boldsymbol{\Delta}_1, \text{vech}'\,\boldsymbol{\Delta}_2\right)', \tag{6.8}$$

and the number of unknown parameters in the BCS structure reduces to $q(q+1)$.

6.3.2 Maximum Likelihood Estimators

Let us assume that the log-likelihood function

$$\begin{aligned} \ln L = \ln L(\boldsymbol{\mu}, \boldsymbol{\Omega}; \boldsymbol{Y}) = &- \frac{nqp}{2}\ln(2\pi) - \frac{n}{2}\ln|\boldsymbol{\Omega}| \\ &- \frac{1}{2}\text{Tr}\left(\boldsymbol{\Omega}^{-1}(\boldsymbol{Y} - \mathbf{1}_n\boldsymbol{\mu}')'(\boldsymbol{Y} - \mathbf{1}_n\boldsymbol{\mu}')\right) \end{aligned} \tag{6.9}$$

for model (6.2) is partially differentiable with respect to each coordinate of the unknown parameter vector $\boldsymbol{\theta}$ related to the respective model, that is, $\boldsymbol{\theta}$ as given in (6.4), (6.7) or (6.8). Using the differentiation rules described by Magnus and Neudecker [16], the derivative of $\ln L$ with respect to $\boldsymbol{\mu}$ is

$$\frac{\partial \ln L}{\partial \boldsymbol{\mu}'} = (\mathbf{1}_n'\boldsymbol{Y} - n\boldsymbol{\mu}')(\boldsymbol{\Omega}^{-1} \otimes \boldsymbol{I}_n).$$

Equating this derivative to zero, we obtain the MLE of $\boldsymbol{\mu}$ as

$$\widehat{\boldsymbol{\mu}} = \frac{1}{n}\boldsymbol{Y}'\mathbf{1}_n, \tag{6.10}$$

which is independent of the covariance parameters, thus it will be the same whether $\boldsymbol{\Omega}$ is unstructured or has a structure. On substituting the value of $\widehat{\boldsymbol{\mu}}$ into (6.9), the log-likelihood function becomes

$$\ln L(\boldsymbol{\Omega}; \boldsymbol{Y}) = -\frac{npq}{2}\ln(2\pi) - \frac{n}{2}\ln|\boldsymbol{\Omega}| - \frac{1}{2}\operatorname{Tr}\left(\boldsymbol{\Omega}^{-1}\boldsymbol{S}\right), \tag{6.11}$$

where $\boldsymbol{S} = \boldsymbol{Y}'\boldsymbol{Q}_n\boldsymbol{Y}$; see, e.g., Filipiak et al. [6]. Note that (6.11) is in fact a profile log-likelihood, which does not depend on the true mean. One can also conclude this property from the commutativity of the space of expectation and the covariance structure $\boldsymbol{\Omega} \otimes \boldsymbol{I}_n$, that is, $\boldsymbol{P}_{\mathrm{I}_{qp}\otimes 1_n}(\boldsymbol{\Omega} \otimes \boldsymbol{I}_n) = (\boldsymbol{\Omega} \otimes \boldsymbol{I}_n)\boldsymbol{P}_{\mathrm{I}_{qp}\otimes 1_n}$.

6.3.2.1 MLE of Unstructured $\boldsymbol{\Omega}$

Differentiating (6.11) with respect to $\boldsymbol{\Omega}$ and equating it to zero, the MLE of $\boldsymbol{\Omega}$ is of the form

$$\widehat{\boldsymbol{\Omega}} = \frac{1}{n}\boldsymbol{S}.$$

This estimator is p.d. and thus invertible only if $n > qp$. Moreover, it is known that $\boldsymbol{S} \sim W_{qp}(\boldsymbol{\Omega}, n-1)$; see, e.g., Kollo and von Rosen [13, Cor. 2.4.3.1].

6.3.2.2 MLE of Separable Structure

Considering the log-likelihood function (6.11) with $\boldsymbol{\Omega} = \boldsymbol{\Psi}_{\mathrm{UN}} \otimes \boldsymbol{\Sigma}$, the MLEs of $\boldsymbol{\Psi}_{\mathrm{UN}}$ and $\boldsymbol{\Sigma}$ satisfy the following system of matrix equations

$$\begin{aligned} nq \ \operatorname{vec}\widehat{\boldsymbol{\Psi}}_{\mathrm{UN}} &= (\boldsymbol{I}_{p^2} \otimes \operatorname{vec}'\widehat{\boldsymbol{\Sigma}}^{-1})(\boldsymbol{I}_p \otimes \boldsymbol{K}_{p,q} \otimes \boldsymbol{I}_q)\operatorname{vec}\boldsymbol{S}, \\ np \ \operatorname{vec}\widehat{\boldsymbol{\Sigma}} &= (\operatorname{vec}'\widehat{\boldsymbol{\Psi}}_{\mathrm{UN}}^{-1} \otimes \boldsymbol{I}_{q^2})(\boldsymbol{I}_p \otimes \boldsymbol{K}_{p,q} \otimes \boldsymbol{I}_q)\operatorname{vec}\boldsymbol{S}; \end{aligned}$$

see, e.g., Filipiak et al. [7]. In order to express these estimators in a more efficient way, one can use the block trace and partial trace operators

$$nq\ \widehat{\boldsymbol{\Psi}}_{\mathrm{UN}} = \operatorname{PTr}_q\left[(\boldsymbol{I}_p \otimes \widehat{\boldsymbol{\Sigma}}^{-1})\boldsymbol{S}\right], \tag{6.12a}$$

$$np\ \widehat{\boldsymbol{\Sigma}} = \operatorname{BTr}_q\left[(\widehat{\boldsymbol{\Psi}}_{\mathrm{UN}}^{-1} \otimes \boldsymbol{I}_q)\boldsymbol{S}\right], \tag{6.12b}$$

where the partial trace operator of a $qp \times qp$ block matrix $\boldsymbol{A}$, partitioned into $q \times q$ blocks, replaces every block by its trace, resulting in a $p \times p$ matrix, while the block trace operator sums all diagonal blocks of $\boldsymbol{A}$, resulting in a $q \times q$ matrix; see Filipiak et al. [9]. The above system has no closed form solution and may be solved, for instance, iteratively using the so-called *flip-flop* algorithm, that is, an initial solution is chosen without loss of generality for $\boldsymbol{\Psi}_{\text{UN}}$ (e.g., $\boldsymbol{I}_p$ or MLE of $\boldsymbol{\Psi}$ assuming $\boldsymbol{\Omega} = \boldsymbol{\Psi} \otimes \boldsymbol{I}_q$) which is used in (6.12b) to compute the first estimate of $\boldsymbol{\Sigma}$, and this is afterwards used in (6.12a) to compute the first estimate of $\boldsymbol{\Psi}_{\text{UN}}$. Second estimates are then computed in the same way starting with the first estimate of $\boldsymbol{\Psi}_{\text{UN}}$, and iterations continue until some convergence criterion is fulfilled, for example, the Frobenius distance between successive Kronecker product estimates is smaller than some given threshold. The problems which arise here are the convergence of this algorithm and the uniqueness of its solution. Recently, Soloveychik and Trushin [27] proved that a sufficient condition for the convergence and for the uniqueness almost surely of the solution of the iterative algorithm to the MLE is that the sample size n should be greater than $p/q + q/p + 1$. Nevertheless, a sample size $n > \max\{p, q\}$ is needed so that the estimators $\widehat{\boldsymbol{\Psi}}_{\text{UN}}$ and $\widehat{\boldsymbol{\Sigma}}$ are p.d. and thus invertible.

6.3.2.3 MLE of Separable Structure with First Component as CS

If $\boldsymbol{\Omega} = \boldsymbol{\Psi}_{\text{CS}} \otimes \boldsymbol{\Sigma}$ in (6.11), Filipiak et al. [6, 7] have shown that the MLEs of ϱ and $\boldsymbol{\Sigma}$ can be obtained as a solution of the following system of equations:

$$\begin{aligned}
&(p-1)^2 c\widehat{\varrho}^3 + (p-1)\{pc + (p-1)a - b\}\widehat{\varrho}^2 + (p-1)\{2a - c\}\widehat{\varrho} + (a-b) = 0,\\
&np\widehat{\boldsymbol{\Sigma}} = \text{BTr}_q\left[(\widehat{\boldsymbol{\Psi}}_{\text{CS}}^{-1} \otimes \boldsymbol{I}_q)\boldsymbol{S}\right]
\end{aligned} \tag{6.13}$$

with

$$a = \text{Tr}\left[\left(\boldsymbol{I}_p \otimes \widehat{\boldsymbol{\Sigma}}^{-1}\right)\boldsymbol{S}\right],\quad b = \text{Tr}\left[\left(\boldsymbol{J}_p \otimes \widehat{\boldsymbol{\Sigma}}^{-1}\right)\boldsymbol{S}\right],\quad c = nqp.$$

Observe that this system does not have any closed form solution, and thus again the flip-flop algorithm may be used to find the MLEs of ϱ and $\boldsymbol{\Sigma}$. Obviously, the estimator $\widehat{\boldsymbol{\Psi}}_{\text{CS}}$ is obtained by substituting $\widehat{\varrho}$ into (6.5), i.e., $\widehat{\boldsymbol{\Psi}}_{\text{CS}} = (1-\widehat{\varrho})\boldsymbol{I}_p + \widehat{\varrho}\boldsymbol{J}_p$. It should be noted that the MLEs given as a numerical solution of (6.13) are equivalent to the MLEs given by Roy and Khattree [21, Eqs. (3) and (6)], who pointed out that the estimator of ϱ is always between $-1/(p-1)$ and 1, and thus the resulting estimator $\widehat{\boldsymbol{\Psi}}_{\text{CS}}$ is a p.d. matrix. Moreover, the estimator $\widehat{\boldsymbol{\Sigma}}$ is positive definite, as the block trace operator of a definite matrix preserves positive definiteness; see Filipiak et al. [9, Lemma 2.4]. However, this positive definiteness property is guaranteed only if $n > q$. Moreover, Filipiak et al. [8] gave two alternative methods of determination of MLEs, and one of them does not require the use of an iterative algorithm.

6.3.2.4 MLE of Separable Structure with First Component as AR(1)

If $\boldsymbol{\Omega} = \boldsymbol{\Psi}_{\text{AR}} \otimes \boldsymbol{\Sigma}$ in (6.11), Filipiak et al. [6, 7] have shown that the MLEs of ϱ and $\boldsymbol{\Sigma}$ can be obtained as a solution of the following system of equations:

$$-2n(p-1)q\widehat{\varrho}^3 + b\widehat{\varrho}^2 + 2[n(p-1)q - a - c]\widehat{\varrho} + b = 0,$$
$$np\widehat{\boldsymbol{\Sigma}} = \text{BTr}_q\left[(\widehat{\boldsymbol{\Psi}}_{\text{AR}}^{-1} \otimes \boldsymbol{I}_q)\boldsymbol{S}\right]$$

with

$$a = \text{Tr}\left[(\boldsymbol{G}_1 \otimes \widehat{\boldsymbol{\Sigma}}^{-1})\boldsymbol{S}\right], \quad b = \text{Tr}\left[(\boldsymbol{G}_2 \otimes \widehat{\boldsymbol{\Sigma}}^{-1})\boldsymbol{S}\right], \quad \text{and} \quad c = \text{Tr}\left[(\boldsymbol{I}_p \otimes \widehat{\boldsymbol{\Sigma}}^{-1})\boldsymbol{S}\right],$$

where $\boldsymbol{G}_1 = \text{diag}(0, \mathbf{1}'_{p-2}, 0)$ and $\boldsymbol{G}_2 = \boldsymbol{C}_p + \boldsymbol{C}'_p$. Even for this system the flip-flop algorithm is needed to find the MLEs $\widehat{\varrho}$ and $\widehat{\boldsymbol{\Sigma}}$. Obviously, the estimator $\widehat{\boldsymbol{\Psi}}_{\text{AR}}$ is obtained by substituting $\widehat{\varrho}$ into (6.6). Similarly as in the case of the previous structure, a sample of size $n > q$ is needed to ensure the positive definiteness of $\widehat{\boldsymbol{\Psi}}_{\text{AR}}$ and $\widehat{\boldsymbol{\Sigma}}$.

6.3.2.5 MLE of BCS Structure

Under the exchangeable structure $\boldsymbol{\Omega} = \boldsymbol{\Gamma}$ in (6.11), Roy et al. [25] have shown that the MLEs of $\boldsymbol{\Gamma}_0$ and $\boldsymbol{\Gamma}_1$ are

$$\widehat{\boldsymbol{\Gamma}}_0 = \frac{1}{p}(\widehat{\boldsymbol{\Delta}}_2 + (p-1)\widehat{\boldsymbol{\Delta}}_1), \quad \widehat{\boldsymbol{\Gamma}}_1 = \frac{1}{p}(\widehat{\boldsymbol{\Delta}}_2 - \widehat{\boldsymbol{\Delta}}_1),$$

where

$$\widehat{\boldsymbol{\Delta}}_1 = \frac{1}{n(p-1)}\text{BTr}_q[(\boldsymbol{Q}_p \otimes \boldsymbol{I}_q)\boldsymbol{S}],$$
$$\widehat{\boldsymbol{\Delta}}_2 = \frac{1}{n}\text{BTr}_q[(\boldsymbol{P}_p \otimes \boldsymbol{I}_q)\boldsymbol{S}].$$

Observe that in this case the estimators are given in an explicit form. Moreover, Roy et al. [23] have shown that

$$\widehat{\boldsymbol{\Delta}}_1 \sim W_q\left(\frac{1}{n(p-1)}\boldsymbol{\Delta}_1, (n-1)(p-1)\right),$$
$$\widehat{\boldsymbol{\Delta}}_2 \sim W_q\left(\frac{1}{n}\boldsymbol{\Delta}_2, n-1\right),$$

which are independent. A sample size $n > q$ is needed for the positive definiteness of $\widehat{\boldsymbol{\Delta}}_1$ and $\widehat{\boldsymbol{\Delta}}_2$.

6.4 Hypotheses and Tests

The hypothesis concerning the structure of the covariance matrix $\boldsymbol{\Omega}$ discussed in this paper is

$$H_0 : \boldsymbol{\Omega} = \boldsymbol{\Omega}_0 \quad \text{versus} \quad H_A : \boldsymbol{\Omega} \text{ unstructured}, \tag{6.14}$$

where $\boldsymbol{\Omega}_0$ is either $\boldsymbol{\Psi}_{\text{UN}} \otimes \boldsymbol{\Sigma}$, $\boldsymbol{\Psi}_{\text{CS}} \otimes \boldsymbol{\Sigma}$, $\boldsymbol{\Psi}_{\text{AR}} \otimes \boldsymbol{\Sigma}$ or $\boldsymbol{\Gamma}$, while $\boldsymbol{\Sigma}$ is assumed to be unstructured in all cases. To test the above hypothesis, we use both LRT and RST, and we compare the behavior of both test statistics. It is known (see, e.g., Rao [19]) that under H_0 both of the tests have asymptotically χ^2 distribution with ν degrees of freedom, where ν is the difference between the number of unknown parameters under the alternative and null hypotheses. Since the exact distributions of the LR and RS test statistics are unknown, with the use of simulation studies, we compare the speed of convergence of the empirical null distribution to the limiting χ^2, the Type I error behavior, as well as the power of both tests.

We show that the LRT statistic can be expressed in terms of the determinant of $\widehat{\boldsymbol{\Omega}}_0^{-1}\widehat{\boldsymbol{\Omega}}$, which is the product of respective eigenvalues, while the RST can be expressed in terms of the trace of $\widehat{\boldsymbol{\Omega}}_0^{-1}\widehat{\boldsymbol{\Omega}}$, that is, the sum of relevant eigenvalues.

6.4.1 Likelihood Ratio Test

The LRT is based on comparison of the maximum likelihoods under the null and alternative hypotheses, i.e.

$$\Lambda = \frac{\max_{H_0} L}{\max_{H_A} L}.$$

This likelihood ratio criterion for testing separability with $\boldsymbol{\Psi}$ either unstructured or assuming a CS or AR(1) structure is given as

$$\Lambda = \left(\frac{|\widehat{\boldsymbol{\Omega}}|}{|\widehat{\boldsymbol{\Psi}}|^q |\widehat{\boldsymbol{\Sigma}}|^p} \right)^{n/2},$$

where $\widehat{\boldsymbol{\Psi}}$ and $\widehat{\boldsymbol{\Sigma}}$ are the MLEs under the respective structures, i.e., $\widehat{\boldsymbol{\Psi}}$ is equal to either $\widehat{\boldsymbol{\Psi}}_{\text{UN}}$, $\widehat{\boldsymbol{\Psi}}_{\text{CS}}$ or $\widehat{\boldsymbol{\Psi}}_{\text{AR}}$. For testing the hypothesis concerning the BCS structure, the criterion Λ has the form

$$\Lambda = \left(\frac{|\widehat{\boldsymbol{\Omega}}|}{|\widehat{\boldsymbol{\Delta}}_1|^{p-1} |\widehat{\boldsymbol{\Delta}}_2|} \right)^{n/2}. \tag{6.15}$$

The exact distribution of Λ is usually unknown, so in practice the LRT statistic,

Table 6.1 Number of degrees of freedom ν of limiting χ^2_ν distribution of the LRT for testing (6.14)

$\boldsymbol{\Omega}_0$	ν
$\boldsymbol{\Psi}_{\text{UN}} \otimes \boldsymbol{\Sigma}$	$\frac{qp(qp+1)}{2} - \frac{p(p+1)+q(q+1)}{2} + 1$
$\boldsymbol{\Psi}_{\text{CS}} \otimes \boldsymbol{\Sigma}$, $\boldsymbol{\Psi}_{\text{AR}} \otimes \boldsymbol{\Sigma}$	$\frac{qp(qp+1)}{2} - \frac{q(q+1)}{2} - 1$
$\boldsymbol{\Gamma}$	$\frac{qp(qp+1)}{2} - q(q+1)$

Table 6.2 The LRT statistics with respect to the null hypothesis $H_0 : \boldsymbol{\Omega} = \boldsymbol{\Omega}_0$

$\boldsymbol{\Omega}_0$	LR
$\boldsymbol{\Psi}_{\text{UN}} \otimes \boldsymbol{\Sigma}$	$nq \ln \lvert\widehat{\boldsymbol{\Psi}}_{\text{UN}}\rvert + np \ln \lvert\widehat{\boldsymbol{\Sigma}}\rvert - n \ln \lvert\widehat{\boldsymbol{\Omega}}\rvert$
$\boldsymbol{\Psi}_{\text{CS}} \otimes \boldsymbol{\Sigma}$	$nq(p-1)\ln(1-\widehat{\varrho}) + nq\ln(1+(p-1)\widehat{\varrho}) +$ $np \ln \lvert\widehat{\boldsymbol{\Sigma}}\rvert - n \ln \lvert\widehat{\boldsymbol{\Omega}}\rvert$
$\boldsymbol{\Psi}_{\text{AR}} \otimes \boldsymbol{\Sigma}$	$nq(p-1)\ln(1-\widehat{\varrho}^2) + np \ln \lvert\widehat{\boldsymbol{\Sigma}}\rvert - n \ln \lvert\widehat{\boldsymbol{\Omega}}\rvert$
$\boldsymbol{\Gamma}$	$n(p-1)\ln \lvert\widehat{\boldsymbol{\Delta}}_1\rvert + n \ln \lvert\widehat{\boldsymbol{\Delta}}_2\rvert - n \ln \lvert\widehat{\boldsymbol{\Omega}}\rvert$

$$\text{LR} = -2 \ln \Lambda$$

is used. It is well known [19] that, under a normality assumption, LR is approximately distributed as χ^2_ν under H_0. The number of degrees of freedom ν for the relevant hypothesis is presented in Table 6.1. It is to be noted that if any of the covariance parameters falls on the boundary of their parameter space, then the asymptotic distribution of LR becomes a mixture of χ^2 distributions, as discussed in Self and Liang [26].

It is easy to see that in both cases, with separability and with BCS, the test statistic $\text{LR} = -2 \ln \Lambda$ can be written as

$$\text{LR} = -n \; \ln |\widehat{\boldsymbol{\Omega}}_0^{-1} \widehat{\boldsymbol{\Omega}}|,$$

which can be simplified to the formulas presented in Table 6.2.

Observe that if we substitute in the above formulas $\widehat{\boldsymbol{\Omega}} = n^{-1}\boldsymbol{S}$, the expression $\ln |\widehat{\boldsymbol{\Omega}}|$ is replaced by $\ln |\boldsymbol{S}| - qp \ln n$. Thus, to be able to compute the LR test statistic a sample size greater than qp is required, otherwise $\boldsymbol{S} = n\widehat{\boldsymbol{\Omega}}$ will be singular.

6.4.2 Rao Score Test

The RST is based upon the Fisher information matrix, which is defined as

$$\boldsymbol{F}(\boldsymbol{\theta}) = -\text{E}\left(\frac{\partial \boldsymbol{s}(\boldsymbol{\theta})}{\partial \boldsymbol{\theta}'}\right) \stackrel{df}{=} \begin{pmatrix} \boldsymbol{F}_{11} & \boldsymbol{F}'_{21} \\ \boldsymbol{F}_{21} & \boldsymbol{F}_{22} \end{pmatrix},$$

where $s(\theta) = (s_1(\theta)', s_2(\theta)')' = \left(\frac{\partial \ln L}{\partial \mu'}, \frac{\partial \ln L}{\partial \operatorname{vech}' \Omega}\right)'$ is the score vector, and F_{11}, F_{21}, and F_{22} are $qp \times qp$, $qp(qp+1)/2 \times qp$, and $qp(qp+1)/2 \times qp(qp+1)/2$ matrices, respectively. The Rao score (RS) assumes that the Fisher information matrix exists and is invertible, and is defined as (see, e.g., Rao [19])

$$\mathrm{RS} = s(\widehat{\theta}_0)' F^{-1}(\widehat{\theta}_0) s(\widehat{\theta}_0),$$

where $\widehat{\theta}_0$ is the MLE of θ_0 under the null hypothesis, i.e., $\widehat{\theta}_0$ is equal to $\widehat{\theta}_{\text{sep}}$, $\widehat{\theta}_{\text{CS}}$, $\widehat{\theta}_{\text{AR}}$ or $\widehat{\theta}_{\text{BCS}}$, depending on the tested structure.

To compute the score vector and Fisher information matrix, the log-likelihood function (6.9) must be differentiated twice with respect to the vector of unknown parameters under the most general model, $\theta = \theta_\Omega$. We obtain the following score vector $s(\theta) = (s_1'(\theta), s_2'(\theta))'$, where

$$s_1(\theta) = (\Omega^{-1} \otimes I_n)(Y'\mathbf{1}_n - n\mu),$$
$$s_2(\theta) = D_{qp}' \left(-\frac{n}{2} \operatorname{vec} \Omega^{-1} + \frac{1}{2}(\Omega^{-1} \otimes \Omega^{-1}) \operatorname{vec}(Y - \mathbf{1}_n\mu')'(Y - \mathbf{1}_n\mu')\right),$$

with D_{qp} being a $qp(qp+1)/2 \times q^2p^2$ duplication matrix; see, e.g., Magnus and Neudecker [16]. Taking the mean value of the derivative of the score vector with respect to $\theta = \theta_\Omega$, the blocks of $F(\theta)$ are

$$F_{11} = n\Omega^{-1}, \qquad F_{21} = \mathbf{0}, \qquad F_{22} = \frac{n}{2} D_{qp}'(\Omega^{-1} \otimes \Omega^{-1}) D_{qp}.$$

Since we are interested in testing only the structure of the covariance matrix, and since $F_{21} = \mathbf{0}$, the RST statistic will be

$$\begin{aligned}\mathrm{RS} = s_2(\widehat{\theta}_0)' F_{22}^{-1}(\widehat{\theta}_0) s_2(\widehat{\theta}_0) s_2(\widehat{\theta}_0) &= \frac{nqp}{2} - \operatorname{Tr}(\widehat{\Omega}_0^{-1} S) + \frac{1}{2n} \operatorname{Tr}(\widehat{\Omega}_0^{-1} S \widehat{\Omega}_0^{-1} S) \\ &= \frac{n}{2} \operatorname{Tr}\left[(\widehat{\Omega}_0^{-1}\widehat{\Omega} - I_{qp})^2\right].\end{aligned}$$

Note that if S is singular, RS can still be computed, and thus the RST can be used even when the sample size is smaller than qp, which is not possible for the LRT. Nevertheless, since the inverse of $\widehat{\Omega}_0^{-1}$ is required, the minimum sample size must be at least $\max\{p, q\} + 1$ when the structure $\Psi_{\text{UN}} \otimes \Sigma$ is tested, and $q + 1$ when the remaining structures are considered.

Observe that since for all considered structures $\operatorname{Tr}(\widehat{\Omega}_0^{-1} S) = nqp$, RS could be written alternatively as

$$\mathrm{RS} = \frac{n}{2} \operatorname{Tr}\left[(\widehat{\Omega}_0^{-1}\widehat{\Omega})^2\right] - \frac{nqp}{2}.$$

6.5 Comparison of Tests

In this section, we use simulation studies to compare the behavior of the LRT and RST statistics' distributions, Type I error and the power of both tests for increasing values of n, p, and q. However, first, we show algebraically that for most of the hypotheses under consideration, the distribution of both statistics does not depend on the true mean and covariance matrix, which makes it possible to fix the true parameters arbitrarily, for example, as a vector of zeros and the identity matrix.

6.5.1 Distribution of Test Statistics Versus True Values of Parameters

It is shown in (6.10) and (6.11) that the MLE of $\boldsymbol{\mu}$ does not depend on the true covariance matrix, and the MLE of the covariance structure does not depend on the true mean. This implies that the LRT and RST statistics distribution, for testing hypothesis (6.14), does not depend on the true mean. Moreover, Lu and Zimmerman [15] showed that the LRT statistic for testing a hypothesis with $\boldsymbol{\Omega}_0 = \boldsymbol{\Psi}_{\mathrm{UN}} \otimes \boldsymbol{\Sigma}$ is invariant with respect to nonsingular linear transformations of the observations, which means that the distribution of the LRT statistic under the null hypothesis of separability with both components unstructured does not depend on the true value of the covariance matrix. An alternative proof of this fact was given by Mitchell et al. [18]. The next two theorems extend this result for all the structures of interest and for the RST statistic. Proofs can be found in Filipiak et al. [7, Th. 3.2] for separable structures, and partially in Roy et al. [25, Th. 3.3] for the BCS structure and the RST. The proof that the distribution of the LRT statistic is not dependent on the true values of unknown parameters will be presented below.

Theorem 6.1 *The distribution of the LRT and RST statistics under the null hypothesis of separability with the first component as UN, CS or AR(1) does not depend on the true values of the unknown parameters $\boldsymbol{\mu}$ or $\boldsymbol{\Sigma}$. Moreover, if $\boldsymbol{\Psi}$ is UN or CS, the distributions of both test statistics do not depend on the true value of $\boldsymbol{\Psi}$.*

Theorem 6.2 *The distribution of the LRT and RST statistics under the null hypothesis of BCS does not depend on the true values of the unknown parameters $\boldsymbol{\mu}$, $\boldsymbol{\Gamma}_0$ or $\boldsymbol{\Gamma}_1$.*

Proof The proof of independence of the distribution of the LRT proceeds in a similar way as that for the RST in Roy et al. [25, Th. 3.3]. Since under the null hypothesis, the observation matrix could be written as $\boldsymbol{Y} = \boldsymbol{E}\boldsymbol{\Gamma}^{1/2}$, where $\boldsymbol{E} \sim N_{n,qp}(\boldsymbol{0}, \boldsymbol{I}_n, \boldsymbol{I}_{qp})$, it is easy to see that

$$\widehat{\boldsymbol{\Omega}} = n^{-1}\boldsymbol{\Gamma}^{1/2}\boldsymbol{E}'\boldsymbol{Q}_n\boldsymbol{E}\boldsymbol{\Gamma}^{1/2} \stackrel{df}{=} \boldsymbol{\Gamma}^{1/2}\widehat{\boldsymbol{\Omega}}_{\mathrm{E}}\boldsymbol{\Gamma}^{1/2},$$

$$\widehat{\boldsymbol{\Delta}}_1 = \boldsymbol{\Delta}_1^{1/2}\widehat{\boldsymbol{\Upsilon}}_1\boldsymbol{\Delta}_1^{1/2}, \quad \text{and} \quad \widehat{\boldsymbol{\Delta}}_2 = \boldsymbol{\Delta}_2^{1/2}\widehat{\boldsymbol{\Upsilon}}_2\boldsymbol{\Delta}_2^{1/2},$$

where none of $\widehat{\boldsymbol{\Omega}}_{\mathrm{E}}$, $\widehat{\boldsymbol{\Upsilon}}_1$ or $\widehat{\boldsymbol{\Upsilon}}_2$ depends on the true values of the unknown parameters. Then for the LRT statistic (6.15) we have

$$\Lambda = \left(\frac{|\boldsymbol{\Gamma}^{1/2}\widehat{\boldsymbol{\Omega}}_{\mathrm{E}}\boldsymbol{\Gamma}^{1/2}|}{|\boldsymbol{\Delta}_1^{1/2}\widehat{\boldsymbol{\Upsilon}}_1\boldsymbol{\Delta}_1^{1/2}|^{p-1}|\boldsymbol{\Delta}_2^{1/2}\widehat{\boldsymbol{\Upsilon}}_2\boldsymbol{\Delta}_2^{1/2}|}\right)^{n/2} = \left(\frac{|\widehat{\boldsymbol{\Omega}}_{\mathrm{E}}|}{|\widehat{\boldsymbol{\Upsilon}}_1|^{p-1}|\widehat{\boldsymbol{\Upsilon}}_2|}\right)^{n/2},$$

which is independent of the unknown parameters. □

6.5.2 Simulation Studies

Extensive simulation studies were performed by Filipiak et al. [6, 7] and Roy et al. [25] to compare the behavior of both LRT and RST statistics for testing four considered structures. Samples of various sizes were generated from a qp-variate normal population $N_{qp}(\mathbf{0}, \boldsymbol{\Omega}_0)$, where $\boldsymbol{\Omega}_0$ is one of $\boldsymbol{\Psi}_{\mathrm{UN}} \otimes \boldsymbol{\Sigma}$, $\boldsymbol{\Psi}_{\mathrm{CS}} \otimes \boldsymbol{\Sigma}$, $\boldsymbol{\Psi}_{\mathrm{AR}} \otimes \boldsymbol{\Sigma}$ or $\boldsymbol{\Gamma}$. The number of generated samples under H_0 for various combinations of parameters p and q was 50,000. Note that in Filipiak et al. [7], *Mathematica* code with the algorithms for the MLEs (as a solution of a system of matrix equations) and for values of both test statistics is provided as supplementary material.

The exact distributions of the LRT and RST statistics are usually unknown. Nevertheless, the empirical null distribution (END) can be computed with the use of Monte Carlo simulations (Rizzo [20]). Similarly as in Filipiak et al. [6, 7] and Roy et al. [25], we present here a comparison of the LRT and RST statistics with respect to the speed of convergence of the END to the limiting χ^2 distribution and the empirical type I error. Moreover, since for a BCS structure the powers of both tests were also compared in Roy et al. [25], we perform similar power simulations for all considered structures in order to present a complete comparison. We shall note that most of the published results were obtained for the nominal significance level 0.01 and for various combinations of experiment parameters. In this paper, the results of simulation studies for parameters $p = q = 3$ and $\varrho = 0.5$ (in the case of $\boldsymbol{\Psi}_{\mathrm{CS}}$ and $\boldsymbol{\Psi}_{\mathrm{AR}}$) are presented to simplify, unify, and clarify the comparison of the tests. Moreover, the nominal significance level is chosen to be 0.05.

Note that the ENDs of the LRT and RST statistics in comparison with the respective limiting χ^2 distributions presented in Figs. 6.2 and 6.3 are based on simulations performed in Filipiak et al. [7, Table A.1] and Filipiak et al. [6, Table 6], while the histograms presented in Fig. 6.4 are reprints of histograms given partially in Filipiak et al. [7, Figs. 1 and 2, second column].

From a practical point of view, one should ask whether and when the quantiles of the limiting distribution can be used to draw conclusions about an experiment, instead of the quantiles of the exact distribution. Thus, in Fig. 6.2, 6.3, 6.4, and 6.5, we present a comparison of the ENDs of the test statistics obtained by simulations

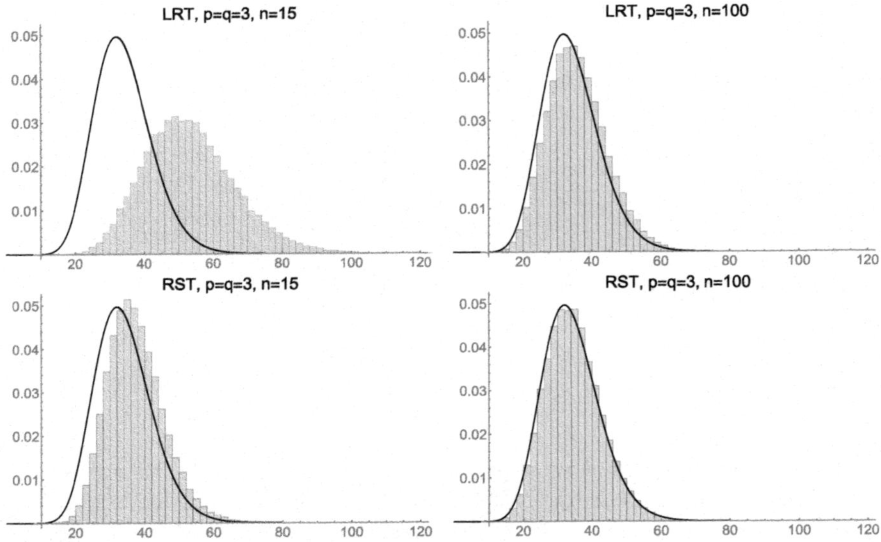

Fig. 6.2 Empirical histogram and the limiting χ^2_{34} distribution for the LRT and RST statistics for testing $\boldsymbol{\Psi}_{\text{UN}} \otimes \boldsymbol{\Sigma}$ with $p = q = 3$ and $n = 15$ and 100 based on 50,000 trials

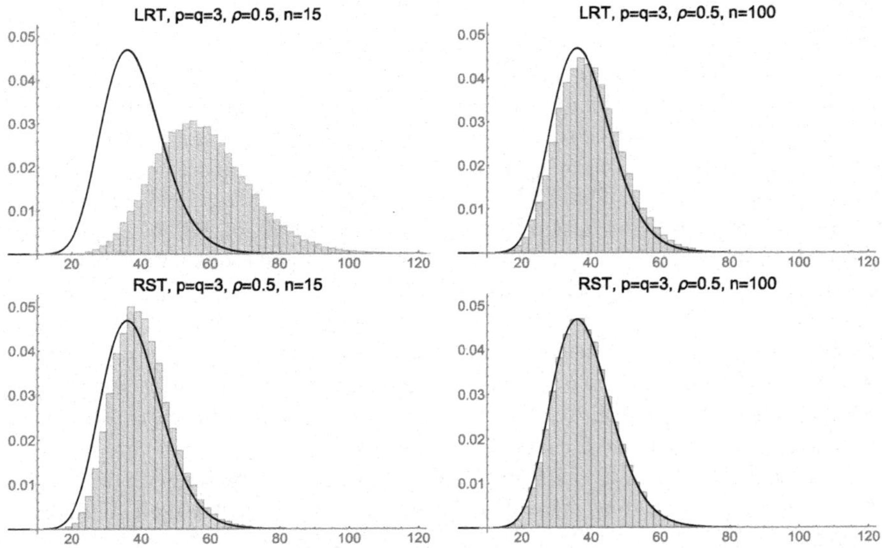

Fig. 6.3 Empirical histogram and the limiting χ^2_{38} distribution for the LRT and RST statistics for testing $\boldsymbol{\Psi}_{\text{CS}} \otimes \boldsymbol{\Sigma}$ with $p = q = 3$, $\varrho = 0.5$ and $n = 15$ and 100 based on 50,000 trials

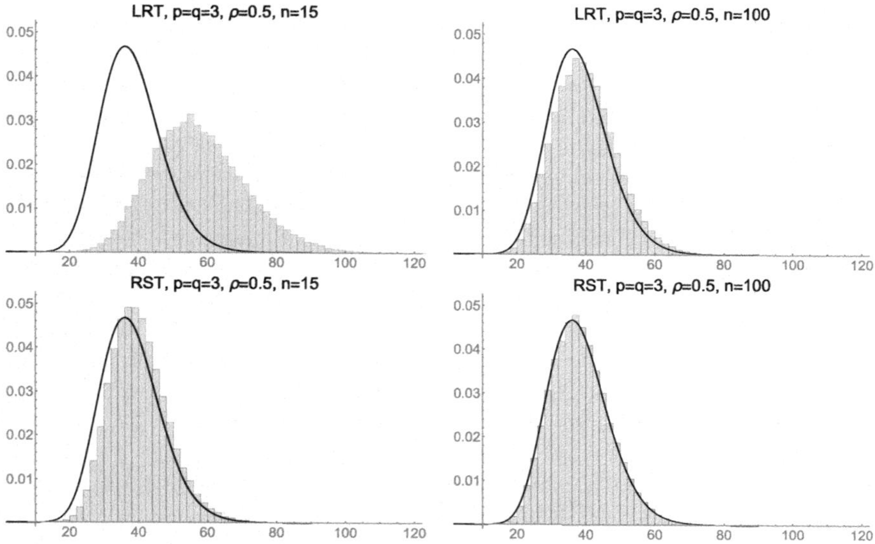

Fig. 6.4 Empirical histogram and the limiting χ^2_{38} distribution for the LRT and RST statistics for testing $\mathbf{\Psi}_{\mathrm{AR}} \otimes \mathbf{\Sigma}$ with $p = q = 3, \varrho = 0.5$ and $n = 15$ and 100 based on 50,000 trials

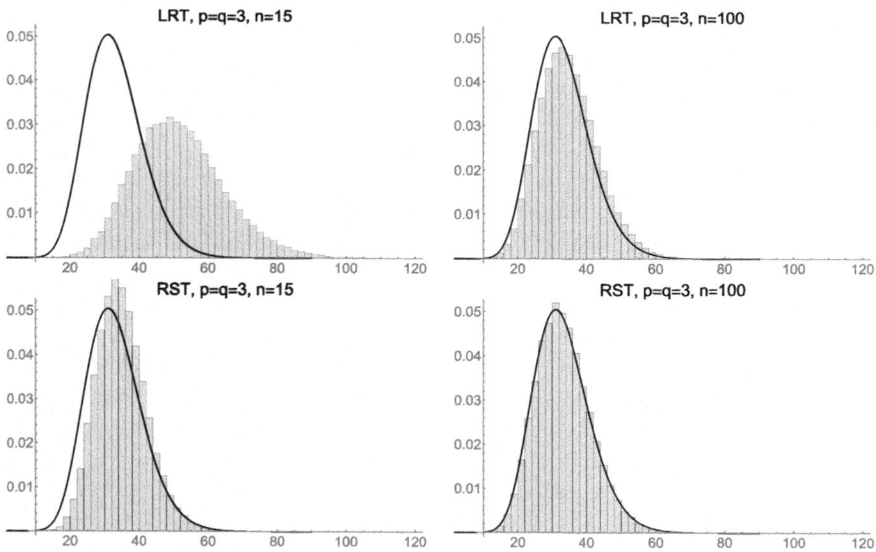

Fig. 6.5 Empirical histogram and the limiting χ^2_{33} distribution for the LRT and RST statistics for testing a BCS structure with $p = q = 3$ and $n = 15$ and 100 based on 50,000 trials

and the limiting χ^2 distribution for small and large sample size, respectively, for the LRT in the first row and for the RST in the second row of each figure. It can be seen that the RST outperforms the LRT in this context, especially for small sample size. Moreover, even for small sample size the END of the RST is so close to the limiting χ^2 distribution that the decision made with the use of critical values of this χ^2 distribution should be the same as with the use of critical values of the END, which is not necessarily true for the LRT; see e.g., the rhesus monkey data example, Table 6.4 in Sect. 6.6. Tables with 90th, 95th, and 99th quantiles of the ENDs for various combinations of experiment parameters can be found in Filipiak et al. [6, Tables 4–6, B1–B6] for testing of the $\boldsymbol{\Psi}_{\text{CS}} \otimes \boldsymbol{\Sigma}$ structure, Filipiak et al. [7, Tables A1–A4 and S3–S5 of Supplementary Material] for testing of the $\boldsymbol{\Psi}_{\text{UN}} \otimes \boldsymbol{\Sigma}$ and $\boldsymbol{\Psi}_{\text{AR}} \otimes \boldsymbol{\Sigma}$ structure, and Roy et al. [25, Tables A1–A4] for testing of the BCS structure. From all these results, we can conclude that for small and moderate sample sizes the END of the RST statistic converges to the limiting χ^2 distribution much faster than the corresponding END of the LRT statistic.

To compare the speed of convergence of Type I error to the nominal significance level when the limiting χ^2 distribution is used, Filipiak et al. [6, 7] and Roy et al. [25] studied the empirical Type I error rates of both test statistics. The results for nominal significance level 0.01 and various combinations of experiment parameters, as well as various values of ϱ in the case of hypotheses concerning $\boldsymbol{\Psi}_{\text{CS}} \otimes \boldsymbol{\Sigma}$ and $\boldsymbol{\Psi}_{\text{AR}} \otimes \boldsymbol{\Sigma}$, appear in the cited papers; see Filipiak et al. [6, Tables 1 and 2], Supplementary Material of Filipiak et al. [7, Tables S1 and S2] and Roy et al. [25, Tables A1–A4].

In Fig. 6.6, in this paper, we present the speed of convergence of the Type I error to the nominal significance level 0.05 for $q = p = 3$. Note that the part of Fig. 6.6 presenting the $\boldsymbol{\Psi}_{\text{CS}} \otimes \boldsymbol{\Sigma}$ structure is based on the empirical Type I error rates presented in Filipiak et al. [6, Table 2].

As expected, the empirical Type I error rate converges to the nominal significance level as the sample size increases. However, the RST statistic performs much better for small and moderate sample size than its counterpart, the LRT statistic, as the achieved significance level is very close to the nominal one even for relatively small sample size. It was also observed in Filipiak et al. [6, Tables 1 and 2] that, for a given sample size, the Type I error rates increase with p for both test statistics.

Finally, let us compare the power of the LRT and RST statistics. Thus, for each structure we generate the data under two alternative hypotheses and verify the rejection rates of the null hypothesis. For that purpose, one can use the quantiles of the limiting χ^2 distribution or the END, as the exact distribution is unknown. Since the simulation study shows differences between the convergence of the END of the LRT and RST statistics to χ^2, especially for small sample size, to be as objective as possible, for power comparison we choose the empirical quantiles.

As the alternative hypotheses, we choose $\boldsymbol{\Omega} = \boldsymbol{\Omega}_0 + \xi \boldsymbol{V}$, where $\boldsymbol{V}$ is a randomly generated positive definite matrix of appropriate size and ξ is a scale parameter (such that $\boldsymbol{\Omega}$ is positive definite), describing the discrepancy between the null and alternative hypotheses. In our considerations, ξ takes the values 0.2 and 1. The results of the power simulations are presented in Table 6.3. Recall that the LRT cannot be performed if the sample size is too small, that is, if n does not exceed qp.

Table 6.3 Powers of the LRT and RST under the nominal significance level $\alpha = 0.05$

	$\xi = 0.2$		$\xi = 1$		$\xi = 0.2$		$\xi = 1$	
n	LRT	RST	LRT	RST	LRT	RST	LRT	RST
	$\Psi_{UN} \otimes \Sigma$				$\Psi_{CS} \otimes \Sigma$			
	$p = 3, q = 2$							
6	–	0.062	–	0.062	–	0.069	–	0.070
10	0.073	0.081	0.075	0.082	0.077	0.087	0.078	0.089
15	0.104	0.108	0.107	0.112	0.113	0.120	0.117	0.124
25	0.178	0.178	0.187	0.186	0.202	0.188	0.214	0.196
50	0.402	0.392	0.425	0.413	0.481	0.440	0.508	0.463
100	0.799	0.785	0.825	0.810	0.884	0.852	0.904	0.874
150	0.953	0.948	0.964	0.960	0.986	0.979	0.991	0.985
200	0.992	0.991	0.995	0.994	0.999	0.998	0.999	0.999
	$p = 3, q = 3$							
6	–	0.117	–	0.120	–	0.127	–	0.131
10	0.095	0.218	0.099	0.226	0.096	0.239	0.100	0.249
15	0.370	0.383	0.398	0.400	0.368	0.412	0.396	0.431
25	0.812	0.728	0.846	0.753	0.812	0.745	0.845	0.770
50	0.999	0.995	1.000	0.997	0.999	0.996	1.000	0.997
100	1.000	1.000	1.000	1.000	1.000	1.000	1.000	1.000
	$p = 5, q = 2$							
6	–	0.108	–	0.111	–	0.191	–	0.196
10	–	0.243	–	0.253	–	0.349	–	0.359
15	0.349	0.448	0.372	0.469	0.441	0.563	0.466	0.580
25	0.826	0.806	0.854	0.828	0.909	0.881	0.927	0.895
50	1.000	0.998	1.000	0.999	1.000	1.000	1.000	1.000
	$\Psi_{AR} \otimes \Sigma$				Γ			
	$p = 3, q = 2$							
6	–	0.067	–	0.067	–	0.059	–	0.060
10	0.077	0.086	0.079	0.087	0.066	0.072	0.068	0.075
15	0.114	0.120	0.118	0.125	0.085	0.091	0.092	0.097
25	0.202	0.192	0.214	0.200	0.128	0.125	0.147	0.139
50	0.481	0.442	0.509	0.466	0.275	0.254	0.331	0.296
100	0.883	0.853	0.904	0.876	0.611	0.563	0.706	0.653
150	0.986	0.979	0.991	0.986	0.840	0.809	0.911	0.887
200	0.999	0.998	0.999	0.999	0.947	0.933	0.979	0.971
	$p = 3, q = 3$							
6	–	0.126	–	0.130	–	0.111	–	0.123
10	0.095	0.231	0.099	0.241	0.081	0.201	0.091	0.236
15	0.357	0.399	0.385	0.418	0.258	0.349	0.332	0.416
25	0.800	0.725	0.834	0.752	0.630	0.642	0.768	0.743
50	0.999	0.994	1.000	0.996	0.983	0.976	0.998	0.994
100	1.000	1.000	1.000	1.000	1.000	1.000	1.000	1.000
	$p = 5, q = 2$							
6	–	0.181	–	0.186	–	0.179	–	0.196
10	–	0.325	–	0.337	–	0.322	–	0.357
15	0.426	0.532	0.450	0.550	0.373	0.517	0.443	0.573
25	0.899	0.861	0.919	0.877	0.845	0.837	0.911	0.887
50	1.000	0.999	1.000	1.000	1.000	0.999	1.000	1.000

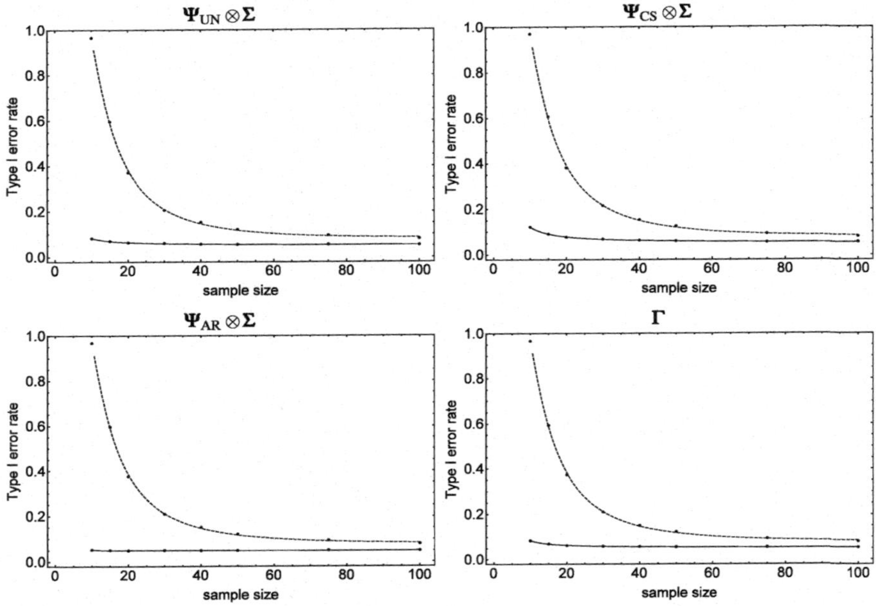

Fig. 6.6 Plots of the Type I error rate as a function of the sample size for the LRT and RST statistics for all hypotheses under consideration with $p = q = 3, \varrho = 0.5$ (in Ψ_{CS} and Ψ_{AR}) and for nominal $\alpha = 0.05$. Plot lines: Dashed — LRT; Solid — RST

As expected, for both tests, the power increases with the sample size and the scale parameter. Note, however, that neither of the tests is uniformly more powerful. In all cases, the power of the RST exceeds that of the LRT for small and moderate sample sizes, while the situation is reversed for higher values of sample size (in our study when $n \geq 25$). Nevertheless, the differences in powers are not as significant as in the case of small samples.

6.6 Real Data Applications

To illustrate the presented methods, we test the hypothesis (6.14) on two real data sets concerning rhesus monkey diet (McKiernan et al. [17]) and a study on epilepsy prevention (Sperling et al. [29]). In the first example, the sample size (number of rhesus monkeys) is large enough to perform both tests: LRT and RST. Therefore, we compare the decisions made from the empirical quantiles of both tests, and in addition, we present a comparison for the inference with the decision based on the quantiles of the limiting χ^2 distribution. Note that, in the second example, the sample size is not large enough to perform the LRT. Thus, only the decision based on the

Table 6.4 Calculated values of the LRT and RST statistics along with the p-values of the limiting χ^2_ν distribution (with degrees of freedom ν) and END for all considered null hypotheses H_0 for the rhesus monkey data set

		$\mathbf{\Psi}_{\text{UN}} \otimes \mathbf{\Sigma}$	$\mathbf{\Psi}_{\text{CS}} \otimes \mathbf{\Sigma}$	$\mathbf{\Psi}_{\text{AR}} \otimes \mathbf{\Sigma}$	$\mathbf{\Gamma}$
LRT	LR	86.71	118.61	106.98	106.55
	p-val: χ^2_ν (ν)	0 (34)	0 (38)	0 (38)	0 (33)
	p-val: END	0.098	0.007	0.023	0.012
RST	RS	49.89	61.39	49.86	51.26
	p-val: χ^2_ν (ν)	0.039 (34)	0.010 (38)	0.094 (38)	0.022 (33)
	p-val: END	0.084	0.017	0.138	0.021

empirical quantiles of the RST is compared with the decision based on the quantiles of the limiting distribution.

Example 6.1 The aim of the study described in McKiernan et al. [17] was to analyze early stage sarcopenia in aging rhesus monkeys, as rhesus monkeys closely model human aging. Thus, the authors collected multivariate longitudinal data over 12 years, where three variables — body weight, fat mass, and estimated upper leg lean muscle mass — were measured in 12 adult male rhesus monkeys from 16 to 22 years of age repeatedly after the 6th, 9th, and 12th year of research. Clearly, in this dataset, $q = 3$ characteristics were measured at $p = 3$ time points for $n = 12$ individuals. The data, originally presented by McKiernan et al. [17, Table 1] and reproduced in Filipiak et al. [7, Table 1], are described in detail in McKiernan et al. [17].

Since, in this data set, there are two sources of variability-characteristics, and time-it is reasonable to assume a separable covariance structure. Moreover, since one of the sources is time, it is also reasonable to assume that the correlation between successive observations decays as the time points become more widely separated; thus the separable structure $\mathbf{\Psi}_{\text{AR}} \otimes \mathbf{\Sigma}$ may be expected.

Nevertheless, we test all the structures discussed in this paper and verify the decisions. The results are presented in Table 6.4 and include:

- first line — values of test statistics for relevant null hypothesis (respectively LR and RS in the first and second part of the table);
- second line — p-values of the test statistics (the cumulative probability beyond the test statistic) of χ^2_ν distribution with respective degrees of freedom ν in parentheses;
- third line — p-values of the test statistics based on the respective END.

Note that the values of LR and RS, as well as the χ^2 p-values, for the first three columns (i.e., for testing the separability with $\mathbf{\Psi}$ either unstructured or structured) can be found in Filipiak et al. [7, Table 4].

Using the limiting χ^2 distribution, the p-values in the case of the LRT statistics are all almost zero, which means that we have to reject H_0 at every usual significance level α. However, using the END, we observe that separability is rejected at $\alpha = 0.1$, while it is not rejected at $\alpha = 0.05$. The structures $\mathbf{\Psi}_{\text{AR}} \otimes \mathbf{\Sigma}$ and BCS are rejected

Table 6.5 Decisions on H_0 under two significance levels (reject: −, not-reject: +)

		$\mathbf{\Psi}_{UN} \otimes \mathbf{\Sigma}$		$\mathbf{\Psi}_{CS} \otimes \mathbf{\Sigma}$		$\mathbf{\Psi}_{AR} \otimes \mathbf{\Sigma}$		$\mathbf{\Gamma}$	
	α	0.01	0.05	0.01	0.05	0.01	0.05	0.01	0.05
LRT	χ^2_ν	−	−	−	−	−	−	−	−
	END	+	+	−	−	+	−	+	−
RST	χ^2_ν	−	+	−	−	+	+	+	−
	END	+	+	+	−	+	+	+	−

at $\alpha = 0.05$, but we fail to reject them at $\alpha = 0.01$. Finally, $\mathbf{\Psi}_{CS} \otimes \mathbf{\Sigma}$ is rejected at $\alpha = 0.01$. If we use the RST statistic to verify the hypothesis H_0, using the limiting χ^2 distribution p-values, we have to reject all of the hypotheses at $\alpha = 0.1$. However, we fail to reject $\mathbf{\Psi}_{AR} \otimes \mathbf{\Sigma}$ at $\alpha = 0.05$ and $\mathbf{\Psi}_{CS} \otimes \mathbf{\Sigma}$ and BCS are not rejected at $\alpha = 0.01$. Using the END, we observe that we fail to reject $\mathbf{\Psi}_{AR} \otimes \mathbf{\Sigma}$ at $\alpha = 0.1$, separability is not rejected at $\alpha = 0.05$ and $\mathbf{\Psi}_{CS} \otimes \mathbf{\Sigma}$ and BCS are not rejected at $\alpha = 0.01$.

Summing up, using the END of the RST statistic, we see that for the rhesus monkey data the hypothesis of the $\mathbf{\Psi}_{AR} \otimes \mathbf{\Sigma}$ structure is not rejected at significance level 0.1, while the structures $\mathbf{\Psi}_{UN} \otimes \mathbf{\Sigma}$, $\mathbf{\Psi}_{CS} \otimes \mathbf{\Sigma}$ and BCS are rejected. Note, however, that the p-values of the limiting χ^2 distribution are much closer to the p-values of END in the RST case, and hence the decisions are similar, which is not the case for the LRT. It can be easily seen from Table 6.5 that the decision based on the RST statistic, when the inference is based either on the END or the limiting distribution, is almost the same for each structure and each significance level under consideration.

Example 6.2 Sperling et al. [29] investigated the function of subcortical nuclei in partial epilepsy to measure the metabolism in the basal ganglia and thalamus. Sixteen patients undergoing surgical evaluation were studied. Eight patients had left and eight patients had right temporal lobe seizure foci. For both groups, the measurements of metabolic rates were taken at $q = 5$ locations of the skull (Frontal, Sensorimotor, Temporal, Parietal, and Occipital) each at $p = 2$ sites (left and right). Since it can be expected that the covariance matrix of the five skull location measurements is the same for both sites, i.e., the covariance matrix will not change if we switch measurements of these sites, the BCS structure may be a good candidate for the covariance structure. Nevertheless, to apply the methods presented in this paper, we again consider all four covariance structures.

Since we have only $n = 8$ patients in each group, only the RST procedure for testing H_0 can be performed. The results are presented in Table 6.6: for both left and right foci the values of the RST statistic are given in the first line, the limiting χ^2_ν distribution p-value with respective degrees of freedom (in parentheses) in the second line, and the END p-value in the third line. Observe that since in this example $p = 2$, there is no difference between the CS and AR(1) structures. Therefore, the results in the second and third column are the same. We note that the results in the last column (i.e., for testing the BCS structure) can be found in Roy et al. [25, Table 2].

Table 6.6 Calculated values and RST statistics along with p-values of the limiting χ^2_ν distribution and the END for all considered null hypotheses H_0 for the epilepsy data

		$\mathbf{\Psi}_{\text{UN}} \otimes \mathbf{\Sigma}$	$\mathbf{\Psi}_{\text{CS}} \otimes \mathbf{\Sigma}$	$\mathbf{\Psi}_{\text{AR}} \otimes \mathbf{\Sigma}$	$\mathbf{\Gamma}$
Left foci	RS	42.76	42.371	42.371	26.645
	p-val: χ^2_ν (ν)	0.274 (38)	0.328 (39)	0.328 (39)	0.374 (25)
	p-val: END	0.67	0.58	0.58	0.78
Right foci	RS	39.161	41.098	41.098	25.634
	p-val: χ^2_ν (ν)	0.416 (38)	0.379 (39)	0.379 (39)	0.427 (25)
	p-val: END	0.78	0.66	0.66	0.91

It can be seen that none of the structures are rejected using either the limiting χ^2 or the END quantiles at any usual significance level. This may be a consequence of the fact that the number of sites p is only 2, which means that, in fact, all of the structures are very similar. Nevertheless, a very important conclusion is that the inferences based on the limiting χ^2 distribution and the END coincide.

6.7 Conclusions

In this paper, two procedures for testing covariance structure were compared.

It can be observed that the RST is a good alternative to the LRT. Its greatest advantage is that it exploits only the null hypothesis, and thus it does not require the assumption $n > qp$ as the LRT does. Moreover, the sample size needed to perform the RST, when the hypothesis concerning separability with one component structured or BCS is tested, does not depend on the number of repeated measures (the minimal sample size for testing these covariance structures is $q + 1$), and thus the test can be applied even for high-dimensional experiments, especially with high numbers of repeated measures.

Another advantage of the RST is the closeness of its test statistic distribution to the limiting χ^2 distribution, even if the sample size is very small. This means that the decisions made with the use of the limiting χ^2 distribution are usually consistent with the decisions made with the use of the END, whose determination is, in general, time-consuming and might not be available. This conclusion also follows from the fact that the empirical Type I error tends to the nominal significance level much faster for the RST than for the LRT, and thus again, for a given significance level, decisions made using the RST (based on either the END or the limiting distribution) should not differ.

The power of the test is another criterion that has been considered. It is observed that neither of the tests is uniformly better. Nevertheless, when the covariance structure of observations from experiments with small or moderate sample size is tested, again the RST outperforms the LRT.

In this paper also sthe dependence of the distribution of the test statistics on the true values of the mean and the covariance matrix is considered. This property depends on the covariance structure; however, both test statistics' distributions have this feature when the null hypothesis concerning separability, or separability with one component structured as CS or BCS, is tested, and both test statistic distributions do not have this attribute under the assumption of separability with one component as AR(1).

Acknowledgements This research is partially supported by Scientific Activities No. 04/43/SBAD/0115 (Katarzyna Filipiak) and by the Slovak Research and Development Agency under contract no. APVV-17-0568 and VEGA MŠ SR 1/0311/18 (Daniel Klein).

References

1. Crowder, M.J., Hand, D.J.: Analysis of Repeated Measures. Chapman & Hall/CRC, Boca Raton, Florida (1990)
2. Devijver, E., Gallopin, M.: Block-diagonal covariance selection for high-dimensional Gaussian graphical models. J. Am. Stat. Assoc. **113**, 306–314 (2018)
3. Dutilleul, P.: The MLE algorithm for the matrix normal distribution. J. Stat. Comput. Simul. **64**, 105–123 (1999)
4. Filipiak, K., Klein, D.: Estimation of parameters under a generalized growth curve model. J. Multivar. Anal. **158**, 73–86 (2017)
5. Filipiak, K., Klein, D.: Approximation with a Kronecker product structure with one component as compound symmetry or autoregression. Linear Algebra Appl. **559**, 11–33 (2018)
6. Filipiak, K., Klein, D., Roy, A.: Score test for a separable covariance structure with the first component as compound symmetric correlation matrix. J. Multivar. Anal. **150**, 105–124 (2016)
7. Filipiak, K., Klein, D., Roy, A.: A comparison of likelihood ratio tests and Rao's score test for three separable covariance matrix structures. Biom. J. **59**, 192–215 (2017)
8. Filipiak, K., Klein, D., Mokrzycka, M.: Estimators comparison of separable covariance structure with one component as compound symmetry matrix. Electron. J. Linear Algebra **33**, 83–98 (2018)
9. Filipiak, K., Klein, D., Vojtková, V.: The properties of partial trace and block trace operators of partitioned matrices. Electron. J. Linear Algebra **33**, 3–15 (2018)
10. Filipiak, K., Klein, D., Markiewicz, A., Mokrzycka, M.: Approximation with a Kronecker product structure with one component as compound symmetry or autoregression via entropy loss function. Linear Algebra Appl. **610**, 625–646 (2021)
11. Gilson, M., Dahmen, D., Moreno-Bote, R., Insabato, A., Helias, M.: The covariance perceptron: a new framework for classification and processing of time series in recurrent neural networks. bioRxiv (2019). https://doi.org/10.1101/562546
12. Kolda, T.G., Bader, B.W.: Tensor decompositions and applications. SIAM Rev. **51**, 455–500 (2009)
13. Kollo, T., von Rosen, D.: Advanced Multivariate Statistics with Matrices. Springer, Dordrecht (2005)
14. Leiva, R.: Linear discrimination with equicorrelated training vectors. J. Multivar. Anal. **98**, 384–409 (2007)
15. Lu, N., Zimmerman, D.: The likelihood ratio test for a separable covariance matrix. Stat. Probab. Lett. **73**, 449–457 (2005)
16. Magnus, J., Neudecker, H.: Symmetry, 0–1 matrices and Jacobians, a review. Econ. Theory **2**, 157–190 (1986)

17. McKiernan, S.H., Colman, R.J., Lopez, M., Beasley, T.M., Weindruch, R., Aiken, J.M.: Longitudinal analysis of early stage sarcopenia in aging rhesus monkeys. Exp. Gerontol. **44**, 170–176 (2009)
18. Mitchell, M., Genton, M., Gumpertz, M.: A likelihood ratio test for separability of covariances. J. Multivar. Anal. **97**, 1025–1043 (2006)
19. Rao, C.R.: Score test: historical review and recent developments. In: Balakrishnan, N., Kannan, N., Nagaijuna, H.N. (eds.) Advances in Ranking and Selection, Multiple Comparisons, and Reliability, pp. 3–20. Birkhäuser, Boston (2005)
20. Rizzo, M.: Statistical Computing with R. Chapman & Hall/CRC, Boca Raton, Florida (2008)
21. Roy, A., Khattree, R.: On implementation of a test for Kronecker product covariance structure for multivariate repeated measures data. Stat. Methodol. **2**, 297–306 (2005)
22. Roy, A., Khattree, R.: Testing the hypothesis of a Kronecker product covariance matrix in multivariate repeated measures data. In: SAS Users Group International, Proc. Stat. Data Anal. Sect., Paper 199-30, 1–11 (2005)
23. Roy, A., Leiva, L., Žežula, I., Klein, D.: Testing the equality of mean vectors for paired doubly multivariate observations in blocked compound symmetric covariance matrix setup. J. Multivar. Anal. **137**, 50–60 (2015)
24. Roy, A., Zmyślony, R., Fonseca, M., Leiva, R.: Optimal estimation for doubly multivariate data in blocked compound symmetric covariance structure. J. Multivar. Anal. **144**, 81–90 (2016)
25. Roy, A., Filipiak, K., Klein, D.: Testing a block exchangeable covariance matrix. Statistics **52**, 393–408 (2018)
26. Self, S.G., Liang, K.: Asymptotic properties of maximum likelihood estimators and likelihood ratio tests under nonstandard conditions. J. Am. Stat. Assoc. **82**, 605–610 (1987)
27. Soloveychik, I., Trushin, D.: Gaussian and robust Kronecker product covariance estimation: existence and uniqueness. J. Multivar. Anal. **149**, 92–113 (2016)
28. Srivastava, M., von Rosen, T., von Rosen, D.: Models with a Kronecker product covariance structure: estimation and testing. Math. Methods Stat. **17**, 357–370 (2008)
29. Sperling, M.R., Gur, R.C., Alavi, A., Gur, R.E., Resnick, S., O'Connor, M.J., Reivich, M.: Subcortical metabolic alterations in partial epilepsy. Epilepsia **31**, 145–155 (1990)

Chapter 7
Testing Equality of Mean Vectors with Block-Circular and Block Compound-Symmetric Covariance Matrices

Carlos A. Coelho

Abstract While the likelihood ratio test (LRT) for the equality of mean vectors when no particular structure is assumed for the covariance matrices is a well-known and well-studied test, the same is not true when some structure, namely a block structure, is assumed for the covariance matrices. In the present work, the author obtains the expressions for the LRT statistics to test the equality of mean vectors when the covariance matrices are assumed to be block-circular or block compound-symmetric and it is shown that actually in most cases the distributions of these statistics have closed finite form representations. For the other cases, families of near-exact distributions are developed and their performance is then numerically assessed. It is shown that these families of near-exact distributions lie very close to the exact distribution, even for very small samples and that they have an asymptotic behavior not only for increasing sample sizes but also for increasing numbers of populations involved and increasing numbers of sets of variables and variables in each set.

7.1 Introduction

The circular or circulant covariance structure is the covariance structure for any finite stretch of a stationary time series (Brillinger [3, Sect. 3.7]; Kedem [11, Sect. 3.2]; Olkin and Press [19]; Pollock [20]). But this structure is also commonly assumed in a number of other situations of interest in many areas as for example in cyclic designs (John [9]), in serially correlated time series (Anderson [1]; Olkin and Press [19]), and in a wealth of other applications (Khattree [10]), and the block version is of interest or appears as a good or a much plausible model in applications in geology and seismology when for example several, say r, variables are measured at m different time points or vertices of a regular polygon. Suppose that in a seismological study we have q different points of interest, in each of which a total of mr continuous variables

C. A. Coelho (✉)
Mathematics Department (DM) and Centro de Matemática e Aplicações (CMA), NOVA School of Science and Technology, NOVA University of Lisbon, Caparica, Portugal
e-mail: cmac@fct.unl.pt

K. Filipiak et al. (eds.), *Multivariate, Multilinear and Mixed Linear Models*, Contributions to Statistics, https://doi.org/10.1007/978-3-030-75494-5_7

are measured. Suppose that these mr variables are indeed a set of r variables which are measured on a set of m points situated at the vertices of a regular polygon whose center is placed on each one of the q points of interest. Suppose also that these polygons are exactly the same for each of these q points of interest and suppose that we take independent samples on each of these q points of interest. Then it would make sense to assume that the r variables measured on each of the mq points that surround the q points of interest have the same variance–covariance matrix in each of these mq points, while for each set of two given different points of the m points surrounding each of the q points of interest, the covariance matrices of the r variables will be only a function of the distance between these two points and will remain the same for the set of q points. Then, if we number consecutively the set of m points around any given point of interest, the variance–covariance matrix for the m sets of r variables measured around each one of the q points of interest will have a block-circular structure. For example, for $m = 5$ and $m = 6$, we would have for the k-th point of interest ($k = 1, \dots, q$) a structure as the one shown in Fig. 7.1, where

$$\boldsymbol{\Sigma}_0^* = \mathrm{Var}(\mathbf{x}_{1k}) = \cdots = \mathrm{Var}(\mathbf{x}_{mk}) \tag{7.1}$$

and, for $i = 1, \dots, m$ and $\ell = 1, \dots, \lfloor m/2 \rfloor$

$$\boldsymbol{\Sigma}_\ell^* = \mathrm{Cov}(\mathbf{x}_{ik}, \mathbf{x}_{\mathrm{mod}^*(i+\ell,m),k}) = \mathrm{Cov}(\mathbf{x}_{ik}, \mathbf{x}_{\mathrm{mod}^*(i+m-\ell,m),k}), \tag{7.2}$$

where $\mathbf{x}_{1k}, \dots, \mathbf{x}_{mk}$ are each an $r \times 1$ random vector with the r random variables measured on each of the m measuring points defined around the k-th point of interest ($k = 1, \dots, q$) and

$$\mathrm{mod}^*(a, b) = \begin{cases} \mathrm{mod}(a, b) & \text{for} \quad \mathrm{mod}(a, b) \neq 0 \\ b & \text{for} \quad \mathrm{mod}(a, b) = 0. \end{cases} \tag{7.3}$$

Let then

$$\mathbf{x}_k = [\mathbf{x}'_{1k}, \dots, \mathbf{x}'_{mk}]'$$

be the random vector of dimension mr for local k ($k = 1, \dots, q$).

Then we have

$$\mathrm{Var}(\mathbf{x}_k) = \boldsymbol{\Sigma},$$

and if we take

$$\mathrm{E}\,(\mathbf{x}_k) = \boldsymbol{\mu}_k,$$

we will be interested in testing the null hypothesis of equality of the q expected value vectors $\boldsymbol{\mu}_k$, that is, the null hypothesis

$$H_0 : \boldsymbol{\mu}_1 = \cdots = \boldsymbol{\mu}_q, \tag{7.4}$$

assuming that $\boldsymbol{\Sigma}$ has a block-circular structure as that in Fig. 7.1, that is, as given by (7.1) and (7.2).

$m = 5$

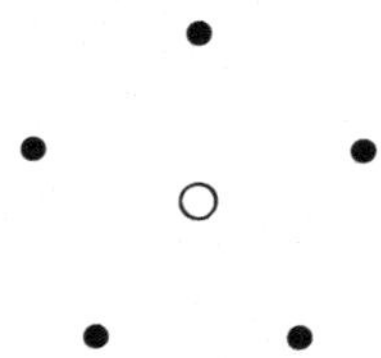

$$\boldsymbol{\Sigma} = \begin{pmatrix} \boldsymbol{\Sigma}_0^* & \boldsymbol{\Sigma}_1^* & \boldsymbol{\Sigma}_2^* & \boldsymbol{\Sigma}_2^* & \boldsymbol{\Sigma}_1^* \\ \boldsymbol{\Sigma}_1^* & \boldsymbol{\Sigma}_0^* & \boldsymbol{\Sigma}_1^* & \boldsymbol{\Sigma}_2^* & \boldsymbol{\Sigma}_2^* \\ \boldsymbol{\Sigma}_2^* & \boldsymbol{\Sigma}_1^* & \boldsymbol{\Sigma}_0^* & \boldsymbol{\Sigma}_1^* & \boldsymbol{\Sigma}_2^* \\ \boldsymbol{\Sigma}_2^* & \boldsymbol{\Sigma}_2^* & \boldsymbol{\Sigma}_1^* & \boldsymbol{\Sigma}_0^* & \boldsymbol{\Sigma}_1^* \\ \boldsymbol{\Sigma}_1^* & \boldsymbol{\Sigma}_2^* & \boldsymbol{\Sigma}_2^* & \boldsymbol{\Sigma}_1^* & \boldsymbol{\Sigma}_0^* \end{pmatrix}$$

$m = 6$

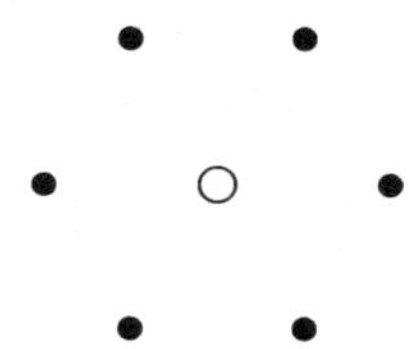

$$\boldsymbol{\Sigma} = \begin{pmatrix} \boldsymbol{\Sigma}_0^* & \boldsymbol{\Sigma}_1^* & \boldsymbol{\Sigma}_2^* & \boldsymbol{\Sigma}_3^* & \boldsymbol{\Sigma}_2^* & \boldsymbol{\Sigma}_1^* \\ \boldsymbol{\Sigma}_1^* & \boldsymbol{\Sigma}_0^* & \boldsymbol{\Sigma}_1^* & \boldsymbol{\Sigma}_2^* & \boldsymbol{\Sigma}_3^* & \boldsymbol{\Sigma}_2^* \\ \boldsymbol{\Sigma}_2^* & \boldsymbol{\Sigma}_1^* & \boldsymbol{\Sigma}_0^* & \boldsymbol{\Sigma}_1^* & \boldsymbol{\Sigma}_2^* & \boldsymbol{\Sigma}_3^* \\ \boldsymbol{\Sigma}_3^* & \boldsymbol{\Sigma}_2^* & \boldsymbol{\Sigma}_1^* & \boldsymbol{\Sigma}_0^* & \boldsymbol{\Sigma}_1^* & \boldsymbol{\Sigma}_2^* \\ \boldsymbol{\Sigma}_2^* & \boldsymbol{\Sigma}_3^* & \boldsymbol{\Sigma}_2^* & \boldsymbol{\Sigma}_1^* & \boldsymbol{\Sigma}_0^* & \boldsymbol{\Sigma}_1^* \\ \boldsymbol{\Sigma}_1^* & \boldsymbol{\Sigma}_2^* & \boldsymbol{\Sigma}_3^* & \boldsymbol{\Sigma}_2^* & \boldsymbol{\Sigma}_1^* & \boldsymbol{\Sigma}_0^* \end{pmatrix}$$

Fig. 7.1 Illustration of a set of $m = 5$ and a set of $m = 6$ points placed at the vertices of a regular polygon centered at the point of interest and the corresponding variance–covariance matrices for the $r \cdot m$ variables measured at the given point of interest ($\circ$ – point of interest; $\bullet$ – one of the m measurement points)

In Sect. 7.2, the estimation process and the likelihood ratio test (LRT) statistic are developed and then in Sect. 7.3, the distribution of the LRT statistic is characterized. It is shown that in most cases, the distribution of the LRT statistic has a finite closed form representation through the EGIG (Exponentiated Generalized Integer Gamma) distribution and a family of near-exact distributions which are asymptotic not only for increasing sample sizes but also for increasing values of r, m, and q is developed for the other cases. In Sect. 7.4 are shown the asymptotic properties of these near-exact distributions and that they indeed lie extremely close to the exact distribution, even for very small sample sizes. Then in Sect. 7.5, the test for equality of mean vectors, with block compound-symmetric covariance matrices, is addressed and in Sect. 7.6, some conclusions are drawn and some concluding remarks are made.

7.2 The Likelihood Ratio Test and Its Statistic

There are a number of different ways in which the LRT statistic to test the null hypothesis in (7.4) may be obtained. One of them is, of course, the common direct way, starting by obtaining the maximum likelihood estimators (MLEs) of $\boldsymbol{\Sigma}$ under the null and the alternative hypotheses and obtaining then the ratio of the maximum of the likelihoods under these two hypotheses. However, given the structure of the covariance matrix $\boldsymbol{\Sigma}$, this is, in this case, a quite long and involved process, and as such we will undertake a different approach.

Let us suppose that $\mathbf{x}_k \sim N_p(\boldsymbol{\mu}_k, \boldsymbol{\Sigma})$ $(k = 1, \ldots, q)$, where $p = r \cdot m$. Then the one that may be the simplest way to obtain the LRT statistic to test H_0 in (7.4) is to start by considering the LRT statistic used to test H_0 in (7.4) when no structure is assumed for $\boldsymbol{\Sigma}$. This statistic is (see, e.g., Kshirsagar [12, Sect. 9.1])

$$\Lambda = \frac{|\mathbf{A}|}{|\mathbf{A} + \mathbf{B}|} \tag{7.5}$$

where

$$\mathbf{A} = \sum_{k=1}^{q} (n_k - 1)\mathbf{S}_k \quad \text{and} \quad \mathbf{B} = \sum_{k=1}^{q} n_k \left(\overline{\mathbf{x}}_k - \overline{\mathbf{x}}\right)\left(\overline{\mathbf{x}}_k - \overline{\mathbf{x}}\right)' \tag{7.6}$$

are, respectively, the "within" and "between" sum of squares and sum of products matrices, where n_k is the size of the sample from $\mathbf{x}_k$ $(k = 1, \ldots, q)$ and $\mathbf{S}_k$ and $\overline{\mathbf{x}}_k$ are, respectively, the sample covariance matrix and mean vector of the sample from $\mathbf{x}_k$, and where we assume the q samples, one from each $\mathbf{x}_k$ $(k = 1, \ldots, q)$ to be independent and where

$$\overline{\mathbf{x}} = \frac{1}{n} \sum_{k=1}^{q} n_k \, \overline{\mathbf{x}}_k$$

is the overall sample mean vector, for

$$n = \sum_{k=1}^{q} n_k .$$

While $\mathbf{A}$ is the MLE of $\boldsymbol{\Sigma}$ under the alternative hypothesis, $\mathbf{A} + \mathbf{B}$ is the MLE of $\boldsymbol{\Sigma}$ under H_0 in (7.4). But, as it is clear from their definitions in (7.6), these are the unstructured MLEs of $\boldsymbol{\Sigma}$, that is, they do not account for the block-circular structure of $\boldsymbol{\Sigma}$.

One can use the LRT statistic in (7.5) to test H_0 in (7.4); however, failure in accounting for the structure of $\boldsymbol{\Sigma}$ will lead to a loss in power.

But then, one may easily argue that in our case the LRT statistic to test H_0 in (7.4), accounting for the block-circulant structure of $\boldsymbol{\Sigma}$, is

$$\Lambda = \frac{|\mathbf{A}^*|}{|\mathbf{A}^* + \mathbf{B}^*|} \tag{7.7}$$

where $\mathbf{A}^*$ and $\mathbf{B}^*$ are now block-circulant matrices, still with $\mathbf{A}^*$ and $\mathbf{A}^* + \mathbf{B}^*$ being the MLEs of $\boldsymbol{\Sigma}$, respectively, under the alternative and the null hypotheses, but now accounting for the block-circular structure of $\boldsymbol{\Sigma}$.

But so, the question now is: "how to find a simple way to obtain $\mathbf{A}^*$ and $\mathbf{B}^*$?".

Let $\boldsymbol{\Sigma}$ be a block-circular matrix with $m \times m$ blocks of dimensions $r \times r$. Then we may write

$$\boldsymbol{\Sigma} = \sum_{i=0}^{m-1} \left(\underset{(m \times m)}{\mathbf{W}_i} \otimes \underset{(r \times r)}{\boldsymbol{\Sigma}_i^*} \right) \tag{7.8}$$

where $\boldsymbol{\Sigma}_0^*$ is an $r \times r$ positive-definite matrix,

$$\boldsymbol{\Sigma}_i^* = \boldsymbol{\Sigma}_{m-i}^*, \quad i = 1, \ldots, m-1 \tag{7.9}$$

are symmetric matrices, and

$$\underset{(m \times m)}{\mathbf{W}_i} = \begin{pmatrix} 0 & \mathbf{I}_{m-i} \\ \mathbf{I}_i & 0 \end{pmatrix}, \quad i = 0, \ldots, m-1 \quad (\text{with } \mathbf{W}_0 = \mathbf{I}_m),$$

where $\mathbf{I}_m$ stands for an identity matrix of order m (see Olkin [18, Sect. 4] and see the representations of $\boldsymbol{\Sigma}$ in Fig. 7.1).

But then, by the invariance property of MLEs, since the matrices $\boldsymbol{\Sigma}_i^*$ $(i = 0, \ldots, m-1)$ in (7.8) are well identified, we may write

$$\widehat{\boldsymbol{\Sigma}} = \sum_{i=0}^{m-1} \left(\underset{(m \times m)}{\mathbf{W}_i} \otimes \underset{(r \times r)}{\widehat{\boldsymbol{\Sigma}}_i^*} \right), \tag{7.10}$$

so that we will take

$$\mathbf{A}^* = \sum_{i=0}^{m-1} \left(\mathbf{W}_i \otimes \mathbf{A}_i^* \right) \quad \text{and} \quad \mathbf{B}^* = \sum_{i=0}^{m-1} \left(\mathbf{W}_i \otimes \mathbf{B}_i^* \right) \tag{7.11}$$

where the question now is: how to find a simple way to obtain $\mathbf{A}_i^*$ and $\mathbf{B}_i^*$ $(i = 0, \ldots, m-1)$?

From an extension of the results of Lv and Huang [13] and also from the results in Lv and Huang [14], as well as from the block spectral decomposition of the matrix $\boldsymbol{\Sigma}$ (see Appendix A), we know that the inverse of a block-circulant matrix is a block-circulant matrix, and as such both $\boldsymbol{\Sigma}$ and $\boldsymbol{\Sigma}^{-1}$ may be written in the form in (7.8). A fact that may also be derived from the simple fact that the space of block-circular matrices is a quadratic space (Seely [22]). We may then use the results in Szatrowski [23, 24] together with the facts that (see (7.8)–(7.10) and Fig. 7.1)

- there are always m blocks $\boldsymbol{\Sigma}_0^*$
- for odd m there are $2m$ of all other blocks
- for even m
  - there are also m blocks $\boldsymbol{\Sigma}_{m/2}^*$
  - and there are $2m$ of all other blocks,

to take

$$\mathbf{A}_0^* = \frac{1}{m} \sum_{j=1}^{m} \widetilde{\mathbf{A}}_{jj}, \quad \mathbf{B}_0^* = \frac{1}{m} \sum_{j=1}^{m} \widetilde{\mathbf{B}}_{jj}, \tag{7.12}$$

and for $i = 1, \ldots, m-1$, for odd m, or for $i = 1, \ldots, m/2-1, m/2+1, \ldots, m$, for even m,

$$\mathbf{A}_i^* = \frac{1}{2m}\sum_{j=1}^{m} \widetilde{\mathbf{A}}_{j,\text{mod}^*(j+i,m)} + \widetilde{\mathbf{A}}_{\text{mod}^*(j+i,m),j}, \tag{7.13}$$

and

$$\mathbf{B}_i^* = \frac{1}{2m}\sum_{j=1}^{m} \widetilde{\mathbf{B}}_{j,\text{mod}^*(j+i,m)} + \widetilde{\mathbf{B}}_{\text{mod}^*(j+i,m),j},$$

with $\mathbf{A}_i^* = \mathbf{A}_{m-i}^*$ and $\mathbf{B}_i^* = \mathbf{B}_{m-i}^*$, for $i = 1, \ldots, m-1$, and for even m

$$\mathbf{A}_{m/2}^* = \frac{1}{m}\sum_{j=1}^{m} \widetilde{\mathbf{A}}_{j,\text{mod}^*(j+m/2,m)}, \quad \mathbf{B}_{m/2}^* = \frac{1}{m}\sum_{j=1}^{m} \widetilde{\mathbf{B}}_{j,\text{mod}^*(j+m/2,m)}, \tag{7.14}$$

where $\text{mod}^*(\cdot,\cdot)$ is given by (7.3) and

$$\widetilde{\mathbf{A}}_{jk} \quad \text{and} \quad \widetilde{\mathbf{B}}_{jk}, \qquad j,k \in \{1,\ldots,m\}$$

denote the $r \times r$ blocks of $\mathbf{A}$ and $\mathbf{B}$ starting at row $r(j-1)+1$ and column $r(k-1)+1$.

7.3 The Distribution of the Likelihood Ratio Statistic

Although (7.7) provides a good way to compute the LRT statistic, we will need to resort to an equivalent formulation in order to be able to derive the distribution of Λ.

Let us consider the matrices

$$\mathbf{A}^{**} = (\boldsymbol{\Gamma}_m \otimes \mathbf{I}_r)\, \mathbf{A}\, (\boldsymbol{\Gamma}_m \otimes \mathbf{I}_r) \quad \text{and} \quad \mathbf{B}^{**} = (\boldsymbol{\Gamma}_m \otimes \mathbf{I}_r)\, \mathbf{B}\, (\boldsymbol{\Gamma}_m \otimes \mathbf{I}_r) \tag{7.15}$$

where Γ_m is an orthogonal symmetrical matrix with elements

$$\gamma_{jk} = \frac{1}{\sqrt{m}}\left[\sin\left(\tfrac{2\pi}{m}(j-1)(k-1)\right) + \cos\left(\tfrac{2\pi}{m}(j-1)(k-1)\right)\right], \quad j,k \in \{1,\ldots,m\}, \tag{7.16}$$

and let $\mathbf{A}_j^{**}$ and $\mathbf{B}_j^{**}$ $(j = 1, \ldots, m)$ denote the j-th diagonal block of dimensions $r \times r$, respectively, of $\mathbf{A}^{**}$ and $\mathbf{B}^{**}$. The matrix $\boldsymbol{\Gamma}_m$ is the matrix used by Olkin and Press [19] to diagonalize circular symmetric matrices and it also diagonalizes all matrices $\mathbf{W}_i + \mathbf{W}_{m-i}$ $(i = 0, \ldots, m)$ (see Olkin [18, Sect. 4]).

Then, going through some algebra (see Appendix B), it is possible to show that, for $m^+ = \lfloor m/2 \rfloor$,

$$|\mathbf{A}^*| = |\mathbf{A}_1^{**}| \left(|\mathbf{A}_{1+m^+}^{**}|\right)^{\mathrm{mod}(m+1,2)} \prod_{j=2}^{m-m^+} \left|\frac{\mathbf{A}_j^{**} + \mathbf{A}_{m-j+2}^{**}}{2}\right|^2 \tag{7.17}$$

and

$$|\mathbf{A}^* + \mathbf{B}^*| = |\mathbf{A}_1^{**} + \mathbf{B}_1^{**}| \left(|\mathbf{A}_{1+m^+}^{**} + \mathbf{B}_{1+m^+}^{**}|\right)^{\mathrm{mod}(m+1,2)} \times \prod_{j=2}^{m-m^+} \left|\frac{\mathbf{A}_j^{**} + \mathbf{B}_j^{**} + \mathbf{A}_{m-j+2}^{**} + \mathbf{B}_{m-j+2}^{**}}{2}\right|^2, \tag{7.18}$$

so that we have for Λ in (7.7),

$$\Lambda = \underbrace{\frac{|\mathbf{A}_1^{**}|}{|\mathbf{A}_1^{**} + \mathbf{B}_1^{**}|}}_{\Lambda_1} \underbrace{\left(\frac{|\mathbf{A}_{1+m^+}^{**}|}{|\mathbf{A}_{1+m^+}^{**} + \mathbf{B}_{1+m^+}^{**}|}\right)^{\mathrm{mod}(m+1,2)}}_{\Lambda_2} \times \underbrace{\prod_{j=2}^{m-m^+} \left(\frac{\left|\mathbf{A}_j^{**} + \mathbf{A}_{m-j+2}^{**}\right|}{\left|\mathbf{A}_j^{**} + \mathbf{B}_j^{**} + \mathbf{A}_{m-j+2}^{**} + \mathbf{B}_{m-j+2}^{**}\right|}\right)^2}_{\Lambda_3}. \tag{7.19}$$

Then we may take into account that, on one hand, in (7.19), the $\mathbf{A}_j^{**}$ and $\mathbf{B}_j^{**}$ are independent, given the independence between $\mathbf{A}$ and $\mathbf{B}$, and that on the other hand, the $\mathbf{A}_j^{**}$ $(j = 1, \ldots, m)$ form a set of m independent matrices and also the $\mathbf{B}_j^{**}$ $(j = 1, \ldots, m)$ form another set of independent matrices. This is so given that we know that for $n = \sum_{k=1}^{q} n_k$,

$$\mathbf{A} \sim W_p(n-q, \mathbf{\Sigma}) \implies \mathbf{A}^{**} \sim W_p\big(n-q, (\boldsymbol{\Gamma}_m \otimes \mathbf{I}_r)\,\mathbf{\Sigma}\,(\boldsymbol{\Gamma}_m \otimes \mathbf{I}_r)\big)$$

and

$$\mathbf{B} \sim W_p(q-1, \mathbf{\Sigma}) \implies \mathbf{B}^{**} \sim W_p\big(q-1, (\boldsymbol{\Gamma}_m \otimes \mathbf{I}_r)\,\mathbf{\Sigma}\,(\boldsymbol{\Gamma}_m \otimes \mathbf{I}_r)\big)$$

with (see Olkin [18, Sect. 4], expressions (4.6) and (4.7))

$$(\boldsymbol{\Gamma}_m \otimes \mathbf{I}_r)\,\mathbf{\Sigma}\,(\boldsymbol{\Gamma}_m \otimes \mathbf{I}_r) = \mathrm{BDiag}\,(\mathbf{\Psi}_1, \mathbf{\Psi}_2, \ldots, \mathbf{\Psi}_m)\,,$$

where, "BDiag$(\cdots)$" stands for a block-diagonal matrix with the diagonal blocks shown inside the parenthesis and where $\mathbf{\Psi}_j$ $(j = 1, \ldots, m)$ are $r \times r$ symmetric matrices (see Olkin [18, Sect. 4]), so that, for $(j = 1, \ldots, m)$,

$$\mathbf{A}_j^{**} \sim W_r(n-q, \mathbf{\Psi}_j) \quad \text{and} \quad \mathbf{B}_j^{**} \sim W_r(q-1, \mathbf{\Psi}_j), \tag{7.20}$$

with $\mathbf{A}_j^{**}$ independent of $\mathbf{A}_{j'}^{**}$, for $j \neq j'$, and $\mathbf{B}_j^{**}$ independent of $\mathbf{B}_{j'}^{**}$.

Furthermore, since for $j = 2, \ldots, m$, $\boldsymbol{\Psi}_j = \boldsymbol{\Psi}_{m-j+2}$ (see Olkin [18, Sect. 4]), we have, for $j = 2, \ldots, m$,

$$\mathbf{A}_j^{**} + \mathbf{A}_{m-j+2}^{**} \sim W_r(2(n-q), \boldsymbol{\Psi}_j)$$

and

$$\mathbf{B}_j^{**} + \mathbf{B}_{m-j+2}^{**} \sim W_r(2(q-1), \boldsymbol{\Psi}_j).$$

As a consequence, the statistics Λ_1, Λ_2, and Λ_3 in (7.19) are independent, with (see, e.g., Coelho and Arnold [6, Chap. 5, Appendix 1])

$$\Lambda_1 \stackrel{d}{\equiv} \Lambda_2 \stackrel{st}{\sim} \prod_{j=1}^{r} Y_j$$

where, "$\stackrel{st}{\sim}$" is to be read as "is stochastically equivalent to" or "has the same distribution as" and where for $n > q + r - 1$,

$$Y_j \sim Beta\left(\frac{n-q+1-j}{2}, \frac{q-1}{2}\right), \qquad j = 1, \ldots, r$$

form a set of r independent RVs, and where r and $q-1$ are interchangeable (see Coelho and Arnold [6, Chap. 5, Appendix 1]), and

$$\Lambda_3 = \prod_{j=2}^{m-m^+} \Lambda_j^* \qquad \text{with} \qquad \Lambda_j^* = \left(\frac{\left|\mathbf{A}_j^{**} + \mathbf{A}_{m-j+2}^{**}\right|}{\left|\mathbf{A}_j^{**} + \mathbf{B}_j^{**} + \mathbf{A}_{m-j+2}^{**} + \mathbf{B}_{m-j+2}^{**}\right|}\right)^2$$

where the $m - m^+ - 1$ statistics Λ_j^* ($j = 2, \ldots, m - m^+$), are also independent. Furthermore, from the distributions of A_j^{**} and B_j^{**} in (7.20), and given that $\boldsymbol{\Psi}_j = \boldsymbol{\Psi}_{m-j+2}$ for $j = 2, \ldots, m$, we have

$$\mathbf{A}_j^{**} + \mathbf{A}_{m-j+2}^{**} \sim W_r(2(n-q), \boldsymbol{\Psi}_j), \quad j = 2, \ldots, m - m^+$$

and

$$\mathbf{B}_j^{**} + \mathbf{B}_{m-j+2}^{**} \sim W_r(2(q-1), \boldsymbol{\Psi}_j), \quad j = 2, \ldots, m - m^+$$

so that (see Coelho and Arnold [6, Sect. 5, Appendix 1])

$$\Lambda_j^* \stackrel{st}{\sim} \prod_{k=1}^{r} (Y_k^*)^2$$

where

$$Y_k^* \sim Beta\left(n - q + \frac{1-k}{2}, q-1\right), \quad k = 1, \ldots, r$$

also form a set of r independent RVs.

But then we can write

$$\Lambda \stackrel{st}{\sim} \left\{ \prod_{j=1}^{r} Y_j \right\}^{1+\text{mod}(m+1,2)} \left\{ \prod_{j=2}^{m-m^+} \prod_{k=1}^{r} (Y_k^*)^2 \right\}$$

where $m^+ = \lfloor m/2 \rfloor$,

$$Y_j \sim Beta\left(\frac{n-q+1-j}{2}, \frac{q-1}{2}\right), \quad j = 1, \ldots, r$$

and

$$Y_k^* \sim Beta\left(n-q+\frac{1-k}{2}, q-1\right), \quad k = 1, \ldots, r,$$

form a set of $r(m-m^+) + 1 + \text{mod}(m+1, 2)$ independent RVs.

Hence, for any q, m, r, and n, the probability density function (p.d.f.) of Λ may be written in terms of the Fox H function (Fox [8]; Mathai [15, Sect. 3.11]; Mathai and Haubold [16, Sect. 1.9]; Mathai and Saxena [17, Sect. 1.1]; Prudnikov et al. [21, Sect. 8.3]), using the notation in Coelho and Arnold [6, Sect. 2.1] as

$$f_\Lambda(z) = \left\{ \prod_{j=1}^{r} \frac{\Gamma\left(\frac{n-j}{2}\right)}{\Gamma\left(\frac{n-q+1-j}{2}\right)} \right\}^{1+\lfloor\lfloor m+1 \rfloor\rfloor} \left\{ \prod_{j=2}^{m-m^+} \prod_{k=1}^{r} \frac{\Gamma\left(n-1+\frac{1-k}{2}\right)}{\Gamma\left(n-q+\frac{1-k}{2}\right)} \right\}$$
$$\times H^{p^*,0}_{p^*,p^*}\left(\begin{array}{c} \left\{ \left\{\left(\frac{n-j-2}{2}, 1\right)\right\}_{j=1:r(1+\lfloor\lfloor m+1\rfloor\rfloor)}, \left\{\left(n-2+\frac{1-k}{2}, 2\right)\right\}_{\substack{j=2:m-m^+\\k=1:r}} \right\} \\ \left\{ \left\{\left(\frac{n-q-1-j}{2}, 1\right)\right\}_{j=1:r(1+\lfloor\lfloor m+1\rfloor\rfloor)}, \left\{\left(n-q-1+\frac{1-k}{2}, 2\right)\right\}_{\substack{j=2:m-m^+\\k=1:r}} \right\} \end{array} \middle| z \right)$$

for $\lfloor\lfloor m+1 \rfloor\rfloor = \text{mod}(m+1, 2)$ and $p^* = r(m-m^+) + \text{mod}(m+1, 2)$, and its cumulative distribution function (c.d.f.) as

$$F_\Lambda(z) = \left\{ \prod_{j=1}^{r} \frac{\Gamma\left(\frac{n-j}{2}\right)}{\Gamma\left(\frac{n-q+1-j}{2}\right)} \right\}^{1+\lfloor\lfloor m+1 \rfloor\rfloor} \left\{ \prod_{j=2}^{m-m^+} \prod_{k=1}^{r} \frac{\Gamma\left(n-1+\frac{1-k}{2}\right)}{\Gamma\left(n-q+\frac{1-k}{2}\right)} \right\}$$
$$\times H^{p^*,1}_{p^*+1,p^*+1}\left(\begin{array}{c} \left\{ (1,1), \left\{\left(\frac{n-j}{2}, 1\right)\right\}_{j=1:r(1+\lfloor\lfloor m+1\rfloor\rfloor)}, \left\{\left(n-1+\frac{1-k}{2}, 2\right)\right\}_{\substack{j=2:m-m^+\\k=1:r}} \right\} \\ \left\{ \left\{\left(\frac{n-q+1-j}{2}, 1\right)\right\}_{j=1:r(1+\lfloor\lfloor m+1\rfloor\rfloor)}, \left\{\left(n-q+\frac{1-k}{2}, 2\right)\right\}_{\substack{j=2:m-m^+\\k=1:r}}, (0,1) \right\} \end{array} \middle| z \right)$$

where

$$H^{m,n}_{p,q}\left(\begin{matrix}\{(a_k,\alpha_k)\}_{k=1:p}\\ \{(b_j,\beta_j)\}_{j=1:q}\end{matrix}\middle|\, z\right)$$
$$=\frac{1}{2\pi\mathrm{i}}\oint_L \frac{\left\{\prod_{j=1}^{m}\Gamma(b_j+\beta_j s)\right\}\left\{\prod_{k=1}^{n}\Gamma(1-a_k-\alpha_k s)\right\}}{\left\{\prod_{j=m+1}^{q}\Gamma(1-b_j-\beta_j s)\right\}\left\{\prod_{k=n+1}^{p}\Gamma(a_k+\alpha_k s)\right\}}\, z^s\, ds$$

represents the Fox H function with arguments $m, n, p, q \in \mathbb{N}_0$ $(m \le q;\ n \le p)$, $z \neq 0$, and any real or complex a_k and b_j and positive reals α_k and β_j $(k = 1, \ldots, p;$ $j = 1, \ldots, q)$, such that for $\nu \in \mathbb{N}_0$, $a_k - (b_j + \nu)\alpha_k/\beta_j \notin \mathbb{N}$ for $k = 1, \ldots, n$ and $j = 1, \ldots, m$.

Although the Fox H function provides a very handy representation for both the p.d.f. and the c.d.f. of Λ, it does not provide a useful means of computation of these functions since even for small values of the parameters q, m, r, and n, any computation will take too much long time if at all possible. As such we will have to resort to a different approach.

7.3.1 The Exact Distribution of Λ for Odd q or Even r

We start by noticing that we can write the h-th moment of Λ_3 as

$$\begin{aligned}\mathrm{E}\left(\Lambda_3^h\right) &= \prod_{j=2}^{m-m^+}\prod_{k=1}^{r}\mathrm{E}\left((Y_k^*)^{2h}\right)\\ &= \prod_{j=2}^{m-m^+}\prod_{k=1}^{r}\frac{\Gamma\left(n-\frac{1}{2}-\frac{k}{2}\right)\Gamma\left(n-q+\frac{1-k}{2}+2h\right)}{\Gamma\left(n-q+\frac{1-k}{2}\right)\Gamma\left(n-\frac{1}{2}-\frac{k}{2}+2h\right)}\\ &= \left\{\prod_{k=1}^{r}\frac{\Gamma\left(n-\frac{1}{2}-\frac{k}{2}\right)\Gamma\left(n-q+\frac{1-k}{2}+2h\right)}{\Gamma\left(n-q+\frac{1-k}{2}\right)\Gamma\left(n-\frac{1}{2}-\frac{k}{2}+2h\right)}\right\}^{m-m^+-1},\end{aligned}$$

which, using the relation

$$\frac{\Gamma(a+n)}{\Gamma(a)}=\prod_{\ell=0}^{n-1}(a+\ell), \qquad (7.21)$$

valid for any real or complex a and positive integer n, may be rewritten as

$$
\begin{aligned}
\mathrm{E}\left(\Lambda_3^h\right) &= \left\{\prod_{k=1}^{r} \frac{\Gamma\left(n-\frac{1}{2}-\frac{k}{2}\right)\,\Gamma\left(n-q+\frac{1-k}{2}+2h\right)}{\Gamma\left(n-q+\frac{1-k}{2}\right)\,\Gamma\left(n-\frac{1}{2}-\frac{k}{2}+2h\right)}\right\}^{m-m^+-1} \\
&= \left\{\prod_{k=1}^{r}\prod_{\ell=0}^{q-2}\left(n-q+\frac{1-k}{2}+\ell\right)\left(n-q+\frac{1-k}{2}+\ell+2h\right)^{-1}\right\}^{m-m^+-1} \\
&= \left\{\prod_{k=1}^{r}\prod_{\ell=0}^{q-2}\left(\frac{n-q+\ell}{2}+\frac{1-k}{4}\right)\left(\frac{n-q+\ell}{2}+\frac{1-k}{4}+h\right)^{-1}\right\}^{m-m^+-1} \\
&= \prod_{j=1}^{r+2q-4}\left(\frac{2n-2q-r+j}{4}\right)^{r_j(m-m^+-1)}\left(\frac{2n-2q-r+j}{4}+h\right)^{-r_j(m-m^+-1)}
\end{aligned}
\tag{7.22}
$$

where

$$
r_j \equiv r(j; r, 2(q-1), 2)
$$

with

$$
r(j; a, b, h^*) = \begin{cases} h_j, & j = 1, \ldots, h^* \\ h_j + r_{j-h^*}, & j = h^*+1, \ldots, a+b-h^* \end{cases} \tag{7.23}
$$

where

$$
h_j \equiv h(j; a, b, h^*) = \begin{cases} +1, & j = 1, \ldots, \min(a,b) \\ \;\;\,0, & j = 1+\min(a,b), \ldots, \max(a,b) \\ -1, & j = 1+\max(a,b), \ldots, a+b-h^*. \end{cases} \tag{7.24}
$$

Then, given that all Gamma functions in (7.22) remain valid when we replace h by any complex value, we may write the characteristic function (c.f.) of $W_3 = -\log \Lambda_3$ as

$$
\begin{aligned}
\Phi_{W_3}(t) &= \mathrm{E}\left(e^{\mathrm{i}tW_3}\right) = \mathrm{E}\left(\Lambda_3^{-\mathrm{i}t}\right) \\
&= \prod_{j=1}^{r+2q-4}\left(\frac{2n-2q-r+j}{4}\right)^{r_j(m-m^+-1)}\left(\frac{2n-2q-r+j}{4}-\mathrm{i}t\right)^{-r_j(m-m^+-1)}
\end{aligned}
$$

which shows that the exact distribution of $W_3 = -\log \Lambda_3$ is a GIG (Generalized Integer Gamma) distribution of depth $2q+r-4$, with shape parameters r_j and rate parameters $\frac{2n-2q-r+j}{4}$ $(j = 1, \ldots, 2q+r-4)$ (see Appendix C and Coelho [4] for the definition of the GIG distribution, its p.d.f., and c.d.f.).

The same result might have been obtained using a modified version of Theorem 3.2 in Coelho and Arnold [6] to encompass the distribution of Λ_3 in the realm of the EGIG distribution, which may be stated as follows.

Theorem 7.1 *For positive integers n_ν, k_ν, and m_ν ($\nu = 1, \dots, m^*$) and for real*

$$a_{\nu\ell} = a_\nu + \ell + \frac{1-\ell}{h^* k_\nu}$$

with $h^* \in \mathbb{N}$ and with $a_\nu > \frac{n_\nu - 1}{h^* k_\nu}$ ($\nu = 1, \dots, m^*$), let

$$Z = \prod_{\nu=1}^{m^*} \prod_{\ell=1}^{n_\nu} \prod_{j=1}^{k_\nu} Y_{\nu\ell j}$$

where

$$Y_{\nu\ell j} \sim Beta\left(a_{\nu\ell} + 1 - \ell - \tfrac{j}{k_\nu}, \tfrac{m_\nu}{k_\nu}\right), \; \nu = 1, \dots, m^*; \; \ell = 1, \dots, n_\nu; \; j = 1, \dots, k_\nu,$$

are independent RVs. Then,

$$Z \stackrel{d}{\equiv} \prod_{\nu=1}^{m^*} \prod_{\ell=1}^{n_\nu} \prod_{j=1+k_\nu(\ell-1)}^{k_\nu \ell} Y^*_{\nu\ell j} \stackrel{d}{\equiv} \prod_{\nu=1}^{m^*} \prod_{\ell=1}^{n_\nu} \left(Y^{**}_{\nu\ell}\right)^{k_\nu} \stackrel{d}{\equiv} \prod_{\nu=1}^{m^*} \prod_{j=1}^{n_\nu + h^*(m_\nu - 1)} e^{-W_{\nu j}} \tag{7.25}$$

where

$$Y^*_{\nu\ell j} \sim Beta\left(a_{\nu\ell} - \frac{j}{k_\nu}, \frac{m_\nu}{k_\nu}\right), \qquad Y^{**}_{\nu\ell} \sim Beta\left((a_{\nu\ell} - \ell)k_\nu, m_\nu\right)$$

and

$$W_{\nu j} \sim \Gamma\left(r_{\nu j}, a_\nu + \frac{j - n_\nu}{h^* k_\nu}\right)$$

for

$$r_{\nu j} \equiv r(j; n_\nu, h^* m_\nu, h^*)$$

where $r(j; a, b, h^*)$ is defined in (7.23)–(7.24) and where $j = 1, ..., n_\nu + h^*(m_\nu - 1)$ and $\nu = 1, \dots, m^*$.

If $a_\nu = a$, $n_\nu = n$, $k_\nu = k$, and $m_\nu = m$ for $\nu = 1, \dots, m^*$, then

$$Z \equiv \prod_{j=1}^{n + h^*(m-1)} e^{-W_j} \tag{7.26}$$

with

$$W_j \sim \Gamma\left(m^* r_j, a + \frac{j - n}{h^* k}\right)$$

where, for $j = 1, \ldots, n + h^*(m-1)$,

$$r_j \equiv r(j; n, h^* m, h^*) \tag{7.27}$$

for the definition of $r(j; a, b, h^*)$ *in (7.23)–(7.24).*

The proof of this theorem follows similar lines to that of Theorem 3.1 in Coelho and Arnold [6] and is shown in Appendix D. Its most significant contribution resides in the fact that the last equivalence in (7.25) shows that the distribution of $W = -\log Z$ is a GIG distribution or, equivalently, that the distribution of Z is an EGIG distribution.

Using the above theorem with $h^* = 2$, $a_\nu = (n-q)/2$, $k_\nu = 2$, $n_\nu = r$, $m_\nu = q - 1$, and $m^* = m - m^+ - 1$, we obtain immediately the distribution of $W_3 = -\log \Lambda_3$ as a GIG distribution, or that of Λ_3.

We will now show that the exact distribution of $W_1 = -\log \Lambda_1$ and $W_2 = -\log \Lambda_2$ for odd q or even r is also GIG distribution and that, as such the exact distribution of $W = -\log \Lambda$ is in this case a GIG distribution, and consequently, that of Λ an EGIG distribution.

The statistics Λ_1 and Λ_2 have the same distribution, and for example, for Λ_1, we may write its h-th moment as

$$\mathrm{E}\left(\Lambda_1^h\right) = \prod_{j=1}^{r} \mathrm{E}\left(Y_j^h\right) = \prod_{j=1}^{r} \frac{\Gamma\left(\frac{n-j}{2}\right)\Gamma\left(\frac{n-q+1-j}{2}+h\right)}{\Gamma\left(\frac{n-q+1-j}{2}\right)\Gamma\left(\frac{n-j}{2}+h\right)}$$

which, for odd q, using (7.21), may then be written as

$$\begin{aligned}\mathrm{E}\left(\Lambda_1^h\right) &= \prod_{j=1}^{r}\prod_{\ell=0}^{\frac{q-1}{2}-1}\left(\frac{n-q+1-j}{2}+\ell\right)\left(\frac{n-q+1-j}{2}+\ell+h\right)^{-1}\\ &= \prod_{j=1}^{r+q-3}\left(\frac{n-r-q+j}{2}\right)^{s_j}\left(\frac{n-r-q+j}{2}+h\right)^{-s_j}\end{aligned}$$

where, for $j = 1, \ldots, r + q - 3$,

$$s_j \equiv r(j; r, q-1, 2) \tag{7.28}$$

for the definition of $r(j; a, b, h^*)$ in (7.23)–(7.24).

As such, we may write the c.f. of $W_1 = -\log \Lambda_1$ as

$$\Phi_{W_1}(t) = \mathrm{E}\left(\Lambda_1^{-\mathrm{i}t}\right) = \prod_{j=1}^{r+q-3}\left(\frac{n-r-q+j}{2}\right)^{s_j}\left(\frac{n-r-q+j}{2}-\mathrm{i}t\right)^{-s_j} \tag{7.29}$$

which is the c.f. of a GIG distribution of depth $r+q-3$, with shape parameters s_j and rate parameters $\frac{n-r-q+j}{2}$ $(j=1,\dots,r+q-3)$.

And it is not hard to show that, since $q-1$ and r are interchangeable in the distributions of Λ_1 and Λ_2 (see Coelho and Arnold [6, Chap. 5, Appendix 1]), a similar result may be obtained when r is even.

Indeed, a similar result for the distribution of Λ_1 and Λ_2 may be obtained by using Theorem 3.2 in Coelho and Arnold [6] with $m^*=1$, and then, for even r

$$a_1=\frac{n-q+1}{2},\quad n_1=r/2,\quad k_1=2,\quad m_1=q-1$$

or, for odd q,

$$a_1=\frac{n-r}{2},\quad n_1=(q-1)/2,\quad k_1=2,\quad m_1=r.$$

But then we may write the c.f. of $W=-\log\Lambda$, for odd q or even r, as

$$\begin{aligned}\Phi_W(t)&=\Phi_{W_1}(t)\,\left\{\Phi_{W_2}(t)\right\}^{\text{mod}(m+1,2)}\,\Phi_{W_3}(t)\\&=\left\{\prod_{j=1}^{r+q-3}\left(\frac{n-r-q+j}{2}\right)^{s_j}\left(\frac{n-r-q+j}{2}-\text{i}t\right)^{-s_j}\right\}^{1+\text{mod}(m+1,2)}\\&\quad\times\prod_{k=1}^{r+2q-4}\left(\frac{2n-2q-r+k}{4}\right)^{r_k(m-m^*-1)}\left(\frac{2n-2q-r+k}{4}-\text{i}t\right)^{-r_k(m-m^*-1)}\end{aligned}$$

which is the c.f. of a GIG distribution, where however some of the rate parameters may be "repeated", so that in order to devise the correct representation for this GIG distribution (where equal rate parameters "are not allowed"), we have to write the c.f. of W as

$$\begin{aligned}\Phi_W(t)=&\prod_{j=1}^{r+q-3}\left(\frac{n-r-q+j}{2}\right)^{s_j^*}\left(\frac{n-r-q+j}{2}-\text{i}t\right)^{-s_j^*}\\&\times\prod_{\substack{k=1+\text{mod}(r,2)\\\text{step }2}}^{r+2q-5}\left(\frac{2n-2q-r+k}{4}\right)^{r_k^*}\left(\frac{2n-2q-r+k}{4}-\text{i}t\right)^{-r_k^*}\\&\times\left(\frac{n-2}{2}\right)^{r^*_{r+2q-4}}\left(\frac{n-2}{2}-\text{i}t\right)^{-r^*_{r+2q-4}}\end{aligned}\tag{7.30}$$

where

$$s_j^*=\begin{cases}s_j(1+\text{mod}(m+1,2)), & j=1,\dots,\lfloor r/2\rfloor\\ s_j(1+\text{mod}(m+1,2))+r^*_{2(j-\lfloor r/2\rfloor)-\text{mod}(r,2)}, & j=\lfloor r/2\rfloor+1,\dots,r+q-3\end{cases}$$

and

$$r_k^* = r_k(m - m^* - 1), \quad k = 1, \ldots, r + 2q - 4.$$

The c.f. in (7.30) shows that the exact distribution of W is for odd q or even r a GIG distribution of depth $r + 2q + \lfloor \frac{r+1}{2} \rfloor - 3$ with rate parameters

$$\left\{ \left\{ \frac{n - r - q + j}{2} \right\}_{j=1:r+q-3}, \left\{ \frac{2n - 2q - r + k}{4} \right\}_{\substack{k=1+\text{mod}(r,2):r+2q-5 \\ \text{step } 2}}, \frac{n-2}{2} \right\}$$

and corresponding shape parameters

$$\left\{ \left\{ s_j^* \right\}_{j=1:r+q-3}, \left\{ r_k^* \right\}_{\substack{k=1+\text{mod}(r,2):r+2q-5 \\ \text{step } 2}}, r_{r+2q-4}^* \right\}.$$

Equivalently, the exact distribution of Λ is, in these cases, an EGIG distribution (Arnold et al. [2]) with the above depth and the above shape and rate parameters. For the definition and expressions of the p.d.f. and c.d.f. of the GIG and EGIG distributions, see Appendix C.

7.3.2 *Near-Exact Distributions for Λ for the Case When q Is Even and r Is Odd*

For even q and odd r, the expressions for the exact p.d.f. and c.d.f. of Λ become non-manageable and the best solution is to resort to the development and use of very sharp near-exact distributions.

These near-exact distributions will be asymptotic not only for increasing sample sizes but also for increasing values of r, m, and q.

For even q and odd r, the distribution of Λ_3 remains the same and in what concerns that of Λ_1 and Λ_2, we may write, for example, the h-th moment of Λ_1 as

$$\mathrm{E}\left(\Lambda_1^h\right) = \left\{ \prod_{j=1}^{r+q-4} \left(\frac{n-r-q+j}{2} \right)^{s_j} \left(\frac{n-r-q+j}{2} + h \right)^{-s_j} \right\} \frac{\Gamma\left(\frac{n-1}{2}\right) \Gamma\left(\frac{n-2}{2} + h\right)}{\Gamma\left(\frac{n-2}{2}\right) \Gamma\left(\frac{n-1}{2} + h\right)} \tag{7.31}$$

where

$$s_j = \begin{cases} h_j^*, & j = 1, 2 \\ h_j^* + s_{j-2}, & j = 3, \ldots, r + q - 4 \end{cases} \tag{7.32}$$

with

$$h_j^* = (\text{\# of elements in } \{r, q-1\} \geq j) - 1, \quad j = 1, \ldots, r+q-4. \tag{7.33}$$

Therefore, for $W_1 = -\log \Lambda_1$, $W_2 = -\log \Lambda_2$, and $W_3 = -\log \Lambda_3$, we may write the exact c.f. of $W = -\log \Lambda$ as

$$\begin{aligned}
\Phi_W(t) &= \Phi_{W_1}(t) \left(\Phi_{W_2}(t)\right)^{\text{mod}(m+1,2)} \Phi_{W_3}(t) \\
&= \underbrace{\begin{aligned}&\prod_{j=1}^{r+q-4} \left(\frac{n-r-q+j}{2}\right)^{s_j^*} \left(\frac{n-r-q+j}{2} - \mathrm{i}t\right)^{-s_j^*} \\
&\times \prod_{\substack{k=2 \\ \text{step } 2}}^{r+2q-5} \left(\frac{2n-2q-r+k}{4}\right)^{r_k^*} \left(\frac{2n-2q-r+k}{4} - \mathrm{i}t\right)^{-r_k^*} \\
&\times \left(\frac{n-3}{2}\right)^{r_{r+2q-6}^*} \left(\frac{n-3}{2} - \mathrm{i}t\right)^{-r_{r+2q-6}^*} \\
&\times \left(\frac{n-2}{2}\right)^{r_{r+2q-4}^*} \left(\frac{n-2}{2} - \mathrm{i}t\right)^{-r_{r+2q-4}^*}\end{aligned}}_{\Phi_{W,1}(t)} \\
&\qquad \times \underbrace{\left(\frac{\Gamma\left(\frac{n-1}{2}\right)\Gamma\left(\frac{n-2}{2} - \mathrm{i}t\right)}{\Gamma\left(\frac{n-2}{2}\right)\Gamma\left(\frac{n-1}{2} - \mathrm{i}t\right)}\right)^{1+\text{mod}(m+1,2)}}_{\Phi_{W,2}(t)} .
\end{aligned}$$

Then, in order to build the near-exact distributions for W and Λ, we will

- keep $\Phi_{W,1}(t)$ unchanged, since it is the c.f. of a GIG distribution, and
- asymptotically approximate $\Phi_{W,2}(t)$, which is the c.f. of the sum of $1 + \text{mod}(m+1, 2)$ $Logbeta\left(\frac{n-2}{2}, \frac{1}{2}\right)$ distributions, with a very sharp asymptotic approximation.

In fact, from the two first expressions in Sect. 5 of Tricomi and Erdélyi [25] and also from expressions (11) and (14) in the same reference, we may write

$$\frac{\Gamma(a - \mathrm{i}t)}{\Gamma(a + b - \mathrm{i}t)} \simeq \sum_{\ell=0}^{\infty} p_\ell(b)(a - \mathrm{i}t)^{-b-\ell}$$

where $p_0(b) = 1$ and for $\ell = 1, 2, \ldots,$

$$p_\ell(b) = \frac{1}{\ell} \sum_{m=0}^{\ell-1} \left(\frac{\Gamma(1-b-m)}{\Gamma(-b-\ell)(\ell-m+1)!} + (-1)^{\ell+m} b^{\ell-m+1} \right) p_m(b),$$

so that, since the h-th moment of a RV X with a $Beta(a, b)$ distribution is given by

$$\mathrm{E}\left(X^h\right)=\frac{\Gamma(a+b)}{\Gamma(a)}\frac{\Gamma(a+h)}{\Gamma(a+b+h)},$$

expression that remains valid for any $h \in \mathbb{C}$, we may write the c.f. of a RV $Y = -\log X$, which will be a RV with a $Logbeta(a, b)$ distribution, as

$$\begin{aligned}\Phi_Y(t) &= \mathrm{E}\left(e^{\mathrm{i}tY}\right)=\mathrm{E}\left(e^{-\mathrm{i}t\log X}\right)=\mathrm{E}\left(X^{-\mathrm{i}t}\right)\\ &=\frac{\Gamma(a+b)}{\Gamma(a)}\frac{\Gamma(a-\mathrm{i}t)}{\Gamma(a+b-\mathrm{i}t)}\simeq\frac{\Gamma(a+b)}{\Gamma(a)}\sum_{\ell=0}^{\infty}p_\ell(b)\,(a-\mathrm{i}t)^{-(b+\ell)}\\ &=\sum_{\ell=0}^{\infty}\underbrace{\frac{\Gamma(a+b)}{\Gamma(a)}\frac{p_\ell(b)}{a^{b+\ell}}}_{p_\ell^*(a,b)}a^{b+\ell}(a-\mathrm{i}t)^{-(b+\ell)},\end{aligned}$$

which is the c.f. of an infinite mixture of $\Gamma(b+\ell, a)$ $(\ell = 0, 1, \dots)$ distributions, with weights $p_\ell^*(\alpha, \beta)$.

As such, since, as noted above, $\Phi_{W,2}(t)$ is the c.f. of a sum of $1+\mathrm{mod}(m+1, 2)$ $Logbeta\left(\frac{n-2}{2}, \frac{1}{2}\right)$ distributions, using the above result, we may approximate $\Phi_{W,2}(t)$ by the c.f. of the sum of $1+\mathrm{mod}(m+1, 2)$ infinite mixtures of $\Gamma\left(\frac{1}{2}+\ell, \frac{n-2}{2}\right)$ distributions, which is an infinite mixture of $\Gamma\left(\frac{1+\mathrm{mod}(m+1,2)}{2}+\ell, \frac{n-2}{2}\right)$ distributions.

This way, using a somewhat heuristic approach, we will approximate $\Phi_{W,2}(t)$ by the c.f. of a finite mixture of $\Gamma\left(\frac{1+\mathrm{mod}(m+1,2)}{2}+\ell, \frac{n-2}{2}\right)$ distributions, or, more precisely, we will, for a given positive integer ν, approximate $\Phi_{W,2}(t)$ by

$$\Phi_2^*(t)=\sum_{\ell=0}^{\nu}\pi_\ell\left(\frac{n-2}{2}\right)^{\frac{1+\mathrm{mod}(m+1,2)}{2}+\ell}\left(\frac{n-2}{2}-\mathrm{i}t\right)^{-\left(\frac{1+\mathrm{mod}(m+1,2)}{2}+\ell\right)}$$

where the weights π_ℓ, $\ell = 0, \dots, \nu-1$ will be determined in such a way that the first ν derivatives of $\Phi_{W,2}(t)$ and $\Phi_2^*(t)$ relative to t, at $t = 0$, are equal, that is, in such a way that

$$\left.\frac{\partial^h}{\partial t^h}\Phi_{W,2}(t)\right|_{t=0}=\left.\frac{\partial^h}{\partial t^h}\Phi_2^*(t)\right|_{t=0},\qquad h=1,\dots,\nu,$$

and with $\pi_\nu = 1 - \sum_{\ell=0}^{\nu-1}\pi_\ell$.

By doing this, one obtains as near-exact c.f. for W

$$
\begin{aligned}
\Phi^*_W(t) &= \Phi_{W,1}(t)\,\Phi^*_2(t)\\
&= \sum_{\ell=0}^{\nu} \pi_\ell \left(\frac{n-2}{2}\right)^{\frac{1+\text{mod}(m+1,2)}{2}+\ell} \left(\frac{n-2}{2}-\text{i}t\right)^{-\left(\frac{1+\text{mod}(m+1,2)}{2}+\ell\right)} \times \Phi_{W,1}(t)\\
&= \sum_{\ell=0}^{\nu} \pi_\ell \left(\frac{n-2}{2}\right)^{r^*_{r+2q-4}+\frac{1+\text{mod}(m+1,2)}{2}+\ell} \left(\frac{n-2}{2}-\text{i}t\right)^{-\left(r^*_{r+2q-4}+\frac{1+\text{mod}(m+1,2)}{2}+\ell\right)}\\
&\quad\times \prod_{j=1}^{r+q-4} \left(\frac{n-r-q+j}{2}\right)^{s^*_j} \left(\frac{n-r-q+j}{2}-\text{i}t\right)^{-s^*_j}\\
&\quad\times \prod_{\substack{k=2\\ \text{step } 2}}^{r+2q-5} \left(\frac{2n-2q-r+k}{4}\right)^{r^*_k} \left(\frac{2n-2q-r+k}{4}-\text{i}t\right)^{-r^*_k}\\
&\quad\times \left(\frac{n-3}{2}\right)^{r^*_{r+2q-6}} \left(\frac{n-3}{2}-\text{i}t\right)^{-r^*_{r+2q-6}}
\end{aligned}
$$

which is the c.f. of a mixture with $\nu+1$ components, each of which is a GNIG distribution in case m is odd, or a GIG distribution in case m is even, in either case with depth $r+2q-4+(r-1)/2$.

The p.d.f.'s and c.d.f.'s of these near-exact distributions for $W=-\log\Lambda$ are, for odd m, respectively, given by (see Appendix C for the definition of the GNIG distribution and its p.d.f. and c.d.f.)

$$
f_W(w) = \sum_{\ell=0}^{\nu} \pi_\ell\, f^{GNIG}\Big(w \,\Big|\, \{u_j\}_{j=1:g}, r^*_{r+2q-4}+\tfrac{1+\text{mod}(m+1,2)}{2}+\ell;\ \{\lambda_j\}_{j=1:g}, \tfrac{n-2}{2};\ g\Big)
$$

and

$$
F_W(w) = \sum_{\ell=0}^{\nu} \pi_\ell\, F^{GNIG}\Big(w \,\Big|\, \{u_j\}_{j=1:g}, r^*_{r+2q-4}+\tfrac{1+\text{mod}(m+1,2)}{2}+\ell;\ \{\lambda_j\}_{j=1:g}, \tfrac{n-2}{2};\ g\Big)
$$

for $w>0$, and for Λ by

$$
\begin{aligned}
f_\Lambda(\lambda) = \sum_{\ell=0}^{\nu} \pi_\ell\, f^{GNIG}\Big(-\log\lambda \,\Big|\, &\{u_j\}_{j=1:g}, r^*_{r+2q-4}+\tfrac{1+\text{mod}(m+1,2)}{2}+\ell;\\
&\{\lambda_j\}_{j=1:g}, \tfrac{n-2}{2};\ g\Big)\,\frac{1}{\lambda}
\end{aligned}
$$

and

$$F_\Lambda(\lambda) \qquad = (\nu+1) - \sum_{\ell=0}^{\nu} \pi_\ell \, F^{GNIG}$$
$$\left(-\log\lambda \,\middle|\, \{u_j\}_{j=1:g}, r^*_{r+2q-4} + \tfrac{1+\text{mod}(m+1,2)}{2} + \ell;\ \{\lambda_j\}_{j=1:g}, \tfrac{n-2}{2}; g\right)$$

for $0 < \lambda < 1$, where $g = r + 2q - 4 + (r-1)/2$,

$$\{u_j\}_{j=1:g} = \left\{ \{s_j^*\}_{j=1:r+q-4}, \{r_k^*\}_{\substack{k=2:r+2q-5\\ \text{step } 2}}, r^*_{r+2q-6}, r^*_{r+2q-4} \right\},$$

and

$$\{\lambda_j\}_{j=1:g} = \left\{ \left\{\frac{n-r-q+j}{2}\right\}_{j=1:r+q-4}, \left\{\frac{2n-2q-r+k}{4}\right\}_{\substack{k=2:r+2q-5\\ \text{step } 2}}, \frac{n-3}{2}, \frac{n-2}{2} \right\},$$

and where the p.d.f. and c.d.f. of the GNIG distribution are to be replaced by the p.d.f. and c.d.f. of the GIG distribution for the case of even m.

We may note that these near-exact distributions match the ν first exact moments of W.

7.4 Numerical Studies for the Near-Exact Distributions

Even though we do not have the expression for the exact c.d.f. of Λ or W, we can measure the distance between the near-exact c.d.f. and the exact c.d.f., through the use of the measure

$$\Delta = \frac{1}{2\pi} \int_{-\infty}^{+\infty} \left| \frac{\Phi_W(t) - \Phi_W^*(t)}{t} \right| dt$$

with

$$\Delta \geq \sup_{w>0} \left| F_W(w) - F_W^*(w) \right| = \sup_{0<\ell<1} \left| F_\Lambda(\ell) - F_\Lambda^+(\ell) \right|,$$

where $\Phi_W(t)$ and $\Phi_W^*(t)$ represent, respectively, the exact and the near-exact c.f.'s of W, and where $F_W(\cdot)$ and $F_\Lambda(\cdot)$ are, respectively, the exact c.d.f. of $W = -\log\Lambda$ and Λ, and $F_W^*(\cdot)$ is the near-exact c.d.f. which corresponds to the c.f. $\Phi_W^*(\cdot)$, with $F_\Lambda^*(\ell) = 1 - F_W^+(-\log\ell)$. In Tables 7.1, 7.2, and 7.3 are shown values of the measure Δ for various values of n, the sample size, r, the number of variables in each subset, q, the number of populations in the study, and m, the number of subsets of variables. In each of these tables, two of these parameters are kept constant, while the sample size and the other parameters evolve in value. This is done in order to allow the reader to evaluate the asymptotic properties of the near-exact distributions developed.

Table 7.1 Values of the measure Δ for increasing values of r and increasing sample sizes, for the near-exact distributions that match ν exact moments of W

n	r	q	m	ν (# of exact moments of W matched)		
				2	6	10
7	3	4	4	$2.83 \cdot 10^{-7}$	$2.30 \cdot 10^{-12}$	$7.40 \cdot 10^{-16}$
57				$1.76 \cdot 10^{-9}$	$2.46 \cdot 10^{-17}$	$9.95 \cdot 10^{-24}$
107				$2.62 \cdot 10^{-10}$	$2.96 \cdot 10^{-19}$	$1.11 \cdot 10^{-26}$
11	7	4	4	$9.33 \cdot 10^{-9}$	$1.47 \cdot 10^{-16}$	$1.26 \cdot 10^{-20}$
61				$3.78 \cdot 10^{-10}$	$8.18 \cdot 10^{-19}$	$6.23 \cdot 10^{-26}$
111				$6.44 \cdot 10^{-11}$	$1.35 \cdot 10^{-20}$	$1.12 \cdot 10^{-28}$
19	15	4	4	$2.19 \cdot 10^{-10}$	$2.72 \cdot 10^{-19}$	$2.28 \cdot 10^{-26}$
69				$7.20 \cdot 10^{-11}$	$1.87 \cdot 10^{-20}$	$1.93 \cdot 10^{-28}$
119				$1.55 \cdot 10^{-11}$	$5.26 \cdot 10^{-22}$	$7.73 \cdot 10^{-31}$
29	25	4	4	$1.39 \cdot 10^{-11}$	$6.46 \cdot 10^{-22}$	$6.08 \cdot 10^{-31}$
79				$1.89 \cdot 10^{-11}$	$8.61 \cdot 10^{-22}$	$1.66 \cdot 10^{-30}$
129				$5.15 \cdot 10^{-12}$	$4.16 \cdot 10^{-23}$	$1.51 \cdot 10^{-32}$

Table 7.2 Values of the measure Δ for increasing values of q and increasing sample sizes, for the near-exact distributions that match ν exact moments of W

n	r	q	m	ν (# of exact moments of W matched)		
				2	6	10
19	15	4	4	$2.19 \cdot 10^{-10}$	$2.72 \cdot 10^{-19}$	$2.28 \cdot 10^{-26}$
69				$7.20 \cdot 10^{-11}$	$1.87 \cdot 10^{-20}$	$1.93 \cdot 10^{-28}$
119				$1.55 \cdot 10^{-11}$	$5.26 \cdot 10^{-22}$	$7.73 \cdot 10^{-31}$
21	15	6	4	$8.08 \cdot 10^{-11}$	$2.53 \cdot 10^{-20}$	$2.59 \cdot 10^{-28}$
71				$2.94 \cdot 10^{-11}$	$2.37 \cdot 10^{-21}$	$7.82 \cdot 10^{-30}$
121				$6.68 \cdot 10^{-12}$	$7.59 \cdot 10^{-23}$	$3.85 \cdot 10^{-32}$
25	15	10	4	$1.94 \cdot 10^{-11}$	$9.49 \cdot 10^{-22}$	$6.77 \cdot 10^{-32}$
75				$9.51 \cdot 10^{-12}$	$1.73 \cdot 10^{-22}$	$1.33 \cdot 10^{-31}$
125				$2.39 \cdot 10^{-12}$	$7.03 \cdot 10^{-24}$	$9.41 \cdot 10^{-34}$

In each table are considered near-exact distributions that match the $\nu = 2, 6$, and 10 first exact moments of $W = -\log \Lambda$, and while from each of the three tables, the asymptotic behavior of the near-exact distributions for increasing sample sizes is clear, from Table 7.1 we may ascertain the asymptotic behavior of the near-exact distributions developed for increasing values of r, the number of variables in each subset, from Table 7.2 the asymptotic behavior of these distributions for increasing values of q, the number of populations involved in the test, and from Table 7.3 the asymptotic behavior for increasing values of m, the number of subsets of variables.

Table 7.3 Values of the measure Δ for increasing values of m and increasing sample sizes, for the near-exact distributions that match ν exact moments of W

n	r	q	m	ν (# of exact moments of W matched)		
				2	6	10
19	15	4	4	$2.19 \cdot 10^{-10}$	$2.72 \cdot 10^{-19}$	$2.28 \cdot 10^{-26}$
69				$7.20 \cdot 10^{-11}$	$1.87 \cdot 10^{-20}$	$1.93 \cdot 10^{-28}$
119				$1.55 \cdot 10^{-11}$	$5.26 \cdot 10^{-22}$	$7.73 \cdot 10^{-31}$
19	15	4	6	$1.81 \cdot 10^{-10}$	$1.55 \cdot 10^{-19}$	$8.14 \cdot 10^{-27}$
69				$4.06 \cdot 10^{-11}$	$4.97 \cdot 10^{-21}$	$2.46 \cdot 10^{-29}$
119				$8.59 \cdot 10^{-12}$	$1.36 \cdot 10^{-22}$	$9.46 \cdot 10^{-32}$
19	15	4	10	$1.33 \cdot 10^{-10}$	$6.64 \cdot 10^{-20}$	$1.81 \cdot 10^{-27}$
69				$1.94 \cdot 10^{-11}$	$9.01 \cdot 10^{-22}$	$1.71 \cdot 10^{-30}$
119				$4.06 \cdot 10^{-12}$	$2.39 \cdot 10^{-23}$	$6.35 \cdot 10^{-33}$

7.5 The Case of Block Compound-Symmetric Matrices

Using a similar approach, we may address rather simply the case of block compound-symmetric covariance matrices.

We say that the matrix $\boldsymbol{\Sigma}$ has a block compound-symmetric (BCS) structure, with m blocks $\boldsymbol{\Sigma}_0$ of dimensions $r \times r$ on the diagonal and off-diagonal blocks $\boldsymbol{\Sigma}_1$, also of dimensions $r \times r$, if

$$\begin{aligned}\boldsymbol{\Sigma} &= \mathbf{I}_m \otimes (\boldsymbol{\Sigma}_0 - \boldsymbol{\Sigma}_1) + \mathbf{J}_m \otimes \boldsymbol{\Sigma}_1 \\ &= \mathbf{I}_m \otimes \boldsymbol{\Sigma}_0 + (\mathbf{J}_m - \mathbf{I}_m) \otimes \boldsymbol{\Sigma}_1\end{aligned}$$

where $\mathbf{I}_m$ is an identity matrix of order m and $\mathbf{J}_m$ an $m \times m$ matrix of 1's, and where $\boldsymbol{\Sigma}_0$ is a positive-definite matrix and $\boldsymbol{\Sigma}_1$ a symmetric matrix such that $\boldsymbol{\Sigma}_0 - \boldsymbol{\Sigma}_1$ and $\boldsymbol{\Sigma}_0 + (m-1)\boldsymbol{\Sigma}_1$ are both positive-definite matrices.

We are now interested in testing a similar null hypothesis to that in (7.4), but now assuming that $\boldsymbol{\Sigma}$ has a BCS structure.

From Appendix B, we may easily see that if all $\mathbf{A}_i^*$ are equal for $i = 1, \ldots, m-1$, that is, if $\mathbf{A}^*$ is BCS, we will have $\widehat{\boldsymbol{\Psi}}_2 = \cdots = \widehat{\boldsymbol{\Psi}}_m$, so that if $\boldsymbol{\Sigma}$ has a BCS structure, then

$$(\boldsymbol{\Gamma}_m \otimes \mathbf{I}_r)\boldsymbol{\Sigma}(\boldsymbol{\Gamma}_m \otimes \mathbf{I}_r) = \boldsymbol{\Psi} = \text{BDiag}(\boldsymbol{\Psi}_1, \underbrace{\boldsymbol{\Psi}_2, \ldots, \boldsymbol{\Psi}_2}_{m-1}) \tag{7.34}$$

where $\boldsymbol{\Psi}_1$ and $\boldsymbol{\Psi}_2$ are positive-definite matrices, actually with $\boldsymbol{\Psi}_1 = \boldsymbol{\Sigma}_0 + (m-1)\boldsymbol{\Sigma}_1$ and $\boldsymbol{\Psi}_2 = \boldsymbol{\Sigma}_0 - \boldsymbol{\Sigma}_1$, and where $\boldsymbol{\Gamma}_m$ is the $m \times m$ orthogonal symmetric matrix that was used in Sect. 7.3, with elements γ_{jk} $(j, k = 1, \ldots, m)$, given by (7.16) and I_r is an identity matrix of order r.

Using a similar procedure to the one used in Appendix A, it is easy to show that also $\boldsymbol{\Sigma}^{-1}$ is a BCS matrix. Thus, we may use the results in Szatrowski [23, 24] in estimating $\boldsymbol{\Sigma}$ and take the MLE of $\boldsymbol{\Sigma}$ given by

$$\widehat{\boldsymbol{\Sigma}} = I_m \otimes \widehat{\boldsymbol{\Sigma}}_0 + (\mathbf{J}_m - \mathbf{I}_m) \otimes \widehat{\boldsymbol{\Sigma}}_1$$

where under H_1

$$\widehat{\boldsymbol{\Sigma}}_{0|_{H_1}} = \frac{1}{m}\sum_{j=1}^{m} \widetilde{\mathbf{A}}_{jj} \quad \text{and} \quad \widehat{\boldsymbol{\Sigma}}_{1|_{H_1}} = \frac{1}{m(m-1)}\sum_{j=1}^{m}\sum_{k=j+1}^{m} (\widetilde{\mathbf{A}}_{jk} + \widetilde{\mathbf{A}}_{kj})$$

and under H_0

$$\widehat{\boldsymbol{\Sigma}}_{0|_{H_1}} = \frac{1}{m}\sum_{j=1}^{m}(\widetilde{\mathbf{A}}_{jj} + \widetilde{\mathbf{B}}_{jj}) \text{ and } \widehat{\boldsymbol{\Sigma}}_{1|_{H_1}} = \frac{1}{m(m-1)}\sum_{j=1}^{m}\sum_{k=j+1}^{m} (\widetilde{\mathbf{A}}_{jk} + \widetilde{\mathbf{A}}_{kj} + \widetilde{\mathbf{B}}_{jk} + \widetilde{\mathbf{B}}_{kj}),$$

where, as before, $\widetilde{\mathbf{A}}_{jk}$ and $\widetilde{\mathbf{B}}_{jk}$ $(j, k = 1, \ldots, m)$, are the $r \times r$ blocks of $\mathbf{A}$ and $\mathbf{B}$ in (7.6), starting at row $r(j-1)+1$ and column $r(k-1)+1$. That is, the MLEs of $\boldsymbol{\Sigma}_0$, respectively, under H_1 and H_0, are the averages of the diagonal blocks of the matrices $\mathbf{A}$ and $\mathbf{A}+\mathbf{B}$, while the MLEs of $\boldsymbol{\Sigma}_1$ are the averages of the off-diagonal blocks of the same matrices.

Then, the LRT statistic to test H_0 in (7.4), assuming the common variance–covariance matrix to be BCS, will be

$$\Lambda = \frac{|\mathbf{A}^*|}{|\mathbf{A}^* + \mathbf{B}^*|}$$

where

$$\mathbf{A}^* = \mathbf{I}_m \otimes \widehat{\boldsymbol{\Sigma}}_{0|_{H_1}} + (\mathbf{J}_m - \mathbf{I}_m) \otimes \widehat{\boldsymbol{\Sigma}}_{1|_{H_1}}$$

and

$$\mathbf{A}^* + \mathbf{B}^* = \mathbf{I}_m \otimes \widehat{\boldsymbol{\Sigma}}_{0|_{H_0}} + (\mathbf{J}_m - \mathbf{I}_m) \otimes \widehat{\boldsymbol{\Sigma}}_{1|_{H_0}}\,.$$

Let us consider the matrices $\mathbf{A}^{**}$ and $\mathbf{B}^{**}$ in (7.15). Then, is possible to show that (see Appendix E)

$$|\mathbf{A}^*| = |\mathbf{A}_1^{**}| \left|\frac{\sum_{j=2}^{m}\mathbf{A}_j^{**}}{m-1}\right|^{m-1} \tag{7.35}$$

and, in a similar manner, that

$$|\mathbf{A}^* + \mathbf{B}^*| = |\mathbf{A}_1^{**} + \mathbf{B}_1^{**}| \left|\frac{\sum_{j=2}^{m}(\mathbf{A}_j^{**} + \mathbf{B}_j^{**})}{m-1}\right|^{m-1},$$

where $\mathbf{A}_j^{**}$ and $\mathbf{B}_j^{**}$ $(j = 1, \ldots, m)$ denote, as before, the j-th diagonal block, respectively, of $\mathbf{A}^{**}$ and $\mathbf{B}^{**}$ of order $r \times r$.

Thus, we may then write

$$\Lambda = \underbrace{\frac{|\mathbf{A}_1^{**}|}{|\mathbf{A}_1^{**} + \mathbf{B}_1^{**}|}}_{\Lambda_1} \underbrace{\left(\frac{\left|\sum_{j=2}^{m} \mathbf{A}_j^{**}\right|}{\left|\sum_{j=2}^{m} (\mathbf{A}_j^{**} + \mathbf{B}_j^{**})\right|} \right)^{m-1}}_{\Lambda_2}, \tag{7.36}$$

where, since we now have

$$\mathbf{A} \sim W_p(n-q, \boldsymbol{\Sigma}) \implies \mathbf{A}^{**} \sim W_p(n-q, \boldsymbol{\Psi})$$

and

$$\mathbf{B} \sim W_p(q-1, \boldsymbol{\Sigma}) \implies \mathbf{B}^{**} \sim W_p(q-1, \boldsymbol{\Psi})$$

for $\boldsymbol{\Psi}$ in (7.34). But so, as before, the diagonal blocks $\mathbf{A}_j^{**}$ and $\mathbf{B}_l^{**}$ are all independent among themselves, now with

$$\begin{aligned} &\mathbf{A}_1^{**} \sim W_r(n-q, \boldsymbol{\Psi}_1), \quad \mathbf{B}_1^{**} \sim W_r(q-1, \boldsymbol{\Psi}_1) \\ &\mathbf{A}_j^{**} \sim W_r(n-q, \boldsymbol{\Psi}_2), \quad \mathbf{B}_j^{**} \sim W_r(q-1, \boldsymbol{\Psi}_2), \quad j = 2, \ldots, m, \end{aligned}$$

and as such with

$$\mathbf{A}_1^{**} + \mathbf{B}_1^{**} \sim W_r(n-1, \boldsymbol{\Psi}_1)$$

and

$$\mathbf{A}_j^{**} + \mathbf{B}_j^{**} \sim W_r(n-1, \boldsymbol{\Psi}_2), \quad j = 2, \ldots, m,$$

and also

$$\sum_{j=2}^{m} \mathbf{A}_j^{**} \sim W_r\left((m-1)(n-q), \boldsymbol{\Psi}_2\right)$$

and

$$\sum_{j=2}^{m} (\mathbf{A}_j^{**} + \mathbf{B}_j^{**}) \sim W_r\left((m-1)(n-1), \boldsymbol{\Psi}_2\right).$$

Hence, we have that

$$\Lambda_1 = \frac{|\mathbf{A}_1^{**}|}{|\mathbf{A}_1^{**} + \mathbf{B}_1^{**}|} \sim \prod_{j=1}^{r} Y_j \tag{7.37}$$

where

$$Y_j \sim Beta\left(\frac{n-q+1-j}{2}, \frac{q-1}{2}\right), \quad j = 1, \ldots, r,$$

are a set of independent RVs, where r and $q-1$ are interchangeable (see Coelho and Arnold [6, Chap. 5, Appendix 1]), and

$$\Lambda_2 = \left(\frac{\left|\sum_{j=2}^{m} \mathbf{A}_j^{**}\right|}{\left|\sum_{j=2}^{m}(\mathbf{A}_j^{**}+\mathbf{B}_j^{**})\right|} \right)^{m-1} \sim \prod_{k=1}^{r} \left(Y_k^*\right)^{m-1} \tag{7.38}$$

where

$$Y_k^* \sim Beta\left(\frac{(m-1)(n-q)}{2}+\frac{1-j}{2}, \frac{(m-1)(q-1)}{2}\right), \quad k=1,\dots,r,$$

also form a set of independent RVs, and where r and $(m-1)(q-1)$ are interchangeable, so that we may also write

$$\Lambda_2 \sim \prod_{k=1}^{(m-1)(q-1)} \left(Y_k^{**}\right)^{m-1}$$

where

$$Y_k^{**} \sim Beta\left(\frac{(m-1)(n-1)+1-k-r}{2}, \frac{r}{2}\right), \quad k=1,\dots,(m-1)(q-1),$$

form a set of independent RVs, and where Λ_1 and Λ_2 are two independent statistics.

7.5.1 The Exact Distribution of Λ in (7.36) for Odd q or Even r

The distribution of the statistic Λ_1 in (7.36) and (7.37) is exactly the same as that of the statistic Λ_1 in (7.19), and as such the c.f. of $W_1 = -\log \Lambda_1$ may be written, for odd q or even r as in (7.29), with s_j given by (7.28), which shows that its distribution is a GIG distribution of depth $r+q-3$ with rate parameters $\frac{n-q}{2}+\frac{j-r}{2}$ $(j=1,\dots,r+q-3)$ and corresponding shape parameters s_j, given by (7.28), a result that, also as already stated in Sect. 7.3.1, may also be obtained from Theorem 3.2 in Coelho and Arnold [6].

In what concerns the distribution of Λ_2 in (7.36) and (7.38), for odd q or odd m, its distribution may be obtained from Theorem 7.1 in Sect. 7.3 with

$$m^*=1,\ a_1=\frac{n-q}{2},\ n_1{=}r,\ k_1{=}m-1,\ m_1=\frac{(m-1)(q-1)}{2},\ h=2$$

and for even r with

$$m^*{=}1,\ a_1{=}\frac{n-1}{2}{-}\frac{r}{2(m-1)},\ n_1=(m-1)(q-1),\ k_1=m-1,\ m_1=\frac{r}{2},\ h=2$$

as an EGIG distribution of depth $r+(m-1)(q-1)-2$, with rate parameters

$$\frac{n-q}{2}+\frac{k-r}{2(m-1)},\quad k=1,\ldots,r+(m-1)(q-1)-2$$

and shape parameters

$$r_k=\begin{cases} h_k, & k=1,2 \\ h_k+r_{k-2}, & k=3,\ldots,r+(m-1)(q-1)-2 \end{cases} \tag{7.39}$$

with

$$h_k=(\#\text{ of elements in}\{r,(m-1)(q-1)\}\geq \mathrm{k})-1 \tag{7.40}$$

for $k=1,\ldots,r+(m-1)(q-1)-2$.

As such, for even r or odd q, the c.f. of $W=-\log\Lambda$, for Λ in (7.36), may be written as

$$\Phi_W(t)=\prod_{j=1}^{r+q-3}\left(\frac{n-r-q+j}{2}\right)^{s_j}\left(\frac{n-r-q+j}{2}+h\right)^{-s_j}$$
$$\times\prod_{k=1}^{r+(m-1)(q-1)-2}\left(\frac{n-q}{2}+\frac{k-r}{2(m-1)}\right)^{r_k}\left(\frac{n-q}{2}+\frac{k-r}{2(m-1)}-\mathrm{i}t\right)^{-r_k} \tag{7.41}$$

which is the c.f. of a GIG distribution. That is, for even r or odd q, the exact distribution of $W=-\log\Lambda$ is a GIG distribution with rate parameters $\frac{n-q}{2}+\frac{j-r}{2}$ ($j=1,\ldots,r+q-3$) and $\frac{n-q}{2}+\frac{k-r}{2(m-1)}$ ($k=1,\ldots,(m-1)(q-1)-2$) and corresponding rate parameters s_j ($j=1,\ldots,r+q-3$) and r_k ($k=1,\ldots,r+(m-1)(q-1)-2$). However, we should be aware that the depth of this GIG distribution is indeed smaller than $r+q-3+r+(m-1)(q-1)-2=2r+mq-m-4$ since some of the rate parameters $\frac{n-q}{2}+\frac{k-r}{2(m-1)}$ match some, or even all, of the rate parameters $\frac{n-q}{2}+\frac{j-r}{2}$ in the first product. More precisely, the rate parameters in the first product in (7.41) will be matched by rate parameters in the second product for $j=j_m=1+\left\lfloor\frac{1+(m-2)r}{m-1}\right\rfloor$ through $j=j_M=q+r-3$, and as such, the rate parameters in the second product that match rate parameters in the first product will be those with

$$k=(m-1)\left(1+\left\lfloor\frac{1+(m-2)r}{m-1}\right\rfloor\right)-r(m-2)$$

through

$$k=(m-1)(q+r-3)-r(m-2)$$

with a step of $m-1$. As such the depth of this GIG distribution that yields the exact distribution of $W=-\log\Lambda$ is indeed equal to $r+(m-1)(q-1)-2+\left\lfloor\frac{1+(m-2)r}{m-1}\right\rfloor$. In fact, using a notation where rate parameters do not appear repeated, we may write $\Phi_W(t)$ as

$$\Phi_W(t)=\prod_{j=1}^{\left\lfloor\frac{1+(m-2)r}{m-1}\right\rfloor}\left(\frac{n-r-q+j}{2}\right)^{s_j}\left(\frac{n-r-q+j}{2}+h\right)^{-s_j}$$
$$\times\prod_{k=1}^{r+(m-1)(q-1)-2}\left(\frac{n-q}{2}+\frac{k-r}{2(m-1)}\right)^{r_k^*}\left(\frac{n-q}{2}+\frac{k-r}{2(m-1)}-\mathrm{i}t\right)^{-r_k^*}$$

where, if we denote by K^* the set

$$\left\{k:k=(m-1)\left(1+\left\lfloor\frac{1+(m-2)r}{m-1}\right\rfloor\right)-r(m-2),\ldots,\right.$$
$$\left.(m-1)(q+r-3)-r(m-2),\ \text{step}\ m-1\right\},$$

then we have

$$r_k^*=\begin{cases}r_k, & \text{for } k\in\{1,\ldots,r+(m-1)(q-1)-2\}\setminus K^*\\ r_k+s_{(k+r(m-2))/(m-1)}, & \text{for } k\in K^*\end{cases}\tag{7.42}$$

for r_k $(k=1,\ldots,r+(m-1)(q-1)-2)$ given by (7.39)–(7.40) and s_j $(j=1,\ldots,r+q-3)$ given by (7.28).

Therefore, the exact distribution of Λ is, in this case, an EGIG distribution of the same depth with the rate and shape parameters above, being thus very easy to handle with any of the available symbolic software.

7.5.2 *Near-Exact Distributions for Λ in (7.36) for Odd r and Even q*

In case r is odd and q is even, the exact distribution of Λ in (7.36) is not manageable and a good choice is to resort to the use of near-exact distributions as we did for the distribution of Λ in (7.19) for the same case.

We should note that even when r is odd and q is even, if m is odd, the exact distribution of $W_2=-\log\Lambda_2$ for Λ_2 in (7.36) and (7.38) is still the GIG distribution in the previous subsection, while the c.f. of $W_1=-\log\Lambda_1$ is given by (7.31), with

h replaced by $-\mathrm{i}t$. Therefore, for r odd, q even, and m odd, we may write the c.f. of $W=-\log\Lambda$ as

$$\Phi_W(t)=\prod_{j=1}^{r+q-4}\left(\frac{n-r-q+j}{2}\right)^{s_j^*}\left(\frac{n-r-q+j}{2}-\mathrm{i}t\right)^{-s_j^*}$$
$$\times\prod_{k=1}^{r+(m-1)(q-1)-2}\left(\frac{n-q}{2}+\frac{k-r}{2(m-1)}\right)^{r_k}\left(\frac{n-q}{2}+\frac{k-r}{2(m-1)}-\mathrm{i}t\right)^{-r_k}$$
$$\times\frac{\Gamma\left(\frac{n-1}{2}\right)}{\Gamma\left(\frac{n-2}{2}\right)}\frac{\Gamma\left(\frac{n-2}{2}-\mathrm{i}t\right)}{\Gamma\left(\frac{n-1}{2}-\mathrm{i}t\right)},$$

where now the s_j are given by (7.32)–(7.33).

However, in the expression above, there is repetition of some of the rate parameters in both products. If one wants to avoid this repetition, we write $\Phi_W(t)$ as

$$\Phi_W(t)=\underbrace{\prod_{j=1}^{\left\lfloor\frac{1+(m-2)r}{m-1}\right\rfloor}\left(\frac{n-r-q+j}{2}\right)^{s_j}\left(\frac{n-r-q+j}{2}+h\right)^{-s_j}\times\prod_{k=1}^{r+(m-1)(q-1)-2}\left(\frac{n-q}{2}+\frac{k-r}{2(m-1)}\right)^{r_k^*}\left(\frac{n-q}{2}+\frac{k-r}{2(m-1)}-\mathrm{i}t\right)^{-r_k^*}}_{\Phi_{W,1}(t)}$$
$$\times\underbrace{\frac{\Gamma\left(\frac{n-1}{2}\right)}{\Gamma\left(\frac{n-2}{2}\right)}\frac{\Gamma\left(\frac{n-2}{2}-\mathrm{i}t\right)}{\Gamma\left(\frac{n-1}{2}-\mathrm{i}t\right)}}_{\Phi_{W,2}(t)},$$

where the r_k^* have a similar definition to the one in (7.42) but now for K^* defined as being the set

$$\left\{k:k=(m-1)\left(1+\left\lfloor\frac{1+(m-2)r}{m-1}\right\rfloor\right)-r(m-2),\ldots,\right.$$
$$\left.(m-1)(q+r-4)-r(m-2),\ \text{step } m-1\right\}.$$

In order to build the near-exact distributions, we will use a procedure in all similar to the one adopted in Sect. 7.3.2, keeping $\Phi_{W,1}(t)$ untouched and asymptotically approximating $\Phi_{W,2}(t)$ by the c.f.

$$\Phi_2^*(t)=\sum_{\ell=0}^{\nu}\pi_\ell\left(\frac{n-2}{2}\right)^{\frac{1}{2}+\ell}\left(\frac{n-2}{2}-\mathrm{i}t\right)^{-(\frac{1}{2}+\ell)}$$

where, similar to what was done in Sect. 7.3.2, the weights π_ℓ, $\ell=0,\dots,\nu-1$, will be determined in such a way that the first ν derivatives of $\Phi_{W,2}(t)$ and $\Phi_2^*(t)$ relative to t, at $t=0$, are equal, that is, in such a way that

$$\frac{\partial^h}{\partial t^h}\Phi_{W,2}(t)\bigg|_{t=0}=\frac{\partial^h}{\partial t^h}\Phi_2^*(t)\bigg|_{t=0},\qquad h=1,\dots,\nu,$$

and with $\pi_{m^*}=1-\sum_{\ell=0}^{\nu}\pi_\ell$.

This procedure will yield near-exact c.f.'s for W of the form

$$\begin{aligned}\Phi_W^*(t)&=\Phi_{W,1}(t)\,\Phi_2^*(t)\\&=\sum_{\ell=0}^{\nu}\pi_\ell\left(\frac{n-2}{2}\right)^{\frac{1}{2}+\ell}\left(\frac{n-2}{2}-\mathrm{i}t\right)^{-\left(\frac{1}{2}+\ell\right)}\\&\quad\times\prod_{j=1}^{\left\lfloor\frac{1+(m-2)r}{m-1}\right\rfloor}\left(\frac{n-r-q+j}{2}\right)^{s_j}\left(\frac{n-r-q+j}{2}+h\right)^{-s_j}\\&\quad\times\prod_{k=1}^{r+(m-1)(q-1)-2}\left(\frac{n-q}{2}+\frac{k-r}{2(m-1)}\right)^{r_k^*}\left(\frac{n-q}{2}+\frac{k-r}{2(m-1)}-\mathrm{i}t\right)^{-r_k^*}\end{aligned}$$

to which correspond a mixture of $\nu+1$ GNIG distributions of depth $r+(m-1)(q-1)-1+\left\lfloor\frac{1+(m-2)r}{m-1}\right\rfloor$, with weights π_ℓ $(\ell=0,\dots,\nu)$, rate parameters $\frac{n-r-q+j}{2}$ $\left(j=1,\dots,\left\lfloor\frac{1+(m-2)r}{m-1}\right\rfloor\right)$ $\frac{n-q}{2}+\frac{k-r}{2(m-1)}$ $(k=1,\dots,r+(m-1)(q-1)-2)$, and $\frac{n-2}{2}$, to which correspond the shape parameters s_j $\left(j=1,\dots,\left\lfloor\frac{1+(m-2)r}{m-1}\right\rfloor\right)$, r_k^* $(k=1,\dots,r+(m-1)(q-1)-2)$, and $\frac{1}{2}+\ell$ $(\ell=0,\dots,\nu)$.

We have to note that for $q=2$, the length of the GNIG distributions will be one less, because then the rate parameter $\frac{n-2}{2}$ will be matched by the r-th rate parameter in the second product, to which will then correspond a shape parameter of $(m-1)/2$ if $m\le r$ or of $(r+1)/2$ if $m>r$.

Similar to what happens with the near-exact distributions developed in Sect. 7.3.2, also these near-exact distributions are asymptotic for increasing sample sizes as well as for increasing values of q, r, and m, yielding approximations of very good quality, with performances of a comparable caliber to the ones developed in Sect. 7.3.2 for the test with block-circular matrices.

In case r is odd and both m and q are even, we will use the duplication formula for the Gamma function, which is the multiplication formula in (7.56) for $n=2$, and (7.21) to write the c.f. of $W_2=-\log\,\Lambda_2$ as

$$
\begin{aligned}
\Phi_{W_2}(t) &= \prod_{k=1}^{(m-1)(q-1)} \frac{\Gamma\left(\frac{(m-1)(n-1)+1-k}{2}\right)}{\Gamma\left(\frac{(m-1)(n-1)+1-k-r}{2}\right)} \frac{\Gamma\left(\frac{(m-1)(n-1)+1-k-r}{2}-(m-1)\mathrm{i}t\right)}{\Gamma\left(\frac{(m-1)(n-1)+1-k}{2}-(m-1)\mathrm{i}t\right)} \\
&= \frac{\Gamma\left(\frac{(m-1)(n-1)}{2}\right)}{\Gamma\left(\frac{(m-1)(n-1)-r}{2}\right)} \frac{\Gamma\left(\frac{(m-1)(n-1)-r}{2}-(m-1)\mathrm{i}t\right)}{\Gamma\left(\frac{(m-1)(n-1)}{2}-(m-1)\mathrm{i}t\right)} \\
&\quad\times \prod_{k=2}^{(m-1)(q-1)} \frac{\Gamma\left(\frac{(m-1)(n-1)+1-k}{2}\right)}{\Gamma\left(\frac{(m-1)(n-1)+1-k-r}{2}\right)} \frac{\Gamma\left(\frac{(m-1)(n-1)+1-k-r}{2}-(m-1)\mathrm{i}t\right)}{\Gamma\left(\frac{(m-1)(n-1)+1-k}{2}-(m-1)\mathrm{i}t\right)} \\
&= \frac{\Gamma\left(\frac{(m-1)(n-1)-r+1}{2}\right)}{\Gamma\left(\frac{(m-1)(n-1)-r}{2}\right)} \frac{\Gamma\left(\frac{(m-1)(n-1)-r}{2}-(m-1)\mathrm{i}t\right)}{\Gamma\left(\frac{(m-1)(n-1)-r+1}{2}-(m-1)\mathrm{i}t\right)} \\
&\quad\times \frac{\Gamma\left(\frac{(m-1)(n-1)}{2}\right)}{\Gamma\left(\frac{(m-1)(n-1)-r+1}{2}\right)} \frac{\Gamma\left(\frac{(m-1)(n-1)-r+1}{2}-(m-1)\mathrm{i}t\right)}{\Gamma\left(\frac{(m-1)(n-1)}{2}-(m-1)\mathrm{i}t\right)} \\
&\quad\times \prod_{\substack{k=2\\ \text{step } 2}}^{(m-1)(q-1)} \frac{\Gamma((m-1)(n-1)-k)}{\Gamma((m-1)(n-1)-k-r)} \frac{\Gamma((m-1)(n-1)-k-r-2(m-1)\mathrm{i}t)}{\Gamma((m-1)(n-1)-k-2(m-1)\mathrm{i}t)} \\
&= \frac{\Gamma\left(\frac{(m-1)(n-1)-r+1}{2}\right)}{\Gamma\left(\frac{(m-1)(n-1)-r}{2}\right)} \frac{\Gamma\left(\frac{(m-1)(n-1)-r}{2}-(m-1)\mathrm{i}t\right)}{\Gamma\left(\frac{(m-1)(n-1)-r+1}{2}-(m-1)\mathrm{i}t\right)} \\
&\quad\times \prod_{\ell=0}^{\frac{r-1}{2}-1} \left(\frac{(m-1)(n-1)-r+1}{2}+\ell\right)\left(\frac{(m-1)(n-1)-r+1}{2}+\ell-(m-1)\mathrm{i}t\right)^{-1} \\
&\quad\times \prod_{\substack{k=2\\ \text{step } 2}}^{(m-1)(q-1)} \prod_{\ell=0}^{r-1} \left(\frac{(m-1)(n-1)-k-r}{2}+\ell\right) \\
&\qquad\qquad\qquad\qquad \times \left(\frac{(m-1)(n-1)-k-r}{2}+\ell-(m-1)\mathrm{i}t\right)^{-1}
\end{aligned}
$$

$$
\begin{aligned}
&= \frac{\Gamma\left(\frac{(m-1)(n-1)-r+1}{2}\right)}{\Gamma\left(\frac{(m-1)(n-1)-r}{2}\right)} \frac{\Gamma\left(\frac{(m-1)(n-1)-r}{2}-(m-1)\mathrm{i}t\right)}{\Gamma\left(\frac{(m-1)(n-1)-r+1}{2}-(m-1)\mathrm{i}t\right)} \\
&\quad\times \prod_{\ell=0}^{\frac{r-1}{2}-1}\left(\frac{n-1}{2}+\frac{1-r}{2(m-1)}+\frac{\ell}{m-1}\right)\left(\frac{n-1}{2}+\frac{1-r}{2(m-1)}+\frac{\ell}{m-1}-\mathrm{i}t\right)^{-1} \\
&\quad\times \prod_{\substack{k=2\\ \text{step } 2}}^{(m-1)(q-1)} \prod_{\ell=0}^{r-1}\left(\frac{n-1}{2}-\frac{k+r}{2}+\frac{\ell}{m-1}\right)\left(\frac{n-1}{2}-\frac{k+r}{2}+\frac{\ell}{m-1}-\mathrm{i}t\right)^{-1} \\
&= \frac{\Gamma\left(\frac{(m-1)(n-1)-r+1}{2}\right)}{\Gamma\left(\frac{(m-1)(n-1)-r}{2}\right)} \frac{\Gamma\left(\frac{(m-1)(n-1)-r}{2}-(m-1)\mathrm{i}t\right)}{\Gamma\left(\frac{(m-1)(n-1)-r+1}{2}-(m-1)\mathrm{i}t\right)} \\
&\quad\times \prod_{\ell=0}^{\frac{r-1}{2}-1}\left(\frac{n-1}{2}+\frac{1-r}{2(m-1)}+\frac{\ell}{m-1}\right)\left(\frac{n-1}{2}+\frac{1-r}{2(m-1)}+\frac{\ell}{m-1}-\mathrm{i}t\right)^{-1} \\
&\quad\times \prod_{j=1}^{r+(m-1)(q-1)-3}\left(\frac{n-1}{2}-\frac{j+2}{2(m-1)}\right)^{r_j}\left(\frac{n-1}{2}-\frac{j+2}{2(m-1)}-\mathrm{i}t\right)^{-r_j} \\
&= \frac{\Gamma\left(\frac{(m-1)(n-1)-r+1}{2}\right)}{\Gamma\left(\frac{(m-1)(n-1)-r}{2}\right)} \frac{\Gamma\left(\frac{(m-1)(n-1)-r}{2}-(m-1)\mathrm{i}t\right)}{\Gamma\left(\frac{(m-1)(n-1)-r+1}{2}-(m-1)\mathrm{i}t\right)} \\
&\quad\times \prod_{j=1}^{r+(m-1)(q-1)-3}\left(\frac{n-1}{2}-\frac{j+2}{2(m-1)}\right)^{r_j^*}\left(\frac{n-1}{2}-\frac{j+2}{2(m-1)}-\mathrm{i}t\right)^{-r_j^*}
\end{aligned} \tag{7.43}
$$

where, for $j = 1, \ldots, r + (m-1)(q-1) - 3$,

$$r_j \equiv (j; r, (m-1)(q-1)-1, 2)$$

for the definition of $r(j; a, b, h^*)$ in (7.23)–(7.24), and

$$
r_j^* = \begin{cases} 1, & j = 0 \\ r_j, & j = 1, \ldots, r-4, \text{ step } 2 \\ r_j + 1, & j = 2, \ldots, r-3, \text{ step } 2 \\ r_j, & j = r-2, \ldots, r+(m-1)(q-1)-3. \end{cases}
$$

From (7.31) and (7.43), we may then write the exact c.f. of $W = -\log \Lambda$ as

$$\Phi_W(t) = \prod_{j=1}^{r+q-4} \left(\frac{n-j-1}{2}-1\right)^{s_{r+q-3-j}} \left(\frac{n-j-1}{2}-1-\mathrm{i}t\right)^{-s_{r+q-3-j}}$$
$$\times \prod_{j=1}^{r+(m-1)(q-1)-3} \left(\frac{n-1}{2}-\frac{j+2}{2(m-1)}\right)^{r_j^*} \left(\frac{n-1}{2}-\frac{j+2}{2(m-1)}-\mathrm{i}t\right)^{-r_j^*}$$
$$\times \frac{\Gamma\left(\frac{n-1}{2}\right)}{\Gamma\left(\frac{n-2}{2}\right)} \frac{\Gamma\left(\frac{n-2}{2}-\mathrm{i}t\right)}{\Gamma\left(\frac{n-1}{2}-\mathrm{i}t\right)}$$
$$\times \frac{\Gamma\left(\frac{(m-1)(n-1)-r+1}{2}\right)}{\Gamma\left(\frac{(m-1)(n-1)-r}{2}\right)} \frac{\Gamma\left(\frac{(m-1)(n-1)-r}{2}-(m-1)\mathrm{i}t\right)}{\Gamma\left(\frac{(m-1)(n-1)-r+1}{2}-(m-1)\mathrm{i}t\right)},$$

which, in order to avoid the repetition of rate parameters between the first and the second product, may be rewritten as

$$\Phi_W(t) = \underbrace{\prod_{j=q-2+\left\lfloor\frac{r-1}{m-1}\right\rfloor}^{r+q-4} \left(\frac{n-j-1}{2}-1\right)^{s_{r+q-3-j}} \left(\frac{n-j-1}{2}-1-\mathrm{i}t\right)^{-s_{r+q-3-j}} \times \prod_{j=1}^{r+(m-1)(q-1)-3} \left(\frac{n-1}{2}-\frac{j+2}{2(m-1)}\right)^{r_j^{**}} \left(\frac{n-1}{2}-\frac{j+2}{2(m-1)}-\mathrm{i}t\right)^{-r_j^{**}}}_{\Phi_{W_1}(t)}$$
$$\underbrace{\times \frac{\Gamma\left(\frac{(m-1)(n-1)-r+1}{2}\right)}{\Gamma\left(\frac{(m-1)(n-1)-r}{2}\right)} \frac{\Gamma\left(\frac{(m-1)(n-1)-r}{2}-(m-1)\mathrm{i}t\right)}{\Gamma\left(\frac{(m-1)(n-1)-r+1}{2}-(m-1)\mathrm{i}t\right)} \times \frac{\Gamma\left(\frac{n-1}{2}\right)}{\Gamma\left(\frac{n-2}{2}\right)} \frac{\Gamma\left(\frac{n-2}{2}-\mathrm{i}t\right)}{\Gamma\left(\frac{n-1}{2}-\mathrm{i}t\right)}}_{\Phi_{W_2}(t)},$$

where the s_j are given by (7.32)–(7.33) and

$$r_j^{**} = \begin{cases} r_j^* + s_{r+q-1-\frac{k+1}{m-1}}, & \text{for } k = 3m-5, \ldots, r+(m-1)(q-1)-3, \text{ step } m-1 \\ r_j^*, & \text{for other values of } k \in \{0, \ldots, r+(m-1)(q-1)-3\}. \end{cases}$$

Then in order to build the near-exact distributions, we will use a procedure in all similar to the ones adopted before, keeping $\Phi_{W,1}(t)$ unchanged and approximating asymptotically $\Phi_{W,2}(t)$, now by the c.f.

$$\Phi_2^*(t) = \sum_{\ell=0}^{\nu} \pi_\ell\, (\lambda)^{1+\ell}\, (\lambda-\mathrm{i}t)^{-(1+\ell)}$$

where now λ will be taken as the harmonic mean of $\frac{n-2}{2}$ and $\frac{n-1}{2}-\frac{r}{2(m-1)}$ and where, also similar to what was done before, the weights π_ℓ, $\ell=0,\ldots,\nu-1$, will be determined in such a way that the first ν derivatives of $\Phi_{W,2}(t)$ and $\Phi_2^*(t)$ relative to t, at $t=0$, are equal, that is, in such a way that

$$\left.\frac{\partial^h}{\partial t^h}\Phi_{W,2}(t)\right|_{t=0}=\left.\frac{\partial^h}{\partial t^h}\Phi_2^*(t)\right|_{t=0}, \quad h=1,\ldots,\nu,$$

taking then $\pi_\nu=1-\sum_{\ell=0}^{\nu-1}\pi_\ell$.

This procedure generates near-exact c.f.'s for W of the form

$$\begin{aligned}\Phi_W^*(t)&=\Phi_{W,1}(t)\,\Phi_2^*(t)\\&=\sum_{\ell=0}^{\nu}\pi_\ell\,(\lambda)^{1+\ell}\,(\lambda-\mathrm{i}t)^{-(1+\ell)}\\&\quad\times\prod_{j=q-2+\left\lfloor\frac{r-1}{m-1}\right\rfloor}^{r+q-4}\left(\frac{n-j-1}{2}-1\right)^{s_{r+q-3-j}}\left(\frac{n-j-1}{2}-1-\mathrm{i}t\right)^{-s_{r+q-3-j}}\\&\quad\times\prod_{j=1}^{r+(m-1)(q-1)-3}\left(\frac{n-1}{2}-\frac{j+2}{2(m-1)}\right)^{r_j^{**}}\left(\frac{n-1}{2}-\frac{j+2}{2(m-1)}-\mathrm{i}t\right)^{-r_j^{**}}\end{aligned}$$

to which corresponds a mixture of $\nu+1$ GIG distributions of depth $2r+(m-1)(q-1)-4-\left\lfloor\frac{r-1}{m-1}\right\rfloor$, with weights π_ℓ ($\ell=0,\ldots,\nu$), rate parameters $\frac{n-j-1}{2}-1$ ($j=q-2+\left\lfloor\frac{r-1}{m-1}\right\rfloor,\ldots,r+q-4$), $\frac{n-1}{2}-\frac{j+2}{2(m-1)}$ ($j=1,\ldots,r+(m-1)(q-1)-3$), and λ, to which correspond the shape parameters $s_{r+q-3-j}$ ($j=q-2+\left\lfloor\frac{r-1}{m-1}\right\rfloor,\ldots,r+q-4$), r_j^{**} ($j=1,\ldots,r+(m-1)(q-1)-3$), and $1+\ell$ ($\ell=0,\ldots,\nu$).

Once again, similar to what happens with other near-exact distributions developed, also these are asymptotic for increasing sample sizes as well as for increasing values of q, r, and m, yielding once again approximations with comparable performances to the other near-exact distributions developed, only with a slightly less marked asymptotic behavior. Slightly better results may be generally obtained if one chooses to use $\lambda=\gamma_1/(\gamma_2-\gamma_1^2)$, where γ_1 and γ_2 represent, respectively, the first and second moments of the distribution that corresponds to $\Phi_{W_2}(t)$.

7.6 Conclusions

The procedures adopted allowed not only for a simple way to obtain the MLEs of the block covariance matrices involved in the tests as well as for a quite simple way to derive the l.r.t statistics and their distributions. It was then possible to show that the exact distribution of these statistics, no matter how complicated they may seem to be at first glance, may have indeed finite form representations in most cases, more precisely, in cases where r, the number of variables in each subgroup of variables, is even or when q, the number of populations involved in the test, is odd. In order to make this task a bit simpler, Theorem 7.1 in Sect. 7.3 was developed, which may be seen as an extended version of Theorem 3.2 in Coelho and Arnold [6]. For the remaining cases, that is, for cases where r is odd and simultaneously q is even, near-exact distributions were developed for the LRT statistics. These distributions are asymptotic not only for increasing sample sizes but also for increasing numbers of variables in each subgroup, that is, for increasing values of r, as well as for increasing values of q and m, which are the number of populations involved in the tests and the number of subgroups of variables. These near-exact distributions besides being asymptotic for increasing values of n, r, q, and m also show very good performances for very small samples, being thus a good choice for the situations where the exact distribution of the LRT statistics is not manageable.

We should also note that for $m = 2$ and $m = 3$ the test in Sect. 7.5, for equality of mean vectors, with block compound-symmetric matrices, is equivalent to the test for equality of mean vectors with block-circular matrices since in these cases the block-circular matrices are indeed also block compound-symmetric matrices.

Acknowledgements The author wants to thank the support provided by the Banach Center of the Institute of Mathematics of the Polish Academy of Sciences and by CMA/UNL (Centro de Matemática e Aplicações, NOVA University of Lisbon) through the project UID/MAT/00297/2019.

Appendix A: The Block Spectral Decomposition of a Block-Circular Matrix and Its Inverse

It is known that the matrix $\boldsymbol{\Sigma}$ is a block-circular matrix with $m \times m$ blocks of dimensions $r \times r$ if and only if

$$(\boldsymbol{\Gamma}_m \otimes \mathbf{I}_r)\boldsymbol{\Sigma}(\boldsymbol{\Gamma}_m \otimes \mathbf{I}_r) = \boldsymbol{\Psi} = \text{BDiag}(\boldsymbol{\Psi}_1, \boldsymbol{\Psi}_2, \ldots, \boldsymbol{\Psi}_m)$$

where $\mathbf{I}_r$ is an identity matrix of order r, $\boldsymbol{\Gamma}_m$ is an $m \times m$ orthogonal symmetric matrix with elements γ_{jk} $(j, k = 1, \ldots, m)$, given by (7.16), and $\boldsymbol{\Psi}_1, \ldots, \boldsymbol{\Psi}_m$ are positive-definite matrices with

$$\boldsymbol{\Psi}_j = \boldsymbol{\Psi}_{m-j+2}, \quad j = 2, \ldots, m.$$

But then, given that $\boldsymbol{\Gamma}_m$ is orthogonal and symmetric, we may write

$$\boldsymbol{\Sigma} = (\boldsymbol{\Gamma}_m \otimes \mathbf{I}_r)\boldsymbol{\Psi}(\boldsymbol{\Gamma}_m \otimes \mathbf{I}_r),$$

and then also we have

$$\boldsymbol{\Sigma}^{-1} = (\boldsymbol{\Gamma}_m \otimes \mathbf{I}_r)\boldsymbol{\Psi}^{-1}(\boldsymbol{\Gamma}_m \otimes \mathbf{I}_r)$$

where

$$\boldsymbol{\Psi}^{-1} = \text{BDiag}(\boldsymbol{\Psi}_1^{-1}, \boldsymbol{\Psi}_2^{-1}, \ldots, \boldsymbol{\Psi}_m^{-1}),$$

and as such $\boldsymbol{\Sigma}^{-1}$ is also a block-circular matrix.

A similar procedure may be used to show that the inverse of a block compound-symmetric matrix is also a block compound-symmetric matrix.

We may note that actually both results may also be directly obtained from the fact that the spaces of both the block-circular and block compound-symmetric structures are quadratic spaces (see Seely [22]).

Appendix B: Proof of Expression (7.17)

We may note that from (7.11), and from the fact that $\mathbf{A}_i^* = \mathbf{A}_{m-i}^*$ for $i = 1, \ldots, m-1$, we may write

$$\mathbf{A}^* = \left(\mathbf{W}_0 \otimes \mathbf{A}_0^*\right) + \sum_{i=1}^{\frac{m-1}{2}} (\mathbf{W}_i + \mathbf{W}_{m-i}) \otimes \mathbf{A}_i^* \tag{7.44}$$

for odd m or as

$$\mathbf{A}^* = \left(\mathbf{W}_0 \otimes \mathbf{A}_0^*\right) + \left(\mathbf{W}_{m/2} \otimes \mathbf{A}_{m/2}^*\right) + \sum_{i=1}^{\frac{m}{2}-1} (\mathbf{W}_i + \mathbf{W}_{m-i}) \otimes \mathbf{A}_i^* \tag{7.45}$$

for even m.

We may also note that since $\mathbf{W}_0 = \mathbf{I}_m$, we have

$$\boldsymbol{\Gamma}_m \mathbf{W}_0 \boldsymbol{\Gamma}_m = \boldsymbol{\Gamma}_m \boldsymbol{\Gamma}_m = \mathbf{I}_m, \tag{7.46}$$

where $\boldsymbol{\Gamma}_m$ is the matrix in (7.15), with elements given by (7.16), and given that for $i = 1, \ldots, m-1$, we have

$$\mathbf{W}_{m-i} = \mathbf{W}_i',$$

so that

$$\mathbf{W}_i + \mathbf{W}_{m-i} = \mathbf{W}_i + \mathbf{W}_i' \qquad (i = 1, \ldots, m-1),$$

are symmetric matrices, and moreover, for $i = 1, \ldots, m-1$,

$$\mathbf{W}_i = \mathbf{W}_1^i$$

so that all matrices $\mathbf{W}_i + \mathbf{W}_{m-i}$ are commutative, and as such may be all diagonalized by the same orthogonal matrix (see Olkin [18, Sect. 2]). This matrix is exactly the matrix $\boldsymbol{\Gamma}_m$, which, given its structure, yields

$$\boldsymbol{\Gamma}_m\left(\mathbf{W}_i+\mathbf{W}_{m-i}\right)\boldsymbol{\Gamma}_m=\boldsymbol{\Delta}_i,\qquad i=1,\ldots,m-1 \tag{7.47}$$

where

$$\boldsymbol{\Delta}_i=\operatorname{diag}\left(\lambda_{i\ell}^{**}\right),\qquad i=1,\ldots,m-1;\ \ell=1,\ldots,m \tag{7.48}$$

with

$$\lambda_{i\ell}^{**}=2\cos\left(2\pi i(\ell-1)/m\right),\qquad i=1,\ldots,m-1;\ \ \ell=1,\ldots,m. \tag{7.49}$$

But then, given the definition of $\mathbf{A}^*$ in (7.44) and (7.45), the matrix $\boldsymbol{\Gamma}_m\otimes\mathbf{I}_r$ is the matrix that block-diagonalizes $\mathbf{A}^*$, with

$$(\boldsymbol{\Gamma}_m\otimes\mathbf{I}_r)\mathbf{A}^*(\boldsymbol{\Gamma}_m\otimes\mathbf{I}_r)=\widehat{\boldsymbol{\Psi}}=\operatorname{BDiag}\left(\widehat{\boldsymbol{\Psi}}_1,\widehat{\boldsymbol{\Psi}}_2,\ldots,\widehat{\boldsymbol{\Psi}}_m\right) \tag{7.50}$$

where

$$\widehat{\boldsymbol{\Psi}}_j=\widehat{\boldsymbol{\Psi}}_{m-j+2},\qquad j=2,\ldots,m$$

and where, for $j=1,\ldots,m$,

$$\widehat{\boldsymbol{\Psi}}_j=\sum_{i=0}^{m-1}\lambda_{ji}\mathbf{A}_i^* \tag{7.51}$$

with

$$\lambda_{ji}=\cos\left(2\pi(j-1)i/m\right),\qquad i=0,\ldots,m-1;\ \ j=1,\ldots,m. \tag{7.52}$$

But then from (7.50) we have

$$|\mathbf{A}^*|=\left|\widehat{\boldsymbol{\Psi}}\right|=\prod_{j=1}^{m}\left|\widehat{\boldsymbol{\Psi}}_j\right|.$$

On the other hand, from (7.12), (7.13), and (7.14), using the definition of BTr (block trace operator) in Filipiak et al. [7], we may write

$$\mathbf{A}_i^*=\frac{1}{2m}\operatorname{BTr}\left(\left(\left(\mathbf{W}_i+\mathbf{W}_{m-i}\right)\otimes\mathbf{I}_r\right)\mathbf{A}\right),\qquad i=0,\ldots,m-1 \tag{7.53}$$

where we take $\mathbf{W}_0=\mathbf{W}_m=\mathbf{I}_m$. Then, if we take

$$\mathbf{A}^{**}=(\boldsymbol{\Gamma}_m\otimes\mathbf{I}_r)\mathbf{A}(\boldsymbol{\Gamma}_m\otimes\mathbf{I}_r),$$

since the matrix $(\boldsymbol{\Gamma}_m\otimes\mathbf{I}_r)$ is orthogonal and symmetric, we may write

$$\mathbf{A}=(\boldsymbol{\Gamma}_m\otimes\mathbf{I}_r)\mathbf{A}^{**}(\boldsymbol{\Gamma}_m\otimes\mathbf{I}_r),$$

so that, using the properties of BTr (see Filipiak et al. [7]) and of the Kronecker product, we have from (7.53), for $i = 0, \ldots, m-1$,

$$\begin{aligned}\mathbf{A}_i^* &= \frac{1}{2m}\mathrm{BTr}\left(((\mathbf{W}_i + \mathbf{W}_{m-i}) \otimes \mathbf{I}_r)(\boldsymbol{\Gamma}_m \otimes \mathbf{I}_r)\mathbf{A}^{**}(\boldsymbol{\Gamma}_m \otimes \mathbf{I}_r)\right)\\ &= \frac{1}{2m}\mathrm{BTr}\left((\boldsymbol{\Gamma}_m \otimes \mathbf{I}_r)((\mathbf{W}_i + \mathbf{W}_{m-i}) \otimes \mathbf{I}_r)(\boldsymbol{\Gamma}_m \otimes \mathbf{I}_r)\mathbf{A}^{**}\right)\\ &= \frac{1}{2m}\mathrm{BTr}\left(((\boldsymbol{\Gamma}_m(\mathbf{W}_i + \mathbf{W}_{m-i})\boldsymbol{\Gamma}_m) \otimes \mathbf{I}_r)\,\mathbf{A}^{**}\right)\end{aligned}$$

where, from (7.46) and (7.47)–(7.49), we have

$$\boldsymbol{\Gamma}_m(\mathbf{W}_i + \mathbf{W}_{m-i})\boldsymbol{\Gamma}_m = \boldsymbol{\Delta}_i^*, \qquad i = 0, \ldots, m-1,$$

with

$$\boldsymbol{\Delta}_i^* = \mathrm{diag}\left(\lambda_{i\ell}^*\right), \qquad i = 0, \ldots, m-1;\ \ell = 1, \ldots, m,$$

where

$$\lambda_{i\ell}^* = 2\cos\left(2\pi i(\ell-1)/m\right), \qquad \ell = 1, \ldots, m;\ i = 0, \ldots, m-1. \tag{7.54}$$

Therefore, we may write for $i = 0, \ldots, m-1$,

$$\mathbf{A}_i^* = \frac{1}{2m}\sum_{\ell=1}^{m}\lambda_{i\ell}^*\mathbf{A}_\ell^{**},$$

where $\mathbf{A}_\ell^{**}$ denotes the ℓ-th diagonal block of dimensions $r \times r$ of $\mathbf{A}^{**}$.

But then, from (7.51), we may write, for $j = 1, \ldots, m$,

$$\widehat{\boldsymbol{\Psi}}_j = \frac{1}{2m}\sum_{i=0}^{m-1}\lambda_{ji}\sum_{\ell=1}^{m}\lambda_{i\ell}^*\mathbf{A}_\ell^{**} = \frac{1}{2m}\sum_{\ell=1}^{m}\left(\sum_{i=0}^{m-1}\lambda_{ji}\lambda_{i\ell}^*\right)\mathbf{A}_\ell^{**}$$

where, from (7.52) and (7.54), we have, for $j = 1$,

$$\sum_{i=0}^{m-1}\lambda_{ji}\lambda_{i\ell}^* = \begin{cases} 2m, & \ell = 1\\ 0, & \ell = 2, \ldots, m\end{cases}$$

for $j = 2, \ldots, m$ for odd m, or for $j = 2, \ldots, m/2, 2 + m/2, \ldots, m$ for even m,

$$\sum_{i=0}^{m-1}\lambda_{ji}\lambda_{i\ell}^* = \begin{cases} m, & \ell = j \text{ or } \ell = m - j + 2\\ 0, & \text{other values of } \ell\end{cases}$$

and for $j = m/2$ for even m

$$\sum_{i=0}^{m-1} \lambda_{ji}\lambda^*_{i\ell} = \begin{cases} 2m, & \ell = m/2 \\ 0, & \text{other values of } \ell \end{cases}$$

so that we have

$$\widehat{\boldsymbol{\Psi}}_1 = \mathbf{A}_1^{**} \qquad \text{and} \qquad \widehat{\boldsymbol{\Psi}}_j = \frac{1}{2}\left(\mathbf{A}_j^{**} + \mathbf{A}_{m-j+2}^{**}\right)$$

for $j = 2, \ldots, m$ for odd m, or for $j = 2, \ldots, m/2, 2 + m/2, \ldots, m$ for even m, and

$$\widehat{\boldsymbol{\Psi}}_{1+m/2} = \mathbf{A}_{1+m/2}^{**}$$

for even m.

We may thus write

$$|\mathbf{A}^*| = |\mathbf{A}_1^{**}| \left(|\mathbf{A}_{1+m^+}^{**}|\right)^{\mathrm{mod}(m+1,2)} \prod_{j=2}^{m-m^+} \left|\frac{\mathbf{A}_j^{**} + \mathbf{A}_{m-j+2}^{**}}{2}\right|^2,$$

which is (7.17).

The equality in expression (7.18) may be shown in a similar manner.

Appendix C: Definition, Notation, and Expression for the p.d.f. and c.d.f. of the GIG and GNIG Distributions

We say that the RV X has a Gamma distribution with shape parameter r (>0) and rate parameter λ (>0), and we will denote this fact by $X \sim \Gamma(r, \lambda)$, if the p.d.f. of X is

$$f_X(x) = \frac{\lambda^r}{\Gamma(r)} e^{-\lambda x} x^{r-1} \qquad (x > 0).$$

Let $Z_j \sim \Gamma(r_j, \lambda_j)$ $j = 1, \ldots, p$ be a set of p independent RVs and consider the RV

$$Z = \sum_{j=1}^{p} Z_j.$$

In case all the $r_j \in \mathbb{N}$, the distribution of Z is what we call a GIG (Generalized Integer Gamma) distribution (Coelho [4]). If all the λ_j are different, Z has a GIG distribution of depth p, with shape parameters r_j and rate parameters λ_j, with p.d.f.

$$f_Z(z) = f^{GIG}\Big(z \,\Big|\, \{r_j\}_{j=1:p}; \{\lambda_j\}_{j=1:p}; p\Big) = K \sum_{j=1}^{p} P_j(z)\, e^{-\lambda_j z} \qquad (z > 0),$$

and c.d.f.

$$F_Z(z) = F^{GIG}\Big(z \,\Big|\, \{r_j\}_{j=1:p}; \{\lambda_j\}_{j=1:p}; p\Big) = 1 - K \sum_{j=1}^{p} P_j^*(z)\, e^{-\lambda_j z} \qquad (z > 0),$$

where

$$K = \prod_{j=1}^{p} \lambda_j^{r_j}, \qquad P_j(z) = \sum_{k=1}^{r_j} c_{j,k}\, z^{k-1}$$

and

$$P_j^*(z) = \sum_{k=1}^{r_j} c_{j,k} (k-1)! \sum_{i=0}^{k-1} \frac{z^i}{i!\, \lambda_j^{k-i}},$$

with

$$c_{j,r_j} = \frac{1}{(r_j - 1)!} \prod_{\substack{i=1 \\ i \neq j}}^{p} (\lambda_i - \lambda_j)^{-r_i}, \quad j = 1, \ldots, p,$$

and, for $k = 1, \ldots, r_j - 1$ and $j = 1, \ldots, p$,

$$c_{j,r_j-k} = \frac{1}{k} \sum_{i=1}^{k} \frac{(r_j - k + i - 1)!}{(r_j - k - 1)!} R(i, j, p) c_{j,r_j-(k-i)},$$

where

$$R(i, j, p) = \sum_{\substack{k=1 \\ k \neq j}}^{p} r_k (\lambda_j - \lambda_k)^{-i}, \quad (i = 1, \ldots, r_j - 1).$$

In case some of the λ_j assume the same value as other λ_j's, the distribution of Z still is a GIG distribution, but in this case with a reduced depth. In this more general case, let $\{\lambda_\ell; \ell = 1, \ldots, g(\leq p)\}$ be the set of different λ_j's and let $\{r_\ell; \ell = 1, \ldots, g(\leq p)\}$ be the set of the corresponding shape parameters, with r_ℓ being the sum of all r_j $(j = 1, \ldots, p)$ which correspond to the λ_j assuming the value λ_ℓ. In this case, Z will have a GIG distribution of depth g, with shape parameters r_ℓ and rate parameters λ_ℓ $(\ell = 1, \ldots, g)$.

If all RVs have Gamma distributions with different rate parameters λ_j and Z_p has a Gamma distribution with a non-integer shape parameter r_p, then we will say that the RV Z has a GNIG (Generalized Near-Integer Gamma) distribution of depth p. The probability density and cumulative distribution functions of Z are, for $z > 0$, respectively, given by Coelho [5]

$$f^{GNIG}(z\,|\,r_1,\ldots,r_{p-1};r_p;\;\lambda_1,\ldots,\lambda_{p-1};\lambda_p;\;p)$$

$$=K\sum_{j=1}^{p-1}e^{-\lambda_j z}\sum_{k=1}^{r_j}\left\{c_{j,k}\frac{\Gamma(k)}{\Gamma(k+r_p)}z^{k+r_p-1}{}_1F_1(r_p,k+r_p,-(\lambda_p-\lambda_j)z)\right\},$$

and

$$F^{GNIG}(z\,|\,r_1,\ldots,r_{p-1};r_p;\;\lambda_1,\ldots,\lambda_{p-1};\lambda_p;\;p)=\frac{\lambda_p^{r_p}\,z^{r_p}}{\Gamma(r_p+1)}{}_1F_1(r_p,r_p+1,-\lambda_p z)$$

$$-K\sum_{j=1}^{p-1}e^{-\lambda_j z}\sum_{k=1}^{r_j}c^*_{j,k}\sum_{i=0}^{k-1}\frac{z^{r_p+i}\lambda_j^i}{\Gamma(r_p+1+i)}{}_1F_1(r_p,r_p+1+i,-(\lambda_p-\lambda_j)z),$$

with $K=\prod_{j=1}^{p}\lambda_j^{r_j}$ and $c^*_{j,k}=\frac{c_{j,k}\Gamma(k)}{\lambda_j^k}$ and where ${}_1F_1(a,b,z)$ represents the Kummer confluent hypergeometric function.

Let Z have a GIG distribution of depth p and let us consider the RV $Z=e^{-W}$. Then the RV Z has what Arnold et al. [2] call an Exponentiated Generalized Integer Gamma (EGIG) distribution of depth p, with p.d.f.

$$f_Z(z)=f^{EGIG}\Big(z\,\Big|\,\{r_j\}_{j=1:p};\{\lambda_j\}_{j=1:p};\,p\Big)$$
$$=f^{GIG}\Big(-\log z\,\Big|\,\{r_j\}_{j=1:p};\{\lambda_j\}_{j=1:p};\,p\Big)\frac{1}{z}\qquad(0<z<1)$$

and c.d.f.

$$F_Z(z)=F^{EGIG}\Big(z\,\Big|\,\{r_j\}_{j=1:p};\{\lambda_j\}_{j=1:p};\,p\Big)$$
$$=1-F^{GIG}\Big(-\log z\,\Big|\,\{r_j\}_{j=1:p};\{\lambda_j\}_{j=1:p};\,p\Big)\qquad(0<z<1).$$

Appendix D: Proof of Theorem 7.1 in Sect. 7.3

Let $W=-\log Z$. Then we may write the c.f. of Z as

$$\Phi_W(t) = E\left(e^{Wit}\right) = E\left(e^{-\mathrm{i}t\log Z}\right) = E\left(Z^{-\mathrm{i}t}\right) = \prod_{\nu=1}^{m^*}\prod_{\ell=1}^{n_\nu}\prod_{j=1}^{k_\nu} E\left(Y_{\nu\ell j}^{-\mathrm{i}t}\right)$$

$$= \prod_{\nu=1}^{m^*}\prod_{\ell=1}^{n_\nu}\prod_{j=1}^{k_\nu} \frac{\Gamma\left(a_\nu+\ell+\frac{1-\ell}{h^*k_\nu}+1-\ell+\frac{m_\nu-j}{k_\nu}\right)\,\Gamma\left(a_\nu+\ell+\frac{1-\ell}{h^*k_\nu}+1-\ell-\frac{j}{k_\nu}-\mathrm{i}t\right)}{\Gamma\left(a_\nu+\ell+\frac{1-\ell}{h^*k_\nu}+1-\ell-\frac{j}{k_\nu}\right)\,\Gamma\left(a_\nu+\ell+\frac{1-\ell}{h^*k_\nu}+1-\ell+\frac{m_\nu-j}{k_\nu}-\mathrm{i}t\right)} \tag{7.55}$$

which, just by reindexing in j, may be written as

$$\Phi_W(t) = \prod_{\nu=1}^{m^*}\prod_{\ell=1}^{n_\nu}\prod_{j=1+k_\nu(\ell-1)}^{k_\nu\ell} \frac{\Gamma\left(a_\nu+\frac{1-\ell}{h^*k_\nu}+1+\frac{m_\nu-j+k_\nu(\ell-1)}{k_\nu}\right)}{\Gamma\left(a_\nu+\frac{1-\ell}{h^*k_\nu}+1-\frac{j+k_\nu(\ell-1)}{k_\nu}\right)}$$

$$\times\frac{\Gamma\left(a_\nu+\frac{1-\ell}{h^*k_\nu}+1-\frac{j+k_\nu(\ell-1)}{k_\nu}-\mathrm{i}t\right)}{\Gamma\left(a_\nu+\frac{1-\ell}{h^*k_\nu}+1+\frac{m_\nu-j+k_\nu(\ell-1)}{k_\nu}-\mathrm{i}t\right)}$$

$$= \prod_{\nu=1}^{m^*}\prod_{\ell=1}^{n_\nu}\prod_{j=1+k_\nu(\ell-1)}^{k_\nu\ell} \frac{\Gamma\left(a_\nu+\frac{1-\ell}{h^*k_\nu}+\frac{m_\nu-j}{k_\nu}\right)\,\Gamma\left(a_\nu+\frac{1-\ell}{h^*k_\nu}-\frac{j}{k_\nu}-\mathrm{i}t\right)}{\Gamma\left(a_\nu+\frac{1-\ell}{h^*k_\nu}-\frac{j}{k_\nu}\right)\,\Gamma\left(a_\nu+\frac{1-\ell}{h^*k_\nu}+\frac{m_\nu-j}{k_\nu}-\mathrm{i}t\right)}$$

$$= \prod_{\nu=1}^{m^*}\prod_{\ell=1}^{n_\nu}\prod_{j=1+k_\nu(\ell-1)}^{k_\nu\ell} \frac{\Gamma\left(a_{\nu\ell}+\frac{m_\nu-j}{k_\nu}\right)\,\Gamma\left(a_{\nu\ell}-\frac{j}{k_\nu}-\mathrm{i}t\right)}{\Gamma\left(a_{\nu\ell}-\frac{j}{k_\nu}\right)\,\Gamma\left(a_{\nu\ell}+\frac{m_\nu-j}{k_\nu}-\mathrm{i}t\right)},$$

which shows the first equivalence in (7.25). Then, from (7.55), rearranging the Gamma arguments and using the multiplication formula for the Gamma function in the second equality in expression (3.5) of Coelho and Arnold [6],

$$\prod_{j=1}^{n}\Gamma\left(z+\frac{j-1}{n}\right) = \prod_{j=1}^{n}\Gamma\left(z+\frac{n-j}{n}\right) = \Gamma(nz)\,(2\pi)^{\frac{n-1}{2}}\,n^{\frac{1}{2}-nz}, \tag{7.56}$$

we may write the c.f. of W as

$$\begin{aligned}\Phi_W(t) &= \prod_{\nu=1}^{m^*}\prod_{\ell=1}^{n_\nu}\prod_{j=1}^{k_\nu} \frac{\Gamma\left(a_\nu+\frac{1-\ell}{h^*k_\nu}+\frac{k_\nu-j}{k_\nu}+\frac{m_\nu}{k_\nu}\right)\,\Gamma\left(a_\nu+\frac{1-\ell}{h^*k_\nu}+\frac{k_\nu-j}{k_\nu}-\mathrm{i}t\right)}{\Gamma\left(a_\nu+\frac{1-\ell}{h^*k_\nu}+\frac{k_\nu-j}{k_\nu}\right)\,\Gamma\left(a_\nu+\frac{1-\ell}{h^*k_\nu}+\frac{k_\nu-j}{k_\nu}+\frac{m_\nu}{k_\nu}-\mathrm{i}t\right)}\\
&= \prod_{\nu=1}^{m^*}\prod_{\ell=1}^{n_\nu} \frac{\Gamma\left(a_\nu k_\nu+\frac{1-\ell}{h^*}+m_\nu\right)\,\Gamma\left(a_\nu k_\nu+\frac{1-\ell}{h^*}-k_\nu\,\mathrm{i}t\right)}{\Gamma\left(a_\nu k_\nu+\frac{1-\ell}{h^*}\right)\,\Gamma\left(a_\nu k_\nu+\frac{1-\ell}{h^*}+m_\nu-k_\nu\,\mathrm{i}t\right)}\\
&= \prod_{\nu=1}^{m^*}\prod_{\ell=1}^{n_\nu} \frac{\Gamma((a_{\nu\ell}-\ell)k_\nu+m_\nu)\,\Gamma((a_{\nu\ell}-\ell)k_\nu-k_\nu\,\mathrm{i}t)}{\Gamma((a_{\nu\ell}-\ell)k_\nu)\,\Gamma((a_{\nu\ell}-\ell)k_\nu+m_\nu-k_\nu\,\mathrm{i}t)},\end{aligned} \tag{7.57}$$

which shows the distribution in the second product in (7.25).

Then, from (7.57), using (7.21), we may write

$$\begin{aligned}\Phi_W(t) &= \prod_{\nu=1}^{m^*}\prod_{\ell=1}^{n_\nu}\prod_{j=0}^{m_\nu-1} ((a_{\nu\ell}-\ell)k_\nu+j)\,((a_{\nu\ell}-\ell)k_\nu+j-k_\nu\,\mathrm{i}t)^{-1}\\
&= \prod_{\nu=1}^{m^*}\prod_{\ell=1}^{n_\nu}\prod_{j=0}^{m_\nu-1} \left(a_{\nu\ell}-\ell+\frac{j}{k_\nu}\right)\left(a_{\nu\ell}-\ell+\frac{j}{k_\nu}-\mathrm{i}t\right)^{-1}\\
&= \prod_{\nu=1}^{m^*}\prod_{\ell=1}^{n_\nu}\prod_{j=0}^{m_\nu-1} \left(a_\nu+\frac{1-\ell}{h^*k_\nu}+\frac{j}{k_\nu}\right)\left(a_\nu+\frac{1-\ell}{h^*k_\nu}+\frac{j}{k_\nu}-\mathrm{i}t\right)^{-1}\\
&= \prod_{\nu=1}^{m^*}\prod_{j=1}^{n_\nu+h^*(m_\nu-1)} \left(a_\nu+\frac{j-n_\nu}{h^*k_\nu}\right)^{r_{\nu j}}\left(a_\nu+\frac{j-n_\nu}{h^*k_\nu}-\mathrm{i}t\right)^{-r_{\nu j}},\end{aligned} \tag{7.58}$$

where, for $j = 1, \ldots, n_\nu + h^*(m_\nu - 1)$ and $\nu = 1, \ldots, m^*$,

$$r_{\nu j} \equiv r(j; n_\nu, h^* m_\nu, h^*)$$

for the definition of $r(j; a, b, h^*)$ in (7.23)–(7.24), and where the last equality is obtained by identifying the different rate parameters in the expression above and their multiplicities, in much the same way that Wald and Brookner [26] identify the different poles and their multiplicities in the Meijer G function representation of p.d.f.'s of distributions of this type.

Expression (7.58) shows that, in case all a_ν are different, the distribution of $W = -\log Z$ is a GIG distribution, and consequently that of Z an EGIG distribution of depth $\sum_{\nu=1}^{m^*} n_\nu + h^*(m_\nu - 1)$, with shape parameters $r_{\nu j}$ and rate parameters $a_\nu + \frac{j-n_\nu}{h^*k_\nu}$, for $\nu = 1, \ldots, m^*$ and $j = 1, \ldots, n_\nu + h^*(m_\nu - 1)$.

In case a_ν, n_ν, k_ν, and m_ν are not function of ν, say with $a_\nu = a$, $n_\nu = n$, $k_\nu = k$, and $m_\nu = m$, we then have for Z, the EGIG distribution depicted in (7.26), which is an EGIG distribution of depth $n + h^*(m - 1)$, with shape parameters $m^* r_j$ and rate parameters $a + \frac{j-n}{h^*k}$, for $j = 1, \ldots, n + h^*(m - 1)$, where the r_j are given by (7.27).

Appendix E: Proof of Expression (7.35)

Actually expression (7.35) is only expression (7.17) for the case where all $\mathbf{A}_i^*$ are equal for $i = 1, \ldots, m-1$, and as such the proof of (7.17) is indeed enough to prove also (7.35).

However, there is a different and simpler way to prove (7.35), if we consider

$$\mathrm{BTr}(\mathbf{A}) = \sum_{j=1}^{m} \mathbf{A}_{jj} \quad \text{and} \quad \mathrm{BSum}(\mathbf{A}) = \sum_{j=1}^{m}\sum_{k=1}^{m} \mathbf{A}_{jk} = \mathrm{BTr}\left[(\mathbf{J}_m \otimes \mathbf{I}_r)\mathbf{A}\right],$$

where $\mathbf{A}_{jk}$ denotes the $r \times r$ block of $\mathbf{A}$ starting at row $r(j-1)+1$ and column $r(k-1)+1$, and as such $\mathbf{A}_{jj}$ denotes the j-th diagonal block of $\mathbf{A}$ of dimensions $r \times r$.

Then, in a proof due to Daniel Klein, with the collaboration of Augustyn Markiewicz and Ivan Žežula, we may write

$$\mathbf{A}_0^* = \widehat{\boldsymbol{\Sigma}}_{0|H_1} = \frac{1}{m}\sum_{j=1}^{m} \mathbf{A}_{jj} = \frac{1}{m}\mathrm{BTr}(\mathbf{A})$$

and

$$\begin{aligned}
\mathbf{A}_1^* = \widehat{\boldsymbol{\Sigma}}_{1|H_1} &= \frac{1}{m(m-1)}\sum_{j=1}^{m}\sum_{k=j+1}^{m}\left(\mathbf{A}_{jk} + \mathbf{A}_{kj}\right) \\
&= \frac{1}{m(m-1)}\left[\mathrm{BSum}(\mathbf{A}) - \mathrm{BTr}(\mathbf{A})\right] \\
&= \frac{1}{m(m-1)}\left\{\mathrm{BTr}\left[(\mathbf{J}_m \otimes \mathbf{I}_r)\mathbf{A}\right] - \mathrm{BTr}(\mathbf{A})\right\} \\
&= \frac{1}{m(m-1)}\mathrm{BTr}\left\{\left[(\mathbf{J}_m - \mathbf{I}_m) \otimes \mathbf{I}_r\right]\mathbf{A}\right\}
\end{aligned}$$

so that, using

$$\mathbf{Q}_m = \mathbf{I}_m - \mathbf{P}_m \qquad \text{with} \qquad \mathbf{P}_m = \frac{1}{m}\mathbf{J}_m,$$

we may write

$$\begin{aligned}\mathbf{A}_0^* - \mathbf{A}_1^* &= \frac{1}{m}\mathrm{BTr}(\mathbf{A}) - \frac{1}{m(m-1)}\mathrm{BTr}\left\{[(\mathbf{J}_m - \mathbf{I}_m)\otimes \mathbf{I}_r]\,\mathbf{A}\right\}\\ &= \frac{1}{m}\mathrm{BTr}\left\{\left[\left(\mathbf{I}_m - \frac{1}{m-1}\mathbf{J}_m + \frac{1}{m-1}\mathbf{I}_m\right)\otimes \mathbf{I}_r\right]\mathbf{A}\right\}\\ &= \frac{1}{m}\mathrm{BTr}\left\{\left[\left(\frac{m}{m-1}\mathbf{I}_m - \frac{1}{m-1}\mathbf{J}_m\right)\otimes \mathbf{I}_r\right]\mathbf{A}\right\}\\ &= \frac{1}{m-1}\mathrm{BTr}\left[(\mathbf{Q}_m \otimes \mathbf{I}_r)\mathbf{A}\right]\end{aligned}$$

and

$$\begin{aligned}\mathbf{A}_0^* + (m-1)\mathbf{A}_1^* &= \frac{1}{m}\mathrm{BTr}(\mathbf{A}) + \frac{1}{m}\mathrm{BTr}\left\{[(\mathbf{J}_m - \mathbf{I}_m)\otimes \mathbf{I}_r]\,\mathbf{A}\right\}\\ &= \frac{1}{m}\mathrm{BTr}\left[(\mathbf{J}_m \otimes \mathbf{I}_r)\mathbf{A}\right] = \mathrm{BTr}\left[(\mathbf{P}_m \otimes \mathbf{I}_r)\mathbf{A}\right].\end{aligned}$$

But then we may write

$$\begin{aligned}\mathbf{A}^* &= \mathbf{I}_m \otimes (\mathbf{A}_0^* - \mathbf{A}_1^*) + \mathbf{J}_m \otimes \mathbf{A}_1^*\\ &= \mathbf{Q}_m \otimes (\mathbf{A}_0^* - \mathbf{A}_1^*) + \mathbf{P}_m \otimes (\mathbf{A}_0^* - \mathbf{A}_1^*) + m\mathbf{P}_m \otimes \mathbf{A}_1^*\\ &= \mathbf{Q}_m \otimes (\mathbf{A}_0^* - \mathbf{A}_1^*) + \mathbf{P}_m \otimes \left(\mathbf{A}_0^* + (m-1)\mathbf{A}_1^*\right),\end{aligned}$$

where

$$\mathbf{Q}_m\mathbf{P}_m = \left(\mathbf{I}_m - \frac{1}{m}\mathbf{J}_m\right)\frac{1}{m}\mathbf{J}_m = \frac{1}{m}\mathbf{J}_m - \frac{1}{m^2}\underbrace{\mathbf{J}_m\mathbf{J}_m}_{=m\mathbf{J}_m} = \frac{1}{m}\mathbf{J}_m - \frac{1}{m}\mathbf{J}_m = 0,$$

with $\mathrm{r}(\mathbf{Q}_m) = m-1$ and $\mathrm{r}(\mathbf{P}_m) = 1$, so that

$$\begin{aligned}|\mathbf{A}^*| &= \left|\mathbf{A}_0^* + (m-1)\mathbf{A}_1^*\right| \left|\mathbf{A}_0^* - \mathbf{A}_1^*\right|^{m-1}\\ &= \left|\mathrm{BTr}\left[(\mathbf{P}_m \otimes \mathbf{I}_r)\mathbf{A}\right]\right| \left|\frac{1}{m-1}\mathrm{BTr}\left((\mathbf{Q}_m \otimes \mathbf{I}_r)\mathbf{A}\right)\right|^{m-1}.\end{aligned}$$

Then, taking

$$\mathbf{A}^{**} = (\boldsymbol{\Gamma}_m \otimes \mathbf{I}_r)\,\mathbf{A}\,(\boldsymbol{\Gamma}_m \otimes \mathbf{I}_r)$$

with $\boldsymbol{\Gamma}_m$ orthogonal and symmetric, we may write

$$\mathbf{A} = (\boldsymbol{\Gamma}_m \otimes \mathbf{I}_r)\,\mathbf{A}^{**}\,(\boldsymbol{\Gamma}_m \otimes \mathbf{I}_r),$$

while, on the other hand, using the properties of BTr (see Filipiak et al. [7]), we may write

$$\begin{aligned}
\mathrm{BTr}\left((\mathbf{P}_m \otimes \mathbf{I}_r)\mathbf{A}\right) &= \mathrm{BTr}\left[(\mathbf{P}_m \otimes \mathbf{I}_r)\,(\boldsymbol{\Gamma}_m \otimes \mathbf{I}_r)\,\mathbf{A}^{**}\,(\boldsymbol{\Gamma}_m \otimes \mathbf{I}_r)\right]\\
&= \mathrm{BTr}\left[(\boldsymbol{\Gamma}_m \otimes \mathbf{I}_r)\,(\mathbf{P}_m \otimes \mathbf{I}_r)\,(\boldsymbol{\Gamma}_m \otimes \mathbf{I}_r)\,\mathbf{A}^{**}\right]\\
&= \mathrm{BTr}\left[(\mathbf{Q}_m^* \otimes \mathbf{I}_r)\mathbf{A}^{**}\right] = \mathbf{A}_1^{**}
\end{aligned}$$

where

$$\mathbf{Q}_m^* = \boldsymbol{\Gamma}_m \mathbf{P}_m \boldsymbol{\Gamma}_m = \left(\begin{array}{c|c} 1 & \mathbf{0}'_{m-1} \\ \hline \mathbf{0}_{m-1} & \mathbf{0}_{(m-1)\times(m-1)} \end{array}\right)$$

and

$$\begin{aligned}
\tfrac{1}{m-1}\mathrm{BTr}\left[(\mathbf{Q}_m \otimes \mathbf{I}_r)\mathbf{A}\right] &= \tfrac{1}{m-1}\mathrm{BTr}\left[(\mathbf{Q}_m \otimes \mathbf{I}_r)\,(\boldsymbol{\Gamma}_m \otimes \mathbf{I}_r)\,\mathbf{A}^{**}\,(\boldsymbol{\Gamma}_m \otimes \mathbf{I}_r)\right]\\
&= \tfrac{1}{m-1}\mathrm{BTr}\left[(\boldsymbol{\Gamma}_m \otimes \mathbf{I}_r)\,(\mathbf{Q}_m \otimes \mathbf{I}_r)\,(\boldsymbol{\Gamma}_m \otimes \mathbf{I}_r)\,\mathbf{A}^{**}\right]\\
&= \tfrac{1}{m-1}\mathrm{BTr}\left[(\mathbf{P}_m^* \otimes \mathbf{I}_r)\mathbf{A}^{**}\right] = \tfrac{1}{m-1}\sum_{j=2}^{m}\mathbf{A}_j^{**}
\end{aligned}$$

where

$$\mathbf{P}_m^* = \boldsymbol{\Gamma}_m \mathbf{Q}_m \boldsymbol{\Gamma}_m = \left(\begin{array}{c|c} 0 & \mathbf{0}'_{m-1} \\ \hline \mathbf{0}_{m-1} & \mathbf{I}_{m-1} \end{array}\right)$$

so that we may indeed write

$$|\mathbf{A}^*| = \left|\mathbf{A}_1^{**}\right| \left|\frac{1}{m-1}\sum_{j=2}^{m}\mathbf{A}_j^{**}\right|^{m-1}.$$

References

1. Anderson, T.W.: The Statistical Analysis of Time Series. Wiley, New York (1972)
2. Arnold, B.C., Coelho, C.A., Marques, F.J.: The distribution of the product of powers of independent uniform random variables—a simple but useful tool to address and better understand the structure of some distributions. J. Multivar. Anal. **113**, 19–36 (2013)
3. Brillinger, D.R.: Time Series-Data Analysis and Theory. SIAM (2001)
4. Coelho, C.A.: The generalized integer Gamma distribution—a basis for distributions in multivariate statistics. J. Multivar. Anal. **64**, 86–102 (1998)
5. Coelho, C.A.: The generalized near-integer Gamma distribution: a basis for "near-exact" approximations to the distribution of statistics which are the product of an odd number of independent Beta random variables. J. Multivar. Anal. **89**, 191–218 (2004)
6. Coelho, C.A., Arnold, B.C.: Finite Forms Representations for Meijer G and Fox H Functions—Applied to Multivariate Likelihood Ratio Tests Using Mathematica®, MAXIMA and R. Lecture Notes in Statistics. Springer (2019)

7. Filipiak, K., Klein, D., Vojtková, E.: The properties of partial trace and block trace operators of partitioned matrices. Electron. J. Linear Algebra **33**, 3–15 (2018)
8. Fox, C.: The G and H functions as symmetrical kernels. Trans. Am. Math. Soc. **98**, 395–429 (1961)
9. John, J.A.: Cyclic Designs. Chapman and Hall, London (1987)
10. Khattree, R.: Multivariate statistical inference involving circulant matrices: a review. In: Gupta, A.K., Girko, V.L. (eds.) Multidimensional Statistical Analysis and Theory of Random Matrices, pp. 101–110. VSP, The Netherlands (1996)
11. Kedem, B.: Time Series Analysis by Higher Order Crossings. IEEE Press Inc., New York (1994)
12. Kshirsagar, A.M.: Multivariate Analysis. Marcel Dekker, New York (1972)
13. Lv, X.-G., Huang, T.-Z.: A note on inversion of Toeplitz matrices. Appl. Math. Lett. **20**, 1189–1193 (2007)
14. Lv, X.-G., Huang, T.-Z.: The inverses of block Toeplitz matrices. J. Math. Article ID 207176, 8 (2013)
15. Mathai, A.M.: A Handbook of Generalized Special Functions for Statistical and Physical Sciences. Oxford University Press, New York (1993)
16. Mathai, A.M., Haubold, H.J.: Special Functions for Applied Scientists. Springer, New York (2008)
17. Mathai, A.M., Saxena, R.K.: The H-function with Applications in Statistics and Other Disciplines. Wiley, New York (1978)
18. Olkin, I.: Testing and estimation for structures which are circularly symmetric in blocks. In: Kabe, D.G., Gupta, R.P. (eds.) Multivariate Statistical Inference, pp. 183–195. North Holland, New York (1973)
19. Olkin, I., Press, S.J.: Testing and estimation for a circular stationary model. Ann. Math. Stat. **40**, 1358–1373 (1969)
20. Pollock, D.S.G.: Circulant matrices and time-series analysis. Int. J. Math. Educ. Sci. Technol. **33**, 213–230 (2002)
21. Prudnikov, A.P., Brychkov, Y.A., Marichev, O.J.: Integrals and Series, Vol. 3: More Special Functions. Gordon and Breach, Newark (1990)
22. Seely, J.: Quadratic subspaces and completeness. Ann. Math. Stat. **42**, 710–721 (1971)
23. Szatrowski, T.D.: Explicit solutions, one iteration convergence and averaging in the multivariate normal estimation problem for patterned means and covariances. Ann. Inst. Stat. Math. **30**, 81–88 (1978)
24. Szatrowski, T.D.: Necessary and sufficient conditions for explicit solutions in the multivariate Normal estimation problem for patterned means and covariances. Ann. Stat. **8**, 802–810 (1980)
25. Tricomi, F.G., Erdélyi, A.: The asymptotic expansion of a ratio of gamma functions. Pac. J. Math. **1**, 133–142 (1951)
26. Wald, A., Brookner, R.J.: On the distribution of Wilks' statistic for testing the independence of several groups of variates. Ann. Math. Stat. **12**, 137–152 (1941)

Chapter 8
Estimation and Testing Hypotheses in Two-Level and Three-Level Multivariate Data with Block Compound Symmetric Covariance Structure

Arkadiusz Kozioł, Anuradha Roy, Roman Zmyślony, Ivan Žežula, and Miguel Fonseca

Abstract This article deals with the estimation and hypotheses testing problems for two-level and three-level multivariate data. The coordinate-free approach is used to prove that the quadratic estimation of covariance parameters is equivalent to linear estimation with a properly defined inner product in the space of symmetric matrices for both two-level and three-level multivariate data. The estimators are shown to be unbiased, sufficient, complete, and consistent. A competitor for the likelihood ratio test on covariance components under linear constraints and the mean vectors are proposed, based on the F distribution. Simulation studies are conducted to see the power of the proposed tests and the proposed methods are implemented with two medical datasets.

A. Kozioł · R. Zmyślony
Faculty of Mathematics, Computer Science and Econometrics, University of Zielona Góra, 65-417 Zielona Góra, Poland
e-mail: A.Koziol@wmie.uz.zgora.pl

R. Zmyślony
e-mail: R.Zmyslony@wmie.uz.zgora.pl

A. Roy (✉)
Department of Management Science and Statistics, The University of Texas at San Antonio, San Antonio, TX 78249, USA
e-mail: Anuradha.Roy@utsa.edu

I. Žežula
Institute of Mathematics, Faculty of Science, P. J. Šafárik University, Košice, Slovakia
e-mail: ivan.zezula@upjs.sk

M. Fonseca
Centro de Matemática e Aplicações (CMA) and Departamento de Matemática, Universidade Nova de Lisboa, Lisbon, Portugal
e-mail: fmig@fct.unl.pt

K. Filipiak et al. (eds.), *Multivariate, Multilinear and Mixed Linear Models*, Contributions to Statistics, https://doi.org/10.1007/978-3-030-75494-5_8

8.1 Introduction

Two-level multivariate data (m dimensional observation vector repeatedly measured at u locations or time points) and three-level multivariate data (m-dimensional observation vector repeatedly measured at u locations and over v time points) are ubiquitous and abundant these days as we can store these complex data easily and inexpensively with the help of modern computing facilities. Since in medical and biomedical sciences small or moderate number of samples are very common, we can quickly arrive at the situation when even basic statistical characteristics needed for inference are not estimable. This can be sometimes circumvented by using some kind of structured variance–covariance matrices, e.g., block compound symmetry (BCS) and double block compound symmetry (DBCS) covariance structures for two-level and three-level datasets, respectively, if the exchangeability or symmetry property is present in the data. If some structure is present in the data, one can use the information to draw better inference to model the data using the structured variance–covariance matrix. The number of unknown parameters in a structured variance–covariance matrix is always much less than the number of unknown parameters in an unstructured variance–covariance matrix. Perlman [23] correctly mentioned that if symmetries are known to be present in the data, then more accurate estimate of the covariance matrix can be obtained, and thus more powerful tests concerning the variance–covariance matrix and the related population mean vectors can be derived. Andersson [1] proposed the general group symmetry variance–covariance model theory, and he named these models as an invariant normal model under normal distribution. He showed that BCS structure is an invariant normal model. Nahtman [22] showed that the random factors in balanced linear models are invariant with respect to marginal permutations of them. Linear models with BCS and DBCS covariance structures as error matrix were developed by Arnold [3] and Roy and Fonseca [26]. In this article, we study the estimation for fixed effects and covariance components along with their optimal properties, based on the fact that both BCS and DBCS covariance matrices for these kinds of models belong to quadratic subspaces. Hypotheses testing procedures of a block diagonal covariance structure with identical blocks as well as the mean vectors are also studied here.

For two-level multivariate observations, u must be greater than 1. For three-level multivariate observations, both u and v must be greater than 1. If either $u = 1$ or $v = 1$, the data become two-level multivariate with BCS covariance structure, and if both $u = 1$ and $v = 1$, the data become traditional multivariate with an unstructured variance–covariance matrix. If $m = 1$ with either $u = 1$ or $v = 1$, the data also become traditional multivariate with compound symmetry covariance structure.

BCS covariance structure was introduced independently by Rao [24, 25] and Wilks [37]. BCS covariance structure was extensively studied by Arnold [2] and Szatrowski [36]. DBCS structure was first studied by Roy and Leiva [28] in the framework of discriminant analysis. Leiva [19] and Leiva and Roy [20] developed classification rules for two-level and three-level multivariate data using BCS and DBCS structures. Hypotheses testing procedures for BCS and DBCS structures are studied by Roy and Leiva [29], Roy et al. [32] and Coelho and Roy [4, 5]. Roy et al. [31] and Koziol et al. [17, 18] studied the optimality properties of both BCS and

DBCS structures using the coordinate-free approach theory. Hypotheses testing of the equality of mean vectors in two populations using the BCS covariance structure was considered by Roy et al. [30] and Žežula et al. [41].

A simple example of a two-level multivariate dataset may be obtained from Johnson and Wichern [14, p. 43], who report data on a study of osteoporosis where an investigator measures the mineral content of three bones (radius, humerus, and ulna, $m = 3$) by photon absorptiometry to examine whether a particular dietary supplement increases bone mineral content and mass in older women. All three measurements are recorded on the dominant and non-dominant sides ($u = 2$) for 25 older women. Another simple example is given in Roy and Leiva [28]. Data come from a clinical trial of the eye disease glaucoma. Measurements of intraocular pressure (IOP) and central corneal thickness (CCT) are obtained from both the eyes (sites), each at three time points at an interval of three months for 30 patients. It is clear that this dataset is a three-level data with $m = 2$, $u = 2$ and $v = 3$. These two examples will be used later in Sects. 8.2.5 and 8.3.5 for illustrative purposes. Our main intention of the analysis of these datasets is to illustrate the proposed methods rather than giving any insight into the datasets.

As mentioned before, the assumption of BCS and DBCS reduces substantially the number of unknown parameters in the model for two-level and three-level datasets; an $vum \times vum$ unstructured variance–covariance matrix has $vum(vum + 1)/2$ unknown parameters, which can be large even for small values of v, u, or m, whereas the same dimensional DBCS covariance structure has only $3m(m + 1)/2$ unknown parameters. This number does not even depend on the number of locations or sites u and the number of time points v. Moreover, both BCS and DBCS covariance structures do not need the repeated measurements to be equally spaced at any level.

Note 1 To simplify reading of this article, we decided to reuse the same notations for different objects having the same function. For example, four different projection matrices are used for two-level and three-level multivariate data, each with unstructured and structured mean vectors. The projection matrices are different in each of the four set-ups, nevertheless being the projection matrix, we exploit the same generic notation $\mathbf{P}$, the projection on the space generated by vector of means, for all of them.

8.2 Two-Level Multivariate Data

In this section, we study the model with BCS covariance structure for two-level multivariate data, i.e., when m dimensional observation vector is repeatedly measured at u locations or time points. It is a model that has several applications in areas such as medicine, biology, and engineering, among many others. We discuss the optimality properties of the model parameters for both unstructured and structured mean vectors. We then compare the two models and illustrate the comparison with the help of a real data example. We finish the section with a hypothesis testing whether the m variables over u sites are independent, and validate and compare the proposed test with some other tests using simulation studies.

8.2.1 Block Compound Symmetric Covariance Structure

The $(um \times um)-$dimensional BCS covariance structure is defined as

$$\boldsymbol{\Gamma} = \begin{pmatrix} \boldsymbol{\Gamma}_0 & \boldsymbol{\Gamma}_1 & \ldots & \boldsymbol{\Gamma}_1 \\ \vdots & \ddots & & \vdots \\ \vdots & & \ddots & \vdots \\ \boldsymbol{\Gamma}_1 & \boldsymbol{\Gamma}_1 & \ldots & \boldsymbol{\Gamma}_0 \end{pmatrix} = \mathbf{I}_u \otimes (\boldsymbol{\Gamma}_0 - \boldsymbol{\Gamma}_1) + \mathbf{J}_u \otimes \boldsymbol{\Gamma}_1.$$

The above BCS structure $\boldsymbol{\Gamma}$ can equivalently be written as follows:

$$\boldsymbol{\Gamma} = \mathbf{I}_u \otimes \boldsymbol{\Gamma}_0 + (\mathbf{J}_u - \mathbf{I}_u) \otimes \boldsymbol{\Gamma}_1, \tag{8.1}$$

where $\mathbf{I}_u$ is the $u \times u$ identity matrix, $\mathbf{1}_u$ is a $u \times 1$ vector of ones, $\mathbf{J}_u = \mathbf{1}_u \mathbf{1}_u'$ is the matrix with all elements equal to one, of size, u and $\otimes$ represents the Kronecker product. We assume $\boldsymbol{\Gamma}_0$ is a positive definite symmetric $m \times m$ matrix, $\boldsymbol{\Gamma}_1$ is a symmetric $m \times m$ matrix, and the constraints $-\frac{1}{u-1}\boldsymbol{\Gamma}_0 \prec \boldsymbol{\Gamma}_1$ and $\boldsymbol{\Gamma}_1 \prec \boldsymbol{\Gamma}_0$, which means that $\boldsymbol{\Gamma}_0 + (u-1)\boldsymbol{\Gamma}_1$ and $\boldsymbol{\Gamma}_0 - \boldsymbol{\Gamma}_1$ are positive definite matrices, so that the $um \times um$ matrix $\boldsymbol{\Gamma}$ is positive definite (for a proof, see Roy and Leiva [29], Lemma 2.1). The $m \times m$ block diagonals $\boldsymbol{\Gamma}_0$ in $\boldsymbol{\Gamma}$ represent the variance–covariance matrix of the m response variables at any given site, whereas the $m \times m$ block off diagonals $\boldsymbol{\Gamma}_1$ in $\boldsymbol{\Gamma}$ represent the covariance matrix of the m response variables between any two sites. We also assume that $\boldsymbol{\Gamma}_0$ is constant for all sites and $\boldsymbol{\Gamma}_1$ is constant for all site pairs.

8.2.2 Estimation in Model with Unstructured Mean Vector

Let $\mathbf{y}_{r,s}$ be a m-variate vector of measurements on the rth individual at the sth site; $r \in \{1, \ldots, n\}$, $s \in \{1, \ldots, u\}$. The n individuals are all independent. Let $\mathbf{y}_r = (\mathbf{y}_{r,1}', \ldots, \mathbf{y}_{r,u}')'$ be the um-variate vector of all measurements corresponding to the rth individual. Finally, let $\mathbf{y}_1, \mathbf{y}_2, \ldots, \mathbf{y}_n$ be a random sample of size n drawn from the population $\mathcal{N}_{um}(\boldsymbol{\mu}, \boldsymbol{\Gamma})$, where, for $r \in \{1, \ldots, n\}$, $\mathrm{E}(\mathbf{y}_r) = \boldsymbol{\mu} \in \mathbb{R}^{um}$ and $\mathrm{D}(\mathbf{y}_r) = \boldsymbol{\Gamma} = \mathbf{I}_u \otimes (\boldsymbol{\Gamma}_0 - \boldsymbol{\Gamma}_1) + \mathbf{J}_u \otimes \boldsymbol{\Gamma}_1$ is assumed to be a $um \times um$ positive definite matrix. Let the matrix $\mathbf{Y}$ represent n independent and identically normally distributed random column vectors with mean vector $\boldsymbol{\mu}$ and variance–covariance matrix $\boldsymbol{\Gamma}$.

In model with unstructured mean vector we assume that the variance–covariance structure is BCS and the mean vector changes over sites. Thus $\boldsymbol{\mu}$ has um components. This model can be written in the following way:

$$\underset{num\times 1}{\mathbf{y}} = \text{vec}\,(\underset{um\times n}{\mathbf{Y}}) \sim \mathcal{N}\left((\mathbf{1}_n \otimes \mathbf{I}_{um})\boldsymbol{\mu}, \mathbf{V}\right), \tag{8.2}$$

where $\mathbf{V} = \mathbf{I}_n \otimes \boldsymbol{\Gamma}$.

Define the projection matrix $\mathbf{P}$ as follows:

$$\mathbf{P} = \frac{1}{n}\mathbf{J}_n \otimes \mathbf{I}_u \otimes \mathbf{I}_m. \tag{8.3}$$

It is clear that $\mathbf{P}$ is an orthogonal projector on the subspace of the mean vector of $\mathbf{y}$. If $\mathbf{I}_n \otimes \mathbf{I}_{um} \in \vartheta = \text{sp}\{\mathbf{V}\}$ (Gnot et al. [10]) it follows that $\mathbf{Py}$ is the best linear unbiased estimator (BLUE) if and only if $\mathbf{P}$ commutes with all covariance matrices $\mathbf{V}$. Therefore, we have the following results.

Result 8.1 *The projection matrix* $\mathbf{P}$ *commutes with the covariance matrix* $\mathbf{V}$*, i.e.,* $\mathbf{PV} = \mathbf{VP}$.

For a proof of the above result, see Result 1 in Roy et al. [31].

Lemma 8.1 *Let* ϑ *denote the subspace spanned by* $\mathbf{V}$*, i.e.,* $\vartheta = \text{sp}\{\mathbf{V}\}$*. Then,* ϑ *is a quadratic subspace, meaning that* ϑ *is a linear space and if* $\mathbf{V} \in \vartheta$ *then* $\mathbf{V}^2 \in \vartheta$ *(see Seely [34] for the definition).*

For a proof of the above lemma, see Roy et al. [31, Lemma 1]. Because orthogonal projector on the space generated by the mean vector commutes with all covariance matrices, there exists BLUEs for each estimable function of mean. Moreover BLUEs are least square estimators (LSEs).

We now construct a basis for the quadratic subspace ϑ. We define

$$\mathbf{A}_{ii} = \mathbf{E}_{ii} \quad \text{and} \quad \mathbf{A}_{ij} = \mathbf{E}_{ij} + \mathbf{E}_{ji}, \quad \text{for } i < j, \text{ and } i, j \in \{1, \ldots, m\},$$

as a basis for symmetric matrices $\boldsymbol{\Gamma}$. The $(m \times m)-$dimensional matrices $\mathbf{E}_{ij}$ has 1 only at the ijth element, and 0 at all other elements. Then it is clear that the basis for diagonal matrices of the form $\mathbf{I}_n \otimes \mathbf{I}_u \otimes \boldsymbol{\Gamma}_0$ is constituted by matrices

$$\mathbf{K}_{ij}^{(0)} = \mathbf{I}_n \otimes \mathbf{I}_u \otimes \mathbf{A}_{ij}, \quad \text{for } i \leq j, \quad j \in \{1, \ldots, m\} \tag{8.4}$$

and the basis for matrices of the form $\mathbf{I}_n \otimes (\mathbf{J}_u - \mathbf{I}_u) \otimes \boldsymbol{\Gamma}_1$ is constituted by matrices

$$\mathbf{K}_{ij}^{(1)} = \mathbf{I}_n \otimes (\mathbf{J}_u - \mathbf{I}_u) \otimes \mathbf{A}_{ij}, \quad \text{for } i \leq j, \quad j \in \{1, \ldots, m\}. \tag{8.5}$$

It is clear from (8.1) that the above basis is orthogonal with respect to the trace of inner product.

Result 8.2 *The complete and minimal sufficient statistics for the mean vector and the variance–covariance matrix are*

$$\left(\mathbf{1}_n' \otimes \mathbf{I}_{um}\right)\mathbf{y} \text{ and } \mathbf{y}'\mathbf{Q}\mathbf{K}_{ij}^{(l)}\mathbf{Q}\mathbf{y}, \quad l \in \{0, 1\},$$

where $\mathbf{Q} = \mathbf{I}_n \otimes \mathbf{I}_u \otimes \mathbf{I}_m - \mathbf{P}$ *and* $\mathbf{P}$ *is given in (8.3), see Fonseca et al. [7], Seely [35] and Zmyślony [39].*

Clearly, $\mathbf{Q}$ is idempotent. Since $\mathbf{P}$ commutes with the covariance matrix of $\mathbf{y}$, then for each parameter of the covariance matrix, there exists best quadratic unbiased estimator (BQUE) if and only if sp$\{\mathbf{QVQ}\}$ is a quadratic subspace (see Zmyślony [38, 39] and Gnot et al. [9–11]) or Jordan algebra (see Jordan et al. [15]), where $\mathbf{V}$ stands for covariance matrix of $\mathbf{y}$. It is clear that if sp$\{\mathbf{V}\}$ is a quadratic subspace and if for each $\mathbf{\Sigma} \in$ sp$\{\mathbf{V}\}$ commutativity $\mathbf{P\Sigma} = \mathbf{\Sigma P}$ holds, then sp$\{\mathbf{QVQ}\}$ = sp$\{\mathbf{QV}\}$ is also a quadratic subspace.

According to the coordinate-free approach, the expectation of $\mathbf{Qyy}'\mathbf{Q}$ can be written as a linear combination of matrices $\mathbf{QK}_{ij}^{(0)}$ and $\mathbf{QK}_{ij}^{(1)}$ with matrices $\boldsymbol{\Gamma}_0$ and $\boldsymbol{\Gamma}_1$, respectively. Note also that the identity covariance operator of $\mathbf{yy}'$ belongs to sp$\{\mathrm{D}(\mathbf{yy}')\}$. It implies that the ordinary best quadratic estimators are LSEs for corresponding matrices $\boldsymbol{\Gamma}_0$ and $\boldsymbol{\Gamma}_1$.

The model formulas and the properties of the best unbiased estimators for mean vector and covariance components are given in the following two theorems.

Theorem 8.1 *Under the model (8.2), the best unbiased estimators of* $\boldsymbol{\mu}$, $\boldsymbol{\Gamma}_0$ *and* $\boldsymbol{\Gamma}_1$ *are, respectively,*

$$\widetilde{\boldsymbol{\mu}} = \frac{1}{n}\sum_{r=1}^{n} \mathbf{y}_r, \tag{8.6}$$

$$\widetilde{\boldsymbol{\Gamma}}_0 = \frac{1}{(n-1)u}\mathbf{C}_0, \tag{8.7}$$

$$\text{and} \quad \widetilde{\boldsymbol{\Gamma}}_1 = \frac{1}{(n-1)u(u-1)}\mathbf{C}_1, \tag{8.8}$$

where matrices $\mathbf{C}_0$ *and* $\mathbf{C}_1$ *are*

$$\mathbf{C}_0 = \sum_{s=1}^{u}\sum_{r=1}^{n}\left(\mathbf{y}_{r,s} - \overline{\mathbf{y}}_{\bullet,s}\right)\left(\mathbf{y}_{r,s} - \overline{\mathbf{y}}_{\bullet,s}\right)',$$

$$\text{and} \quad \mathbf{C}_1 = \sum_{s=1}^{u}\sum_{\substack{s^*=1 \\ s \neq s^*}}^{u}\sum_{r=1}^{n}\left(\mathbf{y}_{r,s} - \overline{\mathbf{y}}_{\bullet,s}\right)\left(\mathbf{y}_{r,s^*} - \overline{\mathbf{y}}_{\bullet,s^*}\right)',$$

with $\overline{\mathbf{y}}_{\bullet,s} = \frac{1}{n}\sum_{r=1}^{n}\mathbf{y}_{r,s}$.

For a proof of the above theorem, see Theorem 1 in Roy et al. [31].

Theorem 8.2 *Estimators given in (8.6), (8.7) and (8.8) are consistent. Moreover, the family of distributions of these estimators is complete.*

For a proof of the above theorem, see Theorem 2 in Roy et al. [31].

8.2.3 Estimation in Model with Structured Mean Vector

Let $\mathbf{y}_1, \mathbf{y}_2, \ldots, \mathbf{y}_n$ be a random sample of size n drawn from the population $\mathcal{N}_{um}(\mathbf{1}_u \otimes \boldsymbol{\mu}, \boldsymbol{\Gamma})$, where, for $r \in \{1, \ldots, n\}$, $\mathrm{E}(\mathbf{y}_r) = \mathbf{1}_u \otimes \boldsymbol{\mu} \in \mathbb{R}^{um}$ and $\mathrm{D}(\mathbf{y}_r) = \boldsymbol{\Gamma} = \mathbf{I}_u \otimes (\boldsymbol{\Gamma}_0 - \boldsymbol{\Gamma}_1) + \mathbf{J}_u \otimes \boldsymbol{\Gamma}_1$ is assumed to be a $um \times um$ positive definite matrix.

In model with structured mean vector we assume that the covariance structure is BCS and the mean vector remains constant over sites. Thus, $\boldsymbol{\mu}$ has m components. This model can be written in the following way

$$\underset{num\times 1}{\mathbf{y}} = \mathrm{vec}\,(\underset{um\times n}{\mathbf{Y}}) \sim \mathcal{N}\left((\mathbf{1}_{nu} \otimes \mathbf{I}_m)\boldsymbol{\mu}, \mathbf{V}\right), \tag{8.9}$$

where $\mathbf{V} = \mathbf{I}_n \otimes \boldsymbol{\Gamma}$. It means that the matrix $\mathbf{Y}$ contains n independent normally distributed random column vectors which are identically distributed with mean vector $\mathbf{1}_u \otimes \boldsymbol{\mu}$ and covariance matrix $\boldsymbol{\Gamma}$.

Define the orthogonal projector on the subspace of the mean vector of $\mathbf{y}$ as follows:

$$\mathbf{P} = \frac{1}{n}\mathbf{J}_n \otimes \frac{1}{u}\mathbf{J}_u \otimes \mathbf{I}_m. \tag{8.10}$$

Just like for the model with unstructured mean vector we need to show that $\mathbf{P}$ commutes with $\mathbf{V}$ and the subspace spanned by $\mathbf{V}$ is a quadratic subspace.

Result 8.3 *The projection matrix* $\mathbf{P}$ *commutes with the covariance matrix* $\mathbf{V}$, *i.e.,* $\mathbf{PV} = \mathbf{VP}$.

For a proof of the above result, see Result 1 in Kozioł et al. [18].

Lemma 8.2 *The subspace* $\vartheta = \mathrm{sp}\{\mathbf{V}\}$ *is a quadratic subspace.*

For a proof of the above lemma, see Lemma 1 in Kozioł et al. [18]. A basis for the quadratic subspace ϑ is the same as in the previous considered case.

Result 8.4 *The complete and minimal sufficient statistics for the mean vector and the variance–covariance matrix are*

$$\left(\mathbf{1}'_{nu} \otimes \mathbf{I}_m\right)\mathbf{y} \text{ and } \mathbf{y}'\mathbf{Q}\boldsymbol{K}_{ij}^{(l)}\mathbf{Q}\mathbf{y}, \quad l \in \{0, 1\},$$

where $\mathbf{Q} = \mathbf{I}_n \otimes \mathbf{I}_u \otimes \mathbf{I}_m - \mathbf{P}$, $\mathbf{P}$ *is given in (8.10),* $\mathbf{K}_{ij}^{(0)}$ *and* $\mathbf{K}_{ij}^{(1)}$ *are given in (8.4) and (8.5), respectively. For more details see Fonseca et al. [7], Seely [35] and Zmyślony [39].*

Now, since $\mathbf{PV} = \mathbf{VP}$, and ϑ is a quadratic space, $\mathbf{Q}\vartheta\mathbf{Q} = \mathbf{Q}\vartheta$ is also a quadratic space. According to the coordinate-free approach, the expectation of $\mathbf{Qyy}'\mathbf{Q}$ can be written as a linear combination of matrices $\mathbf{QK}_{ij}^{(0)}$ and $\mathbf{QK}_{ij}^{(1)}$ with matrices $\boldsymbol{\Gamma}_0$ and $\boldsymbol{\Gamma}_1$, respectively. Note also that identity covariance operator of $\mathbf{yy}'$ belongs to $\mathrm{sp}\{\mathrm{D}(\mathbf{yy}')\}$. It implies that the ordinary best quadratic estimators are LSEs for corresponding matrices $\boldsymbol{\Gamma}_0$ and $\boldsymbol{\Gamma}_1$. They cannot be calculated independently because $\mathbf{QK}_{ij}^{(0)}$ and $\mathbf{QK}_{ij}^{(1)}$ are not orthogonal as we can see.

Theorem 8.3 *Under the model (8.9) the best unbiased estimators of $\boldsymbol{\mu}$, $\boldsymbol{\Gamma}_0$ and $\boldsymbol{\Gamma}_1$ are respectively*

$$\widetilde{\boldsymbol{\mu}} = \frac{1}{nu}\sum_{r=1}^{n}\sum_{s=1}^{u}\mathbf{y}_{r,s}, \tag{8.11}$$

$$\widetilde{\boldsymbol{\Gamma}}_0 = \frac{(n-1)u+1}{(n-1)nu^2}\mathbf{C}_0 + \frac{1}{(n-1)nu^2}\mathbf{C}_1, \tag{8.12}$$

$$\text{and } \widetilde{\boldsymbol{\Gamma}}_1 = \frac{1}{(n-1)nu^2}\mathbf{C}_0 + \frac{nu-1}{(n-1)n(u-1)u^2}\mathbf{C}_1, \tag{8.13}$$

where matrices $\mathbf{C}_0$ *and* $\mathbf{C}_1$ *are*

$$\mathbf{C}_0 = \sum_{s=1}^{u}\sum_{r=1}^{n}\left(\mathbf{y}_{r,s}-\widetilde{\boldsymbol{\mu}}\right)\left(\mathbf{y}_{r,s}-\widetilde{\boldsymbol{\mu}}\right)',$$

$$\text{and } \mathbf{C}_1 = \sum_{s=1}^{u}\sum_{\substack{s^*=1 \\ s\neq s^*}}^{u}\sum_{r=1}^{n}\left(\mathbf{y}_{r,s}-\widetilde{\boldsymbol{\mu}}\right)\left(\mathbf{y}_{r,s^*}-\widetilde{\boldsymbol{\mu}}\right)'.$$

For a proof of the above theorem, see Theorem 1 in Kozioł et al. [18].

Theorem 8.4 *Estimators given in (8.11), (8.12) and (8.13) are consistent. Moreover, the family of distributions of these estimators is complete.*

For a proof of the above theorem, see Theorem 2 in Kozioł et al. [18].

8.2.4 Comparison of BUE in Two Models

In this paragraph we compare variances of estimators of covariance parameters under the two models with unstructured (Mo1) and structured (Mo2) mean vectors as described in the Sects. 8.2.2 and 8.2.3. Let for Mo1, $\widetilde{\sigma}_{[1]ij}^{(0)}$ and $\widetilde{\sigma}_{[1]ij}^{(1)}$ be estimators of covariance parameters in matrices $\boldsymbol{\Gamma}_0$ and $\boldsymbol{\Gamma}_1$, respectively, and for Mo2, $\widetilde{\sigma}_{[2]ij}^{(0)}$ and $\widetilde{\sigma}_{[2]ij}^{(1)}$ be estimators of covariance parameters in matrices $\boldsymbol{\Gamma}_0$ and $\boldsymbol{\Gamma}_1$, respectively. Let us denote $\mathbf{P}_{[1]}$ and $\mathbf{P}_{[2]}$ as orthogonal projectors for unstructured and structured space generated by the respective mean vectors. Moreover, let $\mathbf{Q}_{[1]} = \mathbf{I} - \mathbf{P}_{[1]}$ and $\mathbf{Q}_{[2]} = \mathbf{I} - \mathbf{P}_{[2]}$. Since $R(\mathbf{P}_{[2]}) \subset R(\mathbf{P}_{[1]})$ then it holds $\mathbf{P}_{[1]}\mathbf{P}_{[2]} = \mathbf{P}_{[2]}\mathbf{P}_{[1]} = \mathbf{P}_{[2]}$. This implies that $\mathbf{Q}_{[1]}\mathbf{Q}_{[2]} = \mathbf{Q}_{[2]}\mathbf{Q}_{[1]} = \mathbf{Q}_{[1]}$. One can easily check that the expectation of $\widetilde{\sigma}_{[1]ij}^{(0)}$ and $\widetilde{\sigma}_{[1]ij}^{(1)}$ calculated for Mo1 are unbiased under Mo2.

Alternatively, we can also calculate and present graphically the difference of variances for both models.

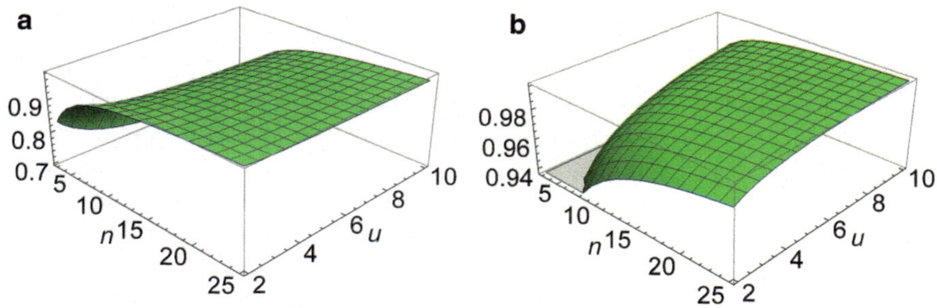

Fig. 8.1 **a** $\frac{\mathrm{D}\left(\widetilde{\sigma}^{(0)}_{[2]ij}\right)}{\mathrm{D}\left(\widetilde{\sigma}^{(0)}_{[1]ij}\right)}$ and **b** $\frac{\mathrm{D}\left(\widetilde{\sigma}^{(1)}_{[2]ij}\right)}{\mathrm{D}\left(\widetilde{\sigma}^{(1)}_{[1]ij}\right)}$

$$\mathrm{D}\left(\widetilde{\sigma}^{(0)}_{[2]ij}\right) - \mathrm{D}\left(\widetilde{\sigma}^{(0)}_{[1]ij}\right) = -\frac{2(u-1)}{(n-1)nu^2}\Big(\mathrm{Tr}\,(\mathbf{A}_{ij}\boldsymbol{\Gamma}_0\mathbf{A}_{ij}\boldsymbol{\Gamma}_0) - 2\mathrm{Tr}\,(\mathbf{A}_{ij}\boldsymbol{\Gamma}_0\mathbf{A}_{ij}\boldsymbol{\Gamma}_1) + \mathrm{Tr}\,(\mathbf{A}_{ij}\boldsymbol{\Gamma}_1\mathbf{A}_{ij}\boldsymbol{\Gamma}_1)\Big).$$

From

$$\mathrm{Tr}\,(\mathbf{A}_{ij}\boldsymbol{\Gamma}_0\mathbf{A}_{ij}\boldsymbol{\Gamma}_0) + \mathrm{Tr}\,(\mathbf{A}_{ij}\boldsymbol{\Gamma}_1\mathbf{A}_{ij}\boldsymbol{\Gamma}_1) > 2\mathrm{Tr}\,(\mathbf{A}_{ij}\boldsymbol{\Gamma}_0\mathbf{A}_{ij}\boldsymbol{\Gamma}_1)$$

we get the following inequality

$$\mathrm{D}\left(\widetilde{\sigma}^{(0)}_{[2]ij}\right) - \mathrm{D}\left(\widetilde{\sigma}^{(0)}_{[1]ij}\right) < 0.$$

Similarly, it is easy to see

$$\mathrm{D}\left(\widetilde{\sigma}^{(1)}_{[2]ij}\right) - \mathrm{D}\left(\widetilde{\sigma}^{(1)}_{[1]ij}\right) = -\frac{2}{(n-1)n(u-1)u^2}\Big(\mathrm{Tr}\,(\mathbf{A}_{ij}\boldsymbol{\Gamma}_0\mathbf{A}_{ij}\boldsymbol{\Gamma}_0) - 2\mathrm{Tr}\,(\mathbf{A}_{ij}\boldsymbol{\Gamma}_0\mathbf{A}_{ij}\boldsymbol{\Gamma}_1) + \mathrm{Tr}\,(\mathbf{A}_{ij}\boldsymbol{\Gamma}_1\mathbf{A}_{ij}\boldsymbol{\Gamma}_1)\Big).$$

From

$$\mathrm{Tr}\,(\mathbf{A}_{ij}\boldsymbol{\Gamma}_0\mathbf{A}_{ij}\boldsymbol{\Gamma}_0) + \mathrm{Tr}\,(\mathbf{A}_{ij}\boldsymbol{\Gamma}_1\mathbf{A}_{ij}\boldsymbol{\Gamma}_1) > 2\mathrm{Tr}\,(\mathbf{A}_{ij}\boldsymbol{\Gamma}_0\mathbf{A}_{ij}\boldsymbol{\Gamma}_1)$$

we get the following inequality

$$\mathrm{D}\left(\widetilde{\sigma}^{(1)}_{[2]ij}\right) - \mathrm{D}\left(\widetilde{\sigma}^{(1)}_{[1]ij}\right) < 0.$$

For more details see Kozioł et al. [18].

For graphical illustration, we fix $\boldsymbol{\Gamma}_0 = \mathbf{I}$ and $\boldsymbol{\Gamma}_1 = \mathbf{0}$, and consider the ratios $\frac{\mathrm{D}\left(\widetilde{\sigma}^{(0)}_{[2]ij}\right)}{\mathrm{D}\left(\widetilde{\sigma}^{(0)}_{[1]ij}\right)}$ (Fig. 8.1a) and $\frac{\mathrm{D}\left(\widetilde{\sigma}^{(1)}_{[2]ij}\right)}{\mathrm{D}\left(\widetilde{\sigma}^{(1)}_{[1]ij}\right)}$ (Fig. 8.1b).

8.2.5 A Real Data Example

This dataset is taken from Johnson and Wichern [14, p. 43]. An investigator measured the mineral content of bones (radius, humerus, and ulna) by photon absorptiometry to examine whether dietary supplements would slow bone loss in 25 older women. Measurements were recorded for three bones on the dominant and non-dominant sides. Thus, the data is doubly multivariate, and clearly $m = 3$ and $u = 2$. We rearrange the variables in the dataset by grouping together the mineral content of the dominant sides of radius, humerus and ulna as the first three variables, that is, the variables in the first location ($u = 1$) and then the mineral contents for the non-dominant side of the same bones ($u = 2$). Using the likelihood ratio test demonstrated in Roy and Leiva [29] we conclude that the data fail to reject the null hypothesis so the covariance structure is of the BCS form (p–value $= 0.5786$). Using the formula (8.11) the unbiased estimate of $\boldsymbol{\mu}$ is

$$\widetilde{\boldsymbol{\mu}} = \begin{pmatrix} 0.83106 & 1.76376 & 0.69912 \end{pmatrix}.$$

Using Theorem 8.3 we say that the above estimate $\widetilde{\boldsymbol{\mu}}$ is BUE for $\boldsymbol{\mu}$. Furthermore, using the formulas (8.12) and (8.13) the unbiased estimates of $\boldsymbol{\Gamma}_0$ and $\boldsymbol{\Gamma}_1$ are

$$\widetilde{\boldsymbol{\Gamma}}_0 = \begin{pmatrix} 0.01234 & 0.02204 & 0.00907 \\ 0.02204 & 0.07559 & 0.01694 \\ 0.00907 & 0.01694 & 0.01105 \end{pmatrix}, \text{ and } \widetilde{\boldsymbol{\Gamma}}_1 = \begin{pmatrix} 0.01025 & 0.01899 & 0.00819 \\ 0.01899 & 0.06610 & 0.01517 \\ 0.00819 & 0.01517 & 0.00810 \end{pmatrix},$$

respectively. Using the above estimates the unbiased estimate of $\boldsymbol{\Gamma}$ is

$$\begin{aligned} \widetilde{\boldsymbol{\Gamma}} &= \mathbf{I}_u \otimes \left(\widetilde{\boldsymbol{\Gamma}}_0 - \widetilde{\boldsymbol{\Gamma}}_1\right) + \mathbf{J}_u \otimes \widetilde{\boldsymbol{\Gamma}}_1 \\ &= \begin{pmatrix} \begin{pmatrix} 0.01234 & 0.02204 & 0.00907 \\ 0.02204 & 0.07559 & 0.01694 \\ 0.00907 & 0.01694 & 0.01105 \end{pmatrix} & \begin{matrix} 0.01025 & 0.01899 & 0.00819 \\ 0.01899 & 0.06610 & 0.01517 \\ 0.00819 & 0.01517 & 0.00810 \end{matrix} \\ \begin{matrix} 0.01025 & 0.01899 & 0.00819 \\ 0.01899 & 0.06610 & 0.01517 \\ 0.00819 & 0.01517 & 0.00810 \end{matrix} & \begin{pmatrix} 0.01234 & 0.02204 & 0.00907 \\ 0.02204 & 0.07559 & 0.01694 \\ 0.00907 & 0.01694 & 0.01105 \end{pmatrix} \end{pmatrix}. \end{aligned}$$

The calculated sum of squares of residuals for the structured constant mean vector model is $SSE_{[2]} = 4.7656172$ and the sum of squares of residuals for the unstructured mean model (Roy et al. [31] and Kozioł et al. [17]) is $SSE_{[1]} = 4.7142896$. By observing the ratio

$$\frac{SSE_{[1]}}{SSE_{[2]}} = 0.9892296,$$

we see that the SSE for the unstructured mean vector model is almost 99% of the SSE for the structured mean vector model, which is expected as the number of parameters in the unstructured mean vector model is more than its counterpart, thus has smaller

SSE. Additionally, the ratio 0.99 also reveals the fact that the SSEs for both the models are very close to each other. So, the model with less number of parameters (structured mean vector model in our case) should be preferred.

8.2.6 Testing Hypotheses in BCS Models

In this section we study a hypothesis testing procedure for identical block diagonal covariance structure versus the BCS covariance structure. Also, we study a test for structure on mean vector where the components of the mean vector remain constant over sites or over time points versus an unstructured mean vector, which can change over sites or over time points, in model with the BCS covariance structure. See Roy and Khattree [27] for a similar mean hypothesis testing in the framework of separable covariance structure.

8.2.6.1 Testing Hypothesis for Covariance Structure

In this section we will test whether the m variables over u sites are independent, i.e., whether the $(um \times um)$−dimensional variance–covariance matrix is a block diagonal covariance structure with u identical blocks with dimension $(m \times m)$, i.e., we will test the following hypothesis, assuming that all elements in $\boldsymbol{\Gamma}_1$ have the same sign

$$H_0 : \boldsymbol{\Gamma}_1 = \mathbf{0} \quad \text{vs.} \quad H_1 : \boldsymbol{\Gamma}_1 \neq \mathbf{0}.$$

For this, we consider the following multivariate normal model

$$\underset{num\times 1}{\mathbf{y}} = \mathrm{vec}\left(\underset{um\times n}{\mathbf{Y}}\right) \sim \mathcal{N}\left((\mathbf{1}_n \otimes \mathbf{I}_{um})\boldsymbol{\mu}, \mathbf{I}_n \otimes \boldsymbol{\Gamma}\right), \tag{8.14}$$

where $\boldsymbol{\mu}$ and $\boldsymbol{\Gamma}$ are unknown, $\boldsymbol{\mu}$ is $um \times 1$ vector and

$$\boldsymbol{\Gamma} = \mathbf{I}_u \otimes \boldsymbol{\Gamma}_0 + (\mathbf{J}_u - \mathbf{I}_u) \otimes \boldsymbol{\Gamma}_1,$$

with $\boldsymbol{\Gamma}_0$ and $\boldsymbol{\Gamma}_1$ are $m \times m$ unknown matrix parameters. The structure of the variance–covariance matrix can be equivalently written as sum of mutually orthogonal matrices, i.e., the product of the matrices is equal to zero

$$\boldsymbol{\Gamma} = \left(\mathbf{I}_u - \frac{1}{u}\mathbf{J}_u\right) \otimes (\boldsymbol{\Gamma}_0 - \boldsymbol{\Gamma}_1) + \frac{1}{u}\mathbf{J}_u \otimes (\boldsymbol{\Gamma}_0 + (u-1)\boldsymbol{\Gamma}_1).$$

From (8.7) and (8.8) in Theorem 8.1 it follows that BUE for $\boldsymbol{\Delta}_0 = \boldsymbol{\Gamma}_0 - \boldsymbol{\Gamma}_1$ and $\boldsymbol{\Delta}_1 = \boldsymbol{\Gamma}_0 + (u-1)\boldsymbol{\Gamma}_1$ are the following

$$\widetilde{\boldsymbol{\Delta}}_0 = \widetilde{\boldsymbol{\Gamma}}_0 - \widetilde{\boldsymbol{\Gamma}}_1,$$
$$\widetilde{\boldsymbol{\Delta}}_1 = \widetilde{\boldsymbol{\Gamma}}_0 + (u-1)\widetilde{\boldsymbol{\Gamma}}_1,$$

because of linearity of parameters. On the other hand the maximum likelihood (ML) estimators for $\boldsymbol{\Delta}_0$ and $\boldsymbol{\Delta}_1$ (see Roy et al. [32]) are given by

$$\widehat{\boldsymbol{\Delta}}_0 = \frac{n-1}{n}\widetilde{\boldsymbol{\Delta}}_0,$$
$$\widehat{\boldsymbol{\Delta}}_1 = \frac{n-1}{n}\widetilde{\boldsymbol{\Delta}}_1.$$

Moreover, from Roy et al. [30] $\widetilde{\boldsymbol{\Delta}}_0$ and $\widetilde{\boldsymbol{\Delta}}_1$ are independent and

$$n(u-1)\widehat{\boldsymbol{\Delta}}_0 \sim \mathcal{W}_m\left(\boldsymbol{\Delta}_0, (n-1)(u-1)\right),$$
$$n\widehat{\boldsymbol{\Delta}}_1 \sim \mathcal{W}_m\left(\boldsymbol{\Delta}_1, n-1\right),$$

where $\mathcal{W}_m(\boldsymbol{\Sigma}, n)$ stands for Wishart distribution with $m \times m$ scale matrix $\boldsymbol{\Sigma}$ and degrees of freedom parameter n. The maximum likelihood for the distribution of (8.14) is given by

$$\ell_1 = \left|\widehat{\boldsymbol{\Delta}}_0\right|^{-\frac{n(u-1)}{2}} \left|\widehat{\boldsymbol{\Delta}}_1\right|^{-\frac{n}{2}} e^{-\frac{num}{2}}.$$

Considering the above forms of $\boldsymbol{\Delta}_0$ and $\boldsymbol{\Delta}_1$ we get the following equivalence

$$H_0 : \boldsymbol{\Gamma}_1 = \mathbf{0} \quad \text{versus} \quad H_1 : \boldsymbol{\Gamma}_1 \neq \mathbf{0} \quad \Leftrightarrow \quad H_0 : \boldsymbol{\Delta}_1 = \boldsymbol{\Delta}_0 \quad \text{versus} \quad H_1 : \boldsymbol{\Delta}_1 \neq \boldsymbol{\Delta}_0.$$

The maximum likelihood under H_0 is given by

$$\ell_0 = \left|(nu)^{-1}\left(n(u-1)\widehat{\boldsymbol{\Delta}}_0 + n\widehat{\boldsymbol{\Delta}}_1\right)\right|^{-\frac{nu}{2}} e^{-\frac{num}{2}}.$$

Finally, the likelihood ratio test is given by

$$\Lambda = \frac{\left|\widehat{\boldsymbol{\Delta}}_0\right|^{\frac{n(u-1)}{2}} \left|\widehat{\boldsymbol{\Delta}}_1\right|^{\frac{n}{2}}}{\left|(nu)^{-1}\left(n(u-1)\widehat{\boldsymbol{\Delta}}_0 + n\widehat{\boldsymbol{\Delta}}_1\right)\right|^{\frac{nu}{2}}}.$$

Now we prove the following.

Lemma 8.3 *If* $\mathbf{W}_1 \sim \mathcal{W}_m(\boldsymbol{\Sigma}, n_1)$ *and* $\mathbf{W}_2 \sim \mathcal{W}_m(\boldsymbol{\Sigma}, n_2)$ *are independent, then for every fixed vector* $\mathbf{x} \neq 0 \in \mathbb{R}^m$

$$T = \frac{n_2 \mathbf{x}'\mathbf{W}_1\mathbf{x}}{n_1 \mathbf{x}'\mathbf{W}_2\mathbf{x}} \sim F_{n_1,n_2}.$$

Proof See Fonseca et al. [8]. □

Under the framework of Michalski and Zmyślony [21], the positive part of $\widetilde{\boldsymbol{\Gamma}}_1$ is given by $\widetilde{\boldsymbol{\Gamma}}_{1+} = \frac{\widetilde{\boldsymbol{\Delta}}_1}{u}$ and negative part is given by $\widetilde{\boldsymbol{\Gamma}}_{1-} = \frac{\widetilde{\boldsymbol{\Delta}}_0}{u}$, so that

$$\widetilde{\boldsymbol{\Gamma}}_1 = \widetilde{\boldsymbol{\Gamma}}_{1+} - \widetilde{\boldsymbol{\Gamma}}_{1-} = \frac{\widetilde{\boldsymbol{\Delta}}_1 - \widetilde{\boldsymbol{\Delta}}_0}{u}.$$

To build a test for the nullity of $\boldsymbol{\Gamma}_1$ we note that the hypothesis is equivalent to $H_0 : \boldsymbol{\Gamma}_{1+} = \boldsymbol{\Gamma}_{1-}$. Thus, we can use Lemma 8.3.

Theorem 8.5 *The test statistic*

$$T = \frac{\mathbf{x}'\widetilde{\boldsymbol{\Gamma}}_{1+}\mathbf{x}}{\mathbf{x}'\widetilde{\boldsymbol{\Gamma}}_{1-}\mathbf{x}} \tag{8.15}$$

is distributed as an F random variable with $(n-1)$ and $(n-1)(u-1)$ degrees of freedom under $H_0 : \boldsymbol{\Gamma}_{1+} = \boldsymbol{\Gamma}_{1-}$ for every $\mathbf{x} \neq \mathbf{0}$.

Proof See Fonseca et al. [8]. □

Under the null hypothesis $H_0 : \boldsymbol{\Gamma}_{1+} = \boldsymbol{\Gamma}_{1-}$ and using $\mathbf{x} = \mathbf{1}_m$, the expectation of the numerator and the denominator of the statistic in (8.15) are equal, while under the alternative hypothesis and assuming that all elements of $\boldsymbol{\Gamma}_1$ are non-negative, the expectation of the numerator is greater than the expectation of the denominator. If the elements in $\boldsymbol{\Gamma}_1$ are non-positive then we reject the null hypothesis when the value of the test statistic is small enough. The next section will show simulation results using $\mathbf{x} = \mathbf{1}_m$, which is to take the sum of the elements of $\boldsymbol{\Gamma}_1$ as null.

One disadvantage of the previous test is that actual value of the test statistic—even if not the distribution—depends on chosen value of $\mathbf{x}$. One can arrive even to the situation when for some $\mathbf{x}_1$ the test is not significant, and for some other $\mathbf{x}_2$ it is. Naturally, if we could choose such $\mathbf{x}_0$ that the value of T would tend to be higher than for other $\mathbf{x}$, we would get more power of the test. Unfortunately, situation is not that simple. The maximizing $\mathbf{x}$ depends on the observed data, and thus the distribution is not F-distribution any more.

Theorem 8.6 *It holds*

$$T_m \stackrel{\text{df}}{=} \max_{\mathbf{x}} T = \lambda_{max}\left(\widetilde{\boldsymbol{\Delta}}_1 \widetilde{\boldsymbol{\Delta}}_0^{-1}\right).$$

The distribution of

$$R = \frac{\frac{1}{u-1}T_m}{1 + \frac{1}{u-1}T_m}$$

is Roy's largest root distribution with parameters m, $(n-1)(u-1)$, and $n-1$ if $u-1 > m$.

Proof See Zmyślony et al. [40]. □

Roy's test (see Roy [33]) does not necessarily have higher power than the F-test. Practical experience from other situations tells us that Roy's other is advantageous only when the largest eigenvalue is substantially larger than the other ones.

Simulation Study

To test the hypotheses

$$H_0 : \mathbf{1}_m' \boldsymbol{\Gamma}_1 \mathbf{1}_m = \mathbf{0} \quad \text{vs.} \quad H_1 : \mathbf{1}_m' \boldsymbol{\Gamma}_1 \mathbf{1}_m \neq \mathbf{0}$$

that are equivalent to the hypotheses described earlier in the section when all elements in $\boldsymbol{\Gamma}_1$ are positive, a simulation study was performed and we compare the power function of one and two-sided F tests with likelihood ratio test (LRT). The chosen parameters were $m = 3$, $u = 5$ and

$$\boldsymbol{\Gamma}_0 = \begin{pmatrix} 0.01221 & 0.02172 & 0.00901 \\ 0.02172 & 0.07492 & 0.01682 \\ 0.00901 & 0.01682 & 0.01108 \end{pmatrix}, \quad \boldsymbol{\Gamma}_1 = \begin{pmatrix} 0.01038 & 0.01931 & 0.00824 \\ 0.01931 & 0.06678 & 0.01529 \\ 0.00824 & 0.01529 & 0.00807 \end{pmatrix},$$

where λ is a multiplier such that $\boldsymbol{\Gamma} = \mathbf{I}_u \otimes \boldsymbol{\Gamma}_0 + (\mathbf{J}_u - \mathbf{I}_u) \otimes \lambda \boldsymbol{\Gamma}_1$ is positive definite. In this case, $-0.280274 < \lambda < 1.12096$. For this instance, we take $\mathbf{x} = \mathbf{1}_3$, which means that the sum of the elements of $\boldsymbol{\Gamma}_1$ is equal to 0 under the alternative hypothesis. Generating 10,000 observation vectors using sample sizes of $n \in \{5, 10, 15, 20\}$ and taking the significance level as 5%, the power of the LRT and one and two sided versions of the F test were compared. The results are shown in Figs. 8.2 and 8.3. In Fig. 8.2 we plot the powers of various tests for $n = 5$ and $n = 10$ in **a** and **b**, respectively. Similarly, in Fig. 8.3 we plot the powers of various tests for $n = 15$ and $n = 20$ in **a** and **b**, respectively.

Consider another case, a very special one, presented in Gąsiorek et al. [13], where $\boldsymbol{\Gamma}_0$ and $\boldsymbol{\Gamma}_1$ are scalars, with $m = 1$. Let $\boldsymbol{\Gamma}_0 = 2$ and $\boldsymbol{\Gamma}_1 = 1$. Additionally, it is assumed that $u = 2$, and parameter n will be one of the values from the set $\{3, 5, 10, 25\}$.

Matrix $\boldsymbol{\Gamma}$ has the following form

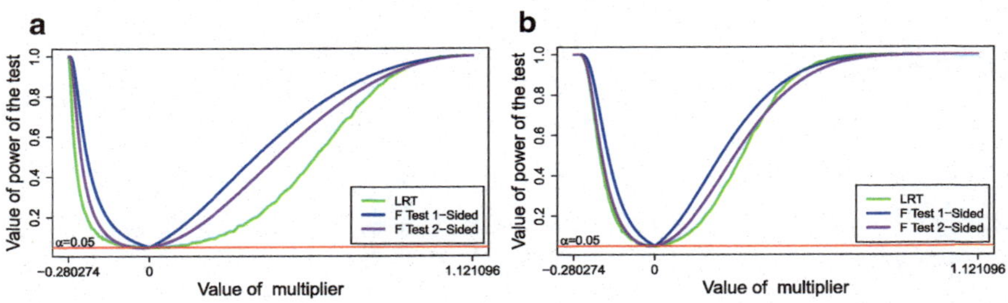

Fig. 8.2 Power comparisons for various tests: **a** $n = 5$ and **b** $n = 10$

$$\boldsymbol{\Gamma} = \begin{pmatrix} 2 & 1 \\ 1 & 2 \end{pmatrix}.$$

From conditions of positive definiteness of matrix $\boldsymbol{\Gamma}$ it is easy to show that values of multiplier λ should be from interval $[-2, 2]$. The results are shown in Figs. 8.4 and 8.5. In Fig. 8.4 we plot the powers of various tests for $n = 3$ and $n = 5$ in **a** and **b**, respectively. Similarly, in Fig. 8.5 we plot the powers of various tests for $n = 10$ and $n = 25$ in **a** and **b** respectively.

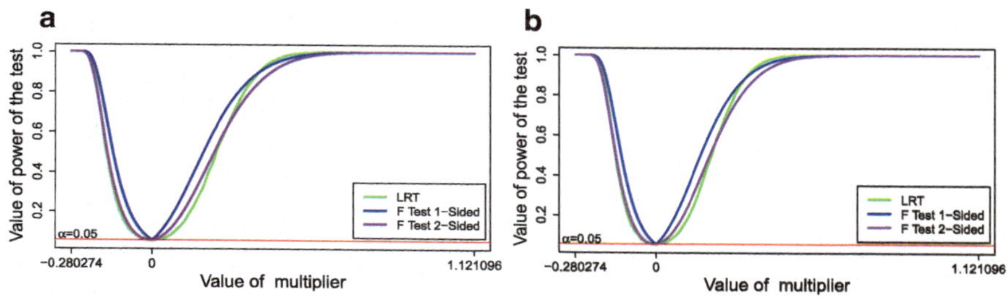

Fig. 8.3 Power comparisons for various tests: **a** $n = 15$ and **b** $n = 20$

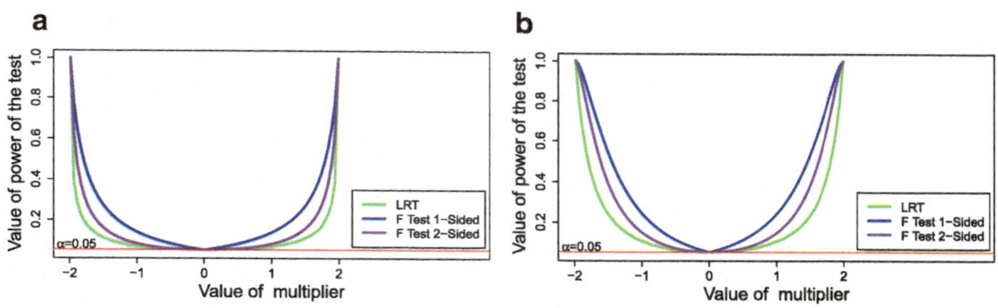

Fig. 8.4 Power comparisons for various tests: **a** $n = 3$ and **b** $n = 5$

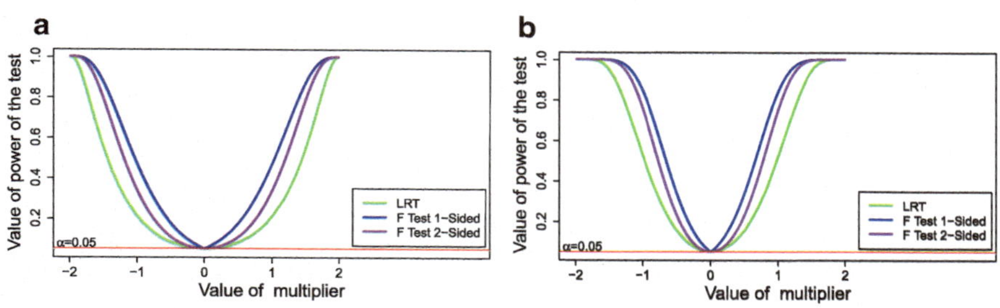

Fig. 8.5 Power comparisons for various tests: **a** $n = 10$ and **b** $n = 25$

8.2.6.2 Testing Hypothesis for Mean Structure

In this section we test the block-wise repetition of the mean structure like the variance–covariance structure. This test in fact will be able to select either of the two models (8.2) and (8.9). We write the null hypothesis as follows:

$$H_0 : \boldsymbol{\mu}_1 = \boldsymbol{\mu}_2 = \ldots = \boldsymbol{\mu}_u.$$

The BUE of the model parameters under H_0 are given in Theorem 8.3. Let $\mathbf{K}_{\mathbf{1}_n}$ be such $n \times (n-1)$ matrix that $\mathbf{K}'_{\mathbf{1}_n}\mathbf{K}_{\mathbf{1}_n} = \mathbf{I}_{n-1}$ and $\mathbf{K}'_{\mathbf{1}_n}\mathbf{1}_n = \mathbf{0}$. Then, hypothesis H_0 can equivalently be expressed as

$$H_0 : \boldsymbol{\mu}_2^{(c)} = \boldsymbol{\mu}_3^{(c)} = \ldots = \boldsymbol{\mu}_u^{(c)} = \mathbf{0}, \quad \text{or} \quad H_0 : \mathbf{M} \stackrel{\text{df}}{=} \sum_{j=2}^{u} \boldsymbol{\mu}_j^{(c)} \boldsymbol{\mu}_j^{(c)'} = \mathbf{0},$$

where $\boldsymbol{\mu}_j^{(c)} = \sqrt{n}\sum_{l=1}^{u} k_{l,j-1}\boldsymbol{\mu}_l$, $j \in \{2, \ldots, u\}$, and $k_{l,j-1}$ are elements of $\mathbf{K}_{\mathbf{1}_u}$. Let us denote $\mathbf{Q}_1 = \left(\frac{1}{\sqrt{u}}\mathbf{1}_u, \mathbf{K}_{\mathbf{1}_u}\right)$, and $\mathbf{Q}_2 = \left(\frac{1}{\sqrt{n}}\mathbf{1}_n, \mathbf{K}_{\mathbf{1}_n}\right)$, that are Helmert matrices, and let $\mathbf{W}_i$, $i \in \{1, \ldots, n\}$ be such $m \times u$ matrices that

$$(\operatorname{vec} \mathbf{W}_1, \ldots, \operatorname{vec} \mathbf{W}_n) = \left(\mathbf{Q}_1' \otimes \mathbf{I}_m\right) \mathbf{Y}\mathbf{Q}_2.$$

The optimal estimator of $\mathbf{M}$ is based on complete sufficient statistics, and it has the following form

$$\widehat{\mathbf{M}} = (u-1)\widehat{\boldsymbol{\Delta}}_{01} - (u-1)\widehat{\boldsymbol{\Delta}}_{00},$$

where

$$\widehat{\boldsymbol{\Delta}}_{01} = \frac{1}{u-1}\sum_{j=2}^{u} \widehat{\boldsymbol{\mu}}_j^{(c)} \widehat{\boldsymbol{\mu}}_j^{(c)'} \quad \text{and} \quad \widehat{\boldsymbol{\Delta}}_{00} = \frac{1}{(n-1)(u-1)} \sum_{i=2}^{n}\sum_{j=2}^{u} \mathbf{w}_j^{(i)} \mathbf{w}_j^{(i)'},$$

with $\widehat{\boldsymbol{\mu}}_j^{(c)} = \sqrt{n}\sum_{l=1}^{u} k_{l,j-1}\widehat{\boldsymbol{\mu}}_l$, while $\mathbf{W}_i = \left(\mathbf{w}_1^{(i)}, \ldots, \mathbf{w}_u^{(i)}\right)$, $i \in \{1, \ldots, n\}$. Then, it holds:

Theorem 8.7 *Under the null hypothesis the test statistic*

$$T = \frac{\mathbf{x}'\widehat{\boldsymbol{\Delta}}_{01}\mathbf{x}}{\mathbf{x}'\widehat{\boldsymbol{\Delta}}_{00}\mathbf{x}}$$

has F distribution with $(u-1)$ *and* $(n-1)(u-1)$ *degrees of freedom for any fixed* $\mathbf{x}$.

Proof See Zmyślony et al. [40]. □

Corollary 8.1 *Since under alternative hypothesis expectation of* $\mathbf{x}'\widehat{\boldsymbol{\Delta}}_{01}\mathbf{x}$ *is bigger than expectation of* $\mathbf{x}'\widehat{\boldsymbol{\Delta}}_{00}\mathbf{x}$*, the null hypothesis is rejected if*

$$T > F_{\alpha,u-1,(n-1)(u-1)}.$$

Using the same principle as in Sect. 8.2.6.1, here also one can see the F-test with arbitrary $\mathbf{x}$ be turned into Roy's test. The only difference is the exchange of role between n and u.

Theorem 8.8 *It holds*

$$T_m \stackrel{\text{df}}{=} \max_{\mathbf{x}} T = \lambda_{max}\left(\widehat{\boldsymbol{\Delta}}_{01}\widehat{\boldsymbol{\Delta}}_{00}^{-1}\right).$$

The distribution of

$$R = \frac{\frac{1}{n-1}T_m}{1+\frac{1}{n-1}T_m}$$

is Roy's largest root distribution with parameters m*,* $(n-1)(u-1)$*, and* $u-1$ *if* $n-1 > m$*.*

The gold standard to which other tests are compared, as in other situations, is the likelihood ratio test:

Theorem 8.9 *Under* H_0 *the LR test statistic is of the form*

$$L = \frac{|\widehat{\boldsymbol{\Delta}}_{00}|}{|\widehat{\boldsymbol{\Delta}}_{00} + \frac{1}{n}\widehat{\boldsymbol{\Delta}}_{01}|},$$

and it has Wilks lambda distribution with parameters m*,* $u-1$*, and* $(n-1)(u-1)$ *if* $n-1 > m$*.*

Proof See Fleiss [6]. □

Zmyślony et al. [40] did a simulation study of the above tests. The Figs. 8.6 and 8.7 show comparison of their powers for $m = 3$ in several different situations. As one can see, each of them can be in a specific situation the best one. Thus, there is no clear clue as to which one to choose for a given dataset.

8.3 Three-Level Multivariate Data

In this section we introduce the DBCS covariance structure for three-level multivariate data, i.e., when m dimensional observation vector is repeatedly measured at u locations and over v time points. We discuss the optimality properties of the model parameters for both unstructured and structured mean vectors. Like two-level

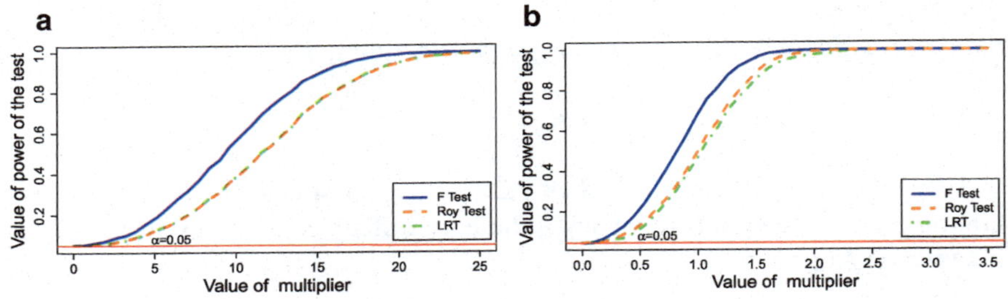

Fig. 8.6 Power comparisons for various tests: **a** $n = 25$, $u = 2$ and **b** $n = 25$, $u = 3$

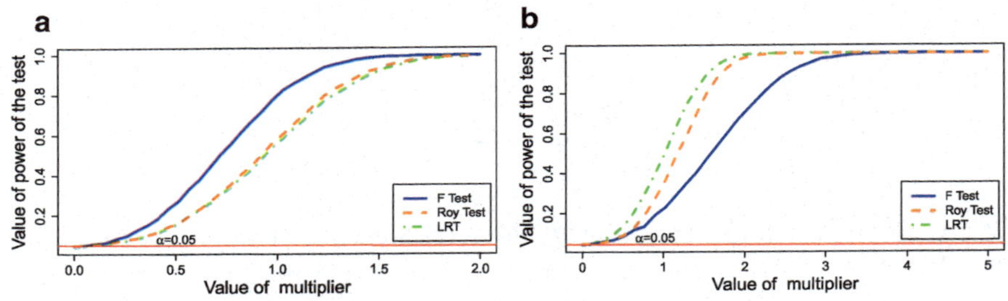

Fig. 8.7 Power comparisons for various tests: **a** $n = 10$, $u = 3$ and **b** $n = 25$, $u = 3$

models, here we also compare the two models and illustrate the comparison with the help of a real data example.

8.3.1 Double Block Compound Symmetry Covariance Structure

The $(vum \times vum)-$dimensional DBCS covariance structure is defined as

$$\boldsymbol{\Gamma} = \begin{pmatrix} \boldsymbol{\Sigma}_0 & \boldsymbol{\Sigma}_1 & \dots & \boldsymbol{\Sigma}_1 \\ \vdots & \ddots & & \vdots \\ \vdots & & \ddots & \vdots \\ \boldsymbol{\Sigma}_1 & \boldsymbol{\Sigma}_1 & \dots & \boldsymbol{\Sigma}_0 \end{pmatrix} = \mathbf{I}_v \otimes (\boldsymbol{\Sigma}_0 - \boldsymbol{\Sigma}_1) + \mathbf{J}_v \otimes \boldsymbol{\Sigma}_1,$$

where

$$\boldsymbol{\Sigma}_0 = \mathbf{I}_u \otimes (\boldsymbol{\Gamma}_0 - \boldsymbol{\Gamma}_1) + \mathbf{J}_u \otimes \boldsymbol{\Gamma}_1, \ \textit{and} \quad \boldsymbol{\Sigma}_1 = \mathbf{J}_u \otimes \boldsymbol{\Gamma}_2.$$

We assume $\mathbf{\Sigma}_0$ is a positive definite symmetric $um \times um$ matrix, and $\mathbf{\Sigma}_1$ is a symmetric $um \times um$ matrix, and $\mathbf{\Sigma}_0 - \mathbf{\Sigma}_1$ and $\mathbf{\Sigma}_0 + (v-1)\mathbf{\Sigma}_1$ are positive definite matrices, so that the $vum \times vum$ matrix $\boldsymbol{\Gamma}$ is positive definite.

We see that the matrix $\boldsymbol{\Gamma}$ is BCS with matrix parameters $\mathbf{\Sigma}_0$ and $\mathbf{\Sigma}_1$, and the matrix $\mathbf{\Sigma}_0$ is BCS with the matrix parameters $\boldsymbol{\Gamma}_0$ and $\boldsymbol{\Gamma}_1$. Because of this doubly BCS nature of this covariance structure $\boldsymbol{\Gamma}$, it is called DBCS covariance structure, and can equivalently be written as follows:

$$\boldsymbol{\Gamma} = \mathbf{I}_v \otimes \mathbf{\Sigma}_0 + (\mathbf{J}_v - \mathbf{I}_v) \otimes \mathbf{\Sigma}_1. \tag{8.16}$$

We can write this doubly exchangeable covariance structure $\boldsymbol{\Gamma}$ in terms of $\boldsymbol{\Gamma}_0$, $\boldsymbol{\Gamma}_1$ and $\boldsymbol{\Gamma}_2$ as

$$\boldsymbol{\Gamma} = \mathbf{I}_v \otimes \mathbf{I}_u \otimes (\boldsymbol{\Gamma}_0 - \boldsymbol{\Gamma}_1) + \mathbf{I}_v \otimes \mathbf{J}_u \otimes (\boldsymbol{\Gamma}_1 - \boldsymbol{\Gamma}_2) + \mathbf{J}_v \otimes \mathbf{J}_u \otimes \boldsymbol{\Gamma}_2,$$

which can equivalently be written as

$$\boldsymbol{\Gamma} = \mathbf{I}_v \otimes \mathbf{I}_u \otimes \boldsymbol{\Gamma}_0 + \mathbf{I}_v \otimes (\mathbf{J}_u - \mathbf{I}_u) \otimes \boldsymbol{\Gamma}_1 + (\mathbf{J}_v - \mathbf{I}_v) \otimes \mathbf{J}_u \otimes \boldsymbol{\Gamma}_2. \tag{8.17}$$

This last form (8.17) will be used to build orthogonal basis with respect to trace of inner product basis for components of the matrix $\boldsymbol{\Gamma}$.

We assume $\boldsymbol{\Gamma}_0$ is a positive definite symmetric $m \times m$ matrix, $\boldsymbol{\Gamma}_1$ and $\boldsymbol{\Gamma}_2$ are symmetric $m \times m$ matrices, and $\boldsymbol{\Gamma}_0 - \boldsymbol{\Gamma}_1$, $\boldsymbol{\Gamma}_0 + (u-1)\boldsymbol{\Gamma}_1 - u\boldsymbol{\Gamma}_2$, $\boldsymbol{\Gamma}_0 + (u-1)\boldsymbol{\Gamma}_1 + (v-1)u\boldsymbol{\Gamma}_2$ are positive definite matrices, so that the $vum \times vum$ matrix $\boldsymbol{\Gamma}$ is positive definite (for a proof, see Lemma 3.1 in Roy and Fonseca [26]). The $m \times m$ diagonal blocks $\boldsymbol{\Gamma}_0$ in $\boldsymbol{\Gamma}$ represent the variance–covariance matrix of the m response variables at any given location and at any given time point, whereas the $m \times m$ off-diagonal blocks $\boldsymbol{\Gamma}_1$ in $\boldsymbol{\Gamma}$ represent the covariance matrix of the m response variables between any two locations and at any given time point. We assume $\boldsymbol{\Gamma}_0$ is constant for all locations and time points, and $\boldsymbol{\Gamma}_1$ is same for all location pairs and for all time points. The $m \times m$ off-diagonal blocks $\boldsymbol{\Gamma}_2$ represent the covariance matrix of the m response variables between any two time points. It is assumed to be the same for any pair of time points, irrespective of the same location or between any two locations.

8.3.2 Estimation in Model with Unstructured Mean Vector

Let $\mathbf{y}_{r,ts}$ be a m-variate vector of measurements on the rth individual at the sth site and at the tth time point; $r \in \{1, \ldots, n\}$, $s \in \{1, \ldots, u\}$, $t \in \{1, \ldots, v\}$. The n individuals are all independent. Let $\mathbf{y}_r = (\mathbf{y}'_{r,11}, \ldots, \mathbf{y}'_{r,vu})'$ be the vum-variate vector of all measurements corresponding to the rth individual. Finally, let $\mathbf{y}_1, \ldots, \mathbf{y}_n$ be a random sample of size n drawn from the population $\mathcal{N}_{vum}(\boldsymbol{\mu}, \boldsymbol{\Gamma})$, where $\boldsymbol{\mu} \in \mathbb{R}^{vum}$ with $\boldsymbol{\mu} = (\boldsymbol{\mu}'_{11}, \ldots, \boldsymbol{\mu}'_{vu})'$ and $\boldsymbol{\Gamma}$ is assumed to be a $vum \times vum$ positive definite matrix.

Optimal properties of unbiased estimators for mean vector and the covariance matrix $\boldsymbol{\Gamma}$ are obtained. Let the data matrix be $\mathbf{Y} = (\mathbf{y}_1, \ldots, \mathbf{y}_n)$. Thus, the model can be written in the following way

$$\underset{num\times 1}{\mathbf{y}} = \operatorname{vec}(\underset{um\times n}{\mathbf{Y}}) \sim \mathcal{N}((\mathbf{1}_n \otimes \mathbf{I}_{vum})\boldsymbol{\mu}, \mathbf{I}_n \otimes \boldsymbol{\Gamma}).$$

This means that n independent random column vectors are identically distributed with $(vum \times vum)$−dimensional variance–covariance matrix

$$\boldsymbol{\Gamma} = \mathbf{I}_v \otimes \mathbf{I}_u \otimes \boldsymbol{\Gamma}_0 + \mathbf{I}_v \otimes (\mathbf{J}_u - \mathbf{I}_u) \otimes \boldsymbol{\Gamma}_1 + (\mathbf{J}_v - \mathbf{I}_v) \otimes \mathbf{J}_u \otimes \boldsymbol{\Gamma}_2.$$

Define the projection matrix $\mathbf{P}$ as follows:

$$\mathbf{P} = \frac{1}{n}\mathbf{J}_n \otimes \mathbf{I}_{vum}, \tag{8.18}$$

and $\mathbf{V} = \mathbf{I}_n \otimes \boldsymbol{\Gamma}_{vum}$ is the covariance matrix of $\mathbf{y}$. It is clear that $\mathbf{P}$ is an orthogonal projector on the subspace of the mean vector of $\mathbf{y}$. If $\mathbf{I}_n \otimes \mathbf{I}_{vum} \in \vartheta = \operatorname{sp}\{\mathbf{V}\}$, from Gnot [12] it follows that $\mathbf{Py}$ is the best linear unbiased estimator (BLUE) if and only if $\mathbf{P}$ commutes with all covariance matrices $\mathbf{V}$. Therefore, we have the following lemmas.

Lemma 8.4 *The projection matrix* $\mathbf{P}$ *commutes with the covariance matrix* $\mathbf{V}$*, i.e.,* $\mathbf{PV} = \mathbf{VP}$*, where* $\mathbf{V} = \mathbf{I}_n \otimes \boldsymbol{\Gamma}$*, the covariance matrix of* $\mathbf{y}$.

For a proof of the above lemma, see Result 1 in Kozioł et al. [17].

Lemma 8.5 *Let* ϑ *denote the subspace spanned by* $\mathbf{V}$*, i.e.,* $\vartheta = \operatorname{sp}\{\mathbf{V}\}$*. Then,* ϑ *is a quadratic subspace. That is,* ϑ *is a linear space and if* $\mathbf{V} \in \vartheta$ *then* $\mathbf{V}^2 \in \vartheta$ *(see Seely [34] for the definition).*

Now, because orthogonal projector on the space generated by the mean vector commutes with all covariances matrices, there exists BLUEs for each estimable function of mean. Moreover, BLUEs are LSEs.

Likewise, BLUEs are LSEs in view of Lemma 8.4. Thus, $\widetilde{\boldsymbol{\mu}}$ is the unique solution of the following normal equation

$$(\mathbf{1}_n \otimes \mathbf{I}_{vum})'(\mathbf{1}_n \otimes \mathbf{I}_{vum})\boldsymbol{\mu} = (\mathbf{1}_n \otimes \mathbf{I}_{vum})'\mathbf{y}$$

or

$$n\mathbf{I}_{vum}\boldsymbol{\mu} = (\mathbf{I}_{vum}, \mathbf{I}_{vum}, \ldots, \mathbf{I}_{vum})\mathbf{y},$$

which means that

$$\widetilde{\boldsymbol{\mu}} = \frac{1}{n}\sum_{r=1}^{n}\mathbf{y}_r.$$

Let $\mathbf{Q} = \mathbf{I}_n \otimes \mathbf{I}_v \otimes \mathbf{I}_u \otimes \mathbf{I}_m - \mathbf{P}$. So, $\mathbf{Q}$ is idempotent. Now, since $\mathbf{PV} = \mathbf{VP}$, and ϑ is a quadratic space, $\mathbf{Q}\vartheta\mathbf{Q} = \mathbf{Q}\vartheta$ is also a quadratic space. We now construct a basis for this quadratic subspace ϑ. We define

$$\mathbf{A}_{ii} = \mathbf{E}_{ii} \quad \text{and} \quad \mathbf{A}_{ij} = \mathbf{E}_{ij} + \mathbf{E}_{ji}, \quad \text{for } i < j,\ j \in \{1, \ldots, m\},$$

as a basis for symmetric matrices $\boldsymbol{\Gamma}$. It is clear that the basis for diagonal matrices of the form $\mathbf{I}_n \otimes \mathbf{I}_v \otimes \mathbf{I}_u \otimes \boldsymbol{\Gamma}_0$ is constituted by matrices

$$\mathbf{K}_{ij}^{(0)} = \mathbf{I}_n \otimes \mathbf{I}_v \otimes \mathbf{I}_u \otimes \mathbf{A}_{ij}, \quad \text{for } i \le j,\ j \in \{1, \ldots, m\}, \tag{8.19}$$

the basis for matrices of the form $\mathbf{I}_n \otimes \mathbf{I}_v \otimes (\mathbf{J}_u - \mathbf{I}_u) \otimes \boldsymbol{\Gamma}_1$ is constituted by matrices

$$\mathbf{K}_{ij}^{(1)} = \mathbf{I}_n \otimes \mathbf{I}_v \otimes (\mathbf{J}_u - \mathbf{I}_u) \otimes \mathbf{A}_{ij}, \quad \text{for } i \le j,\ j \in \{1, \ldots, m\} \tag{8.20}$$

and the basis for matrices of the form $\mathbf{I}_n \otimes (\mathbf{J}_v - \mathbf{I}_v) \otimes \mathbf{J}_u \otimes \boldsymbol{\Gamma}_2$ is constituted by matrices

$$\mathbf{K}_{ij}^{(2)} = \mathbf{I}_n \otimes (\mathbf{J}_v - \mathbf{I}_v) \otimes \mathbf{J}_u \otimes \mathbf{A}_{ij}, \quad \text{for } i \le j,\ j \in \{1, \ldots, m\}. \tag{8.21}$$

It is clear from (8.16) that above basis is orthogonal with respect to trace of inner product.

Lemma 8.6 *The complete and minimal sufficient statistics for the mean vector and the variance–covariance matrix are*

$$(\mathbf{1}_n' \otimes \mathbf{I}_{vum})\mathbf{y} \quad \text{and} \quad \mathbf{y}'\mathbf{Q}\mathbf{K}_{ij}^{(l)}\mathbf{Q}\mathbf{y}, \quad l \in \{0, 1, 2\},$$

where $\mathbf{Q} = \mathbf{I}_{nvum} - \mathbf{P}$ and $\mathbf{P}$ is given in (8.18), see Fonseca et al. [7], Zmyślony [39] and Seely [35].

Theorem 8.10 *Under the above model the unbiased estimators of $\boldsymbol{\mu}$, $\boldsymbol{\Gamma}_0$, $\boldsymbol{\Gamma}_1$ and $\boldsymbol{\Gamma}_2$ are respectively*

$$\widetilde{\boldsymbol{\mu}} = \frac{1}{n}\sum_{r=1}^{n} \mathbf{y}_r, \tag{8.22}$$

$$\widetilde{\boldsymbol{\Gamma}}_0 = \frac{1}{(n-1)vu}\mathbf{C}_0 = \frac{1}{(n-1)vu}\sum_{t=1}^{v}\sum_{s=1}^{u}\sum_{r=1}^{n}\left(\mathbf{y}_{r,ts} - \overline{\mathbf{y}}_{\bullet,ts}\right)\left(\mathbf{y}_{r,ts} - \overline{\mathbf{y}}_{\bullet,ts}\right)', \tag{8.23}$$

$$\begin{aligned}\widetilde{\boldsymbol{\Gamma}}_1 &= \frac{1}{(n-1)vu\,(u-1)}\mathbf{C}_1 \\ &= \frac{1}{(n-1)vu\,(u-1)}\sum_{t=1}^{v}\sum_{s=1}^{u}\sum_{\substack{s^*=1 \\ s\neq s^*}}^{u}\sum_{r=1}^{n}\left(\mathbf{y}_{r,ts^*} - \overline{\mathbf{y}}_{\bullet,ts^*}\right)\left(\mathbf{y}_{r,ts} - \overline{\mathbf{y}}_{\bullet,ts}\right)',\end{aligned} \tag{8.24}$$

and

$$\widetilde{\boldsymbol{\Gamma}}_2 = \frac{1}{(n-1)v\,(v-1)\,u^2}\mathbf{C}_2$$
$$= \frac{1}{(n-1)v\,(v-1)\,u^2}\sum_{t=1}^{v}\sum_{s=1}^{u}\sum_{t\neq t^*=1}^{v}\sum_{s^*=1}^{u}\sum_{r=1}^{n}\left(\mathbf{y}_{r,t^*s^*}-\overline{\mathbf{y}}_{\bullet,t^*s^*}\right)\left(\mathbf{y}_{r,ts}-\overline{\mathbf{y}}_{\bullet,ts}\right)', \tag{8.25}$$

where $\mathbf{y}_{\bullet,ts} = \frac{1}{n}\sum_{r=1}^{n}\mathbf{y}_{r,ts}$, *for* $t \in \{1,\ldots,v\}$, $s \in \{1,\ldots,u\}$. *The estimators*

$$\widetilde{\boldsymbol{\mu}} = \frac{1}{n}\sum_{r=1}^{n}\mathbf{y}_r,$$

and

$$\widetilde{\boldsymbol{\Gamma}} = \mathbf{I}_v \otimes \mathbf{I}_u \otimes (\widetilde{\boldsymbol{\Gamma}}_0 - \widetilde{\boldsymbol{\Gamma}}_1) + \mathbf{I}_v \otimes \mathbf{J}_u \otimes (\widetilde{\boldsymbol{\Gamma}}_1 - \widetilde{\boldsymbol{\Gamma}}_2) + \mathbf{J}_v \otimes \mathbf{J}_u \otimes \widetilde{\boldsymbol{\Gamma}}_2,$$

are best unbiased estimators (BUE) for $\boldsymbol{\mu}$ *and* $\boldsymbol{\Gamma}$ *respectively.*

For a proof, see Theorem 2 in Kozioł et al. [17].

Theorem 8.11 *Estimators given in (8.22), (8.23), (8.24) and (8.25) are consistent. Moreover, the family of distributions of these estimators is complete.*

For a proof of the above theorem, see Theorem 3 in Kozioł et al. [17].

8.3.3 Estimation in Model with Structured Mean Vector

Let $\mathbf{y}_1, \mathbf{y}_2, \ldots, \mathbf{y}_n$ be a random sample of size n drawn from the population $\mathcal{N}_{vum}(\mathbf{1}_{vu} \otimes \boldsymbol{\mu}, \boldsymbol{\Gamma})$, where $\mathrm{E}(\mathbf{y}_r) = \mathbf{1}_{vu} \otimes \boldsymbol{\mu} \in \mathbb{R}^{vum}$ and $\mathrm{D}(\mathbf{y}_r) = \boldsymbol{\Gamma} = \mathbf{I}_v \otimes \mathbf{I}_u \otimes (\boldsymbol{\Gamma}_0 - \boldsymbol{\Gamma}_1) + \mathbf{I}_v \otimes \mathbf{J}_u \otimes (\boldsymbol{\Gamma}_1 - \boldsymbol{\Gamma}_2) + \mathbf{J}_v \otimes \mathbf{J}_u \otimes \boldsymbol{\Gamma}_2$ is assumed to be a $vum \times vum$ positive definite matrix.

In model with structured mean vector we assume that the covariance structure is DBCS and the mean vector remains constant over sites and over time points. Now $\boldsymbol{\mu}$ has m components. This model can be written in the following way

$$\underset{nvum\times 1}{\mathbf{y}} = \mathrm{vec}\,(\underset{vum\times n}{\mathbf{Y}}) \sim \mathcal{N}\left((\mathbf{1}_{nvu} \otimes \mathbf{I}_m)\boldsymbol{\mu}, \mathbf{V}\right), \tag{8.26}$$

where $\mathbf{V} = \mathbf{I}_n \otimes \boldsymbol{\Gamma}$. It means that matrix $\mathbf{Y}$ contains n independent normally distributed random column vectors which are identically distributed with mean vector $\mathbf{1}_{vu} \otimes \boldsymbol{\mu}$ and covariance matrix $\boldsymbol{\Gamma}$. Define the orthogonal projector on the subspace of the mean vector of $\mathbf{y}$ as follows:

$$\mathbf{P} = \frac{1}{n}\mathbf{J}_n \otimes \frac{1}{v}\mathbf{J}_v \otimes \frac{1}{u}\mathbf{J}_u \otimes \mathbf{I}_m. \tag{8.27}$$

Just like for the model with unstructured mean vector we need to show that $\mathbf{P}$ commutes with $\mathbf{V}$ and the subspace spanned by $\mathbf{V}$ is a quadratic subspace.

Result 8.5 *The projection matrix* $\mathbf{P}$ *commutes with the covariance matrix* $\mathbf{V}$*, i.e.,* $\mathbf{PV} = \mathbf{VP}$.

For a proof of the above result, see Result 1 in Kozioł [16].

Lemma 8.7 *The subspace* $\vartheta = \text{sp}\{\mathbf{V}\}$ *is a quadratic subspace.*

For a proof of the above lemma, see Lemma 8.4.

A basis for the quadratic subspace ϑ is the same as in the previous considered case. Let $\mathbf{Q} = \mathbf{I}_n \otimes \mathbf{I}_v \otimes \mathbf{I}_u \otimes \mathbf{I}_m - \mathbf{P}$. So, $\mathbf{Q}$ is idempotent.

Result 8.6 *The complete and minimal sufficient statistics for the mean vector and the variance–covariance matrix are*

$$\left(\mathbf{1}'_{nvu} \otimes \mathbf{I}_m\right)\mathbf{y} \text{ and } \mathbf{y}'\mathbf{Q}\mathbf{K}_{ij}^{(l)}\mathbf{Q}\mathbf{y}, \quad l \in \{0, 1, 2\}$$

where $\mathbf{Q} = \mathbf{I}_n \otimes \mathbf{I}_v \otimes \mathbf{I}_u \otimes \mathbf{I}_m - \mathbf{P}$*,* $\mathbf{P}$ *is given in (8.27),* $\mathbf{K}_{ij}^{(0)}$*,* $\mathbf{K}_{ij}^{(1)}$ *and* $\mathbf{K}_{ij}^{(2)}$ *are given in (8.19), (8.20) and (8.21), respectively. For more details see Fonseca et al. [7], Seely [35] and Zmyślony [39].*

Now, since $\mathbf{PV} = \mathbf{VP}$, and ϑ is a quadratic space, $\mathbf{Q}\vartheta\mathbf{Q} = \mathbf{Q}\vartheta$ is also a quadratic space. According to the coordinate-free approach, the expectation of $\mathbf{Qyy}'\mathbf{Q}$ can be written as a linear combination of matrices $\mathbf{QK}_{ij}^{(0)}$, $\mathbf{QK}_{ij}^{(1)}$ and $\mathbf{QK}_{ij}^{(2)}$ with matrices $\boldsymbol{\Gamma}_0$, $\boldsymbol{\Gamma}_1$ and $\boldsymbol{\Gamma}_2$, respectively. Note also that identity covariance operator of $\mathbf{yy}'$ belongs to $\text{sp}\{\text{D}(\mathbf{yy}')\}$. It implies that the ordinary best quadratic estimators are LSEs for corresponding matrices $\boldsymbol{\Gamma}_0$, $\boldsymbol{\Gamma}_1$ and $\boldsymbol{\Gamma}_2$. They cannot be calculated independently because $\mathbf{QK}_{ij}^{(0)}$, $\mathbf{QK}_{ij}^{(1)}$ and $\mathbf{QK}_{ij}^{(2)}$ are not orthogonal as we can see.

Theorem 8.12 *Under the model (8.26) the best unbiased estimators of* $\boldsymbol{\mu}$*,* $\boldsymbol{\Gamma}_0$*,* $\boldsymbol{\Gamma}_1$ *and* $\boldsymbol{\Gamma}_2$ *are respectively*

$$\widetilde{\boldsymbol{\mu}} = \frac{1}{nvu}\sum_{r=1}^{n}\sum_{t=1}^{v}\sum_{s=1}^{u}\mathbf{y}_{r,ts}, \tag{8.28}$$

$$\widetilde{\boldsymbol{\Gamma}}_0 = \frac{(n-1)vu+1}{n(n-1)v^2u^2}\mathbf{C}_0 + \frac{1}{n(n-1)v^2u^2}(\mathbf{C}_1 + \mathbf{C}_2), \tag{8.29}$$

$$\widetilde{\boldsymbol{\Gamma}}_1 = \frac{(n-1)vu+u-1}{n(n-1)v^2u^2(u-1)}\mathbf{C}_1 + \frac{1}{n(n-1)v^2u^2}(\mathbf{C}_0 + \mathbf{C}_2), \tag{8.30}$$

$$\text{and } \widetilde{\boldsymbol{\Gamma}}_2 = \frac{nv-1}{n(n-1)v^2(v-1)u^2}\mathbf{C}_2 + \frac{1}{n(n-1)v^2u^2}(\mathbf{C}_0 + \mathbf{C}_1), \tag{8.31}$$

where matrices $\mathbf{C}_0$ *and* $\mathbf{C}_1$ *are*

$$\mathbf{C}_0=\sum_{t=1}^{v}\sum_{s=1}^{u}\sum_{r=1}^{n}\left(\mathbf{y}_{r,ts}-\widetilde{\boldsymbol{\mu}}\right)\left(\mathbf{y}_{r,ts}-\widetilde{\boldsymbol{\mu}}\right)',$$

$$\mathbf{C}_1=\sum_{t=1}^{v}\sum_{s=1}^{u}\sum_{\substack{s^*=1\\ s\neq s^*}}^{u}\sum_{r=1}^{n}\left(\mathbf{y}_{r,ts^*}-\widetilde{\boldsymbol{\mu}}\right)\left(\mathbf{y}_{r,ts}-\widetilde{\boldsymbol{\mu}}\right)',$$

and $$\mathbf{C}_2=\sum_{t=1}^{v}\sum_{\substack{t^*=1\\ t\neq t^*}}^{v}\sum_{s=1}^{u}\sum_{s^*=1}^{u}\sum_{r=1}^{n}\left(\mathbf{y}_{r,t^*s^*}-\widetilde{\boldsymbol{\mu}}\right)\left(\mathbf{y}_{r,ts}-\widetilde{\boldsymbol{\mu}}\right)'.$$

For a proof of the above theorem, see Theorem 2 in Kozioł [16].

Theorem 8.13 *Estimators given in (8.28), (8.29), (8.30) and (8.31) are consistent. Moreover, the family of distributions of these estimators is complete.*

For a proof of the above theorem, see Theorem 3 in Kozioł [16].

8.3.4 Comparison of BUE in Two Models

Consider two models, Mo1 and Mo2, both with a DBCS covariance structure, in which the first one has the unstructured mean vector $\mathbf{1}_n\otimes\boldsymbol{\mu}$ (where $\boldsymbol{\mu}$ has vum components) and the second one has the structured mean vector $\mathbf{1}_{nvu}\otimes\boldsymbol{\mu}$ (where $\boldsymbol{\mu}$ has m components). See Sects. 8.3.2 and 8.3.3. In this section we compare variances of estimators of the covariance parameters under models Mo1, $\widetilde{\sigma}^{(0)}_{[1]ij}$, $\widetilde{\sigma}^{(1)}_{[1]ij}$ and $\widetilde{\sigma}^{(2)}_{[1]ij}$, and Mo2, $\widetilde{\sigma}^{(0)}_{[2]ij}$, $\widetilde{\sigma}^{(1)}_{[2]ij}$ and $\widetilde{\sigma}^{(2)}_{[2]ij}$. Denote $\mathbf{P}_{[1]}$ and $\mathbf{P}_{[2]}$ respectively be the orthogonal projectors for unstructured and structured space generated by the corresponding mean vectors. Moreover, let $\mathbf{Q}_{[1]}=\mathbf{I}-\mathbf{P}_{[1]}$ and $\mathbf{Q}_{[2]}=\mathbf{I}-\mathbf{P}_{[2]}$. Since $R(\mathbf{P}_{[2]})\subset R(\mathbf{P}_{[1]})$ then matrices $\mathbf{P}_{[1]}\mathbf{P}_{[2]}=\mathbf{P}_{[2]}\mathbf{P}_{[1]}=\mathbf{P}_{[2]}$. This implies that $\mathbf{Q}_{[1]}\mathbf{Q}_{[2]}=\mathbf{Q}_{[2]}\mathbf{Q}_{[1]}=\mathbf{Q}_{[1]}$. One can easily check that the expectation of $\widetilde{\sigma}^{(0)}_{[1]ij}$, $\widetilde{\sigma}^{(1)}_{[1]ij}$ and $\widetilde{\sigma}^{(2)}_{[1]ij}$ calculated for Mo1 are unbiased under Mo2. Alternatively, it can also calculate and present graphically the difference of variances for both models.

$$\begin{aligned}
\mathrm{D}\left(\widetilde{\sigma}^{(0)}_{[2]ij}\right)-\mathrm{D}\left(\widetilde{\sigma}^{(0)}_{[1]ij}\right)&=-\frac{2}{(n-1)nu^2v^2}\Big[(u-1)\mathrm{Tr}\left(\mathbf{A}_{ij}\boldsymbol{\Gamma}_0\mathbf{A}_{ij}\boldsymbol{\Gamma}_0\right)\\
&-2(u-1)\mathrm{Tr}\left(\mathbf{A}_{ij}\boldsymbol{\Gamma}_0\mathbf{A}_{ij}\boldsymbol{\Gamma}_1\right)+(u-1)\mathrm{Tr}\left(\mathbf{A}_{ij}\boldsymbol{\Gamma}_1\mathbf{A}_{ij}\boldsymbol{\Gamma}_1\right)\\
&+(v-1)u\mathrm{Tr}\left(\mathbf{A}_{ij}\boldsymbol{\Gamma}_0\mathbf{A}_{ij}\boldsymbol{\Gamma}_0\right)-2(v-1)u\mathrm{Tr}\left(\mathbf{A}_{ij}\boldsymbol{\Gamma}_0\mathbf{A}_{ij}\boldsymbol{\Gamma}_2\right)\\
&+(v-1)u\mathrm{Tr}\left(\mathbf{A}_{ij}\boldsymbol{\Gamma}_2\mathbf{A}_{ij}\boldsymbol{\Gamma}_2\right)+(v-1)u(u-1)\mathrm{Tr}\left(\mathbf{A}_{ij}\boldsymbol{\Gamma}_1\mathbf{A}_{ij}\boldsymbol{\Gamma}_1\right)\\
&-2(v-1)u(u-1)\mathrm{Tr}\left(\mathbf{A}_{ij}\boldsymbol{\Gamma}_1\mathbf{A}_{ij}\boldsymbol{\Gamma}_2\right)+(v-1)u(u-1)\mathrm{Tr}\left(\mathbf{A}_{ij}\boldsymbol{\Gamma}_2\mathbf{A}_{ij}\boldsymbol{\Gamma}_2\right)\Big],
\end{aligned}$$

thus $\mathrm{D}\left(\widetilde{\sigma}^{(0)}_{[2]ij}\right) - \mathrm{D}\left(\widetilde{\sigma}^{(0)}_{[1]ij}\right) < 0$ if

$$\begin{aligned}
&\mathrm{Tr}\,(\mathbf{A}_{ij}\boldsymbol{\Gamma}_0\mathbf{A}_{ij}\boldsymbol{\Gamma}_0) + \mathrm{Tr}\,(\mathbf{A}_{ij}\boldsymbol{\Gamma}_1\mathbf{A}_{ij}\boldsymbol{\Gamma}_1) > 2\mathrm{Tr}\,(\mathbf{A}_{ij}\boldsymbol{\Gamma}_0\mathbf{A}_{ij}\boldsymbol{\Gamma}_1),\\
&\mathrm{Tr}\,(\mathbf{A}_{ij}\boldsymbol{\Gamma}_0\mathbf{A}_{ij}\boldsymbol{\Gamma}_0) + \mathrm{Tr}\,(\mathbf{A}_{ij}\boldsymbol{\Gamma}_2\mathbf{A}_{ij}\boldsymbol{\Gamma}_2) > 2\mathrm{Tr}\,(\mathbf{A}_{ij}\boldsymbol{\Gamma}_0\mathbf{A}_{ij}\boldsymbol{\Gamma}_2),\\
\text{and}\quad &\mathrm{Tr}\,(\mathbf{A}_{ij}\boldsymbol{\Gamma}_1\mathbf{A}_{ij}\boldsymbol{\Gamma}_1) + \mathrm{Tr}\,(\mathbf{A}_{ij}\boldsymbol{\Gamma}_2\mathbf{A}_{ij}\boldsymbol{\Gamma}_2) > 2\mathrm{Tr}\,(\mathbf{A}_{ij}\boldsymbol{\Gamma}_1\mathbf{A}_{ij}\boldsymbol{\Gamma}_2)
\end{aligned}$$

which holds for any fixed $\boldsymbol{\Gamma}_0, \boldsymbol{\Gamma}_1, \boldsymbol{\Gamma}_2$. Similarly, after calculations we get that

$$\begin{aligned}
&\mathrm{D}\left(\widetilde{\sigma}^{(1)}_{[2]ij}\right) - \mathrm{D}\left(\widetilde{\sigma}^{(1)}_{[1]ij}\right) =\\
&-\frac{2}{(n-1)n(u-1)u^2v^2}\Big[(1-(u-2)u(v-1))\,\mathrm{Tr}\,(\mathbf{A}_{ij}\boldsymbol{\Gamma}_0\mathbf{A}_{ij}\boldsymbol{\Gamma}_0)\\
&-2\,(1-(u-2)u(v-1))\,\mathrm{Tr}\,(\mathbf{A}_{ij}\boldsymbol{\Gamma}_0\mathbf{A}_{ij}\boldsymbol{\Gamma}_1)\\
&+(1-(u-2)u(v-1))\,\mathrm{Tr}\,(\mathbf{A}_{ij}\boldsymbol{\Gamma}_1\mathbf{A}_{ij}\boldsymbol{\Gamma}_1)\\
&+(v-1)u(u-1)\mathrm{Tr}\,(\mathbf{A}_{ij}\boldsymbol{\Gamma}_0\mathbf{A}_{ij}\boldsymbol{\Gamma}_0) - 2(v-1)u(u-1)\mathrm{Tr}\,(\mathbf{A}_{ij}\boldsymbol{\Gamma}_0\mathbf{A}_{ij}\boldsymbol{\Gamma}_2)\\
&+(v-1)u(u-1)\mathrm{Tr}\,(\mathbf{A}_{ij}\boldsymbol{\Gamma}_2\mathbf{A}_{ij}\boldsymbol{\Gamma}_2) + (v-1)u(u-1)^2\mathrm{Tr}\,(\mathbf{A}_{ij}\boldsymbol{\Gamma}_1\mathbf{A}_{ij}\boldsymbol{\Gamma}_1)\\
&-2(v-1)u(u-1)^2\mathrm{Tr}\,(\mathbf{A}_{ij}\boldsymbol{\Gamma}_1\mathbf{A}_{ij}\boldsymbol{\Gamma}_2) + (v-1)u(u-1)^2\mathrm{Tr}\,(\mathbf{A}_{ij}\boldsymbol{\Gamma}_2\mathbf{A}_{ij}\boldsymbol{\Gamma}_2)\Big],
\end{aligned}$$

thus $\mathrm{D}\left(\widetilde{\sigma}^{(1)}_{[2]ij}\right) - \mathrm{D}\left(\widetilde{\sigma}^{(1)}_{[1]ij}\right) < 0$ if

$$\begin{aligned}
&\mathrm{Tr}\,(\mathbf{A}_{ij}\boldsymbol{\Gamma}_0\mathbf{A}_{ij}\boldsymbol{\Gamma}_0) + \mathrm{Tr}\,(\mathbf{A}_{ij}\boldsymbol{\Gamma}_1\mathbf{A}_{ij}\boldsymbol{\Gamma}_1) > 2\mathrm{Tr}\,(\mathbf{A}_{ij}\boldsymbol{\Gamma}_0\mathbf{A}_{ij}\boldsymbol{\Gamma}_1),\\
&\mathrm{Tr}\,(\mathbf{A}_{ij}\boldsymbol{\Gamma}_0\mathbf{A}_{ij}\boldsymbol{\Gamma}_0) + \mathrm{Tr}\,(\mathbf{A}_{ij}\boldsymbol{\Gamma}_2\mathbf{A}_{ij}\boldsymbol{\Gamma}_2) > 2\mathrm{Tr}\,(\mathbf{A}_{ij}\boldsymbol{\Gamma}_0\mathbf{A}_{ij}\boldsymbol{\Gamma}_2),\\
\text{and}\quad &\mathrm{Tr}\,(\mathbf{A}_{ij}\boldsymbol{\Gamma}_1\mathbf{A}_{ij}\boldsymbol{\Gamma}_1) + \mathrm{Tr}\,(\mathbf{A}_{ij}\boldsymbol{\Gamma}_2\mathbf{A}_{ij}\boldsymbol{\Gamma}_2) > 2\mathrm{Tr}\,(\mathbf{A}_{ij}\boldsymbol{\Gamma}_1\mathbf{A}_{ij}\boldsymbol{\Gamma}_2)
\end{aligned}$$

which holds for any fixed $\boldsymbol{\Gamma}_0, \boldsymbol{\Gamma}_1, \boldsymbol{\Gamma}_2$. Now,

$$\begin{aligned}
&\mathrm{D}\left(\widetilde{\sigma}^{(2)}_{[2]ij}\right) - \mathrm{D}\left(\widetilde{\sigma}^{(2)}_{[1]ij}\right) = -\frac{2}{(n-1)nu^2(v-1)v^2}\Big[(1-u)\mathrm{Tr}\,(\mathbf{A}_{ij}\boldsymbol{\Gamma}_0\mathbf{A}_{ij}\boldsymbol{\Gamma}_0)\\
&-2(1-u)\mathrm{Tr}\,(\mathbf{A}_{ij}\boldsymbol{\Gamma}_0\mathbf{A}_{ij}\boldsymbol{\Gamma}_1) + (1-u)\mathrm{Tr}\,(\mathbf{A}_{ij}\boldsymbol{\Gamma}_1\mathbf{A}_{ij}\boldsymbol{\Gamma}_1)\\
&+u\mathrm{Tr}\,(\mathbf{A}_{ij}\boldsymbol{\Gamma}_0\mathbf{A}_{ij}\boldsymbol{\Gamma}_0) - 2u\mathrm{Tr}\,(\mathbf{A}_{ij}\boldsymbol{\Gamma}_0\mathbf{A}_{ij}\boldsymbol{\Gamma}_2)\\
&+u\mathrm{Tr}\,(\mathbf{A}_{ij}\boldsymbol{\Gamma}_2\mathbf{A}_{ij}\boldsymbol{\Gamma}_2) + u(u-1)\mathrm{Tr}\,(\mathbf{A}_{ij}\boldsymbol{\Gamma}_1\mathbf{A}_{ij}\boldsymbol{\Gamma}_1)\\
&-2u(u-1)\mathrm{Tr}\,(\mathbf{A}_{ij}\boldsymbol{\Gamma}_1\mathbf{A}_{ij}\boldsymbol{\Gamma}_2) + u(u-1)\mathrm{Tr}\,(\mathbf{A}_{ij}\boldsymbol{\Gamma}_2\mathbf{A}_{ij}\boldsymbol{\Gamma}_2)\Big],
\end{aligned}$$

thus $\mathrm{D}\left(\widetilde{\sigma}^{(2)}_{[2]ij}\right) - \mathrm{D}\left(\widetilde{\sigma}^{(2)}_{[1]ij}\right) < 0$ if

$$\begin{aligned}
&\mathrm{Tr}\,(\mathbf{A}_{ij}\boldsymbol{\Gamma}_0\mathbf{A}_{ij}\boldsymbol{\Gamma}_0) + \mathrm{Tr}\,(\mathbf{A}_{ij}\boldsymbol{\Gamma}_1\mathbf{A}_{ij}\boldsymbol{\Gamma}_1) > 2\mathrm{Tr}\,(\mathbf{A}_{ij}\boldsymbol{\Gamma}_0\mathbf{A}_{ij}\boldsymbol{\Gamma}_1),\\
&\mathrm{Tr}\,(\mathbf{A}_{ij}\boldsymbol{\Gamma}_0\mathbf{A}_{ij}\boldsymbol{\Gamma}_0) + \mathrm{Tr}\,(\mathbf{A}_{ij}\boldsymbol{\Gamma}_2\mathbf{A}_{ij}\boldsymbol{\Gamma}_2) > 2\mathrm{Tr}\,(\mathbf{A}_{ij}\boldsymbol{\Gamma}_0\mathbf{A}_{ij}\boldsymbol{\Gamma}_2),\\
\text{and}\quad &\mathrm{Tr}\,(\mathbf{A}_{ij}\boldsymbol{\Gamma}_1\mathbf{A}_{ij}\boldsymbol{\Gamma}_1) + \mathrm{Tr}\,(\mathbf{A}_{ij}\boldsymbol{\Gamma}_2\mathbf{A}_{ij}\boldsymbol{\Gamma}_2) > 2\mathrm{Tr}\,(\mathbf{A}_{ij}\boldsymbol{\Gamma}_1\mathbf{A}_{ij}\boldsymbol{\Gamma}_2)
\end{aligned}$$

which holds for any fixed $\boldsymbol{\Gamma}_0, \boldsymbol{\Gamma}_1, \boldsymbol{\Gamma}_2$. For graphical illustration of these differences, we fix $\boldsymbol{\Gamma}_0 = \mathbf{I}$ and $\boldsymbol{\Gamma}_1 = \boldsymbol{\Gamma}_2 = \mathbf{0}$. For each figure, values for n are chosen from 3 to 25 and for u and v from 2 to 10. For the plot of n and u, v is treated as constant and $v = 2$. Similarly, for the plot of n and v, u is treated as constant and $u = 2$.

All Figs. 8.8, 8.9 and 8.10 reveal the fact that differences between variances of estimators for $\sigma_{ij}^{(0)}$, $\sigma_{ij}^{(1)}$ and $\sigma_{ij}^{(2)}$ in models Mo2 and Mo1 are negative and if $n \to \infty$

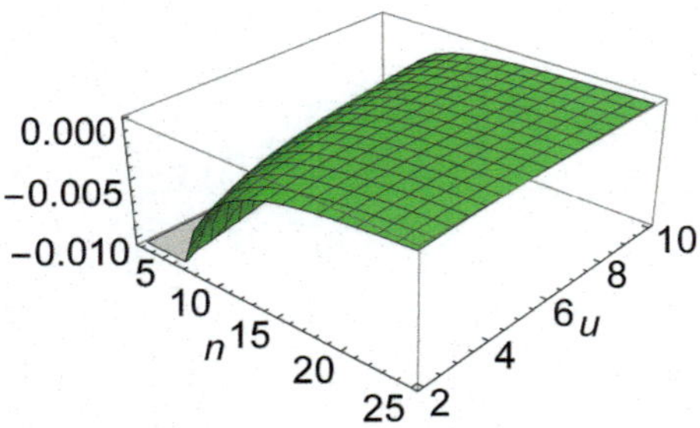

Fig. 8.8 $\mathrm{D}\left(\widetilde{\sigma}_{[2]ij}^{(0)}\right) - \mathrm{D}\left(\widetilde{\sigma}_{[1]ij}^{(0)}\right)$ for n and u, plotting separate figure for parameters n and v is redundant because difference is symmetric with respect to u and v

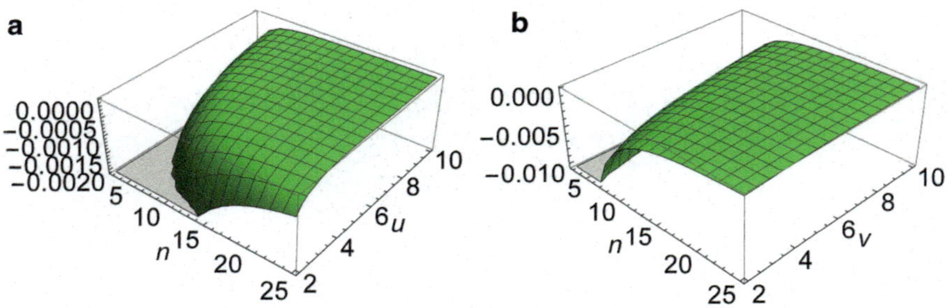

Fig. 8.9 **a** $\mathrm{D}\left(\widetilde{\sigma}_{[2]ij}^{(1)}\right) - \mathrm{D}\left(\widetilde{\sigma}_{[1]ij}^{(1)}\right)$ for n and u, and **b** $\mathrm{D}\left(\widetilde{\sigma}_{[2]ij}^{(1)}\right) - \mathrm{D}\left(\widetilde{\sigma}_{[1]ij}^{(1)}\right)$ for n and v

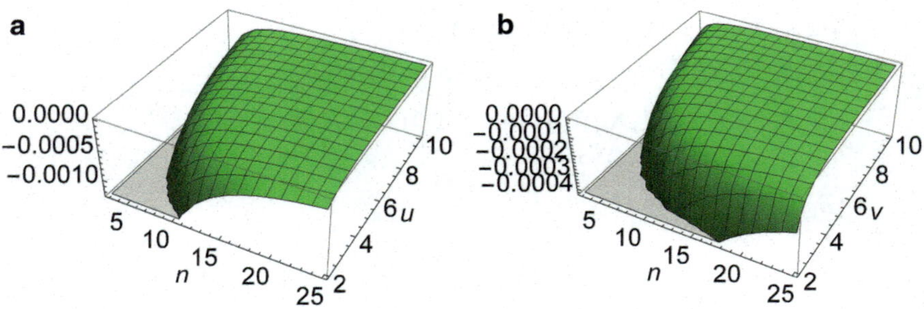

Fig. 8.10 **a** $\mathrm{D}\left(\widetilde{\sigma}_{[2]ij}^{(2)}\right) - \mathrm{D}\left(\widetilde{\sigma}_{[1]ij}^{(2)}\right)$ for n and u, and **b** $\mathrm{D}\left(\widetilde{\sigma}_{[2]ij}^{(2)}\right) - \mathrm{D}\left(\widetilde{\sigma}_{[1]ij}^{(2)}\right)$ for n and v

they $\rightarrow 0$, thus variances of estimators for sigmas in Mo2 are smaller than corresponding variances of estimators for sigmas in Mo1.

8.3.5 A Real Data Example

Results of our proposed methods are applied to the Glaucoma dataset that is described in the Introduction. Using the formula (8.22) presented in Theorem 8.10, the (2×1) dimensional partitioned mean vector for different $s \in \{1, 2\}$, and for different $t \in \{1, 2, 3\}$ are presented in Table 8.1.

Using Theorem 8.10 we say that the above estimate $\widetilde{\boldsymbol{\mu}}$ is BLUE for $\boldsymbol{\mu}$. Additionally, using the formulas in Theorem 8.10 the unbiased estimates $\widetilde{\boldsymbol{\Gamma}}_0$, $\widetilde{\boldsymbol{\Gamma}}_1$ and $\widetilde{\boldsymbol{\Gamma}}_2$ are

$$\widetilde{\boldsymbol{\Gamma}}_0 = \begin{pmatrix} 12.230 & 12.061 \\ 12.061 & 426.155 \end{pmatrix} \widetilde{\boldsymbol{\Gamma}}_1 = \begin{pmatrix} 5.826 & 6.939 \\ 6.939 & 164.156 \end{pmatrix}, \text{ and } \widetilde{\boldsymbol{\Gamma}}_2 = \begin{pmatrix} 3.528 & 9.268 \\ 9.268 & 288.684 \end{pmatrix},$$

respectively. Using the above estimates the unbiased estimate of $\boldsymbol{\Gamma}$ (rounded to two decimal places) is

$$\widetilde{\boldsymbol{\Gamma}} = \mathbf{I}_v \otimes \mathbf{I}_u \otimes \widetilde{\boldsymbol{\Gamma}}_0 + \mathbf{I}_v \otimes (\mathbf{J}_u - \mathbf{I}_u) \otimes \widetilde{\boldsymbol{\Gamma}}_1 + (\mathbf{J}_v - \mathbf{I}_v) \otimes \mathbf{J}_u \otimes \widetilde{\boldsymbol{\Gamma}}_2 =$$

$$\left(\begin{array}{cccc|cccc|cccc}
\left(\begin{smallmatrix}12.23 & 12.06\\ 12.06 & 426.16\end{smallmatrix}\right) & & 5.83 & 6.94 & 3.53 & 9.27 & 3.53 & 9.27 & 3.53 & 9.27 & 3.53 & 9.27 \\
 & & 6.94 & 164.16 & 9.27 & 288.68 & 9.27 & 288.68 & 9.27 & 288.68 & 9.27 & 288.68 \\
5.83 & 6.94 & \left(\begin{smallmatrix}12.23 & 12.06\\ 12.06 & 426.16\end{smallmatrix}\right) & & 3.53 & 9.27 & 3.53 & 9.27 & 3.53 & 9.27 & 3.53 & 9.27 \\
6.94 & 164.16 & & & 9.27 & 288.68 & 9.27 & 288.68 & 9.27 & 288.68 & 9.27 & 288.68 \\
\hline
3.53 & 9.27 & 3.53 & 9.27 & \left(\begin{smallmatrix}12.23 & 12.06\\ 12.06 & 426.16\end{smallmatrix}\right) & & 5.83 & 6.94 & 3.53 & 9.27 & 3.53 & 9.27 \\
9.27 & 288.68 & 9.27 & 288.68 & & & 6.94 & 164.16 & 9.27 & 288.68 & 9.27 & 288.68 \\
3.53 & 9.27 & 3.53 & 9.27 & 5.83 & 6.94 & \left(\begin{smallmatrix}12.23 & 12.06\\ 12.06 & 426.16\end{smallmatrix}\right) & & 3.53 & 9.27 & 3.53 & 9.27 \\
9.27 & 288.68 & 9.27 & 288.68 & 6.94 & 164.16 & & & 9.27 & 288.68 & 9.27 & 288.68 \\
\hline
3.53 & 9.27 & 3.53 & 9.27 & 3.53 & 9.27 & 3.53 & 9.27 & \left(\begin{smallmatrix}12.23 & 12.06\\ 12.06 & 426.16\end{smallmatrix}\right) & & 5.83 & 6.94 \\
9.27 & 288.68 & 9.27 & 288.68 & 9.27 & 288.68 & 9.27 & 288.68 & & & 6.94 & 164.16 \\
3.53 & 9.27 & 3.53 & 9.27 & 3.53 & 9.27 & 3.53 & 9.27 & 5.83 & 6.94 & \left(\begin{smallmatrix}12.23 & 12.06\\ 12.06 & 426.16\end{smallmatrix}\right) & \\
9.27 & 288.68 & 9.27 & 288.68 & 9.27 & 288.68 & 9.27 & 288.68 & 6.94 & 164.16 & &
\end{array}\right).$$

Table 8.1 The (2×1) dimensional partitioned mean vector

t	s	$\widetilde{\boldsymbol{\mu}}_{ts}$
1	1	$(24.333, 527.367)'$
1	2	$(23.567, 534.633)'$
2	1	$(20.233, 525.333)'$
2	2	$(19.567, 532.500)'$
3	1	$(19.233, 527.133)'$
3	2	$(18.933, 534.867)'$

Using Theorems 8.10 we say that the above estimate $\widetilde{\boldsymbol{\Gamma}}$ is the best unbiased, consistent and complete estimate of the DBCS covariance structure $\boldsymbol{\Gamma}$.

8.4 Conclusions

Estimates of variance–covariance matrices are needed for the principal component analysis and factor analysis, and are also needed in varieties of regression analysis that treat the multiple dependent variables in a dataset. Thus, optimal estimation of variance–covariance matrices is very important property in any data analysis. The results in this article demonstrate the optimality of estimates for both fixed effects and variance–covariance matrices using the coordinate-free approach theory for two-level and three-level datasets.

It thus provides a valuable alternative to maximum likelihood estimation, taking as a base for estimation the algebraic structure of the model. Another significant property is the unbiasedness of the proposed estimates, a property that maximum likelihood normally does not have.

Acknowledgements The authors would like to thank the two anonymous reviewers for their careful reading and valuable suggestions of an earlier version of this article that led to the much-improved version of the article. The work of Ivan Žežula was supported by the Slovak Research and Development Agency under contract no. APVV-17-0568, and by grant VEGA MŠ SR 1/0311/18. The work of Miguel Fonseca was partially supported by the Fundação para a Ciência e a Tecnologia (Portuguese Foundation for Science and Technology) through the project UIDB/00297/2020 (Centro de Matemática e Aplicações).

References

1. Andersson, S.: Invariant normal models. Ann. Stat. **3**, 132–154 (1975)
2. Arnold, S.F.: Applications of products to the generalized compound symmetry problem. Ann. Stat. **3**, 227–233 (1976)
3. Arnold, S.F.: Linear models with exchangeably distributed errors. J. Am. Stat. Assoc. **74**, 194–199 (1979)
4. Coelho, C.A., Roy, A.: Testing the hypothesis of a block compound symmetric covariance matrix for elliptically contoured distributions. Test **26**(2), 308–330 (2017)
5. Coelho, C.A., Roy, A.: Testing the hypothesis of a doubly exchangeable covariance matrix. Metrika **83**, 45–68 (2020)
6. Fleiss, J.L.: Assessing the accuracy of multivariate observations. J. Am. Stat. Assoc. **61**(314), 403–412 (1966)
7. Fonseca, M., Mexia, J.T., Zmyślony, R.: Least squares and generalized least squares in models with orthogonal block structure. J. Stat. Plan. Inference **140**, 1346–1352 (2010)
8. Fonseca, M., Zmyślony, R., Kozioł, A.: Testing hypotheses of covariance structure in multivariate data. Electronic J. Linear Algebra **33**, 53–62 (2018)
9. Gnot, S., Klonecki, W., Zmyślony, R.: Uniformly minimum variance unbiased estimation in Euclidean vector space. Bull. Acad. Pol. Sci. **XXIV(4)**, 281–286 (1976)

10. Gnot, S., Klonecki, W., Zmyślony, R.: Best linear plus quadratic estimation of parameters in mixed linear models. Appl. Math. (Warsaw) **XV(4)**, 455–462 (1977)
11. Gnot, S., Klonecki, W., Zmyślony, R.: Uniformly minimum variance unbiased estimation in various classes of estimators. Statistics **8**(2), 199–210 (1977)
12. Gnot, S.: Locally best linear estimation in Euclidean vector spaces. In: Klonecki, W., Kozek, A., Rosiński, J. (eds.) Mathematical Statistics and Probability Theory. Lecture Notes in Statistics, vol. 2. Springer, New York, NY (1980)
13. Gąsiorek, E., Michalski, A., Zmyślony, R.: Tests of independence of normal random variables with known and unknown variance ratio. Discuss. Math. Probab. Stat. **20**, 233–247 (2000)
14. Johnson, R.A., Wichern, D.W.: Applied Multivariate Statistical Analysis, 6th edn. Pearson Prentice Hall, Englewood Cliffs, NJ (2007)
15. Jordan, P., von Neumann, J., Wigner, E.: On an algebraic generalization of the quantum mechanical formalism. Ann. Math. **35**(1), 29–64 (1934)
16. Kozioł, A.: Best unbiased estimates for parameters of three-level multivariate data with doubly exchangeable covariance structure and structured mean vector. Discuss. Math. Probab. Stat. **36**(1–2), 93–113 (2016)
17. Kozioł, A., Roy, A., Zmyślony, R., Leiva, R., Fonseca, M.: Best unbiased estimates for parameters of three-level multivariate data with doubly exchangeable covariance structure. Linear Algebr. Appl. **535**, 87–104 (2017)
18. Kozioł, A., Roy, A., Zmyślony, R., Leiva, R., Fonseca, M.: Free-coordinate estimation for doubly multivariate data. Linear Algebr. Appl. **547**, 217–239 (2018)
19. Leiva, R.: Linear discrimination with equicorrelated training vectors. J. Multivariate Anal. **98**, 384–409 (2007)
20. Leiva, R., Roy, A.: Linear discrimination for three-level multivariate data with separable additive mean vector and doubly exchangeable covariance structure. Comput. Stat. Data Anal. **56**(6), 1644–1661 (2012)
21. Michalski, A., Zmyślony, R.: Testing hypotheses for variance components in mixed linear models. Statistics **27**(3–4), 297–310 (1996)
22. Nahtman, T.: Marginal permutation invariant covariance matrices with applications to linear models. Linear Algebr. Appl. **417**, 183–210 (2006)
23. Perlman, M.D.: Group symmetry covariance models. Stat. Sci. **2**, 421–425 (1987)
24. Rao, C.R.: Familial correlations or the multivariate generalizations of the intraclass correlation. Curr. Sci. **14**, 66–67 (1945)
25. Rao, C.R.: Discriminant functions for genetic differentiation and selection. Sankhyā **12**, 229–246 (1953)
26. Roy, A., Fonseca, M.: Linear models with doubly exchangeable distributed errors. Comm. Stat. Theory Methods **41**, 2545–2569 (2012)
27. Roy, A., Khattree, R.: Tests for mean and covariance structures relevant in repeated measures based discriminant analysis. J. Appl. Stat. Sci. **12**(2), 91–104 (2003)
28. Roy, A., Leiva, R.: Discrimination with jointly equicorrelated multi-level multivariate data. Adv. Data Anal. Classif. **1**(3), 175–199 (2007)
29. Roy, A., Leiva, R.: Estimating and testing a structured covariance matrix for three-level multivariate data. Comm. Stat. Theory Methods **40**, 1945–1963 (2011)
30. Roy, A., Leiva, R., Žežula, I., Klein, D.: Testing of equality of mean vectors for paired doubly multivariate observations in blocked compound symmetric covariance matrix setup. J. Multivariate Anal. **137**, 50–60 (2015)
31. Roy, A., Zmyślony, R., Fonseca, M., Leiva, R.: Optimal estimation for doubly multivariate data in blocked compound symmetric covariance structure. J. Multivariate Anal. **144**, 81–90 (2016)
32. Roy, A., Filipiak, K., Klein, D.: Testing a block exchangeable covariance matrix. Statistics **52**(2), 393–408 (2018)
33. Roy, S.N.: On a heuristic method of test construction and its use in multivariate analysis. Ann. Math. Stat. **24**, 220–238 (1953)
34. Seely, J.F.: Quadratic subspaces and completeness. Ann. Math. Stat. **42**(2), 710–721 (1971)

35. Seely, J.F.: Minimal sufficient statistics and completeness for multivariate normal families. Sankhyā A **39**, 170–185 (1977)
36. Szatrowski T.H.: Estimation and testing for block compound symmetry and other patterned covariance matrices with linear and non-linear structure. Technical Report No. 107, Department of Statistics, Stanford University (1976)
37. Wilks, S.S.: Sample criteria for testing equality of means, equality of variances, and equality of covariances in a normal multivariate distribution. Ann. Math. Stat. **17**(3), 257–281 (1946)
38. Zmyślony, R.: On estimation of parameters in linear models. Appl. Math. (Warsaw) **XV(3)**, 271–276 (1976)
39. Zmyślony, R.: Completeness for a family of normal distributions. Math. Stat. Banach Center Publ. **6**, 355–357 (1980)
40. Zmyślony, R., Žežula, I., Kozioł, A.: Application of Jordan algebra for testing hypotheses about structure of mean vector in model with block compound symmetric covariance structure. Electronic J. Linear Algebr. **33**, 41–52 (2018)
41. Žežula, I., Klein, D., Roy, A.: Testing of multivariate repeated measures data with block exchangeable covariance structure. Test **27**(2), 360–378 (2018)

Chapter 9
Testing of Multivariate Repeated Measures Data with Block Exchangeable Covariance Structure

Ivan Žežula, Daniel Klein, and Anuradha Roy

Abstract The article contains a review of hypothesis testing of mean vectors in normal populations for multivariate repeated measures data on q response variables at p sites or time points with the block exchangeable covariance matrix structure. Some simulation studies are performed to compare the power of the tests.

9.1 Introduction

Multivariate data often naturally form matrix structures, especially repeated measures data. Naturally, these data are correlated. For example, analysis of multivariate repeated measures data needs to take into account the correlations among the measurements of q different variables as well as the correlations among measurements taken at p different sites or time points. If we simply take all correlations or covariances as unknown parameters, the number of the second-order parameters can be really high and we need a large number of observations to estimate all of them. However, in many situations, the design of the experiment produces a special structure of the correlations. If this is the case, realistic special structure assumption can substantially reduce the number of unknown parameters and, through the reduction of the required number of observations, also the cost of the whole experiment.

We will consider a linear model for matrix-valued random variable $\mathbf{X}$, which is suitable for multivariate repeated measures data or doubly multivariate data:

$$\mathbf{X} = \mathbf{M} + \mathbf{E}, \tag{9.1}$$

I. Žežula (✉) · D. Klein
Institute of Mathematics, Faculty of Science, P. J. Šafárik University, Košice, Slovakia
e-mail: ivan.zezula@upjs.sk

D. Klein
e-mail: daniel.klein@upjs.sk

A. Roy
Department of Management Science and Statistics, The University of Texas at San Antonio, San Antonio, TX 78249, USA
e-mail: Anuradha.Roy@utsa.edu

K. Filipiak et al. (eds.), *Multivariate, Multilinear and Mixed Linear Models*, Contributions to Statistics, https://doi.org/10.1007/978-3-030-75494-5_9

where $\mathbf{M}$ is a $q \times p$ location center matrix, which may depend on the explaining variables, and $\mathbf{E}$ is the $q \times p$ error matrix. The independent and identically distributed $q \times p$ matrix observations $\mathbf{X}_1, \ldots, \mathbf{X}_n$ may then come from a multivariate hierarchical model with two hierarchy levels (p classes in the upper level; q response variables in the lower level). Students in classes, patients in hospitals, and repeated measurements on subjects are examples of these models. We will stick to multivariate repeated measures terminology with n individuals, q variables, and p measurement repetitions. For a review of literature on the topic, see, e.g., Žežula et al. [15].

An example of such multivariate data is a study of cerebral metabolism in epileptic patients by Sperling [9]. The metabolic rate of glucose was measured at 16 locations in the brain (8 on the right-hand side and 8 on the left-hand side) by PE tomography. Clearly, the data are doubly multivariate with $q = 8$ and $p = 2$. The sample consisted of 18 normal control subjects, 8 patients with a right brain hemisphere focus of the epilepsy, and 8 patients with a left brain hemisphere focus of the epilepsy.

Classical multivariate analysis assumes both unstructured mean and unstructured variance–covariance matrix $\mathbf{\Sigma} = \mathrm{D}\,(\mathrm{vec}\,\mathbf{E})$. In the case of model (9.1), the number of unknown variance–covariance parameters to be estimated is $pq(pq+1)/2$, which can be quite large.

However, in a certain situation, exchangeability can be assumed. There are two types of exchangeability—exchangeability of mean values and/or exchangeability of errors. Exchangeability of mean values means that all q-dimensional columns of $\mathbf{X}$ are exchangeable, i.e., $\mathrm{E}\,(\mathbf{X}) = \mathbf{M} = \boldsymbol{\xi}\mathbf{1}_p'$ (mean does not change over time/space). Exchangeability of errors means that all q-dimensional columns of $\mathbf{E}$ are exchangeable with respect to variance and covariance, i.e., they all have the same variance matrix and pairwise the same covariance matrices. Structure of the variance matrix of vec $(\mathbf{E})$ is then called block exchangeable or block compound symmetry (BCS) variance structure:

$$\mathrm{D}\,(\mathrm{vec}\,\mathbf{E}) = \mathbf{\Sigma} = \begin{pmatrix} \mathbf{\Sigma}_0 & \mathbf{\Sigma}_1 & \ldots & \mathbf{\Sigma}_1 \\ \mathbf{\Sigma}_1 & \mathbf{\Sigma}_0 & \ldots & \mathbf{\Sigma}_1 \\ \vdots & \vdots & \ddots & \vdots \\ \mathbf{\Sigma}_1 & \mathbf{\Sigma}_1 & \ldots & \mathbf{\Sigma}_0 \end{pmatrix} = \mathbf{I}_p \otimes (\mathbf{\Sigma}_0 - \mathbf{\Sigma}_1) + \mathbf{J}_p \otimes \mathbf{\Sigma}_1 \,. \tag{9.2}$$

Thus, each column of $\mathbf{E}$ has the same variance matrix $\mathbf{\Sigma}_0$, and any two different columns have the same covariance matrix $\mathbf{\Sigma}_1$. Here $\mathbf{I}_p$ is the $p \times p$ identity matrix, $\mathbf{J}_p = \mathbf{1}_p\mathbf{1}_p'$, and $\mathbf{1}_p$ is the $p \times 1$ vector of ones. It is clear that $p \geq 2$ is needed for the block exchangeable variance structure.

In the next, we will assume that the $q \times q$ matrix $\mathbf{\Sigma}_0$ is positive definite (denoted by $\mathbf{\Sigma}_0 > 0$), and the $q \times q$ symmetric matrix $\mathbf{\Sigma}_1$ must satisfy $\mathbf{\Sigma}_0 - \mathbf{\Sigma}_1 > 0$ and $\mathbf{\Sigma}_0 + (p-1)\mathbf{\Sigma}_1 > 0$. This guarantees the positive definiteness of $\mathbf{\Sigma}$ (for the proof see, e.g., Arnold [1]).

The two types of exchangeability do not go always together. We can feel in our example of cerebral metabolism that it can be reasonable to expect the same variance and covariance matrices at different sites, but one cannot expect the same metabolic activity at different sides of the brain. That is why we assume an unstructured mean $\mathbf{M}$ in the basic model, but we do assume BCS of the variance matrix.

BCS variance structure is a realistic assumption in many multivariate data. It has been studied most extensively by Arnold [2] and Szatrowski [10]. Its desirability lies mostly in the substantially reduced number of unknown parameters (only $q(q+1)$ variance–covariance parameters), which results in the reliability of estimates and higher test power. Moreover, the number of unknown parameters does not even depend on p, and this variance structure does not need equal spacing of the repeated measurements over time.

The aim of the paper is to make a review of the basic tests for the mean: one-sample test, paired-sample test, and two-sample test under the assumption of block exchangeable variance matrix of the data. In the two-sample case, we assume the equality of BCS variance matrices in the two populations.

Old literature on the topic contained mostly only some special cases, e.g., BCS combined with exchangeable mean structure, and relied on asymptotic methods and approximations of their distributions (see, e.g., Wilks [14], Tukey and Wilks [12], Votaw [13], Geisser [5], Szatrowski [10], Arnold [3], Szatrowski [11]). However, asymptotic methods—even if asymptotically optimal—can have severe disadvantages in small samples. Recent literature (see citations in Sects. 9.3, 9.4, and 9.5, and literature cited therein) introduced method using exact distribution based on the Mahalanobis distance. These are much more effective especially for small sample sizes. Increased computer power makes the resulting distributions practically applicable.

9.2 Preliminaries

Hypothesis testing of mean vector in a doubly multivariate framework is much more difficult than in a multivariate framework as the number of parameters increases with the increase of p. In this article, we present several test procedures for the mean matrix $\mathbf{M}$ (or matrices in the case of more populations involved) in a doubly multivariate setup using BCS covariance structure as defined in (9.2). These should cover all basic mean testing cases met in practice. Moreover, we compare different approaches to the testing.

We will assume multivariate normality throughout the paper. The following notation we use as equivalent:

$$\mathbf{X} \sim N_{q\times p}(\mathbf{M}, \boldsymbol{\Sigma}) \quad \text{or} \quad \mathbf{x} = \text{vec}(\mathbf{X}) \sim N_{pq}(\boldsymbol{\mu}, \boldsymbol{\Sigma}),$$

where $\text{E}(\mathbf{x}) = \boldsymbol{\mu} = \text{vec}(\mathbf{M}) = \left(\boldsymbol{\mu}_1', \ldots, \boldsymbol{\mu}_p'\right)'$ and $\text{D}(\mathbf{x}) = \boldsymbol{\Sigma} > 0$ as defined in (9.2).[1] We will assume that $p \geq 2$ and $q \geq 1$ in the rest of the article.

[1] Note that notation $N_{q\times p}$ is usually used for matrix-valued random variable only if the variance matrix of vec $(\mathbf{X})$ can be decomposed into product $\boldsymbol{\Gamma} \otimes \boldsymbol{\Psi}$ (say), and then the two matrices $\boldsymbol{\Gamma}$ and

Further, we will use the following notation for orthogonal projector matrices: For any matrix $\mathbf{A}$, $\mathbf{P_A} = \mathbf{A}(\mathbf{A}'\mathbf{A})^{+}\mathbf{A}'$ denotes the orthogonal projector onto its column space, and $\mathbf{Q_A} = \mathbf{I} - \mathbf{P_A}$ the orthogonal projector on its orthogonal complement, where $(\mathbf{A}'\mathbf{A})^{+}$ is the Moore–Penrose inverse of $\mathbf{A}'\mathbf{A}$. For the sake of simplicity, the matrices $\mathbf{P}_{\mathbf{J}_n}$ and $\mathbf{Q}_{\mathbf{J}_n}$ will be denoted by $\mathbf{P}_n$ and $\mathbf{Q}_n$, respectively.

Using this notation, it is easy to observe that the BCS structure of $\boldsymbol{\Sigma}$ can also be written in the form of two mutually orthogonal components:

$$\boldsymbol{\Sigma} = \mathbf{P}_p \otimes \boldsymbol{\Delta}_2 + \mathbf{Q}_p \otimes \boldsymbol{\Delta}_1 ,$$

where $\boldsymbol{\Delta}_1 = \boldsymbol{\Sigma}_0 - \boldsymbol{\Sigma}_1 > 0$ and $\boldsymbol{\Delta}_2 = \boldsymbol{\Sigma}_0 + (p-1)\boldsymbol{\Sigma}_1 > 0$.

9.3 One-Sample Test

9.3.1 Orthogonal Decomposition Solution

Let $\mathbf{X}_1, \dots, \mathbf{X}_n$ be a random sample from $N_{q\times p}(\mathbf{M}, \boldsymbol{\Sigma})$, where $\boldsymbol{\Sigma}$ has the BCS structure (9.2), and $\overline{\mathbf{X}}$ be the sample mean. Equivalently, we will write vectorized form of the sample as $\mathbf{x}_1, \dots, \mathbf{x}_n$ i.i.d. $N_{pq}(\boldsymbol{\mu}, \boldsymbol{\Sigma})$ and $\overline{\mathbf{x}}$ for the sample mean. We want to test the hypothesis

$$H_0 : \mathbf{M} = \mathbf{M}_0 \quad \text{against} \quad H_1 : \mathbf{M} \neq \mathbf{M}_0. \tag{9.3}$$

Let $\underset{\sim}{\mathbf{X}} = (\mathbf{x}_1, \dots, \mathbf{x}_n)$ be the $(pq \times n)$-dimensional data matrix.

It is easy to see that the $(pq \times pq)$-dimensional sample variance–covariance matrix $\mathbf{S}$ can be expressed in the form

$$\mathbf{S} = \frac{1}{n-1}\underset{\sim}{\mathbf{X}}\mathbf{Q}_n\underset{\sim}{\mathbf{X}}' = \begin{pmatrix} \mathbf{S}_{11} & \mathbf{S}_{12} & \dots & \mathbf{S}_{1p} \\ \mathbf{S}_{21} & \mathbf{S}_{22} & \dots & \mathbf{S}_{2p} \\ \vdots & \vdots & \ddots & \vdots \\ \mathbf{S}_{p1} & \mathbf{S}_{p2} & \dots & \mathbf{S}_{pp} \end{pmatrix},$$

where $\mathbf{S}_{ij} = \frac{1}{n-1}\underset{\sim}{\mathbf{X}}^*_{\bullet i}\mathbf{Q}_n\underset{\sim}{\mathbf{X}}^{*\prime}_{\bullet j}$, $i, j \in \{1, \dots, p\}$, and $\underset{\sim}{\mathbf{X}}^*_{\bullet i}$, $i \in \{1, \dots, p\}$, is the data matrix at the ith site or time point. The matrix $\mathbf{S}$ is an unbiased estimator of $\boldsymbol{\Sigma}$, and it is known that $\mathbf{S}$ has the Wishart distribution with $n-1$ degrees of freedom and covariance matrix $\frac{1}{n-1}\boldsymbol{\Sigma}$, i.e., $\mathbf{S} \sim W_{pq}\left(n-1, \frac{1}{n-1}\boldsymbol{\Sigma}\right)$. Therefore, $\mathrm{E}\left[\mathbf{S}_{ij}\right] = \boldsymbol{\Sigma}_{1-\delta_{ij}}$, where $\delta_{ij} = 1$ if $i = j$, and $\delta_{ij} = 0$ if $i \neq j$. It is natural to use the following unbiased estimators of the variance and covariance matrices $\boldsymbol{\Sigma}_0$ and $\boldsymbol{\Sigma}_1$:

$\boldsymbol{\Psi}$ are given as separate parameters of the distribution. However, if the decomposition of this type is not possible, some authors—including us—simply give $\boldsymbol{\Sigma}$ as the second matrix parameter of the distribution.

$$\widehat{\boldsymbol{\Sigma}}_0 = \frac{1}{p}\sum_{i=1}^{p}\mathbf{S}_{ii}\,, \quad \text{and } \widehat{\boldsymbol{\Sigma}}_1 = \frac{1}{p(p-1)}\sum_{\substack{i=1\\ i\neq j}}^{p}\sum_{j=1}^{p}\mathbf{S}_{ij}\,.$$

Then, the estimator $\widehat{\boldsymbol{\Sigma}}$ can also be written as the sum of two orthogonal parts

$$\widehat{\boldsymbol{\Sigma}} = \mathbf{P}_p \otimes \widehat{\boldsymbol{\Delta}}_2 + \mathbf{Q}_p \otimes \widehat{\boldsymbol{\Delta}}_1\,,$$

where $\widehat{\boldsymbol{\Delta}}_1 = \widehat{\boldsymbol{\Sigma}}_0 - \widehat{\boldsymbol{\Sigma}}_1$ and $\widehat{\boldsymbol{\Delta}}_2 = \widehat{\boldsymbol{\Sigma}}_0 + (p-1)\widehat{\boldsymbol{\Sigma}}_1$. Since $\mathbf{P}_p\mathbf{Q}_p = \mathbf{0}$, the inverse of $\widehat{\boldsymbol{\Sigma}}$ can be written as $\widehat{\boldsymbol{\Sigma}}^{-1} = \mathbf{P}_p \otimes \widehat{\boldsymbol{\Delta}}_2^{-1} + \mathbf{Q}_p \otimes \widehat{\boldsymbol{\Delta}}_1^{-1}$. This form enables us to find explicit formula of the Mahalanobis type test statistic for testing the Hypothesis (9.3):

$$\begin{aligned} D^2 &= n\left(\overline{\mathbf{x}} - \boldsymbol{\mu}_0\right)'\left[\mathbf{P}_p \otimes \widehat{\boldsymbol{\Delta}}_2^{-1} + \mathbf{Q}_p \otimes \widehat{\boldsymbol{\Delta}}_1^{-1}\right]\left(\overline{\mathbf{x}} - \boldsymbol{\mu}_0\right) = \\ &= n\mathrm{Tr}\left[\left(\overline{\mathbf{X}} - \mathbf{M}_0\right)'\widehat{\boldsymbol{\Delta}}_2^{-1}\left(\overline{\mathbf{X}} - \mathbf{M}_0\right)\mathbf{P}_p\right] + n\mathrm{Tr}\left[\left(\overline{\mathbf{X}} - \mathbf{M}_0\right)'\widehat{\boldsymbol{\Delta}}_1^{-1}\left(\overline{\mathbf{X}} - \mathbf{M}_0\right)\mathbf{Q}_p\right], \end{aligned}$$

where $\boldsymbol{\mu}_0 = \operatorname{vec}\mathbf{M}_0$.

The distribution of this D^2 test statistic is not Hotelling's T^2, since the estimator $\widehat{\boldsymbol{\Sigma}}$ does not follow the Wishart distribution. The distribution is provided by the following theorem.

Theorem 9.1 *If $n-1 > q$, then under H_0 it holds*

$$D^2 \sim T_0^2(q;\, 1, n-1) \oplus T_0^2(q;\, p-1, (n-1)(p-1)),$$

where $\oplus$ denotes the convolution operation and T_0^2 denotes the Lawley–Hotelling trace distribution.

Proof See Žežula et al. [15]. □

Unfortunately, there is no simple way for obtaining critical values of this convolution. However, Lawley-Hotelling trace distribution is usually approximated by F-distribution (see McKeon [6]). In fact, $T_0^2(q;\, 1, n-1)$ is equal to usual Hotelling $T^2_{q,n-1}$, which is equivalent to $\frac{(n-1)q}{n-q}F(q, n-q)$. The second term, $T_0^2(q;\, p-1, (n-1)(p-1))$, can be approximated by

$$\frac{(n-1)(p-1)^2 q}{np-n-p-q}\cdot\frac{b-2}{b}F((p-1)q, b),$$

where

$$b = 4 + \frac{(pq-q+2)(np-n-p-q-2)(np-n-p-q+1)}{(np-n-p)(p+q)-(q-1)(q+2)}\,.$$

Therefore, we can approximate the distribution of D^2 by the convolution of the two (one exact and other approximating) F-distributions, where its critical values can be obtained by the method of Dyer [4]. It is interesting, from a practical point of view, to compare these critical values with those obtained by simulation.

9.3.2 Canonical Transformation Solution

Since $H_0 : \boldsymbol{\mu} = \boldsymbol{\mu}_0$ is equivalent to $H_0 : \mathbf{Z}\boldsymbol{\mu} = \mathbf{Z}\boldsymbol{\mu}_0$ for any non-singular matrix $\mathbf{Z}$, we can use $\mathbf{Z} = \mathbf{H}_p \otimes \mathbf{I}_q$, where $\mathbf{H}_p$ is an orthogonal matrix with the first row proportional to a vector of all ones. Using this, we get $\mathbf{y} = (\mathbf{y}_1', \ldots, \mathbf{y}_p')' = \mathbf{Z}\mathbf{x} \sim N_{pq}(\boldsymbol{\nu}, \boldsymbol{\Omega})$ where $\boldsymbol{\nu} = \mathbf{Z}\boldsymbol{\mu}$, and according to Roy and Fonseca [7]

$$\boldsymbol{\Omega} = \mathbf{Z}\boldsymbol{\Sigma}\mathbf{Z}' = \begin{pmatrix} \boldsymbol{\Delta}_2 & \mathbf{0} \\ \mathbf{0} & \mathbf{I}_{p-1} \otimes \boldsymbol{\Delta}_1 \end{pmatrix}.$$

We see that this is the canonical transformation, since the $q \times 1$ component vectors $\mathbf{y}_i$, $i \in \{1, \ldots, p\}$, are mutually independent. Even though the estimator $\widehat{\boldsymbol{\Omega}} = \begin{pmatrix} \widehat{\boldsymbol{\Delta}}_2 & \mathbf{0} \\ \mathbf{0} & \mathbf{I}_{p-1} \otimes \widehat{\boldsymbol{\Delta}}_1 \end{pmatrix}$ does not have a Wishart distribution, the following lemma holds.

Lemma 9.1 *Distributions of* $(n-1)(p-1)\widehat{\boldsymbol{\Delta}}_1$ *and* $(n-1)\widehat{\boldsymbol{\Delta}}_2$ *are independent, and*

$$(n-1)(p-1)\widehat{\boldsymbol{\Delta}}_1 \sim W_q\left((n-1)(p-1), \boldsymbol{\Delta}_1\right),$$
$$(n-1)\widehat{\boldsymbol{\Delta}}_2 \sim W_q\left(n-1, \boldsymbol{\Delta}_2\right).$$

Proof See Theorem 1 in Roy et al. [8]. □

For testing the hypothesis (9.3), Žežula et al. [15] proposed a Mahalanobis type test statistic in the transformed model (similar to Hotelling's T^2) named Block T^2

$$BT^2 = n\left(\overline{\mathbf{x}} - \boldsymbol{\mu}_0\right)' \mathbf{Z}' \begin{pmatrix} \widehat{\boldsymbol{\Delta}}_2^{-1} & \mathbf{0} \\ \mathbf{0} & \mathbf{P}_{p-1} \otimes \widehat{\boldsymbol{\Delta}}_1^{-1} \end{pmatrix} \mathbf{Z}\left(\overline{\mathbf{x}} - \boldsymbol{\mu}_0\right).$$

It is just a small modification of the method of Roy et al. [8] developed originally for the paired samples test.

Theorem 9.2 *If* $n - 1 > q$, *then under* H_0 *it holds*

$$BT^2 \sim T^2_{q,n-1} \oplus T^2_{q,(n-1)(p-1)},$$

where $T^2_{a,b}$ *denotes the Hotelling* T^2 *distribution with appropriate parameters.*

Proof See Žežula et al. [15]. □

However, such a test statistic is not independent of the choice of the orthogonal matrix $\mathbf{H}_p$ which is involved in $\mathbf{Z}$. Even if for a given data matrix $\mathbf{X}$ the distribution of BT^2 is the same for two different $\mathbf{H}_p^1$ and $\mathbf{H}_p^2$, it can be easily verified that the particular value of the test statistic is not the same with the exception of $p = 2$. As a consequence, a different power can be achieved by using different $\mathbf{H}_p$'s. In order to achieve the highest power, one should use the matrix $\mathbf{H}_p$ which gives the highest value of BT^2. Unfortunately, such $\mathbf{H}_p$ depends on the data observed. The following lemma shows a connection between the maximum of BT^2 and D^2.

Lemma 9.2 *It holds*

$$\max_{\mathbf{H}_p} BT^2 \leq D^2,$$

where the maximization is over all orthogonal matrices $\mathbf{H}_p$ with the first row proportional to a vector of all ones.

Proof This is a simplified version of the proof by Žežula et al. [15].

Let us consider a given fixed data matrix $\mathbf{X}$ and let us write $\mathbf{H}_p = (\mathbf{h}_1, \mathbf{G})'$, where $\mathbf{h}_1 = \frac{1}{\sqrt{p}}\mathbf{1}_p$ and $\mathbf{G} = \left(\mathbf{h}_2, \ldots, \mathbf{h}_p\right)$. The orthogonality of $\mathbf{H}_p$ immediately implies

$$\mathbf{G}'\mathbf{G} = \mathbf{I}_{p-1}, \quad \text{and} \quad \mathbf{G}\mathbf{G}' = \mathbf{Q}_p. \tag{9.4}$$

Since $\mathbf{h}_1\mathbf{h}_1' = \mathbf{P}_p$, it is easy to derive that

$$\mathbf{Z}' \begin{pmatrix} \widehat{\boldsymbol{\Delta}}_2^{-1} & \mathbf{0} \\ \mathbf{0} & \mathbf{P}_{p-1} \otimes \widehat{\boldsymbol{\Delta}}_1^{-1} \end{pmatrix} \mathbf{Z} = \mathbf{P}_p \otimes \widehat{\boldsymbol{\Delta}}_2^{-1} + \mathbf{G}\mathbf{P}_{p-1}\mathbf{G}' \otimes \widehat{\boldsymbol{\Delta}}_1^{-1}.$$

It follows that

$$\left(\mathbf{P}_p \otimes \widehat{\boldsymbol{\Delta}}_2^{-1} + \mathbf{Q}_p \otimes \widehat{\boldsymbol{\Delta}}_1^{-1}\right) - \mathbf{Z}' \begin{pmatrix} \widehat{\boldsymbol{\Delta}}_2^{-1} & \mathbf{0} \\ \mathbf{0} & \mathbf{P}_{p-1} \otimes \widehat{\boldsymbol{\Delta}}_1^{-1} \end{pmatrix} \mathbf{Z} = \mathbf{G}\mathbf{Q}_{p-1}\mathbf{G}' \otimes \widehat{\boldsymbol{\Delta}}_1^{-1},$$

and

$$D^2 - BT^2 = n\,(\overline{\mathbf{x}} - \boldsymbol{\mu}_0)' \left(\mathbf{G}\mathbf{Q}_{p-1}\mathbf{G}' \otimes \widehat{\boldsymbol{\Delta}}_1^{-1}\right) (\overline{\mathbf{x}} - \boldsymbol{\mu}_0).$$

Since $\mathbf{Q}_{p-1}$ is a projector and therefore p.s.d., it follows that $\mathbf{G}\mathbf{Q}_{p-1}\mathbf{G}' \otimes \widehat{\boldsymbol{\Delta}}_1^{-1}$ is a p.s.d. matrix with probability 1 for any $\mathbf{G}$ satisfying (9.4).

Thus, $D^2 - BT^2 \geq 0$. □

Lemma 9.3 *There exists $\mathbf{H}_p$ satisfying (9.4) for which $\max_{\mathbf{H}_p} BT^2 = D^2$ only if $p = 2$ (for any q) or $q = 1$ (for any p).*

Proof See Žežula et al. [15]. □

As Lemma 9.2 implies, optimal choice of $\mathbf{H}_p$ which maximizes BT^2 depends on data matrix $\mathbf{X}$. As a consequence, distribution of $BTR = \max_{\mathbf{H}_p} BT^2$ is no longer that of BT^2 with fixed $\mathbf{H}_p$. The proof of this Lemma also implies that BT^2 can be equivalently expressed as

$$BT^2 = n\mathrm{Tr}\left[\left(\overline{\mathbf{X}} - \mathbf{M}_0\right)' \widehat{\boldsymbol{\Delta}}_2^{-1} \left(\overline{\mathbf{X}} - \mathbf{M}_0\right) \mathbf{P}_p\right] + \frac{n}{p-1}\mathbf{1}'_{p-1}\mathbf{G}' \left(\overline{\mathbf{X}} - \mathbf{M}_0\right)' \widehat{\boldsymbol{\Delta}}_1^{-1} \left(\overline{\mathbf{X}} - \mathbf{M}_0\right) \mathbf{G}\mathbf{1}_{p-1}.$$

This form helps us to find the maximum. These results have not yet been published.

Theorem 9.3

$$BTR = n\mathrm{Tr}\left[\left(\overline{\mathbf{X}} - \mathbf{M}_0\right)' \widehat{\boldsymbol{\Delta}}_2^{-1} \left(\overline{\mathbf{X}} - \mathbf{M}_0\right) \mathbf{P}_p\right] + n\lambda_1\left[\left(\overline{\mathbf{X}} - \mathbf{M}_0\right)' \widehat{\boldsymbol{\Delta}}_1^{-1} \left(\overline{\mathbf{X}} - \mathbf{M}_0\right) \mathbf{Q}_p\right].$$

If $n-1 > q$, then the distribution of BTR under H_0 is

$$T^2_{q,n-1} \oplus R\left(q, \tfrac{1}{2}(p-q-2), \tfrac{1}{2}((n-1)(p-1)-q-1)\right),$$

where $R(d, \nu_1, \nu_2)$ denotes (unstandardized) Roy's largest root distribution.

Proof Since $\mathbf{Q}_p\mathbf{G} = \mathbf{G}$ and $\left\|\mathbf{G}\mathbf{1}_{p-1}\right\|^2 = p-1$, the second term is equal to

$$\frac{n}{p-1}\mathbf{1}'_{p-1}\mathbf{G}'\mathbf{Q}_p \left(\overline{\mathbf{X}} - \mathbf{M}_0\right)' \widehat{\boldsymbol{\Delta}}_1^{-1} \left(\overline{\mathbf{X}} - \mathbf{M}_0\right) \mathbf{Q}_p\mathbf{G}\mathbf{1}_{p-1} \leq n\lambda_1\left[\mathbf{Q}_p \left(\overline{\mathbf{X}} - \mathbf{M}_0\right)' \widehat{\boldsymbol{\Delta}}_1^{-1} \left(\overline{\mathbf{X}} - \mathbf{M}_0\right) \mathbf{Q}_p\right],$$

where $\lambda_1[\mathbf{A}]$ is the largest eigenvalue of $\mathbf{A}$. Since non-zero eigenvalues of matrix products $\mathbf{AB}$ and $\mathbf{BA}$ are the same, the result follows. Equality is attained for such $\mathbf{G}$ that $\mathbf{u}_1 = \mathbf{G}\mathbf{1}_{p-1}$ is the eigenvector corresponding to λ_1. □

Thus, we see that the first convolution terms in the distributions of D^2, BT^2, and BTR are the same. The difference is made by the second term, where $T^2_{q,(n-1)(p-1)}$ changes to

$$R\left(q, \tfrac{1}{2}(p-q-2), \tfrac{1}{2}((n-1)(p-1)-q-1)\right) \quad \text{or} \quad T_0^2(q; p-1, (n-1)(p-1)).$$

Remark 9.1 It is worth noting that a minimal sample size needed for the application of the above tests is $n \geq q+1$ regardless of p, while the computation of standard Hotelling's T^2 test statistic (if we ignore the special variance structure) requires $n \geq pq+1$. We can see that not only we have much less parameters to estimate in such a reduced model—and thus have more stability of the estimators for the same data—but also that the proposed methods are applicable in more situations.

9.3.3 Exchangeable Mean Structure

Sometimes, it can be assumed that mean value remains constant over time or space. It means that $\boldsymbol{\mu} = \mathbf{1}_p \otimes \boldsymbol{\xi}_q$. Using similar method as above, one can easily arrive at the estimator

$$\widehat{\boldsymbol{\xi}} = \frac{1}{np}\left(\mathbf{1}_p' \otimes \mathbf{I}_q\right) \mathbb{X}\mathbf{1}_n \sim N_q\left(\boldsymbol{\xi}, \frac{1}{np}\boldsymbol{\Delta}_2\right).$$

Since $\mathbf{Q}_n\mathbf{1}_n = \mathbf{0}$, $\mathbb{X}\mathbf{Q}_n\mathbb{X}'$ and $\mathbb{X}\mathbf{1}_n$ are independent. It follows that $\mathbf{S}$ and $\widehat{\boldsymbol{\xi}}$ are independent, and also $\widehat{\boldsymbol{\Delta}}_2$ and $\widehat{\boldsymbol{\xi}}$. As a result, we can construct directly the Hotelling T^2 statistic for testing $H_0 : \boldsymbol{\xi} = \boldsymbol{\xi}_0$ against $H_1 : \boldsymbol{\xi} \neq \boldsymbol{\xi}_0$. Under H_0, it holds

$$T^2 = np\left(\widehat{\boldsymbol{\xi}} - \boldsymbol{\xi}_0\right)' \widehat{\boldsymbol{\Delta}}_2^{-1} \left(\widehat{\boldsymbol{\xi}} - \boldsymbol{\xi}_0\right) \sim T^2_{q,n-1}.$$

This statistic is very similar to the first component of D^2, BT^2, and BTR statistics.

9.4 Paired Samples Test

Let us have random sample $(\mathbf{X}_{11}, \mathbf{X}_{12}), \ldots, (\mathbf{X}_{n1}, \mathbf{X}_{n2})$ of doubly multivariate data measured before and after a treatment on the same individuals. We want to test the effect of the treatment, which can be reformulated as testing zero difference of the corresponding means. Let us assume that $\mathbf{D}_i = \mathbf{X}_{1i} - \mathbf{X}_{2i} \sim N_{q\times p}(\mathbf{M}_D, \boldsymbol{\Sigma})\ \forall i$, where $\boldsymbol{\Sigma}$ has the BCS structure (9.2).

Applying the results from Sect. 9.3 to $\mathbf{D}_1, \ldots, \mathbf{D}_n$, we obtain all three test statistics of

$$H_0 : \mathbf{M}_D = \mathbf{0} \quad \text{against} \quad H_1 : \mathbf{M}_D \neq \mathbf{0}.$$

The first one was BT^2, which was derived in Roy et al. [8]. If $n - 1 > q$, then

$$\begin{aligned} BT^2_D &= n\overline{\mathbf{d}}'\mathbf{Z}' \begin{pmatrix} \widehat{\boldsymbol{\Delta}}_2^{-1} & 0 \\ 0 & \mathbf{P}_{p-1} \otimes \widehat{\boldsymbol{\Delta}}_1^{-1} \end{pmatrix} \mathbf{Z}\overline{\mathbf{d}} \\ &= n\mathrm{Tr}\left[\overline{\mathbf{D}}'\widehat{\boldsymbol{\Delta}}_2^{-1}\overline{\mathbf{D}}\mathbf{P}_p\right] + \frac{n}{p-1}\mathbf{1}_{p-1}'\mathbf{G}'\overline{\mathbf{D}}'\widehat{\boldsymbol{\Delta}}_1^{-1}\overline{\mathbf{D}}\mathbf{G}\mathbf{1}_{p-1} \\ &\qquad \sim T^2_{q,n-1} \oplus T^2_{q,(n-1)(p-1)}, \end{aligned}$$

where $\mathbf{G}$ satisfies (9.4).

The next possibility is the D^2 statistic, which was introduced in Žežula et al. [15]. If $n - 1 > q$, then

$$D_D^2 = n\overline{\mathbf{d}}'\left[\mathbf{P}_p \otimes \widehat{\boldsymbol{\Delta}}_2^{-1} + \mathbf{Q}_p \otimes \widehat{\boldsymbol{\Delta}}_1^{-1}\right]\overline{\mathbf{d}} = n\mathrm{Tr}\left[\overline{\mathbf{D}}'\widehat{\boldsymbol{\Delta}}_2^{-1}\overline{\mathbf{D}}\mathbf{P}_p\right] + n\mathrm{Tr}\left[\overline{\mathbf{D}}'\widehat{\boldsymbol{\Delta}}_1^{-1}\overline{\mathbf{D}}\mathbf{Q}_p\right]$$
$$\sim T_0^2(q; 1, n-1) \oplus T_0^2(q; p-1, (n-1)(p-1)),$$

where T_0^2 is the Lawley–Hotelling trace distribution.

The last test statistic, not yet published, naturally is

$$BTR_D = n\mathrm{Tr}\left[\overline{\mathbf{D}}'\widehat{\boldsymbol{\Delta}}_2^{-1}\overline{\mathbf{D}}\mathbf{P}_p\right] + n\lambda_1\left[\overline{\mathbf{D}}'\widehat{\boldsymbol{\Delta}}_1^{-1}\overline{\mathbf{D}}\mathbf{Q}_p\right]$$
$$\sim T_{q,n-1}^2 \oplus R\left(q, \tfrac{1}{2}(p-q-2), \tfrac{1}{2}((n-1)(p-1)-q-1)\right).$$

9.5 Two-Sample Test

One-sample test is important to have, but a two-sample test is much more often the procedure we need. We now derive it using the results of the previous sections. We will consider only the case with a common variance matrix for both samples.

Let $\mathbf{X}_{11}, \ldots, \mathbf{X}_{1n}$ be a random sample from $N_{q\times p}(\mathbf{M}_1, \boldsymbol{\Sigma})$ with sample mean $\overline{\mathbf{X}}_1$, and $\mathbf{X}_{21}, \ldots, \mathbf{X}_{2m}$ be a random sample from $N_{q\times p}(\mathbf{M}_2, \boldsymbol{\Sigma})$ with sample mean $\overline{\mathbf{X}}_2$, and let $\boldsymbol{\Sigma}$ have BCS structure.

Equivalently, we will write vectorized form of the samples as $\mathbf{x}_{11}, \ldots, \mathbf{x}_{1n}$ i.i.d. $N_{pq}(\boldsymbol{\mu}_1, \boldsymbol{\Sigma})$, and $\mathbf{x}_{21}, \ldots, \mathbf{x}_{2m}$ i.i.d. $N_{pq}(\boldsymbol{\mu}_2, \boldsymbol{\Sigma})$, with sample means $\overline{\mathbf{x}}_1$ and $\overline{\mathbf{x}}_2$. We test the hypothesis

$$H_0: \mathbf{M}_1 = \mathbf{M}_2 \quad \text{against} \quad H_1: \mathbf{M}_1 \neq \mathbf{M}_2.$$

Let $\underset{\sim}{\mathbf{X}_1} = (\mathbf{x}_{11}, \ldots, \mathbf{x}_{1n})$ and $\underset{\sim}{\mathbf{X}_2} = (\mathbf{x}_{21}, \ldots, \mathbf{x}_{2m})$ be the data matrices. We know that sample means $\overline{\mathbf{X}}_1$ and $\overline{\mathbf{X}}_2$ are independent of variance matrices estimators $\mathbf{S}_1 = \frac{1}{n-1}\underset{\sim}{\mathbf{X}_1}\mathbf{Q}_n\underset{\sim}{\mathbf{X}_1}'$ and $\mathbf{S}_2 = \frac{1}{m-1}\underset{\sim}{\mathbf{X}_2}\mathbf{Q}_m\underset{\sim}{\mathbf{X}_2}'$, and thus also independent of the pooled estimator

$$\mathbf{S}^{\mathrm{pl}} = \frac{1}{n+m-2}\left((n-1)\mathbf{S}_1 + (m-1)\mathbf{S}_2\right) = \left(\mathbf{S}_{ij}^{\mathrm{pl}}\right)_{i,j=1}^{p}.$$

As in the one-sample case, we define estimators

$$\widehat{\boldsymbol{\Sigma}}_0^{\mathrm{pl}} = \tfrac{1}{p}\textstyle\sum_{i=1}^{p}\mathbf{S}_{ii}^{\mathrm{pl}}, \quad \widehat{\boldsymbol{\Sigma}}_1^{\mathrm{pl}} = \frac{1}{p(p-1)}\sum_{\substack{i=1\\ i\neq j}}^{p}\sum_{j=1}^{p}\mathbf{S}_{ij}^{\mathrm{pl}},$$
$$\widehat{\boldsymbol{\Delta}}_1^{\mathrm{pl}} = \widehat{\boldsymbol{\Sigma}}_0^{\mathrm{pl}} - \widehat{\boldsymbol{\Sigma}}_1^{\mathrm{pl}} \quad \text{and} \quad \widehat{\boldsymbol{\Delta}}_2^{\mathrm{pl}} = \widehat{\boldsymbol{\Sigma}}_0^{\mathrm{pl}} + (p-1)\widehat{\boldsymbol{\Sigma}}_1^{\mathrm{pl}}.$$

Then it holds:

Lemma 9.4

$$(n+m-2)(p-1)\widehat{\boldsymbol{\Delta}}_1^{\mathrm{pl}} \sim W_q\left((n+m-2)(p-1), \boldsymbol{\Delta}_1\right),$$
$$(n+m-2)\widehat{\boldsymbol{\Delta}}_2^{\mathrm{pl}} \sim W_q\left(n+m-2, \boldsymbol{\Delta}_2\right),$$

and the two random matrices are independent.

Proof See Žežula et al. [15]. □

Thus, all three test statistics can again be constructed.

Theorem 9.4 *If* **G** *satisfies (9.4), then under* H_0 *it holds*

$$\begin{aligned} BT_{2S}^2 &= \frac{nm}{n+m}\left(\overline{\mathbf{x}}_1-\overline{\mathbf{x}}_2\right)'\mathbf{Z}'\begin{pmatrix}\left(\widehat{\boldsymbol{\Delta}}_2^{\mathrm{pl}}\right)^{-1} & \mathbf{0}\\ \mathbf{0} & \mathbf{P}_{p-1}\otimes\left(\widehat{\boldsymbol{\Delta}}_1^{\mathrm{pl}}\right)^{-1}\end{pmatrix}\mathbf{Z}\left(\overline{\mathbf{x}}_1-\overline{\mathbf{x}}_2\right)\\ &= \frac{nm}{n+m}\mathrm{Tr}\left[\left(\overline{\mathbf{X}}_1-\overline{\mathbf{X}}_2\right)'\widehat{\boldsymbol{\Delta}}_2^{-1}\left(\overline{\mathbf{X}}_1-\overline{\mathbf{X}}_2\right)\mathbf{P}_p\right]\\ &\quad+\frac{nm}{(p-1)(n+m)}\mathbf{1}_{p-1}'\mathbf{G}'\left(\overline{\mathbf{X}}_1-\overline{\mathbf{X}}_2\right)'\widehat{\boldsymbol{\Delta}}_1^{-1}\left(\overline{\mathbf{X}}_1-\overline{\mathbf{X}}_2\right)\mathbf{G}\mathbf{1}_{p-1}\\ &\qquad\qquad\sim T_{q,n+m-2}^2\oplus T_{q,(n+m-2)(p-1)}^2. \end{aligned}$$

Proof See Žežula et al. [15]. The second form is a new one, but trivially follows from Sect. 9.3. □

Theorem 9.5 *Under* H_0 *it holds*

$$\begin{aligned} D_{2S}^2 &= \frac{nm}{n+m}\left(\overline{\mathbf{x}}_1-\overline{\mathbf{x}}_2\right)'\left[\mathbf{P}_p\otimes\left(\widehat{\boldsymbol{\Delta}}_2^{\mathrm{pl}}\right)^{-1}+\mathbf{Q}_p\otimes\left(\widehat{\boldsymbol{\Delta}}_1^{\mathrm{pl}}\right)^{-1}\right]\left(\overline{\mathbf{x}}_1-\overline{\mathbf{x}}_2\right)\\ &= \frac{nm}{n+m}\mathrm{Tr}\left[\left(\overline{\mathbf{X}}_1-\overline{\mathbf{X}}_2\right)'\left(\widehat{\boldsymbol{\Delta}}_2^{\mathrm{pl}}\right)^{-1}\left(\overline{\mathbf{X}}_1-\overline{\mathbf{X}}_2\right)\mathbf{P}_p\right]\\ &\quad+\frac{nm}{n+m}\mathrm{Tr}\left[\left(\overline{\mathbf{X}}_1-\overline{\mathbf{X}}_2\right)'\left(\widehat{\boldsymbol{\Delta}}_1^{\mathrm{pl}}\right)^{-1}\left(\overline{\mathbf{X}}_1-\overline{\mathbf{X}}_2\right)\mathbf{Q}_p\right]\\ &\quad\sim T_0^2(q;1,n+m-2)\oplus T_0^2(q;p-1,(n+m-2)(p-1)). \end{aligned}$$

Proof See Žežula et al. [15]. The second form is a new one, also following from results of Sect. 9.3. □

As before, we can use the fact that $T_{q,n+m-2}^2$ is equivalent to

$$\frac{n+m-q-1}{q(n+m-2)}F(q,n+m-q-1),$$

and $T_0^2(q;p-1,(n+m-2)(p-1))$ can be approximated by

$$\frac{(n+m-2)(p-1)^2 q}{(n+m-2)(p-1)-q-1} \cdot \frac{c-2}{c} F((p-1)q, c),$$

where

$$c = 4 + \frac{(pq-q+2)[(n+m-2)(p-1)-q-3][(n+m-2)(p-1)-q]}{(np+mp-2p-n-m+1)(p+q)-(q-1)(q+2)}.$$

Theorem 9.6 *Under H_0 it holds*

$$\begin{aligned} BTR_{2S} &= \frac{nm}{n+m} \mathrm{Tr} \left[\left(\overline{\mathbf{X}}_1 - \overline{\mathbf{X}}_2\right)' \left(\widehat{\boldsymbol{\Delta}}_2^{\mathrm{pl}}\right)^{-1} \left(\overline{\mathbf{X}}_1 - \overline{\mathbf{X}}_2\right) \mathbf{P}_p \right] \\ &+ \frac{nm}{n+m} \lambda_1 \left[\left(\overline{\mathbf{X}}_1 - \overline{\mathbf{X}}_2\right)' \left(\widehat{\boldsymbol{\Delta}}_1^{\mathrm{pl}}\right)^{-1} \left(\overline{\mathbf{X}}_1 - \overline{\mathbf{X}}_2\right) \mathbf{Q}_p \right] \\ &\sim T^2_{q,n+m-2} \oplus R\left(q, \tfrac{1}{2}(p-q-2), \tfrac{1}{2}((n+m-2)(p-1)-q-1)\right). \end{aligned}$$

Proof The proof follows from results of Sect. 9.3. □

Remark 9.2 The only condition needed on sample size in order to have the test applicable is $n+m-2 \geq q$, i.e., $n+m \geq q+2$. Again, the value of p adds no other requirement on the sample size.

9.5.1 Exchangeable Means Structure

If both samples have also exchangeable mean structure, i.e., $\boldsymbol{\mu}_1 = \mathbf{1}_p \otimes \boldsymbol{\xi}_1$, $\boldsymbol{\mu}_2 = \mathbf{1}_p \otimes \boldsymbol{\xi}_2$, we can test the hypothesis $H_0 \colon \boldsymbol{\xi}_1 = \boldsymbol{\xi}_2$ using the same method as in Sect. 9.3.3. We obtain the test statistic

$$T^2 = \frac{nmp}{n+m} \left(\widehat{\boldsymbol{\xi}}_1 - \widehat{\boldsymbol{\xi}}_2\right)' \left(\widehat{\boldsymbol{\Delta}}_2^{\mathrm{pl}}\right)^{-1} \left(\widehat{\boldsymbol{\xi}}_1 - \widehat{\boldsymbol{\xi}}_2\right) \sim T^2_{q,n+m-2}.$$

9.6 Simulation Study

Power comparisons of the tests were done using simulations in Wolfram Mathematica environment and it is published here for the first time. Every single simulation was based on 50 000 samples. The Helmert matrix was used for H_p in BT^2 statistic. The distribution of every sample was exactly that of H_0 but shifted by adding a non-zero mean value. The multiplicative constants were chosen in such a way that we can observe a reasonable increase in power over the investigated range of sample sizes.

Parameters of the tests:

- all combinations of $p \in \{3, 5, 8\}$ and $q \in \{2, 5, 10\}$;

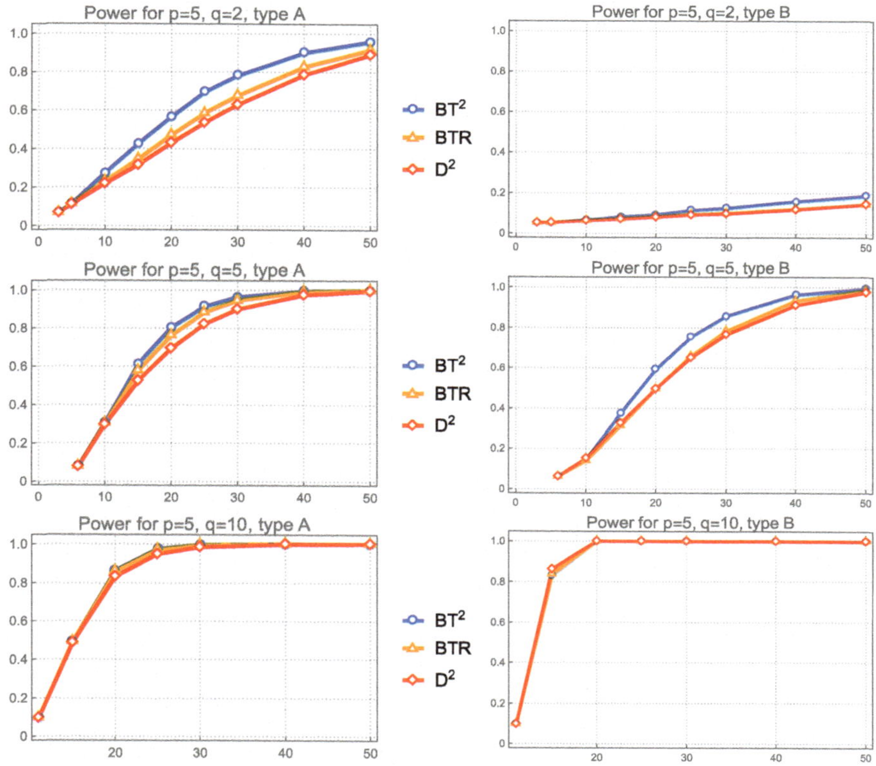

Fig. 9.1 Increase of power with q (type A and B)

- four types of mean value:
  - A: $\boldsymbol{\mu} = 0.5 \cdot \mathbf{1}_{pq}$ (uniform mean shift);
  - B: $\boldsymbol{\mu} = 0.02 \cdot (1, 2, \ldots, pq)'$ (slanted mean shift);
  - C: $\boldsymbol{\mu} = 0.4 \cdot (\mathbf{1}_q', \mathbf{0}_{(p-1)q}')'$ (mean shift only in $\boldsymbol{\Delta}_2$-area);
  - D: $\boldsymbol{\mu} = 0.35 \cdot (\mathbf{0}_q', \mathbf{1}_{(p-1)q}')'$ (mean shift only in $\boldsymbol{\Delta}_1$-area);
- sample sizes $n \in \{q + 1, 5, 10, 15, 20, 25, 30, 40, 50\}$ (5 and/or 10 were omitted if they were less than $q + 1$).

9.6.1 Comparisons for Fixed Type of Alternative

See Figs. 9.1, 9.2, 9.3, and 9.4.

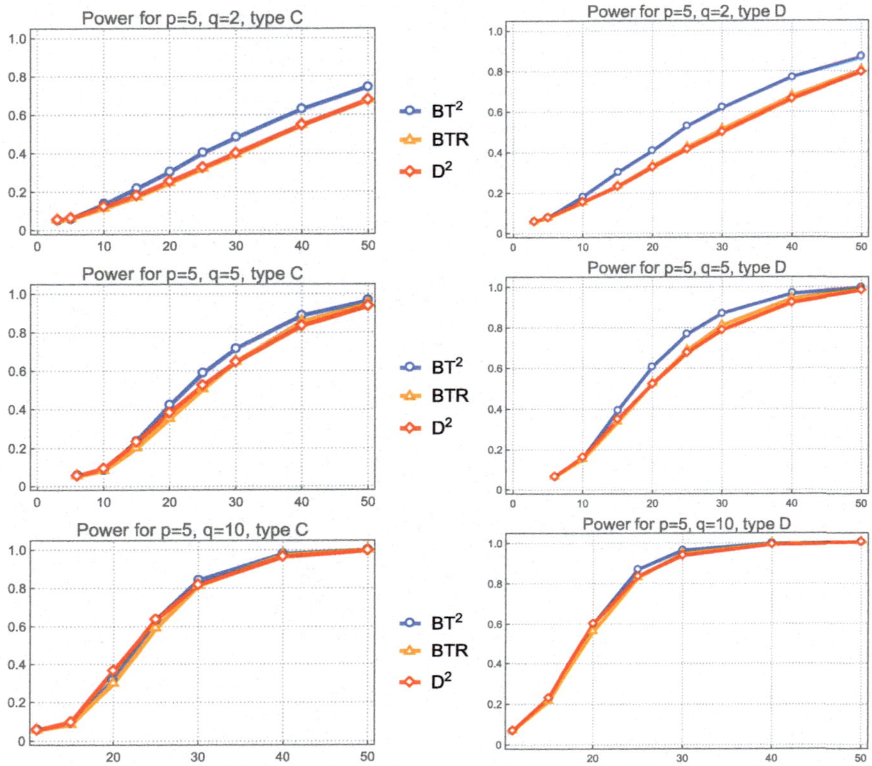

Fig. 9.2 Increase of power with q (type C and D)

9.6.2 *Comparisons for Individual Tests*

See Figs. 9.5, 9.6, 9.7, 9.8, 9.9 and 9.10.

Two major conclusions can be made: Even if BT^2 (with Helmert matrix) can be viewed as non-optimal, it has the highest power in the vast majority of investigated cases. However, the differences between BT^2, BTR, and D^2 are very little in most cases. The power of the tests depends much more on q (response dimension) than on p (number of sites/times).

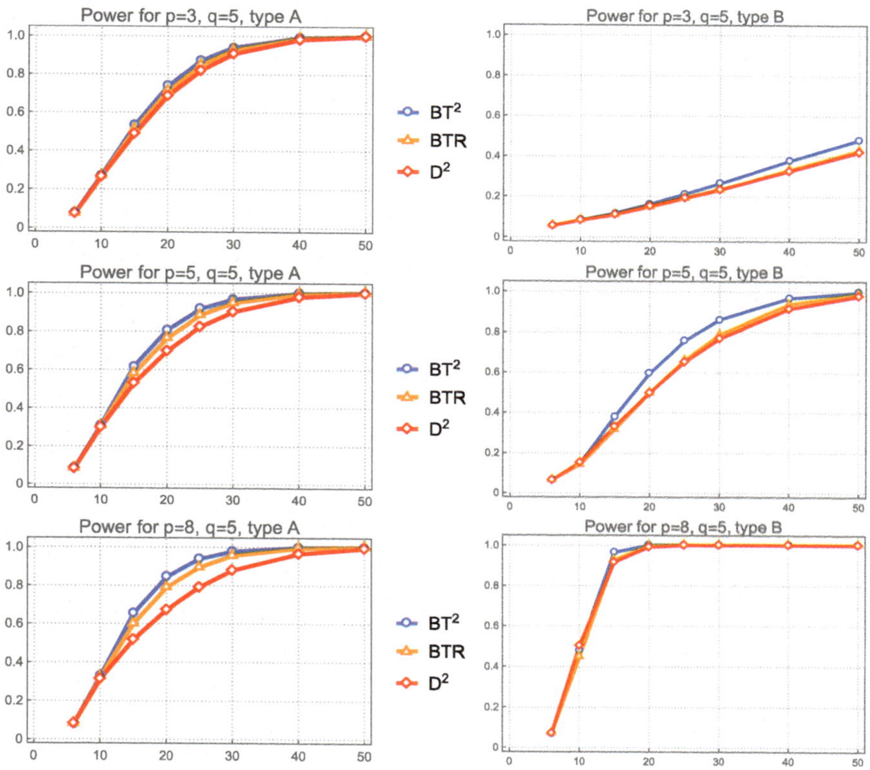

Fig. 9.3 Increase of power with p (type A and B)

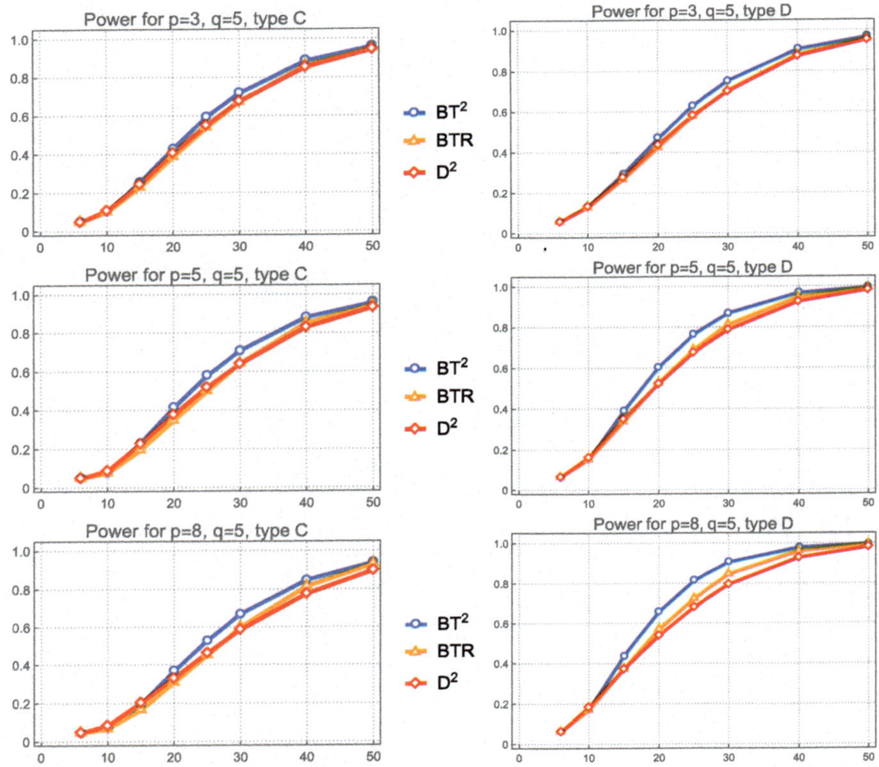

Fig. 9.4 Increase of power with p (type C and D)

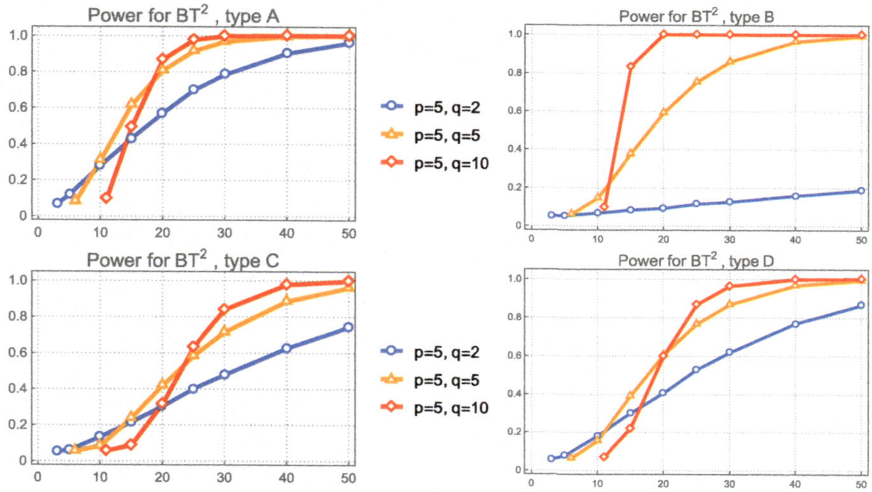

Fig. 9.5 Increase of power with q for BT^2

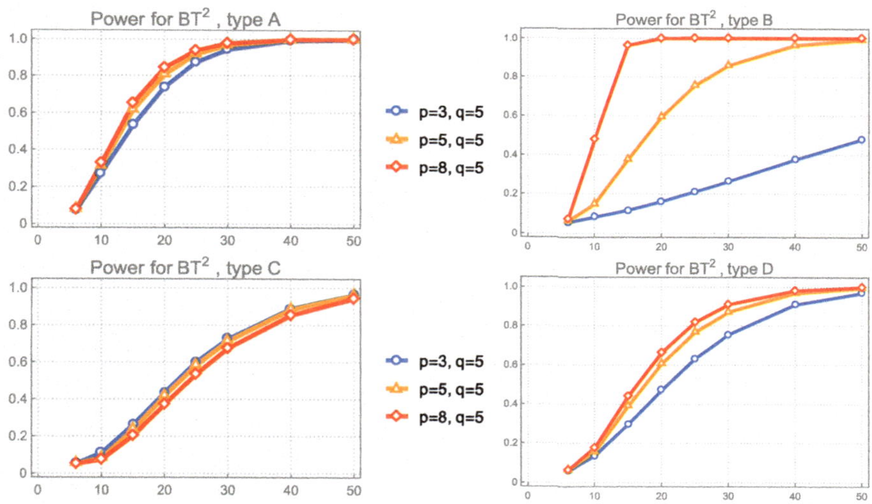

Fig. 9.6 Increase of power with p for BT^2

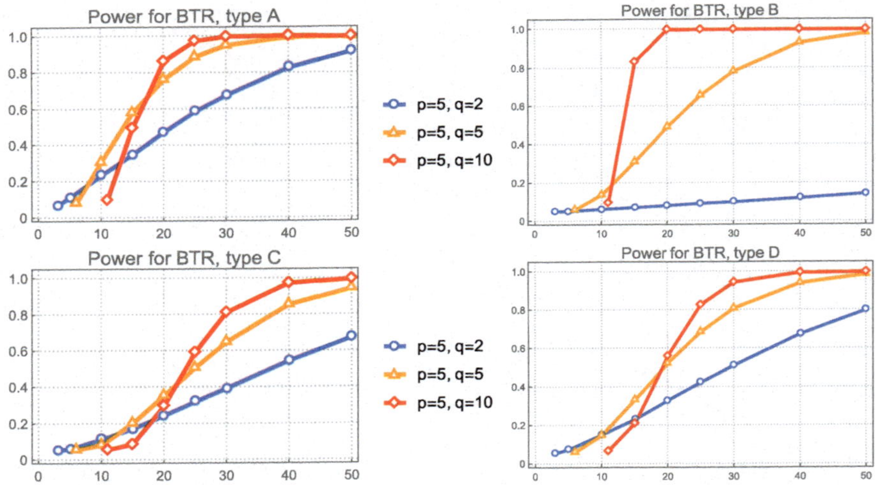

Fig. 9.7 Increase of power with q for BTR

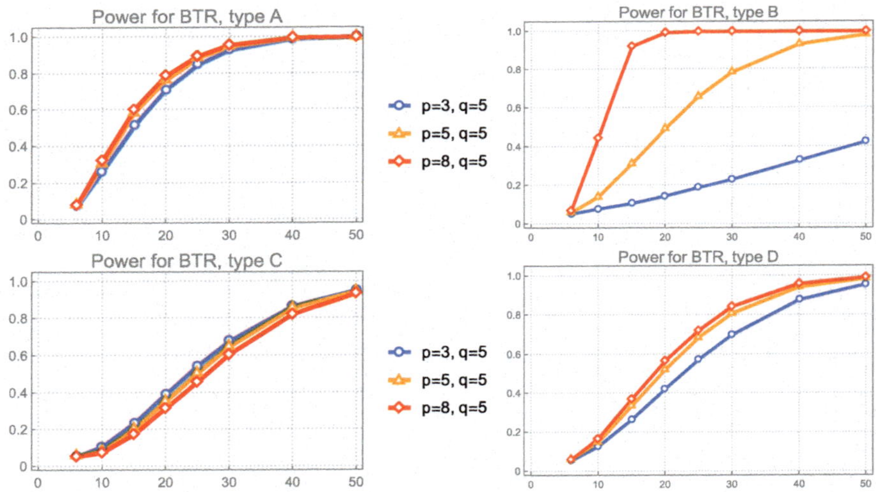

Fig. 9.8 Increase of power with p for BTR

9.7 Concluding Remarks

Overview of matrix mean testing in a simple linear model with BCS variance structure is given in this article. Such a structure is a realistic assumption in many cases and substantially reduces the number of estimated parameters. All basic tests are provided. The simulation study shows that all methods recently proposed work satis-

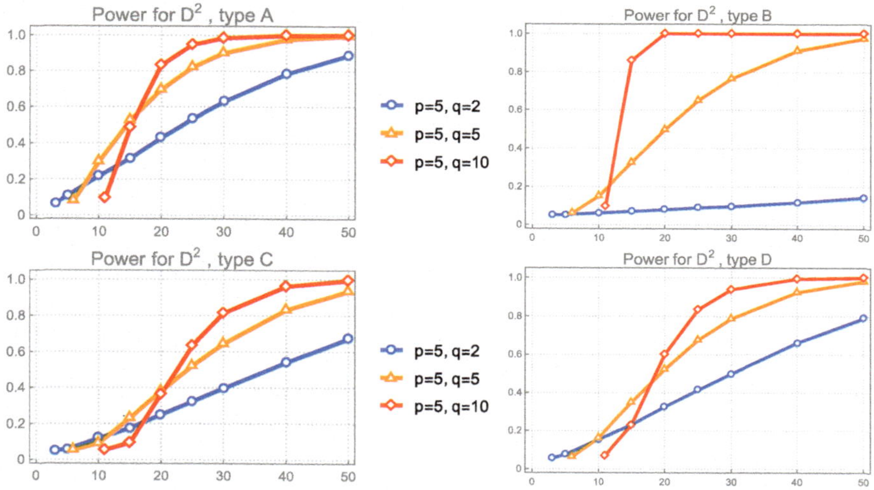

Fig. 9.9 Increase of power with q for D^2

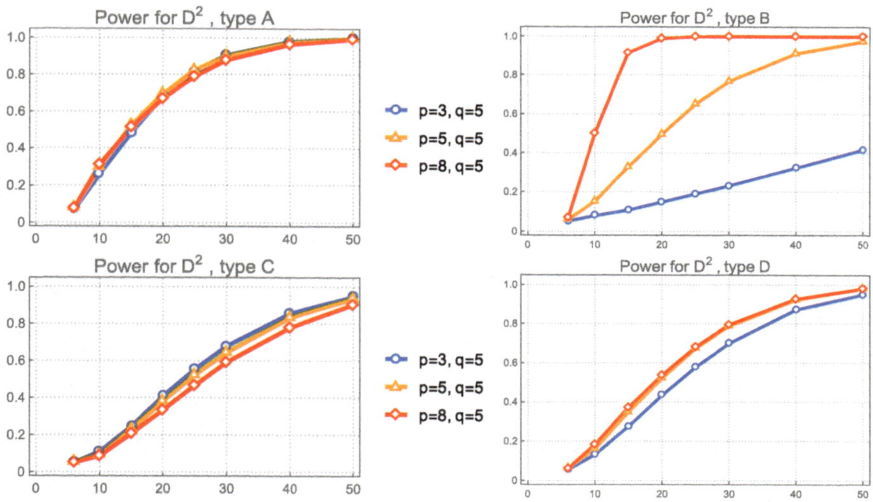

Fig. 9.10 Increase of power with p for D^2

factorily and there are no big differences between their powers. The best performance in the majority of cases provides BT^2 test with the Helmert matrix.

Acknowledgements Žežula's and Klein's research was supported by the Slovak Research and Development Agency under the Contract No. APVV-17-0568 and by grant VEGA MŠ SR 1/0311/18.

References

1. Arnold, S.F.: Application of the theory of products of problems to certain patterned covariance matrices. Ann. Stat. **1**(4), 682–699 (1973)
2. Arnold, S.F.: Applications of products to the generalized compound symmetry problem. Ann. Stat. **3**, 227–233 (1976)
3. Arnold, S.F.: Linear models with exchangeably distributed errors. J. Am. Stat. Assoc. **74**(365), 194–199 (1979)
4. Dyer, D.: The convolution of generalized F distributions. J. Am. Stat. Assoc. **77**(377), 184–189 (1982)
5. Geisser, S.: Multivariate Analysis of Variance for a Special Covariance Case. J. Am. Stat. Assoc. **58**(303), 660–669 (1963)
6. McKeon, J.J.: F-approximations to the distribution of Hotelling's T_0^2. Biometrika **61**(2), 381–383 (1974)
7. Roy, A., Fonseca, M.: Linear Models with Doubly Exchangeable Distributed Errors. Comm. Stat. Theory Methods **41**(13), 2545–2569 (2012)
8. Roy, A., Leiva, R., Žežula, I., Klein, D.: Testing the equality of mean vectors for paired doubly multivariate observations in blocked compound symmetric covariance matrix setup. J. Multivariate Anal **137**, 50–60 (2015)
9. Sperling, M.R., Gur, R.C., Alavi, A., Gur, R.E., Resnick, S., O'Connor, M.J., Reivich, M.: Subcortical metabolic alterations in partial epilepsy. Epilepsia **31**(2), 145–155 (1990)
10. Szatrowski, T.H.: Estimation and Testing for Block Compound Symmetry and Other Patterned Covariance Matrices With Linear and Non-linear Structure. Technical Report No. OLK NSF 107, Stanford University, Department of Statistics (1976)
11. Szatrowski, T.H.: Testing and estimation in the block compound symmetry problem. J. Educ. Stat. **7**(1), 3–18 (1982)
12. Tukey, J.W., Wilks, S.S.: Approximation of the distribution of the product of beta variables by a single beta variable. Ann. Math. Stat. **17**, 318–324 (1946)
13. Votaw, D.F., Jr.: Testing compound symmetry in a normal multivariate distribution. Ann. Math. Stat. **19**(4), 447–473 (1948)
14. Wilks, S.S.: Sample criteria for testing equality of means, equality of variances, and equality of covariances in a normal multivariate distribution. Ann. Math. Stat. **17**(3), 257–281 (1946)
15. Žežula, I., Klein, D., Roy, A.: Testing of multivariate repeated measures data with block exchangeable covariance structure. Test **27**(2), 360–378 (2018)

Chapter 10
On a Simplified Approach to Estimation in Experiments with Orthogonal Block Structure

Radosław Kala

Abstract Experiments with the orthogonal block structure form a wide class of designs having, under the assumption of full randomization, the dispersion matrix of a special spectral form with unknown variance components. In the paper, it is shown how the known estimation procedures of both the treatment parameters and variance components can be simplified. The approach proposed is direct, quite general, and mainly uses the technique of orthogonal projection.

10.1 Introduction

The analysis of observations following from planned experiments is based on a model which must take into account many elements. In general, the model concerning a set of n observations forming the random vector $\mathbf{y}$ determines the expectation of $\mathbf{y}$, $\mathrm{E}(\mathbf{y})$, and the dispersion matrix of $\mathbf{y}$, $\mathrm{D}(\mathbf{y})$. Usually, the expectation of $\mathbf{y}$ is related to a set of v treatments through an $n \times v$ design matrix $\mathbf{X}$. This relation is expressed in the form

$$\mathrm{E}(\mathbf{y}) = \boldsymbol{\mu} = \mathbf{X}\boldsymbol{\beta},$$

where $\boldsymbol{\beta}$ is a v-dimensional vector of unknown treatment means. The column space of $\mathbf{X}$, $\mathscr{C}(\mathbf{X})$, will be called the expectation subspace. It is assumed here that $\mathbf{X}$ is of full column rank and fulfills the condition

$$\mathbf{X}\mathbf{1}_v = \mathbf{1}_n, \tag{10.1}$$

where $\mathbf{1}_n$ is the n-dimensional vector of ones, which, in what follows, will be denoted without n subscript. This condition means that the vector $\mathbf{1}$ is contained in $\mathscr{C}(\mathbf{X})$. Moreover, if $\mathbf{X}$ is a binary matrix, then the property (10.1) implies that each observation is related only with one treatment mean.

R. Kala (✉)
Department of Mathematical and Statistical Methods, Poznań University of Life Sciences, Poznań, Poland
e-mail: kalar@up.poznan.pl

K. Filipiak et al. (eds.), *Multivariate, Multilinear and Mixed Linear Models*, Contributions to Statistics, https://doi.org/10.1007/978-3-030-75494-5_10

In the case of experiments conducted with full randomization of experimental material (units), the dispersion matrix has to reflect the consequences of such random procedures. The final form of D (**y**) depends on the systems of blocking of units as well as the presence of random errors of individual measurements. If the experimental material possesses Orthogonal Block Structure (see, e.g., Nelder [17, 18]), the dispersion matrix takes the form

$$\mathrm{D}\,(\mathbf{y}) = \mathbf{V} = \sigma_0^2\mathbf{C}_0 + \sigma_1^2\mathbf{C}_1 + \cdots + \sigma_t^2\mathbf{C}_t,$$

where $\sigma_0^2, \sigma_1^2, \ldots, \sigma_t^2$ are unknown but positive variance components, while $\mathbf{C}_0, \mathbf{C}_1, \ldots, \mathbf{C}_t$ are known mutually orthogonal (in the standard sense) $n \times n$ projectors and such that

$$\mathbf{C}_0 + \mathbf{C}_1 + \cdots + \mathbf{C}_t = \mathbf{I}.$$

Of course, such a dispersion matrix is non-singular, which ensures that the model $\{\mathbf{y}, \mathbf{X}\boldsymbol{\beta}, \mathbf{V}\}$ is statistically consistent for any observed vector $\mathbf{y}$ (see, e.g., Rao [22, p. 297]). Each projector $\mathbf{C}_i$ if applied to $\mathbf{y}$ leads to the so-called stratum submodel. They are singular and have the forms $\{\mathbf{C}_i\mathbf{y}, \mathbf{C}_i\mathbf{X}\boldsymbol{\beta}, \sigma_i^2\mathbf{C}_i\}$, $i \in \{0, 1, \ldots, t\}$. Each vector $\mathbf{C}_i\mathbf{y}$ represents a part of information contained in $\mathbf{y}$ which is uncorrelated with the other parts. In each stratum submodel, the expectation subspace $\mathscr{C}(\mathbf{C}_i\mathbf{X})$ is the invariant subspace of the corresponding dispersion matrix, which ensures the possibility of obtaining the Best Linear Unbiased Estimate (BLUE) of its expectation by a solution of the simple least squares normal equations. A complete review of results concerning such equality is presented in the paper by Baksalary et al. [4]. However, the BLUE of $\boldsymbol{\mu} = \mathbf{X}\boldsymbol{\beta}$ in the overall model $\{\mathbf{y}, \mathbf{X}\boldsymbol{\beta}, \mathbf{V}\}$ does not exist as long as the variance components are unknown.

In this paper, some simplifications of the estimation procedures of both the treatment parameters and variance components are proposed. In the next section, some remarks about the forms of dispersion matrix of three standard experiments are discussed, giving the basis for the aforementioned simplifications. Section three is devoted to the development of the alternative projectors producing the BLUE of $\boldsymbol{\mu}$ or only of the set of all standard treatment contrasts, however, under the assumption of known variance components. The next section shows the simplifications concerning the system of non-linear equations, originally initiated by Nelder [19] and later supplemented by Houtman and Speed [11], which provide the estimates of variance components indispensable to form of the so-called empirical BLUE. The solutions of these equations coincide with those following from the REML method due to Patterson and Thompson [21] and those obtainable by the iterated MINQUE procedure of Rao [23]. It is worth noticing that the approach proposed here is free from such specific assumptions as of the general balance of the design, used by Nelder [19], or the normality distribution of observations, exploited by Patterson and Thompson [20]. The final section is devoted to some remarks concerning the possibility of extension of the model assumptions and to some summarizing conclusions.

10.2 Three Basic Experiments

The simplest experiment with the OBS property is any proper block design, i.e., the design with equal block sizes (see, e.g., Caliński and Kageyama [6]). The dispersion matrix of such design reflecting the randomization of blocks and independently of units in each block takes the form

$$\mathbf{V} = \sigma_0^2\mathbf{C}_0 + \sigma_1^2\mathbf{C}_1 + \sigma_2^2\mathbf{C}_2,$$

where

$$\mathbf{C}_0 = n^{-1}\mathbf{1}\mathbf{1}',\ \mathbf{C}_1 = \mathbf{P}_B - n^{-1}\mathbf{1}\mathbf{1}',\ \mathbf{C}_2 = \mathbf{I} - \mathbf{P}_B = \mathbf{P}_2,$$

while $\mathbf{P}_B$, denoted also as $\mathbf{P}_1$, is the orthogonal projector on the subspace corresponding to blocks. It is easy to notice that not only $\mathbf{C}_0 + \mathbf{C}_1 + \mathbf{C}_2 = \mathbf{I}$, but also $\mathbf{P}_1 + \mathbf{P}_2 = \mathbf{I}$. Moreover, in this type of the design the projectors $\mathbf{P}_1$ and $\mathbf{P}_2$ are also mutually orthogonal.

Another experiment with the OSB property is any row and column design (see, e.g., Bailey [1, 2]). Its dispersion matrix derived by full randomization is composed of four projectors,

$$\mathbf{V} = \sigma_0^2\mathbf{C}_0 + \sigma_1^2\mathbf{C}_1 + \sigma_2^2\mathbf{C}_2 + \sigma_3^2\mathbf{C}_3,$$

where

$$\mathbf{C}_0 = n^{-1}\mathbf{1}\mathbf{1}',\ \ \mathbf{C}_1 = \mathbf{P}_R - n^{-1}\mathbf{1}\mathbf{1}',\ \ \mathbf{C}_2 = \mathbf{P}_C - n^{-1}\mathbf{1}\mathbf{1}',$$
$$\mathbf{C}_3 = \mathbf{I} - \mathbf{P}_R - \mathbf{P}_C + n^{-1}\mathbf{1}\mathbf{1}' = \mathbf{P}_3 + n^{-1}\mathbf{1}\mathbf{1}'.$$

The operators $\mathbf{P}_R (= \mathbf{P}_1)$ and $\mathbf{P}_C (= \mathbf{P}_2)$ project on the subspaces related to rows and columns, respectively. These subspaces are not disjoint because $\mathbf{P}_1\mathbf{C}_0 = \mathbf{P}_2\mathbf{C}_0 = \mathbf{C}_0$. The subspace $\mathscr{C}(\mathbf{C}_0)$ is also contained in $\mathscr{C}(\mathbf{P}_3)$, but in this case, $\mathbf{P}_3\mathbf{C}_0 = -\mathbf{C}_0$. Nevertheless, again $\mathbf{P}_1 + \mathbf{P}_2 + \mathbf{P}_3 = \mathbf{I}$.

The designs with nested blocks, as considered by Caliński [5] (see also Caliński and Kageyama [7], Kala [13]; or Caliński and Łacka [8]), will play as the third example. The dispersion matrix of such designs can be expressed as

$$\mathbf{V} = \sigma_0^2\mathbf{C}_0 + \sigma_1^2\mathbf{C}_1 + \sigma_2^2\mathbf{C}_2 + \sigma_3^2\mathbf{C}_3,$$

where

$$\mathbf{C}_0 = n^{-1}\mathbf{1}\mathbf{1}',\ \ \mathbf{C}_1 = \mathbf{P}_S - n^{-1}\mathbf{1}\mathbf{1}',\ \ \mathbf{C}_2 = \mathbf{P}_B - \mathbf{P}_S = \mathbf{P}_2,$$
$$\mathbf{C}_3 = \mathbf{I} - \mathbf{P}_B = \mathbf{P}_3.$$

The operators $\mathbf{P}_S (= \mathbf{P}_1)$ and $\mathbf{P}_B$ project on the subspaces of superblocks and blocks, respectively. The subspace of superblocks, $\mathscr{C}(\mathbf{P}_S)$, is contained in $\mathscr{C}(\mathbf{P}_B)$, i.e., $\mathbf{P}_B\mathbf{P}_S = \mathbf{P}_S$. Moreover, $\mathbf{P}_S\mathbf{C}_0 = \mathbf{C}_0$ and $\mathbf{P}_B\mathbf{C}_0 = \mathbf{C}_0$. Finally, as in the previous two examples, $\mathbf{P}_1 + \mathbf{P}_2 + \mathbf{P}_3 = \mathbf{I}$.

The above examples illustrate two basic operations, known as nesting and crossing, used when organizing the experimental material. Mixing these two procedure, one may build up more complex classifications of experimental units which with full randomization leads to a wide class of experiments with orthogonal block structure (see, e.g., Nelder [17]).

The dispersion matrix corresponding to all such designs contains the projector $\mathbf{C}_0$ which plays a specific role. It appears that each other projector $\mathbf{C}_i$ is equal to $\mathbf{P}_i$, to $\mathbf{P}_i - \mathbf{C}_0$, or to $\mathbf{P}_i + \mathbf{C}_0$. Therefore, the projector $\mathbf{C}_0$ can be considered as a correction of projectors $\mathbf{P}_i$ with respect to the mutual orthogonality of projectors $\mathbf{C}_i$.

On the other hand, the projector $\mathbf{C}_0$ leads to the submodel of the form $\{\overline{y}\mathbf{1}, \mu\mathbf{1}, (\sigma_0^2/n)\mathbf{11}'\}$, which is called the total experimental area stratum. This stratum is defective because it represents only the general mean over all observations, $\overline{y} = n^{-1}\mathbf{1}'\mathbf{y}$, and, due to the lack of degrees of freedom, does not give any base to estimate σ_0^2. Therefore, the question arises, how to eliminate this defective component from considerations?

To this aim, let $\mathbf{S}$ be the orthogonal projector on the complement of $\mathscr{C}(\mathbf{C}_0)$, i.e., $\mathbf{S} = \mathbf{I} - \mathbf{C}_0$. Then

$$\mathbf{SC}_i = \mathbf{C}_i \text{ and } \mathbf{SP}_i = \mathbf{C}_i, \quad i \in \{1, 2, \ldots, t\}. \tag{10.2}$$

The further considerations are conducted under assumption that these properties hold together with the equality:

$$\mathbf{P}_1 + \mathbf{P}_2 + \cdots + \mathbf{P}_t = \mathbf{I}. \tag{10.3}$$

It is also worth noticing that the projector $\mathbf{S}$ leads to the stratum submodel of the form

$$\{\mathbf{Sy}, \mathbf{SX}\boldsymbol{\beta}, \mathbf{V}_{\#}\}, \tag{10.4}$$

where

$$\mathbf{V}_{\#} = \mathbf{SVS} = \sigma_1^2\mathbf{C}_1 + \sigma_2^2\mathbf{C}_2 + \cdots + \sigma_t^2\mathbf{C}_t \tag{10.5}$$

is a singular dispersion matrix. This submodel collects information contained in all strata except the information provided by $\mathbf{C}_0\mathbf{y}$. Actually, (10.4) represents the stratum submodel where all treatment contrasts $\mathbf{SX}\boldsymbol{\beta}$ can be estimated.

It should be emphasized here that the linear statistic $\mathbf{Sy}$ is linearly sufficient for all treatment contrasts. It means (see, e.g., Baksalary and Kala [3], Kala et al. [15]) that the BLUE of $\mathbf{SX}\boldsymbol{\beta}$ in the original model $\{\mathbf{y}, \mathbf{X}\boldsymbol{\beta}, \mathbf{V}\}$ and in the transformed model (10.4) are the same.

10.3 Various Representations of the BLUE

Under the assumption that all variance components are known, the BLUE of $\mathbf{X}\boldsymbol{\beta}$ in the model $\{\mathbf{y}, \mathbf{X}\boldsymbol{\beta}, \mathbf{V}\}$ is delivered by the statistic $\mathbf{P}_V\mathbf{y}$, where

$$\mathbf{P}_V = \mathbf{X}(\mathbf{X}'\mathbf{V}^{-1}\mathbf{X})^{-1}\mathbf{X}'\mathbf{V}^{-1}.$$

The operator $\mathbf{P}_V$ is the $\mathbf{V}^{-1}$-orthogonal projector on $\mathscr{C}(\mathbf{X})$, i.e., it fulfills two conditions:

$$\mathbf{P}_V\mathbf{X} = \mathbf{X}, \quad \mathbf{V}^{-1}\mathbf{P}_V = \mathbf{P}_V'\mathbf{V}^{-1}.$$

In case of the model (10.4), which is singular, but fulfills the condition $\mathscr{C}(\mathbf{SX}) \subset \mathscr{C}(\mathbf{V}_\#)$, the corresponding projector can be expressed as

$$\mathbf{P}_\# = \mathbf{SX}(\mathbf{X}'\mathbf{SV}_\#^+\mathbf{SX})^+\mathbf{X}'\mathbf{SV}_\#^+,$$

where

$$\mathbf{V}_\#^+ = \sigma_1^{-2}\mathbf{C}_1 + \sigma_2^{-2}\mathbf{C}_2 + \cdots + \sigma_t^{-2}\mathbf{C}_t$$

is the Moore–Penrose inverse of the dispersion matrix given in (10.5). It should be noticed that, in view of the first equality in (10.2),

$$\mathbf{SV}_\#^+ = \mathbf{V}_\#^+ = \mathbf{V}_\#^+\mathbf{S} = \mathbf{V}^{-1}\mathbf{S} = \mathbf{V}^{-1} - \sigma_0^{-2}\mathbf{C}_0. \tag{10.6}$$

In consequence, the projector $\mathbf{P}_\#$ can be written as

$$\mathbf{P}_\# = \mathbf{SX}(\mathbf{X}'\mathbf{V}_\#^+\mathbf{X})^+\mathbf{X}'\mathbf{V}_\#^+.$$

Moreover, the projector $\mathbf{P}_\#$ fulfills the equations:

$$\mathbf{P}_\#\mathbf{SX} = \mathbf{SX}, \quad \mathbf{V}^{-1}\mathbf{P}_\# = \mathbf{P}_\#'\mathbf{V}^{-1},$$

i.e., it is the $\mathbf{V}^{-1}$-orthogonal projector on $\mathscr{C}(\mathbf{SX})$.

It appears that there is a direct relation between $\mathbf{P}_V$ and $\mathbf{P}_\#$. For showing this, first observe that in view of (10.6) and (10.1), the following equality:

$$\mathbf{C}_0 = \sigma_0^2\mathbf{C}_0\mathbf{V}^{-1} = \sigma_0^2 n^{-1}\mathbf{11}_v'\mathbf{X}'\mathbf{V}^{-1}, \tag{10.7}$$

holds. In consequence,

$$\mathbf{C}_0\mathbf{P}_V = \sigma_0^2 n^{-1}\mathbf{11}_v'\mathbf{X}'\mathbf{V}^{-1}\mathbf{X}(\mathbf{X}'\mathbf{V}^{-1}\mathbf{X})^{-1}\mathbf{X}'\mathbf{V}^{-1} = \mathbf{C}_0, \tag{10.8}$$

which implies that

$$\mathbf{SP}_V = \mathbf{P}_V - \mathbf{C}_0. \tag{10.9}$$

On the other hand, it follows from (10.1) that $\mathbf{P}_V\mathbf{C}_0 = \mathbf{C}_0$. This together with (10.9) shows that the projectors $\mathbf{S}$ and $\mathbf{P}_V$ commute, i.e., $\mathbf{SP}_V = \mathbf{P}_V\mathbf{S}$. In consequence, the product $\mathbf{SP}_V$ is also the projector, $(\mathbf{SP}_V)^2 = \mathbf{SP}_V$. It projects on $\mathscr{C}(\mathbf{SX})$, $(\mathbf{SP}_V)\mathbf{SX} = \mathbf{SP}_V\mathbf{X} = \mathbf{SX}$, and is $\mathbf{V}^{-1}$-orthogonal,

$$\mathbf{V}^{-1}(\mathbf{SP}_V) = \mathbf{V}^{-1}\mathbf{P}_V - \mathbf{V}^{-1}\mathbf{C}_0 = \mathbf{P}_V'\mathbf{V}^{-1} - \mathbf{C}_0\mathbf{V}^{-1} = (\mathbf{SP}_V)'\mathbf{V}^{-1}.$$

The same properties also has a projector $\mathbf{P}_{\#}$. Therefore,

$$\mathbf{SP}_V = \mathbf{P}_V - \mathbf{C}_0 = \mathbf{P}_{\#}. \tag{10.10}$$

In the consideration above, the stratum related to $\mathbf{C}_0$ was eliminated from the initial model by the projection of $\mathbf{y}$ on $\mathscr{C}(\mathbf{S})$. Now, some other possibility of eliminating $\mathbf{C}_0$ will be demonstrated.

Using three possible relations between $\mathbf{C}_i$ and projectors $\mathbf{P}_i$, let $\mathcal{I}_+$ and $\mathcal{I}_-$ denote two sets of indexes related to projectors of the form $\mathbf{C}_i = \mathbf{P}_i \pm \mathbf{C}_0$, i.e., $\mathcal{I}_+ = \{i : \mathbf{C}_i = \mathbf{P}_i + \mathbf{C}_0\}$ and $\mathcal{I}_- = \{i : \mathbf{C}_i = \mathbf{P}_i - \mathbf{C}_0\}$. Finally, let

$$\alpha = \sum_{i\in\mathcal{I}_-} \sigma_i^{-2} - \sum_{i\in\mathcal{I}_+} \sigma_i^{-2}.$$

Then, the following decomposition of $\mathbf{V}^{-1}$ appears,

$$\mathbf{V}^{-1} = \sum_{i=0}^{t} \sigma_i^{-2}\mathbf{C}_i = \mathbf{V}_* + (\sigma_0^{-2} - \alpha)\mathbf{C}_0, \tag{10.11}$$

where

$$\mathbf{V}_* = \sum_{i=1}^{t} \sigma_i^{-2}\mathbf{P}_i.$$

In view of (10.3), the matrix $\mathbf{V}_*$ is positive definite, but, in general, is not the inverse of $\sum_{i=1}^{t} \sigma_i^2 \mathbf{P}_i$, which holds for the proper block designs. Anyway, one can define the projector

$$\mathbf{P}_* = \mathbf{X}(\mathbf{X}'\mathbf{V}_*\mathbf{X})^{-1}\mathbf{X}'\mathbf{V}_*.$$

To exhibit its properties, let us start from the equality

$$\mathbf{V}_{\#}^{+} = \mathbf{SV}^{-1} = \mathbf{SV}_*, \tag{10.12}$$

being obvious in view of (10.2). On the other hand, the equality (10.11) implies

$$\sigma_0^{-2}\mathbf{C}_0 = \mathbf{C}_0\mathbf{V}^{-1} = \mathbf{C}_0\mathbf{V}_* + (\sigma_0^{-2} - \alpha)\mathbf{C}_0,$$

which shows that

$$\mathbf{C}_0\mathbf{V}_* = \alpha\mathbf{C}_0, \tag{10.13}$$

or, again in view of (10.1), that

$$\mathbf{C}_0 = \alpha^{-1} n^{-1} \mathbf{1}\mathbf{1}_v' \mathbf{X}'\mathbf{V}_*. \tag{10.14}$$

The last equality is similar to that in (10.7). Thus, it is not surprising that $\mathbf{C}_0\mathbf{P}_* = \mathbf{C}_0$ and, in consequence, that $\mathbf{SP}_* = \mathbf{P}_* - \mathbf{C}_0$. This, together with obvious equality $\mathbf{P}_*\mathbf{C}_0 = \mathbf{C}_0$, shows that $\mathbf{SP}_* = \mathbf{P}_*\mathbf{S}$. In result, $(\mathbf{SP}_*)^2 = \mathbf{SP}_*$ and $\mathbf{SP}_*\mathbf{SX} = \mathbf{SP}_*\mathbf{X} = \mathbf{SX}$. Moreover, from (10.12) and (10.13), it follows that

$$\mathbf{V}^{-1}(\mathbf{SP}_*) = \mathbf{SV}_*\mathbf{P}_* = \mathbf{V}_*\mathbf{P}_* - \alpha\mathbf{C}_0\mathbf{P}_* = \mathbf{V}_*\mathbf{P}_* - \alpha\mathbf{C}_0$$

is a symmetric matrix. In consequence, $\mathbf{SP}_*$ is the $\mathbf{V}^{-1}$-orthogonal projector on $\mathscr{C}(\mathbf{SX})$ and thus, it is the same as projector $\mathbf{P}_\#$,

$$\mathbf{SP}_* = \mathbf{P}_* - \mathbf{C}_0 = \mathbf{P}_\#. \tag{10.15}$$

Taking into account the last equality and that in (10.10), the conclusion arises that projectors $\mathbf{P}_V$ and $\mathbf{P}_*$ are equal. Practically, it means that the BLUE for $\mathbf{X}\boldsymbol{\beta}$ in the model $\{\mathbf{y}, \mathbf{X}\boldsymbol{\beta}, \mathbf{V}\}$ is delivered by $\mathbf{P}_V\mathbf{y}$, or by $\mathbf{P}_*\mathbf{y}$, whereas projectors $\mathbf{P}_\#$, $\mathbf{SP}_V$, and $\mathbf{SP}_*$ provide the BLUE for all treatment contrasts $\mathbf{SX}\boldsymbol{\beta}$ in the transformed model $\{\mathbf{Sy}, \mathbf{SX}\boldsymbol{\beta}, \mathbf{SVS}\}$ which coincide with the BLUE for $\mathbf{SX}\boldsymbol{\beta}$ in the original model $\{\mathbf{y}, \mathbf{X}\boldsymbol{\beta}, \mathbf{V}\}$. Finally, it should be stressed here that these conclusions are true for any values of variance components. This means, in particular, that $\mathbf{P}_V$ is equal to $\mathbf{P}_*$ for arbitrary choice of σ_0^2 provided that $\sigma_0^2 > 0$.

10.4 Estimation of Variance Components

In the considerations of the previous section it was assumed that all variance components are known. It is not the case in practice. Thus, some estimates of these parameters are indispensable. In a simple fixed model, where only one variance component is needed, the standard estimation approach is based on equating an appropriate quadratic form of $\mathbf{y}$ with its expectation. In the case of the model $\{\mathbf{y}, \mathbf{X}\boldsymbol{\beta}, \mathbf{V}\}$, the suitable quadratic form is based on the length of the vector of the so-called residuals under the inner product defined by $\mathbf{V}^{-1}$. The vector of residuals can be expressed as

$$\mathbf{y} - \mathbf{P}_V\mathbf{y} = \mathbf{Ry},$$

where $\mathbf{R} = \mathbf{I} - \mathbf{P}_V$ is the projector on the $\mathbf{V}^{-1}$-orthogonal complement of the expectation subspace $\mathscr{C}(\mathbf{X})$. Taking into account the form of $\mathbf{V}^{-1}$ given in (10.11), the residual sum of squares can be written as

$$Q = \mathbf{y}'\mathbf{R}'\mathbf{V}^{-1}\mathbf{Ry} = \sum_{i=0}^{t} \sigma_i^{-2}\mathbf{y}'\mathbf{R}'\mathbf{C}_i\mathbf{Ry}. \tag{10.16}$$

The expectation of Q can also be expressed in a similar form,

$$\mathrm{E}\,(Q) = \sum_{i=0}^{t} \sigma_i^{-2} \mathrm{E}\,(\mathbf{y}'\mathbf{R}'\mathbf{C}_i\mathbf{R}\mathbf{y}) = \sum_{i=0}^{t} \mathrm{Tr}\,(\mathbf{C}_i\mathbf{R}). \tag{10.17}$$

It is so, because $\mathrm{E}\,(\mathbf{y}'\mathbf{R}'\mathbf{C}_i\mathbf{R}\mathbf{y}) = \sigma_i^2\mathrm{Tr}\,(\mathbf{C}_i\mathbf{R})$. The comparison of the corresponding terms on the right-hand sides in (10.16) and (10.17) leads directly to the estimating equations, first established by Nelder [19] under the additional assumption of general balance of the design (for details see Houtman and Speed [11] or Kala [14]). They take the following form:

$$\mathbf{y}'\mathbf{R}'\mathbf{C}_i\mathbf{R}\mathbf{y} = \widehat{\sigma}_i^2\mathrm{Tr}\,(\mathbf{C}_i\mathbf{R}), \quad i \in \{0, 2, \ldots, t\}. \tag{10.18}$$

which also coincide, as it was observed by Patterson and Thompson [20], with those following from their maximum likelihood approach under the assumption that $\mathbf{y}$ has a multivariate normal distribution. These equations can be simplified. The first equation in (10.18) is satisfied by any $\widehat{\sigma}_0^2$. It is so, because, in view of (10.8),

$$\mathbf{C}_0\mathbf{R} = \mathbf{C}_0 - \mathbf{C}_0\mathbf{P}_V = \mathbf{0}. \tag{10.19}$$

This means that the first equation in (10.18) can be eliminated and, simultaneously, in the other equations, each product $\mathbf{C}_i\mathbf{R}$ can be replaces by $\mathbf{P}_i\mathbf{R}$. In result, the simplified equations (10.18) take the form

$$\mathbf{y}'\mathbf{R}'\mathbf{P}_i\mathbf{R}\mathbf{y} = \widehat{\sigma}_i^2\mathrm{Tr}\,(\mathbf{P}_i\mathbf{R}), \quad i \in \{1, 2, \ldots, t\}. \tag{10.20}$$

The solution of these equations can not be obtained directly because in the residual projector $\mathbf{R} = \mathbf{I} - \mathbf{P}_V = \mathbf{I} - \mathbf{P}_*$ all variance components are involved. However, having the vector $\mathbf{y}$ of data established, the solution of (10.20) with respect to $\widehat{\sigma}_i^2$, $i \in \{1, 2, \ldots, t\}$, can be obtained iteratively starting with a set of initial values for variance components. The simplest starting point is to set all $\widehat{\sigma}_i^2$ as equal to one. This choice means that the initial residual projector $\mathbf{R}$ coincides with $\mathbf{I} - \mathbf{P}$, where $\mathbf{P}$ is the simple least squares operator, $\mathbf{P} = \mathbf{X}(\mathbf{X}'\mathbf{X})^{-1}\mathbf{X}'$. In the next step, a revised set of estimates can be obtained by solving the equations (10.20). This leads to a revised residual projector $\mathbf{R}$, and so on. The iteration procedure is continued until its convergence, which is usually achieved after a few steps (see, e.g., Caliński and Łacka [8], Caliński and Siatkowski [9, 10]).

It is worth noticing that for any fixed residual projector each equation in (10.20), treated separately, leads to a quadratic, unbiased, even, and transition invariant estimator of corresponding σ_i^2. Under these conditions (see: Kackar and Harville [12], Jiang [16]), the so-called empirical BLUE, i.e., the statistic $\mathbf{P}_*\mathbf{y}$ with $\mathbf{V}_*$ in which all the unknown variance components are replaced by their estimates, being the solutions of (10.20), is unbiased for $\mathbf{X}\boldsymbol{\beta}$.

Moreover, summing up the equations (10.18) premultiplied by $\widehat{\sigma}_i^{-2}, i \in \{0, \ldots, t\}$, respectively, and using (10.3) together with the equality $\mathbf{P}_i\mathbf{R} = \mathbf{P}_i\mathbf{S}\mathbf{R} = \mathbf{C}_i\mathbf{R}$, implied by (10.19), leads to the equality

$$\mathbf{y}'\mathbf{R}'\widehat{\mathbf{V}}^{-1}\mathbf{R}\mathbf{y} = \mathrm{Tr}\,(\mathbf{R}) = n - \mathrm{r}(\mathbf{X}),$$

where $\widehat{\mathbf{V}}$ denotes $\mathbf{V}$ with the variance components replaced by the exact solutions of the equation (10.18). Therefore, the so-called mean square error corresponding to the overall model $\{\mathbf{y}, \mathbf{X}\boldsymbol{\beta}, \widehat{\mathbf{V}}\}$ is equal to one. This property may be utilized to control the convergence of iteration process when solving the (10.20) or (10.18).

In case of the Nelder equation (10.18), there is some freedom in choosing $\widehat{\sigma_0^2}$. If this variance component is set as zero, then the empirical projector will be operating as $\mathbf{SP}_V$ or $\mathbf{P}_{\#}$, i.e., the statistics $\mathbf{SP}_V\mathbf{y}$, $\mathbf{P}_V\mathbf{Sy}$ or $\mathbf{P}_{\#}\mathbf{y}$ will be producing the same estimates of all treatment contrasts represented by $\mathbf{SX}\boldsymbol{\beta}$. If, however, $\widehat{\sigma_0^2}$ will be replaced by any positive scalar, then the empirical projector $\mathbf{P}_V$ will be operating as $\mathbf{P}_*$, i.e., they both, if applied to $\mathbf{y}$ will be providing the same empirical BLUE of $\mathbf{X}\boldsymbol{\beta}$.

10.5 Some Final Comments and Conclusions

The basic model assumptions concern the design matrix $\mathbf{X}$. The most limiting is the condition (10.1). Fortunately, it can be simply weakened allowing designs with many sets of treatments and their interactions. In such designs, the $n \times v$ matrix $\mathbf{X}$ is not of full column rank, the matrices $(\mathbf{X}'\mathbf{V}^{-1}\mathbf{X})$ or $(\mathbf{X}'\mathbf{V}_*\mathbf{X})$ are singular, and the condition (10.1) is not satisfied.

Let us assume, therefore, that the rank of $\mathbf{X}$ is less than v, which number is now only loosely related with the single set of treatments, and let the condition (10.1) be replaced by the equality

$$\mathbf{Xk}_v = \mathbf{1},$$

for some v-dimensional vector $\mathbf{k}_v$. So, the condition that the vector $\mathbf{1}$ belongs to the expectation subspace is preserved.

Under these conditions, the projectors considered in the previous sections can be expressed as

$$\mathbf{P}_V = \mathbf{X}(\mathbf{X}'\mathbf{V}^{-1}\mathbf{X})^-\mathbf{X}'\mathbf{V}^{-1} \quad \text{and} \quad \mathbf{P}_* = \mathbf{X}(\mathbf{X}'\mathbf{V}_*\mathbf{X})^-\mathbf{X}'\mathbf{V}_*, \tag{10.21}$$

where $\mathbf{A}^-$ stands for any g-inverse of $\mathbf{A}$ (see, e.g., Rao and Rao [24]). On the other hand, the key equations (10.7) and (10.14) now can be replaced by

$$\mathbf{C}_0 = \sigma_0^2 n^{-1}\mathbf{1k}_v'\mathbf{X}'\mathbf{V}^{-1} \text{ and } \mathbf{C}_0 = \alpha^{-1}n^{-1}\mathbf{1k}_v'\mathbf{X}'\mathbf{V}_*,$$

respectively. In consequence, the conditions (10.10) and (10.15) hold with projectors as defined in (10.21). This is sufficient to repeat the general conclusion that the statistics $\mathbf{SP}_V\mathbf{y} = \mathbf{P}_V\mathbf{Sy} = (\mathbf{SP}_*\mathbf{y} = \mathbf{P}_*\mathbf{Sy})$ with variance components replaced by the solutions of the equations (10.20) provides the same estimate of $\mathbf{SX}\boldsymbol{\beta}$, while $\mathbf{P}_V\mathbf{y}(= \mathbf{P}_*\mathbf{y})$ provides the same estimate of $\mathbf{X}\boldsymbol{\beta}$.

Summarizing the above considerations, it can be stated that the defective total experimental area stratum, related with projector $\mathbf{C}_0$ can be eliminated from the model without any loss in estimation of the variance components and estimation of the estimable linear treatment functions by the use of the empirical BLUE projectors $\mathbf{P}_*$ and $\mathbf{P}_\#$. On the other hand, the Nelder equations can be established directly by the standard approach based on comparing the residual sum of squares with its expectation, without any need of the general balance property. Moreover, they can be simplified by reducing their number and replacing all projectors $\mathbf{C}_i$ by the corresponding projectors $\mathbf{P}_i$, which reduces computational burden.

Acknowledgements The author is grateful to the referee for valuable comments and suggestions substantially improving the presentation of this paper.

References

1. Bailey, R.A.: Strata for randomized experiments (with discussion). J. R. Stat. Soc. Ser. B. Stat. Methodol. **53**, 27–78 (1991)
2. Bailey, R.A.: General balance: artificial theory or practical relevance. In: Caliński, T., Kala, R. (eds.) Proceedings of the International Conference on Linear Statistical Inference LINSTAT'93, pp. 171–184. Kluver Acad. Publ. (1994)
3. Baksalary, J.K., Kala, R.: Linear sufficiency with respect to a given vector of parametric functions. J. Stat. Plan. Infer. **14**, 331–338 (1986)
4. Baksalary, J.K., Puntanen, S., Styan, G.P.S.: On T.W. Anderson's contributions to solving the problem of when the ordinary least-squares estimator is best linear unbiased and to characterizing rank additivity of matrices. In: Styan, G.P.S. (ed.) The Collected Papers of T.W. Anderson: 1943–1985, pp. 1579–1591. Wiley, New York (1990)
5. Caliński, T.: Recovery of inter-block information when the experiment is in nested block design. Biomet. Lett. **34**, 9–26 (1997)
6. Caliński, T., Kageyama, S.: The randomization model for experiments in bock designs and the recovery of inter-block information. J. Stat. Plan. Infer. **52**, 359–374 (1996)
7. Caliński, T., Kageyama, S.: Block Designs: A Randomization Approach, Vol I: Analysis. Springer, New York (2000)
8. Caliński, T., Łacka, A.: On combining information in generally balanced nested block designs. Comm. Stat. Theory Methods **43**, 954–974 (2014)
9. Caliński, T., Siatkowski, I.: On a new approach to the analysis of variance for experiments with orthogonal block structure. I. Experiments in proper block designs. Biomet. Lett. **54**, 91–122 (2017)
10. Caliński, T., Siatkowski, I.: On a new approach to the analysis of variance for experiments with orthogonal block structure II experiments in nested block designs. Biomet. Lett. **55**, 147–178 (2018)
11. Houtman, A.M., Speed, T.P.: Balance in designed experiments with orthogonal block structure. Ann. Math. Stat. **11**, 1069–1085 (1983)
12. Kackar, R.N., Harville, D.A.: Approximations for standard errors of estimators of fixed and random effects in mixed linear models. J. Amer. Stat. Assoc. **79**, 853–862 (1984)
13. Kala, R.: On nested block designs geometry. Stat. Pap. **50**, 805–815 (2009)
14. Kala, R.: A new look at combining information in experiments with orthogonal block structure. In: Ahmed, S.E., Carvalho, F., Puntanen, S. (eds.) Matrices, Statistics and Big Data. Proc. of the 25th International Workshop on Matrices and Statistics, IWMS-2016, pp. 35–49. Springer (2019)
15. Kala, R., Puntanen, S., Tian, Y.: Some notes on linear sufficiency. Stat. Pap. **58**, 1–17 (2017)
16. Jiang, J.: On unbiasedness of the empirical BLUE and BLUP. Stat. Probab. Lett. **41**, 19–24 (1999)

17. Nelder, J.A.: The analysis of randomized experiments with orthogonal block structure. I. Block structure and the null analysis of variance. Proc. Roy. Soc. Lond. Ser. A **283**, 147–162 (1965)
18. Nelder, J.A.: The analysis of randomized experiments with orthogonal block structure. II. Treatment structure and the general analysis of variance. Proc. Roy. Soc. Lond. Ser. A **283**, 163–178 (1965)
19. Nelder, J.A.: The combination of information in generally balanced designs. J. R. Stat. Soc. Ser. B. Stat. Methodol. **30**, 303–311 (1968)
20. Patterson, H.D., Thompson, R.: Recovery of inter-block information when the block sizes are unequal. Biometrika **58**, 545–554 (1971)
21. Patterson, H.D., Thompson, R.: Maximum likelihood estimation of components of variance. In: Corsten, L.C.A., Postelnicu, T. (eds.) Proceedings 8th International Biometric Conference, pp. 197–207. Bucuresti, Editura Academiei (1975)
22. Rao, C.R.: Linear Statistical Inference and its Applications, 2nd edn. Wiley, New York (1973)
23. Rao, C.R.: MINQUE theory and its relation to ML and MML estimation of variance components. Sankhyā B **41**, 138–153 (1979)
24. Rao, C.R., Rao, M.B.: Matrix Algebra and Its Applications to Statistics and Econometrics. World Scientific, Singapore (2004)

Chapter 11
A Review of the Linear Sufficiency and Linear Prediction Sufficiency in the Linear Model with New Observations

Stephen J. Haslett, Jarkko Isotalo, Radosław Kala, Augustyn Markiewicz, and Simo Puntanen

Abstract We consider the general linear model $\mathbf{y} = \mathbf{X}\boldsymbol{\beta} + \boldsymbol{\varepsilon}$, denoted as $\mathscr{M} = \{\mathbf{y}, \mathbf{X}\boldsymbol{\beta}, \mathbf{V}\}$, supplemented with the new unobservable random vector $\mathbf{y}_*$, coming from $\mathbf{y}_* = \mathbf{X}_*\boldsymbol{\beta} + \boldsymbol{\varepsilon}_*$, where the covariance matrix of $\mathbf{y}_*$ is known as well as the cross-covariance matrix between $\mathbf{y}_*$ and $\mathbf{y}$. A linear statistic $\mathbf{Fy}$ is called linearly sufficient for $\mathbf{X}_*\boldsymbol{\beta}$ if there exists a matrix $\mathbf{A}$ such that $\mathbf{AFy}$ is the best linear unbiased estimator, BLUE, for $\mathbf{X}_*\boldsymbol{\beta}$. The concept of linear sufficiency with respect to a predictable random vector is defined in the corresponding way but considering the best linear unbiased predictor, BLUP instead of BLUE. In this paper, we consider the linear sufficiency of $\mathbf{Fy}$ with respect to $\mathbf{y}_*$, $\mathbf{X}_*\boldsymbol{\beta}$, and $\boldsymbol{\varepsilon}_*$. We also apply our results into the linear mixed model. The concept of linear sufficiency was essentially introduced in early 1980s by Baksalary, Kala, and Drygas. Recently, several papers providing further properties of the linear sufficiency have been published by the present authors. Our aim is to provide an easy-to-read review of recent results and while doing that,

S. J. Haslett (✉)
Research School of Finance, Actuarial Studies and Statistics,
The Australian National University, Canberra, Australia

Centre for Public Health Research, Massey University, Wellington, New Zealand

School of Fundamental Sciences, Massey University, Palmerston North, New Zealand
e-mail: stephen.haslett@anu.edu.au; s.j.haslett@massey.ac.nz

J. Isotalo
Faculty of Information Technology and Communication Sciences, Tampere University,
Tampere, Finland
e-mail: jarkko.isotalo@tuni.fi

R. Kala · A. Markiewicz
Department of Mathematical and Statistical Methods, Poznań University of Life Sciences,
Poznań, Poland
e-mail: kalar@up.poznan.pl

A. Markiewicz
e-mail: augustyn.markiewicz@up.poznan.pl

S. Puntanen
Faculty of Information Technology and Communication Sciences, Tampere University,
Tampere, Finland
e-mail: simo.puntanen@tuni.fi

K. Filipiak et al. (eds.), *Multivariate, Multilinear and Mixed Linear Models*,
Contributions to Statistics, https://doi.org/10.1007/978-3-030-75494-5_11

we go through some basic concepts related to linear sufficiency. As a review paper, we do not provide many proofs, instead our goal is to explain and clarify the central results.

11.1 Preliminaries and Introduction to the Models

In this section, we introduce the notation to be used and briefly go through the various versions of linear models that we are interested in. Also, we present some handy matrix-algebraic tools that will be needed later on.

The symbol $\mathbb{R}^{m\times n}$ denotes the set of $m \times n$ real matrices, while $\mathbf{A}'$, $\mathbf{A}^-$, $\mathbf{A}^+$, $\mathscr{C}(\mathbf{A})$, $\mathscr{C}(\mathbf{A})^\perp$, $\mathrm{r}(\mathbf{A})$, and $\mathscr{N}(\mathbf{A})$ denote, respectively, the transpose, a generalized inverse, the Moore–Penrose inverse, the column space, the orthogonal complement of the column space, rank, and the null space of the matrix $\mathbf{A}$. If $\mathbf{G}$ satisfies $\mathbf{AGA} = \mathbf{A}$, then we denote $\mathbf{G} = \mathbf{A}^-$. The Moore–Penrose inverse $\mathbf{A}^+$ is defined as a unique matrix satisfying the following four conditions:

$$\mathbf{AA}^+\mathbf{A} = \mathbf{A}\,, \quad \mathbf{A}^+\mathbf{AA}^+ = \mathbf{A}^+, \quad (\mathbf{AA}^+)' = \mathbf{AA}^+, \quad (\mathbf{A}^+\mathbf{A})' = \mathbf{A}^+\mathbf{A}\,.$$

By $(\mathbf{A} : \mathbf{B})$ we denote the columnwise partitioned matrix with $\mathbf{A}_{a\times b}$ and $\mathbf{B}_{a\times c}$ as submatrices. By $\mathbf{A}^\perp$ we denote any matrix satisfying $\mathscr{C}(\mathbf{A}^\perp) = \mathscr{C}(\mathbf{A})^\perp = \mathscr{N}(\mathbf{A}')$. Notice that if $\mathbf{A} \in \mathbb{R}^{a\times b}$, then $\mathbf{A}^\perp \in \mathbb{R}^{a\times d}$, where $d \geq a - \mathrm{r}(\mathbf{A})$. Notation $\mathbf{P}_\mathbf{A} = \mathbf{AA}^+ = \mathbf{A}(\mathbf{A}'\mathbf{A})^-\mathbf{A}'$ stands for the orthogonal projector (with respect to the standard inner product) onto the column space $\mathscr{C}(\mathbf{A})$, and so $\mathbf{P}_{\mathbf{A}'} = \mathbf{A}^+\mathbf{A}$. The orthogonal projector onto $\mathscr{C}(\mathbf{A})^\perp$ is denoted as $\mathbf{Q}_\mathbf{A} = \mathbf{I}_a - \mathbf{P}_\mathbf{A}$, where $\mathbf{I}_a$ refers to the $a \times a$ identity matrix and a is the number of rows of $\mathbf{A}$. It appears convenient to use the short notation

$$\mathbf{M} = \mathbf{I}_n - \mathbf{P}_\mathbf{X}\,,$$

where $\mathbf{X}_{n\times p}$ refers the matrix determining the expectation subspace in the linear model. One handy choice (for its symmetry and idempotence) for $\mathbf{X}^\perp$ is $\mathbf{M}$.

The concept of nonnegative definite matrices plays an important role in statistics. Formally, a symmetric $n \times n$ matrix $\mathbf{A}$ is said to be nonnegative definite (or positive semidefinite), denoted as $\mathbf{A} \in \mathrm{NND}_n$, if

$$\mathbf{x}'\mathbf{Ax} \geq 0 \ \text{ for all } \mathbf{x} \in \mathbb{R}^n, \quad \text{or equivalently,} \quad \mathbf{A} = \mathbf{C}'\mathbf{C} \ \text{ for some } \mathbf{C}\,.$$

Such matrices can be partially ordered. If $\mathbf{A}, \mathbf{B} \in \mathrm{NND}_n$ and simultaneously $\mathbf{B} - \mathbf{A} \in \mathrm{NND}_n$, then $\mathbf{A}$ is said to be below $\mathbf{B}$ in the Löwner partial order. This fact will be denoted as $\mathbf{A} \leq_\mathrm{L} \mathbf{B}$.

Next, we shortly describe the various statistical models that we are interested in.

(a) *General linear model, $\mathscr{M}$.* Our main interest lies in the general linear model

$$\mathbf{y} = \mathbf{X}\boldsymbol{\beta} + \boldsymbol{\varepsilon}, \quad \text{or shortly } \mathscr{M} = \{\mathbf{y}, \mathbf{X}\boldsymbol{\beta}, \mathbf{V}\}, \tag{11.1}$$

where $\mathbf{y}$ is an observable n-dimensional random vector, $\mathbf{X}_{n\times p}$ is a known model matrix, $\boldsymbol{\beta}$ is a $p \times 1$ vector of unknown parameters, $\boldsymbol{\varepsilon}$ is an unobservable vector of random errors with expectation $\mathrm{E}(\boldsymbol{\varepsilon}) = \mathbf{0}$, and covariance matrix $\mathrm{Cov}(\boldsymbol{\varepsilon}) = \mathbf{V}$. The nonnegative definite covariance matrix $\mathbf{V}$ is known and can be singular.

(b) *Partitioned linear model, $\mathscr{M}_{12}$.* One special case of $\mathscr{M}$ is the partitioned linear model $\mathbf{y} = \mathbf{X}_1\boldsymbol{\beta}_1 + \mathbf{X}_2\boldsymbol{\beta}_2 + \boldsymbol{\varepsilon}$, or shortly denoted

$$\mathscr{M}_{12} = \{\mathbf{y},\ \mathbf{X}\boldsymbol{\beta},\ \mathbf{V}\} = \{\mathbf{y},\ \mathbf{X}_1\boldsymbol{\beta}_1 + \mathbf{X}_2\boldsymbol{\beta}_2,\ \mathbf{V}\}. \tag{11.2}$$

In addition to the *full* model $\mathscr{M}_{12}$, we will consider the *small* models $\mathscr{M}_i = \{\mathbf{y},\ \mathbf{X}_i\boldsymbol{\beta}_i,\ \mathbf{V}\}$, $i = 1, 2$, and the *reduced* model

$$\mathscr{M}_{12\cdot 2} = \{\mathbf{M}_2\mathbf{y},\ \mathbf{M}_2\mathbf{X}_1\boldsymbol{\beta}_1,\ \mathbf{M}_2\mathbf{V}\mathbf{M}_2\},$$

which is obtained by premultiplying the model $\mathscr{M}_{12}$ by $\mathbf{M}_2 = \mathbf{I}_n - \mathbf{P}_{\mathbf{X}_2}$. Transformation by $\mathbf{M}_2$ eliminates the "nuisance parameter" $\boldsymbol{\beta}_2$, as some authors say.

(c) *Transformed model, $\mathscr{T}$.* Premultiplying the model $\mathscr{M}$ by an $f \times n$ matrix $\mathbf{F}$ yields the transformed model

$$\mathbf{F}\mathbf{y} = \mathbf{F}\mathbf{X}\boldsymbol{\beta} + \mathbf{F}\boldsymbol{\varepsilon}\,, \quad \text{or shortly } \mathscr{T} = \{\mathbf{F}\mathbf{y}, \mathbf{F}\mathbf{X}\boldsymbol{\beta}, \mathbf{F}\mathbf{V}\mathbf{F}'\}\,.$$

The reduced model $\mathscr{M}_{12\cdot 2}$ is, of course, one example of the transformed models. The transformed model $\mathscr{T}$ will play a crucial role in our considerations. Loosely described, one of the main questions in this article will be the following: does the multiplication of the response $\mathbf{y}$ by $\mathbf{F}$ keep the estimation/prediction "undisturbed," that is, do we lose anything essential as a consequence? Notice that in $\mathscr{M}$ the response $\mathbf{y}$ is n-dimensional while in $\mathscr{T}$ the response $\mathbf{F}\mathbf{y}$ is f-dimensional. In principle, the number of rows in $\mathbf{F}$, f, can be greater than, less than, or equal to n but intuitively it seems clear that the rows of $\mathbf{F}$ could be chosen linearly independent, see (11.28). In the partitioned model (11.2), the reduction into $\mathscr{M}_{12\cdot 2}$ has been done by $\mathbf{M}_2$ which has n rows but another essentially identical reduction could be carried out by choosing $\mathbf{F}' = \mathbf{X}_2^{\perp}$, where the columns of $n \times r_2$ matrix $\mathbf{X}_2^{\perp}$ would span $\mathscr{C}(\mathbf{M}_2)$, $\mathrm{r}(\mathbf{M}_2) = r_2$.

(d) *Linear model with new observations, $\mathscr{M}_*$.* Let $\mathbf{y}_*$ denote a $q \times 1$ unobservable random vector containing new observations. The new observations are assumed to be generated from

$$\mathbf{y}_* = \mathbf{X}_*\boldsymbol{\beta} + \boldsymbol{\varepsilon}_*\,, \tag{11.3}$$

where $\mathbf{X}_*$ is a known $q \times p$ matrix, $\boldsymbol{\beta}$ is the same vector of fixed but unknown parameters as in $\mathscr{M}$, and $\boldsymbol{\varepsilon}_*$ is a q-dimensional random error vector. We further assume that

$$\mathrm{E}\begin{pmatrix}\mathbf{y}\\ \mathbf{y}_*\end{pmatrix} = \begin{pmatrix}\mathbf{X}\boldsymbol{\beta}\\ \mathbf{X}_*\boldsymbol{\beta}\end{pmatrix} = \begin{pmatrix}\mathbf{X}\\ \mathbf{X}_*\end{pmatrix}\boldsymbol{\beta}, \quad \mathrm{Cov}\begin{pmatrix}\mathbf{y}\\ \mathbf{y}_*\end{pmatrix} = \begin{pmatrix}\mathbf{V} & \mathbf{V}_{12}\\ \mathbf{V}_{21} & \mathbf{V}_{22}\end{pmatrix} = \boldsymbol{\Gamma},$$

where $\boldsymbol{\Gamma} \in \mathrm{NND}_{n+q}$ is known. We denote this setup shortly as

$$\mathscr{M}_* = \left\{ \begin{pmatrix}\mathbf{y}\\ \mathbf{y}_*\end{pmatrix}, \begin{pmatrix}\mathbf{X}\\ \mathbf{X}_*\end{pmatrix}\boldsymbol{\beta}, \begin{pmatrix}\mathbf{V} & \mathbf{V}_{12}\\ \mathbf{V}_{21} & \mathbf{V}_{22}\end{pmatrix} \right\}. \tag{11.4}$$

We call $\mathscr{M}_*$ "the linear model with new observations." Of course, the word "new" need not be taken here literally. Our main interest in $\mathscr{M}_*$ lies in predicting $\mathbf{y}_*$ on the basis of observable $\mathbf{y}$, but we will also be interested in predicting $\boldsymbol{\varepsilon}_*$. Notice the key role of the (cross-)covariance matrix $\mathrm{Cov}(\mathbf{y}, \mathbf{y}_*) = \mathbf{V}_{12} \in \mathbb{R}^{n\times q}$.

(e) *Transformed linear model with new observations,* $\mathscr{T}_*$. Suppose we transform $\mathscr{M}$ into $\mathscr{T}$ and do the prediction of the new observations with the "help" of $\mathbf{Fy}$. Corresponding to $\mathscr{M}_*$, we have now the following setup:

$$\mathscr{T}_* = \left\{ \begin{pmatrix}\mathbf{Fy}\\ \mathbf{y}_*\end{pmatrix}, \begin{pmatrix}\mathbf{FX}\\ \mathbf{X}_*\end{pmatrix}\boldsymbol{\beta}, \begin{pmatrix}\mathbf{FVF}' & \mathbf{FV}_{12}\\ \mathbf{V}_{21}\mathbf{F}' & \mathbf{V}_{22}\end{pmatrix} \right\}.$$

(f) *Mixed linear model,* $\mathscr{L}$. One application of the model $\mathscr{M}_*$ is the linear mixed model

$$\mathbf{y} = \mathbf{X}\boldsymbol{\beta} + \mathbf{Zu} + \mathbf{e}, \quad \text{or shortly,} \quad \mathscr{L} = \{\mathbf{y}, \mathbf{X}\boldsymbol{\beta} + \mathbf{Zu}, \mathbf{D}, \mathbf{R}, \mathbf{S}\},$$

where $\mathbf{X}_{n\times p}$ and $\mathbf{Z}_{n\times q}$ are known matrices, $\boldsymbol{\beta} \in \mathbb{R}^p$ is a vector of unknown fixed effects, $\mathbf{u}$ is an unobservable vector (q elements) of random effects with $\mathrm{E}(\mathbf{u}) = \mathbf{0}$, $\mathrm{Cov}(\mathbf{u}) = \mathbf{D}$, $\mathrm{Cov}(\mathbf{e}, \mathbf{u}) = \mathbf{S}$, and $\mathrm{E}(\mathbf{e}) = \mathbf{0}$, $\mathrm{Cov}(\mathbf{e}) = \mathbf{R}$. (Often in applications $\mathbf{S} = \mathbf{0}$.) In this situation, we have

$$\mathrm{Cov}\begin{pmatrix}\mathbf{e}\\ \mathbf{u}\end{pmatrix} = \begin{pmatrix}\mathbf{R} & \mathbf{S}\\ \mathbf{S}' & \mathbf{D}\end{pmatrix}, \quad \mathrm{Cov}\begin{pmatrix}\mathbf{y}\\ \mathbf{u}\end{pmatrix} = \begin{pmatrix}\boldsymbol{\Sigma} & \mathbf{ZD} + \mathbf{S}\\ (\mathbf{ZD} + \mathbf{S})' & \mathbf{D}\end{pmatrix},$$

$$\boldsymbol{\Sigma} = \mathrm{Cov}(\mathbf{y}) = \mathrm{Cov}(\mathbf{Zu} + \mathbf{e}) = \mathbf{ZDZ}' + \mathbf{R} + \mathbf{ZS}' + \mathbf{SZ}'.$$

In Sect. 11.9, it is shown how the mixed model can be expressed as a version of the model with new observations.

Our matrix expressions will use generalized inverses heavily and in this context it is essential to know whether the expressions are independent of the choice of the generalized inverses involved. Lemma 2.2.4 of Rao and Mitra [76] gives the condition under which the matrix product $\mathbf{AB}^-\mathbf{C}$ is invariant with respect to the choice of $\mathbf{B}^-$.

Lemma 11.1 *For nonnull matrices* $\mathbf{A}$ *and* $\mathbf{C}$ *the following holds:*

(a) $\mathbf{AB}^{-}\mathbf{C} = \mathbf{AB}^{+}\mathbf{C}$ *for all* $\mathbf{B}^{-} \iff \mathscr{C}(\mathbf{C}) \subset \mathscr{C}(\mathbf{B})$ *and* $\mathscr{C}(\mathbf{A}') \subset \mathscr{C}(\mathbf{B}')$,
(b) $\mathbf{AA}^{-}\mathbf{C} = \mathbf{C}$ *for some (and hence for all)* $\mathbf{A}^{-} \iff \mathscr{C}(\mathbf{C}) \subset \mathscr{C}(\mathbf{A})$.

Suppose that the matrix equation

$$\mathbf{YB} = \mathbf{A} \tag{11.5}$$

is solvable for $\mathbf{Y}$, i.e., $\mathscr{C}(\mathbf{A}') \subset \mathscr{C}(\mathbf{B}')$. Then it is well known, see, e.g., Rao and Mitra [76, p. 24] and Ben-Israel and Greville [16, p. 52], that the general solution $\mathbf{Y}_0$ to (11.5) can be written as

$$\mathbf{Y}_0 = \{\text{one solution to (11.5)}\} + \mathbf{E}_0\mathbf{Q}_{\mathbf{B}}\,, \quad \text{where } \mathbf{E}_0 \text{ is free to vary.}$$

There is one special class of matrices worth particular attention and that is the set $\mathscr{W}$ of nonnegative definite matrices defined as

$$\mathscr{W} = \left\{\mathbf{W} \in \mathbb{R}^{n\times n} : \mathbf{W} = \mathbf{V} + \mathbf{XUU}'\mathbf{X}',\ \mathscr{C}(\mathbf{W}) = \mathscr{C}(\mathbf{X} : \mathbf{V})\right\}. \tag{11.6}$$

In (11.6) $\mathbf{U}$ can be any matrix comprising p rows as long as $\mathscr{C}(\mathbf{W}) = \mathscr{C}(\mathbf{X} : \mathbf{V})$ is satisfied. One obvious choice is $\mathbf{U} = \mathbf{I}_p$. In particular, if $\mathscr{C}(\mathbf{X}) \subset \mathscr{C}(\mathbf{V})$, we can choose $\mathbf{U} = \mathbf{0}$. Occasionally, we may use the notation $\mathscr{W}_{\mathscr{A}}$ to indicate that the model $\mathscr{A}$, say, is under consideration. We will also use the phrase "$\mathbf{W}_{\mathscr{A}}$ is a $\mathbf{W}$-matrix under the model $\mathscr{A}$." The set $\mathscr{W}$ in (11.6) is, of course, $\mathscr{W}_{\mathscr{M}}$. The following result is easy to confirm:

$$\mathbf{W} \in \mathscr{W}_{\mathscr{M}} \implies \mathbf{FWF}' \in \mathscr{W}_{\mathscr{T}}\,. \tag{11.7}$$

We will later utilize some particular properties of $\mathscr{W}$ and of the corresponding extended set

$$\mathscr{W}^{\#} = \left\{\mathbf{W} \in \mathbb{R}^{n\times n} : \mathbf{W} = \mathbf{V} + \mathbf{XUX}',\ \mathscr{C}(\mathbf{W}) = \mathscr{C}(\mathbf{X} : \mathbf{V})\right\}.$$

Above $\mathbf{U} \in \mathbb{R}^{p\times p}$ is free to vary subject to condition $\mathscr{C}(\mathbf{W}) = \mathscr{C}(\mathbf{X} : \mathbf{V})$. Notice that $\mathbf{W}$ belonging to $\mathscr{W}^{\#}$ is not necessarily nonnegative definite and it can be nonsymmetric. Lemma 11.2 collects together some important properties of the class $\mathscr{W}^{\#}$.

Lemma 11.2 *Let* $\mathbf{V}$ *be an* $n \times n$ *nonnegative definite matrix, let* $\mathbf{X}$ *be an* $n \times p$ *matrix, and define* $\mathbf{W}$ *as* $\mathbf{W} = \mathbf{V} + \mathbf{XUX}'$, *where* $\mathbf{U}$ *is a* $p \times p$ *matrix. Then the following statements are equivalent:*

(a) $\mathscr{C}(\mathbf{X} : \mathbf{V}) = \mathscr{C}(\mathbf{W})$,
(b) $\mathscr{C}(\mathbf{X}) \subset \mathscr{C}(\mathbf{W})$,
(c) $\mathbf{X}'\mathbf{W}^{-}\mathbf{X}$ *is invariant for any choice of* $\mathbf{W}^{-}$,
(d) $\mathscr{C}(\mathbf{X}'\mathbf{W}^{-}\mathbf{X}) = \mathscr{C}(\mathbf{X}')$ *for any choice of* $\mathbf{W}^{-}$,

(e) $\mathbf{X}(\mathbf{X}'\mathbf{W}^-\mathbf{X})^-\mathbf{X}'\mathbf{W}^-\mathbf{X} = \mathbf{X}$ *for any choices of* $\mathbf{W}^-$ *and* $(\mathbf{X}'\mathbf{W}^-\mathbf{X})^-$.

Moreover, each of these statements is equivalent also to $\mathscr{C}(\mathbf{X} : \mathbf{V}) = \mathscr{C}(\mathbf{W}')$, and hence to the statements (b)–(e) by replacing $\mathbf{W}$ with $\mathbf{W}'$. Observe that obviously $\mathscr{C}(\mathbf{W}) = \mathscr{C}(\mathbf{W}')$ and that the invariance properties in (d) and (e) concern also the choice of $\mathbf{W} \in \mathscr{W}^{\#}$. For further properties of $\mathscr{W}^{\#}$, see, e.g., Baksalary et al. [13, Th. 2], Baksalary and Mathew [12, Th. 2], Harville [36, p. 468], and Puntanen et al. [72, Sect. 12.3].

For the following lemma, see, e.g., Isotalo et al. [48], Puntanen et al. [72, Prop. 15.2], and Markiewicz and Puntanen [64, Sect. 4].

Lemma 11.3 *Consider the partitioned linear model* $\mathscr{M} = \{\mathbf{y}, (\mathbf{X}_1 : \mathbf{X}_2)\boldsymbol{\beta}, \mathbf{V}\}$, *let* $\mathbf{W} = \mathbf{V} + \mathbf{X}\mathbf{U}\mathbf{U}'\mathbf{X}' \in \mathscr{W}$ *and denote* $\mathbf{M}_2 = \mathbf{I}_n - \mathbf{P}_{\mathbf{X}_2}$ *and*

$$\dot{\mathbf{M}} = \mathbf{M}(\mathbf{M}\mathbf{V}\mathbf{M})^-\mathbf{M}, \quad \dot{\mathbf{M}}_{2W} = \mathbf{M}_2(\mathbf{M}_2\mathbf{W}\mathbf{M}_2)^-\mathbf{M}_2.$$

Then the following equalities hold:

(a) $\mathbf{X}(\mathbf{X}'\mathbf{W}^-\mathbf{X})^-\mathbf{X}'\mathbf{W}^+ = \mathbf{P}_\mathbf{W} - \mathbf{V}\mathbf{M}(\mathbf{M}\mathbf{V}\mathbf{M})^-\mathbf{M}\mathbf{P}_\mathbf{W} = \mathbf{P}_\mathbf{W} - \mathbf{V}\dot{\mathbf{M}}\mathbf{P}_\mathbf{W}$,
(b) $\mathbf{X}(\mathbf{X}'\mathbf{W}^-\mathbf{X})^-\mathbf{X}' = \mathbf{W} - \mathbf{W}\mathbf{M}(\mathbf{M}\mathbf{V}\mathbf{M})^-\mathbf{M}\mathbf{W} = \mathbf{V} - \mathbf{V}\dot{\mathbf{M}}\mathbf{V} + \mathbf{X}\mathbf{U}\mathbf{U}'\mathbf{X}'$,
(c) $\mathbf{X}_2(\mathbf{X}_2'\mathbf{W}^-\mathbf{X}_2)^-\mathbf{X}_2'\mathbf{W}^+ = \mathbf{P}_\mathbf{W} - \mathbf{W}\mathbf{M}_2(\mathbf{M}_2\mathbf{W}\mathbf{M}_2)^-\mathbf{M}_2\mathbf{P}_\mathbf{W} = \mathbf{P}_\mathbf{W} - \mathbf{W}\dot{\mathbf{M}}_{2\mathbf{W}}\mathbf{P}_\mathbf{W}$,
(d) $\mathbf{X}_2(\mathbf{X}_2'\mathbf{W}^-\mathbf{X}_2)^-\mathbf{X}_2' = \mathbf{W} - \mathbf{W}\mathbf{M}_2(\mathbf{M}_2\mathbf{W}\mathbf{M}_2)^-\mathbf{M}_2\mathbf{W}$,
(e) $\mathbf{W}\dot{\mathbf{M}}_{2\mathbf{W}}\mathbf{X}_1 = [\mathbf{I}_n - \mathbf{X}_2(\mathbf{X}_2'\mathbf{W}^-\mathbf{X}_2)^-\mathbf{X}_2'\mathbf{W}^-]\mathbf{X}_1$.

As noted by Isotalo et al. [48, p. 1439], the matrix $\dot{\mathbf{M}} = \mathbf{M}(\mathbf{M}\mathbf{V}\mathbf{M})^-\mathbf{M}$ is not necessarily unique with respect to the choice of the generalized inverse $(\mathbf{M}\mathbf{V}\mathbf{M})^-$. It is unique if and only if $\mathbb{R}^n = \mathscr{C}(\mathbf{X} : \mathbf{V})$. However, for example, $\mathbf{V}\dot{\mathbf{M}}\mathbf{P}_\mathbf{W}$ is unique. It is noteworthy that using the Moore–Penrose inverse the following holds:

$$\mathbf{M}(\mathbf{M}\mathbf{V}\mathbf{M})^+\mathbf{M} = (\mathbf{M}\mathbf{V}\mathbf{M})^+\mathbf{M} = \mathbf{M}(\mathbf{M}\mathbf{V}\mathbf{M})^+ = (\mathbf{M}\mathbf{V}\mathbf{M})^+. \tag{11.8}$$

If $\mathbf{V}$ is positive definite and $\mathbf{Z} \in \{\mathbf{X}^\perp\}$, then

$$\dot{\mathbf{M}} = \mathbf{M}(\mathbf{M}\mathbf{V}\mathbf{M})^-\mathbf{M} = \mathbf{Z}(\mathbf{Z}'\mathbf{V}\mathbf{Z})^-\mathbf{Z}' = \mathbf{V}^{-1} - \mathbf{V}^{-1}\mathbf{X}(\mathbf{X}'\mathbf{V}^{-1}\mathbf{X})^-\mathbf{X}'\mathbf{V}^{-1}.$$

The following result concerning the matrix $\mathbf{F}'\mathbf{Q}_{\mathbf{FX}}$, where $\mathbf{Q}_{\mathbf{FX}} = \mathbf{I}_f - \mathbf{P}_{\mathbf{FX}}$, is useful for our considerations. For the proof, see Rao and Mitra [76, Complement 7, p. 118]. For related results, see also Markiewicz and Puntanen [61, 64, Sect. 2, Lemma 3].

Lemma 11.4 *Suppose that* $\mathbf{F}$ *is an* $f \times n$ *matrix. Then*

$$\mathscr{C}(\mathbf{F}'\mathbf{Q}_{\mathbf{FX}}) = \mathscr{C}(\mathbf{F}') \cap \mathscr{C}(\mathbf{M}),$$

and denoting $\mathbf{N} = \mathbf{P}_{\mathbf{F}'\mathbf{Q}_{\mathbf{FX}}} = \mathbf{P}_{\mathscr{C}(\mathbf{F}')\cap\mathscr{C}(\mathbf{M})}$, *we have*

$$\mathbf{M}\mathbf{F}'\mathbf{Q}_{\mathbf{FX}} = \mathbf{F}'\mathbf{Q}_{\mathbf{FX}}, \quad \mathbf{N} = \mathbf{M}\mathbf{N} = \mathbf{N}\mathbf{M}.$$

We assume the model $\mathscr{M}$ to be consistent in the sense that $\mathbf{y}$ lies in $\mathscr{C}(\mathbf{X} : \mathbf{V})$ with probability 1, see, e.g., Baksalary et al. [14], Groß [31]. Hence, we assume that under the model $\mathscr{M}$ the observed numerical value of $\mathbf{y}$ satisfies

$$\mathbf{y} \in \mathscr{C}(\mathbf{X} : \mathbf{V}) = \mathscr{C}(\mathbf{X} : \mathbf{V}\mathbf{X}^{\perp}) = \mathscr{C}(\mathbf{X} : \mathbf{V}\mathbf{M}) = \mathscr{C}(\mathbf{X}) \oplus \mathscr{C}(\mathbf{V}\mathbf{M}),$$

where "$\oplus$" refers to the direct sum, implying that

$$\mathscr{C}(\mathbf{X}) \cap \mathscr{C}(\mathbf{V}\mathbf{X}^{\perp}) = \{\mathbf{0}\}. \tag{11.9}$$

For the equality $\mathscr{C}(\mathbf{X} : \mathbf{V}) = \mathscr{C}(\mathbf{X} : \mathbf{V}\mathbf{M})$, we refer to Rao [75, Lemma 2.1]. There is a related decomposition, see, e.g., Puntanen et al. [72, Th. 8], that is worth mentioning in this context: for any conformable matrices $\mathbf{A}$ and $\mathbf{B}$ we have

$$\mathscr{C}(\mathbf{A} : \mathbf{B}) = \mathscr{C}(\mathbf{A} : \mathbf{Q}_{\mathbf{A}}\mathbf{B}), \text{ and thereby } \mathbf{P}_{(\mathbf{A}:\mathbf{B})} = \mathbf{P}_{\mathbf{A}} + \mathbf{P}_{\mathbf{Q}_{\mathbf{A}}\mathbf{B}}.$$

Thus, we can obtain the following useful results for the partitioned linear model; for the part (b) in Lemma 11.5, see the rank rule of the matrix product of Marsaglia and Styan [66, Corollary 6.2].

Lemma 11.5 *Consider* $\mathbf{X} = (\mathbf{X}_1 : \mathbf{X}_2)$ *and let* $\mathbf{M}_2 = \mathbf{I}_n - \mathbf{P}_{\mathbf{X}_2}$. *Then*

(a) $\mathbf{M} = \mathbf{I}_n - \mathbf{P}_{(\mathbf{X}_1:\mathbf{X}_2)} = \mathbf{I}_n - (\mathbf{P}_{\mathbf{X}_2} + \mathbf{P}_{\mathbf{M}_2\mathbf{X}_1}) = \mathbf{M}_2\mathbf{Q}_{\mathbf{M}_2\mathbf{X}_1} = \mathbf{Q}_{\mathbf{M}_2\mathbf{X}_1}\mathbf{M}_2$,
(b) $\mathrm{r}(\mathbf{M}_2\mathbf{X}_1) = \mathrm{r}(\mathbf{X}_1) - \dim \mathscr{C}(\mathbf{X}_1) \cap \mathscr{C}(\mathbf{X}_2)$.

Let $\mathbf{A}$ and $\mathbf{B}$ be arbitrary $m \times n$ matrices. Then, in the consistent linear model $\mathscr{M}$, the estimators $\mathbf{Ay}$ and $\mathbf{By}$ are said to be equal with probability 1 if

$$\mathbf{Ay} = \mathbf{By} \quad \text{for all } \mathbf{y} \in \mathscr{C}(\mathbf{X} : \mathbf{V}) = \mathscr{C}(\mathbf{W}), \tag{11.10}$$

where $\mathbf{W} \in \mathscr{W}$. Thus, if $\mathbf{A}$ and $\mathbf{B}$ satisfy (11.10), then $\mathbf{A} - \mathbf{B} = \mathbf{C}\mathbf{Q}_{\mathbf{W}}$ for some matrix $\mathbf{C}$. When talking about the equality of estimators like $\mathbf{Ay} = \mathbf{By}$, we often drop off the phrase "with probability 1."

The following lemma collects together some equivalent expressions for (11.10). For part (d) of Lemma 11.6, see Groß and Trenkler [34, Th. 1].

Lemma 11.6 *Let* $\mathbf{A}$ *and* $\mathbf{B}$ *be* $m \times n$ *matrices. Then under the model* $\mathscr{M}$ *the identity* $\mathbf{Ay} = \mathbf{By}$ *holds with probability* 1 *if and only if any of the following equivalent conditions hold:*

(a) $\mathbf{AX} = \mathbf{BX}$ *and* $\mathbf{AV} = \mathbf{BV}$,
(b) $\mathbf{AX} = \mathbf{BX}$ *and* $\mathbf{AVM} = \mathbf{BVM}$,
(c) $\mathbf{AX} = \mathbf{BX}$ *and* $\mathrm{Cov}(\mathbf{Ay} - \mathbf{By}) = \mathbf{0}$,
(d) $\mathbf{AX} = \mathbf{BX}$, $\mathrm{Cov}(\mathbf{Ay}) = \mathrm{Cov}(\mathbf{By})$, $2\mathrm{Cov}(\mathbf{Ay}) = \mathrm{Cov}(\mathbf{Ay}, \mathbf{By}) + \mathrm{Cov}(\mathbf{By}, \mathbf{Ay})$.

One more notational matter is worth mentioning. Let $\mathbf{A}$ be a nonnegative definite $n \times n$ matrix with $\mathrm{r}(\mathbf{A}) = r$ and let $\mathbf{A} = \mathbf{Q}\boldsymbol{\Lambda}\mathbf{Q}'$ be its eigenvalue decomposition in terms of nonzero eigenvalues; here $\boldsymbol{\Lambda}$ is an $r \times r$ diagonal matrix of the nonzero eigenvalues of $\mathbf{A}$. Then $\mathbf{A}^{1/2}$ refers to the nonnegative definite square root of $\mathbf{A}$, i.e., $\mathbf{A}^{1/2} = \mathbf{Q}\boldsymbol{\Lambda}^{1/2}\mathbf{Q}'$. Notation $\mathbf{A}^{+1/2} = \mathbf{Q}\boldsymbol{\Lambda}^{-1/2}\mathbf{Q}'$ refers to $(\mathbf{A}^{+})^{1/2}$ or equivalently $(\mathbf{A}^{1/2})^{+}$. Notice that $\mathbf{A}^{1/2}\mathbf{A}^{+1/2} = \mathbf{Q}\mathbf{Q}' = \mathbf{A}\mathbf{A}^{+} = \mathbf{P}_{\mathbf{A}}$.

As regards the structure of this paper, below is an outline. As a review paper, we do not provide many proofs; instead, our goal is to explain and clarify the relevant central results.

- Section 11.2: We go through some basic properties of the best linear unbiased estimators and predictors, BLUPs and BLUEs, and introduce the well-known fundamental BLUP and BLUE equations. When dealing with linear predictors, our aim is to predict $\mathbf{y}_*$ and $\boldsymbol{\varepsilon}_*$, and when considering linear estimators, we are interested in estimating $\boldsymbol{\mu} = \mathbf{X}\boldsymbol{\beta}$ or some estimable parametric function $\boldsymbol{\mu}_* = \mathbf{X}_*\boldsymbol{\beta}$.
- Section 11.3: We recall some known conditions for the linear sufficiency. When dealing with the predictors, we may sometimes use the term "linearly prediction sufficient" instead of the phrase "linearly sufficient."
- Section 11.4: We consider the transformed model $\mathscr{T} = \{\mathbf{F}\mathbf{y}, \mathbf{F}\mathbf{X}\boldsymbol{\beta}, \mathbf{F}\mathbf{V}\mathbf{F}'\}$ and introduce various properties of the linear sufficiency using this transformed model. Explicit expressions for the BLUEs and BLUPs under the original and under the transformed model are given.
- Section 11.5: We shortly discuss the concept of relative linear sufficiency, introduced by Kala et al. [52, Sect. 5].
- Section 11.6: We explore the coincidence of the multipliers of $\mathbf{y}$ providing the BLUEs for $\mathbf{X}_*\boldsymbol{\beta}$ under $\mathscr{M}$ and under $\mathscr{T}$. If every representation of the BLUE of $\mathbf{X}_*\boldsymbol{\beta}$ under $\mathscr{T}$ is BLUE also under $\mathscr{M}$, we can denote
$$\{\mathrm{BLUE}(\mathbf{X}_*\boldsymbol{\beta}) \mid \mathscr{T}\} \subset \{\mathrm{BLUE}(\mathbf{X}_*\boldsymbol{\beta}) \mid \mathscr{M}\}. \tag{11.11}$$
Proposition 11.17 provides a new interesting characterization for the equality in (11.11).
- Section 11.7: We consider, in the spirit of Markiewicz and Puntanen [64], a partitioned linear model $\mathscr{M}_{12} = \{\mathbf{y}, \mathbf{X}_1\boldsymbol{\beta}_1 + \mathbf{X}_2\boldsymbol{\beta}_2, \mathbf{V}\}$. Particular attention is paid on the situation when the transformation matrix is $\mathbf{M}_2 = \mathbf{I}_n - \mathbf{P}_{\mathbf{X}_2}$, so that the transformed model is $\mathscr{M}_{12\cdot 2} = \{\mathbf{M}_2\mathbf{y}, \mathbf{M}_2\mathbf{X}_1\boldsymbol{\beta}_1, \mathbf{M}_2\mathbf{V}\mathbf{M}_2\}$.
- Section 11.8: We explore the mutual relations of linear sufficiencies. In addition, we go through some interesting connections between the covariance matrices of the BLUPs and the linear sufficiencies. We also comment on the upper bounds of the Euclidean distance between the BLUPs when the prediction is based on the original model $\mathscr{M}$ and when it is based on the transformed model $\mathscr{T}$.
- Section 11.9: In this section, we consider the linear mixed model in the spirit of Isotalo et al. [44] and Haslett et al. [38]. The key feature here is the fact that linear mixed model can be interpreted as a special case of the model with new observations.

- Section 11.10: We give necessary and sufficient conditions that $\mathbf{Fy}$ continues to be linearly sufficient for $\mathbf{X}_*\boldsymbol{\beta}$ (or $\boldsymbol{\varepsilon}_*$) under the misspecified model $\underline{\mathscr{M}}$. The misspecification concerns the covariance part of the setup.

11.2 BLUEs and BLUPs

In this section, we go through some basic properties of the best linear unbiased estimators and predictors in the frames of the general linear model. We will talk about estimation (and estimators) when our main focus is in the fixed parametric function, $\mathbf{X}_*\boldsymbol{\beta}$, say, and about prediction (and predictors), when the main interest lies in predicting a random vector like $\mathbf{y}_*$ which is believed to be generated by $\mathbf{y}_* = \mathbf{X}_*\boldsymbol{\beta} + \boldsymbol{\varepsilon}_*$ in the frames of the model $\mathscr{M}_*$. Not everybody agrees with this division between estimators and predictors, see Robinson [77, Sect. 1].

A linear statistic $\mathbf{By}$ is said to be linear unbiased estimator (LUE) for $\boldsymbol{\mu}_* = \mathbf{X}_*\boldsymbol{\beta}$ in $\mathscr{M}$ if its expectation is equal to $\boldsymbol{\mu}_*$, i.e.,

$$\mathrm{E}\,(\mathbf{By} - \boldsymbol{\mu}_*) = (\mathbf{BX} - \mathbf{X}_*)\boldsymbol{\beta} = \mathbf{0} \quad \text{for all } \boldsymbol{\beta} \in \mathbb{R}^p,$$

which happens if and only if $\mathbf{X}_*' = \mathbf{X}'\mathbf{B}'$. When $\mathscr{C}(\mathbf{X}_*') \subset \mathscr{C}(\mathbf{X}')$ holds, the linear parametric function $\boldsymbol{\mu}_* = \mathbf{X}_*\boldsymbol{\beta}$ is said to be estimable in $\mathscr{M}$. The LUE $\mathbf{By}$ is the best linear unbiased estimator, BLUE, of estimable $\mathbf{X}_*\boldsymbol{\beta}$ if $\mathbf{By}$ has the smallest covariance matrix in the Löwner sense among all LUEs of $\mathbf{X}_*\boldsymbol{\beta}$:

$$\mathrm{Cov}(\mathbf{By}) \leq_{\mathrm{L}} \mathrm{Cov}(\mathbf{B}_{\#}\mathbf{y}) \quad \text{for all } \mathbf{B}_{\#} : \mathbf{B}_{\#}\mathbf{X} = \mathbf{X}_*. \tag{11.12}$$

Correspondingly, the linear predictor $\mathbf{Ay}$ is said to be unbiased for $\mathbf{y}_*$ if the expected prediction error is zero, i.e., $\mathrm{E}\,(\mathbf{y}_* - \mathbf{Ay}) = \mathbf{0}$ for all $\boldsymbol{\beta} \in \mathbb{R}^p$, which happens if and only if $\mathbf{X}_*' = \mathbf{X}'\mathbf{A}'$. When $\mathscr{C}(\mathbf{X}_*') \subset \mathscr{C}(\mathbf{X}')$ holds, we will say that $\mathbf{y}_*$ is predictable under $\mathscr{M}_*$, that is, $\mathbf{y}_*$ is predictable whenever $\mathbf{X}_*\boldsymbol{\beta}$ is estimable. Now a linear unbiased predictor $\mathbf{Ay}$ is the best linear unbiased predictor, BLUP, for $\mathbf{y}_*$, if we have the Löwner ordering

$$\mathrm{Cov}(\mathbf{y}_* - \mathbf{Ay}) \leq_{\mathrm{L}} \mathrm{Cov}(\mathbf{y}_* - \mathbf{A}_{\#}\mathbf{y}) \quad \text{for all } \mathbf{A}_{\#} : \mathbf{A}_{\#}\mathbf{X} = \mathbf{X}_*. \tag{11.13}$$

Notice that

- in (11.12) we minimize the covariance matrix of the estimator subject to the unbiasedness of the estimation, while
- in (11.13) we are minimizing the covariance matrix of the prediction error subject to the unbiasedness of the prediction.

Consider then the BLUP of $\boldsymbol{\varepsilon}_*$. Obviously $\mathbf{Dy}$ is an unbiased predictor for $\boldsymbol{\varepsilon}_*$ if and only if $\mathbf{DX} = \mathbf{0}$, i.e., $\mathbf{D} = \mathbf{LM}$ for some $\mathbf{L}$. Thus, the unbiased $\mathbf{Dy}$ is the BLUP for $\boldsymbol{\varepsilon}_*$ if and only if

$$\mathrm{Cov}(\boldsymbol{\varepsilon}_* - \mathbf{D}\mathbf{y}) \leq_{\mathrm{L}} \mathrm{Cov}(\boldsymbol{\varepsilon}_* - \mathbf{D}_{\#}\mathbf{y}) \quad \text{for all } \mathbf{D}_{\#} : \mathbf{D}_{\#}\mathbf{X} = \mathbf{0},$$

or equivalently,

$$\mathrm{Cov}(\boldsymbol{\varepsilon}_* - \mathbf{D}\mathbf{y}) \leq_{\mathrm{L}} \mathrm{Cov}(\boldsymbol{\varepsilon}_* - \mathbf{L}\mathbf{M}\mathbf{y}) \quad \text{for all } \mathbf{L} \in \mathbb{R}^{q \times n}. \tag{11.14}$$

For Proposition 11.1, characterizing the BLUE, see, e.g., Rao [74, p. 282], and Drygas [25, p. 55], Kala [50, Th. 3.1], and the BLUP, see, e.g., Christensen [22, p. 294], and Isotalo and Puntanen [46, p. 1015]. For part (c), see Isotalo et al. [44, Th. 3.1]. For the general reviews of the BLUP-properties, see, e.g., Robinson [77], Searle [79], Tian [82, 83] and Haslett and Puntanen [39].

Proposition 11.1 *Consider the linear model with new observations defined as $\mathscr{M}_*$ in (11.4), where $\mathscr{C}(\mathbf{X}_*') \subset \mathscr{C}(\mathbf{X}')$, i.e., $\mathbf{y}_*$ is predictable.*

(a) *The linear predictor* $\mathbf{Ay}$ *is the* BLUP *for* $\mathbf{y}_*$ *if and only if*

$$\mathbf{A}(\mathbf{X} : \mathbf{V}\mathbf{X}^{\perp}) = (\mathbf{X}_* : \mathbf{V}_{21}\mathbf{X}^{\perp}). \tag{11.15}$$

(b) *The linear estimator* $\mathbf{By}$ *is the* BLUE *of* $\boldsymbol{\mu}_* = \mathbf{X}_*\boldsymbol{\beta}$ *if and only if*

$$\mathbf{B}(\mathbf{X} : \mathbf{V}\mathbf{X}^{\perp}) = (\mathbf{X}_* : \mathbf{0}). \tag{11.16}$$

In particular, $\mathbf{Cy}$ *is the* BLUE *for* $\boldsymbol{\mu} = \mathbf{X}\boldsymbol{\beta}$ *if and only if*

$$\mathbf{C}(\mathbf{X} : \mathbf{V}\mathbf{X}^{\perp}) = (\mathbf{X} : \mathbf{0}). \tag{11.17}$$

(c) *The linear predictor* $\mathbf{Dy}$ *is the* BLUP *for* $\boldsymbol{\varepsilon}_*$ *if and only if*

$$\mathbf{D}(\mathbf{X} : \mathbf{V}\mathbf{X}^{\perp}) = (\mathbf{0} : \mathbf{V}_{21}\mathbf{X}^{\perp}). \tag{11.18}$$

Equations (11.15) and (11.16) are sometimes called the fundamental BLUP and BLUE equations, respectively. It is noteworthy that equations (11.15) and (11.16) are solvable for $\mathbf{A}$ and $\mathbf{B}$, respectively, if and only if $\mathbf{X}_*\boldsymbol{\beta}$ is estimable while (11.17) and (11.18) are always solvable for $\mathbf{C}$ and $\mathbf{D}$, respectively. If $\mathbf{V}_{12} = \mathbf{0}$, then (11.15) and (11.16) become the same and the $\mathrm{BLUP}(\mathbf{y}_*)$ is the $\mathrm{BLUE}(\mathbf{X}_*\boldsymbol{\beta})$ and the $\mathrm{BLUP}(\boldsymbol{\varepsilon}_*) = \mathbf{0}$.

Putting (11.16) and (11.18) together yields

$$\begin{pmatrix} \mathbf{B} \\ \mathbf{D} \end{pmatrix} (\mathbf{X} : \mathbf{V}\mathbf{X}^{\perp}) = \begin{pmatrix} \mathbf{X}_* & \mathbf{0} \\ \mathbf{0} & \mathbf{V}_{21}\mathbf{X}^{\perp} \end{pmatrix},$$

which implies that

$$(\mathbf{B} + \mathbf{D})(\mathbf{X} : \mathbf{V}\mathbf{X}^{\perp}) = (\mathbf{X}_* : \mathbf{V}_{21}\mathbf{X}^{\perp}),$$

and thereby $(\mathbf{B}+\mathbf{D})\mathbf{y}$ is a BLUP for $\mathbf{y}_*$ and then we have the following result, see, e.g., Isotalo et al. [44, Sect. 3].

Proposition 11.2 *Under the linear model $\mathscr{M}_*$, where $\mathbf{y}_*$ is predictable, the following decomposition holds (with probability 1):*

$$\mathrm{BLUP}(\mathbf{y}_*) = \mathrm{BLUE}(\mathbf{X}_*\boldsymbol{\beta}) + \mathrm{BLUP}(\boldsymbol{\varepsilon}_*), \quad \textit{or shortly,}\ \widetilde{\mathbf{y}}_* = \widetilde{\boldsymbol{\mu}}_* + \widetilde{\boldsymbol{\varepsilon}}_*.$$

Let us define the sets $\{\mathbf{P}_{\mathbf{y}_*|\mathscr{M}}\}$, $\{\mathbf{P}_{\mathbf{X}_*|\mathscr{M}}\}$, and $\{\mathbf{P}_{\boldsymbol{\varepsilon}_*|\mathscr{M}}\}$ as follows:

$$\begin{aligned}
\mathbf{A} \in \{\mathbf{P}_{\mathbf{y}_*|\mathscr{M}_*}\} &\iff \mathbf{A}(\mathbf{X} : \mathbf{VM}) = (\mathbf{X}_* : \mathbf{V}_{21}\mathbf{X}^{\perp}),\\
\mathbf{B} \in \{\mathbf{P}_{\mathbf{X}_*|\mathscr{M}}\} &\iff \mathbf{B}(\mathbf{X} : \mathbf{VM}) = (\mathbf{X}_* : \mathbf{0}),\\
\mathbf{D} \in \{\mathbf{P}_{\boldsymbol{\varepsilon}_*|\mathscr{M}_*}\} &\iff \mathbf{D}(\mathbf{X} : \mathbf{VM}) = (\mathbf{0} : \mathbf{V}_{21}\mathbf{X}^{\perp}).
\end{aligned} \tag{11.19}$$

Using Lemma 11.2 we can obtain, for example, the following well-known solutions to (11.16) and (11.17):

$$\mathbf{G}_* := \mathbf{X}_*(\mathbf{X}'\mathbf{W}^-\mathbf{X})^-\mathbf{X}'\mathbf{W}^- \in \{\mathbf{P}_{\mathbf{X}_*|\mathscr{M}}\}, \tag{11.20a}$$
$$\mathbf{G} := \mathbf{X}(\mathbf{X}'\mathbf{W}^-\mathbf{X})^-\mathbf{X}'\mathbf{W}^- \in \{\mathbf{P}_{\mathbf{X}|\mathscr{M}}\}, \tag{11.20b}$$

where $\mathbf{W} \in \mathscr{W}$ and we can freely choose the generalized inverses involved. Alternative solutions to (11.19) are, for example,

$$\mathbf{B}_1 := \mathbf{X}_*(\mathbf{X}'\mathbf{W}^-\mathbf{X})^-\mathbf{X}'\mathbf{W}^+, \quad \mathbf{B}_2 := (\mathbf{X}_* : \mathbf{0})(\mathbf{X} : \mathbf{VM})^+.$$

Notice that $\mathbf{B}_1$ is unique with respect to choice of $\mathbf{W}^-$ and $(\mathbf{X}'\mathbf{W}^-\mathbf{X})^-$. In (11.20a)–(11.20b), it is actually enough if $\mathbf{W}$ belongs to the extended set $\mathscr{W}^\#$ but for simplicity we deal here with $\mathscr{W}$. The *general* solutions for $\mathbf{B}$ and $\mathbf{C}$ in (11.17) and (11.16), respectively, can be expressed, for example, as

$$\mathbf{P}_{\mathbf{X}_*|\mathscr{M}} = \mathbf{G}_* + \mathbf{E}\mathbf{Q}_{\mathbf{W}}, \quad \mathbf{P}_{\mathbf{X}|\mathscr{M}} = \mathbf{G} + \mathbf{E}_1\mathbf{Q}_{\mathbf{W}},$$

where $\mathbf{E} \in \mathbb{R}^{q\times n}$ and $\mathbf{E}_1 \in \mathbb{R}^{n\times n}$ are free to vary and $\mathbf{Q}_{\mathbf{W}} = \mathbf{I}_n - \mathbf{P}_{\mathbf{W}}$. It is worth emphasizing that the matrix $\mathbf{G}$ in (11.20b) is a kind of extended version of the projector: it is a projector onto $\mathscr{C}(\mathbf{X})$ along $\mathscr{C}(\mathbf{VM})$. The same concerns any member of the class $\{\mathbf{P}_{\mathbf{X}|\mathscr{M}}\}$, i.e., any matrix of the form $\mathbf{G} + \mathbf{E}_1\mathbf{Q}_{\mathbf{W}}$. For generalized projectors, see, e.g., Rao [75], Kala [50], and Puntanen et al. [72, Sect. 2.5].

Under the consistency of $\mathscr{M}$, for a given $\mathbf{y} \in \mathscr{C}(\mathbf{X} : \mathbf{V})$, we can write $\mathbf{y}$ as

$$\mathbf{y} = \mathbf{Xa} + \mathbf{VMb} \quad \text{for some } \mathbf{a} \in \mathbb{R}^p \text{ and } \mathbf{b} \in \mathbb{R}^n,$$

where $\mathbf{Xa}$ and $\mathbf{VMb}$ are unique. Thus, in view of Lemma 11.2, the observed numerical value of the BLUE is uniquely $\mathbf{By} = \mathbf{X}_*\mathbf{a}$, even though $\mathbf{B} \in \{\mathbf{P}_{\mathbf{X}_*|\mathscr{M}}\}$ may not be unique; $\mathbf{B}$ is unique if and only if $\mathscr{C}(\mathbf{W}) = \mathbb{R}^n$. The properties of the BLUE deserve particular attention when $\mathscr{C}(\mathbf{W}) = \mathbb{R}^n$ does not hold: then there is an infinite number of multipliers $\mathbf{B}$ such that $\mathbf{By}$ is BLUE.

By Lemma 11.3 and taking into account the consistency of the model $\mathscr{M}$, we obtain, for $\mathbf{W} = \mathbf{V} + \mathbf{XUU'X'} \in \mathscr{W}$,

$$\widetilde{\boldsymbol{\mu}} = \mathrm{BLUE}(\mathbf{X}\boldsymbol{\beta}) = \mathbf{X}(\mathbf{X'W^-X})^-\mathbf{X'W^-y} = [\mathbf{I}_n - \mathbf{VM}(\mathbf{MVM})^-\mathbf{M}]\mathbf{y}, \quad (11.21)$$

which further gives the BLUE's residual

$$\mathbf{y} - \mathrm{BLUE}(\mathbf{X}\boldsymbol{\beta}) = \mathbf{VM}(\mathbf{MVM})^-\mathbf{My}, \quad (11.22)$$

and the covariance matrix of $\widetilde{\boldsymbol{\mu}}$, cf. part (b), Lemma 11.3,

$$\mathrm{Cov}\,(\widetilde{\boldsymbol{\mu}}) = \mathbf{X}(\mathbf{X'W^-X})^-\mathbf{X'} - \mathbf{XUU'X} = \mathbf{V} - \mathbf{VM}(\mathbf{MVM})^-\mathbf{MV}. \quad (11.23)$$

From (11.18), we observe that $\mathbf{Dy}$ is the BLUP for $\boldsymbol{\varepsilon}_*$ if $\mathbf{D} = \mathbf{LM}$ for some matrix $\mathbf{L} \in \mathbb{R}^{q\times n}$ such that $\mathbf{LMVM} = \mathbf{V}_{21}\mathbf{M}$, from which one solution to $\mathbf{L}$ is $\mathbf{L} = \mathbf{V}_{21}\mathbf{M}(\mathbf{MVM})^-$ yielding the following expression:

$$\mathrm{BLUP}(\boldsymbol{\varepsilon}_*) = \mathbf{Dy} = \mathbf{V}_{21}\mathbf{M}(\mathbf{MVM})^-\mathbf{My} = \mathbf{V}_{21}\dot{\mathbf{M}}\mathbf{y},$$

where $\dot{\mathbf{M}} = \mathbf{M}(\mathbf{MVM})^-\mathbf{M}$. By (11.21) and (11.22), we have, for example, the following further representations, see Haslett et al. [37, Th. 2]:

$$\begin{aligned}\mathrm{BLUP}(\boldsymbol{\varepsilon}_*) &= \mathbf{V}_{21}\mathbf{M}(\mathbf{MVM})^-\mathbf{My} = \mathbf{V}_{21}\mathbf{V}^-\mathbf{VM}(\mathbf{MVM})^-\mathbf{My} \\ &= \mathbf{V}_{21}\mathbf{W}^-\mathbf{WM}(\mathbf{MVM})^-\mathbf{My} = \mathbf{V}_{21}\mathbf{V}^-[\mathbf{y} - \mathrm{BLUE}(\mathbf{X}\boldsymbol{\beta})] \\ &= \mathbf{V}_{21}\mathbf{V}^-(\mathbf{I}_n - \mathbf{G})\mathbf{y} = \mathbf{V}_{21}\mathbf{W}^-(\mathbf{I}_n - \mathbf{G})\mathbf{y},\end{aligned}$$

where $\mathbf{G} = \mathbf{X}(\mathbf{X'W^-X})^-\mathbf{X'W^-}$ and $\mathbf{V}_{21}\mathbf{V}^-\mathbf{V} = \mathbf{V}_{21}\mathbf{W}^-\mathbf{W} = \mathbf{V}_{21}$.

If $\mathbf{V}$ is positive definite and $\mathrm{r}(\mathbf{X}) = p$, as in Goldberger [28], we obtain

$$\begin{aligned}\mathrm{BLUP}(\mathbf{y}_*) &= \mathrm{BLUE}(\mathbf{X}_*\boldsymbol{\beta}) + \mathrm{BLUP}(\boldsymbol{\varepsilon}_*) \\ &= \mathbf{X}_*\widetilde{\boldsymbol{\beta}} + \mathbf{V}_{21}\mathbf{V}^{-1}\left(\mathbf{y} - \mathbf{X}\widetilde{\boldsymbol{\beta}}\right) \\ &= \mathbf{X}_*\widetilde{\boldsymbol{\beta}} + \mathbf{V}_{21}\mathbf{M}(\mathbf{MVM})^-\mathbf{My},\end{aligned}$$

where $\widetilde{\boldsymbol{\beta}} = (\mathbf{X'V^{-1}X})^{-1}\mathbf{X'V^{-1}y}$. Apparently Goldberger [28] was the first to use the term best linear unbiased predictor.

11.3 Conditions for Linear Sufficiency

Let us formally define the concept of linear sufficiency as done by Baksalary and Kala [9, 10] and Drygas [26].

Definition 11.1 Suppose that $\boldsymbol{\mu}_* = \mathbf{X}_*\boldsymbol{\beta}$, where $\mathbf{X}_* \in \mathbb{R}^{q\times p}$ is estimable under the model $\mathscr{M} = \{\mathbf{y}, \mathbf{X}\boldsymbol{\beta}, \mathbf{V}\}$. Then a linear statistic $\mathbf{Fy}$, where $\mathbf{F} \in \mathbb{R}^{f\times n}$ is called linearly sufficient for $\boldsymbol{\mu}_*$ if there exists a matrix $\mathbf{A} \in \mathbb{R}^{q\times f}$ such that $\mathbf{AFy}$ is the BLUE for $\boldsymbol{\mu}_*$, that is, there exists a matrix $\mathbf{A}$ such that

$$\mathbf{AF}(\mathbf{X} : \mathbf{VM}) = (\mathbf{X}_* : \mathbf{0}).$$

Of course, $\mathbf{Fy}$ is linearly sufficient for $\boldsymbol{\mu} = \mathbf{X}\boldsymbol{\beta}$, if there exists a matrix $\mathbf{A} \in \mathbb{R}^{n\times f}$ such that $\mathbf{AF}(\mathbf{X} : \mathbf{VM}) = (\mathbf{X} : \mathbf{0})$.

The concept of linear prediction sufficiency is defined in the corresponding way:

Definition 11.2 Let $\mathbf{y}_* = \mathbf{X}_*\boldsymbol{\beta} + \boldsymbol{\varepsilon}_*$ be predictable under the model $\mathscr{M}_*$, i.e., $\mathscr{C}(\mathbf{X}_*') \subset \mathscr{C}(\mathbf{X}')$. Then $\mathbf{Fy}$ is called linearly prediction sufficient for $\mathbf{y}_*$ if there exists a matrix $\mathbf{A}$ such that $\mathbf{AFy}$ is the BLUP for $\mathbf{y}_*$, that is, there exists a matrix $\mathbf{A}$ such that

$$\mathbf{AF}(\mathbf{X} : \mathbf{VM}) = (\mathbf{X}_* : \mathbf{V}_{21}\mathbf{M}).$$

Moreover, $\mathbf{Fy}$ is linearly prediction sufficient for $\boldsymbol{\varepsilon}_*$ if there exists a matrix $\mathbf{A}$ such that $\mathbf{AFy}$ is the BLUP for $\boldsymbol{\varepsilon}_*$, that is, there exists a matrix $\mathbf{A}$ such that

$$\mathbf{AF}(\mathbf{X} : \mathbf{VM}) = (\mathbf{0} : \mathbf{V}_{21}\mathbf{M}).$$

Sometimes we will use the phrases "BLUE-sufficient" and "BLUP-sufficient" when dealing with estimation and with prediction, respectively, and the short notations like $\mathbf{Fy} \in \mathscr{S}(\mathbf{y}_*)$, $\mathbf{Fy} \in \mathscr{S}(\mathbf{X}_*\boldsymbol{\beta})$, or $\mathbf{Fy} \in \mathscr{S}(\boldsymbol{\varepsilon}_*)$. However, the division into BLUE-sufficiency vs. BLUP-sufficiency is not necessary and we can simply refer to linear sufficiency of $\mathbf{Fy}$ with respect to $\mathbf{y}_*$, $\mathbf{X}_*\boldsymbol{\beta}$ or $\boldsymbol{\varepsilon}_*$. Thus, we have

$$\begin{aligned}
\mathscr{S}(\mathbf{y}_*) &= \{\mathbf{Fy} : \mathbf{AF}(\mathbf{X} : \mathbf{VM}) = (\mathbf{X}_* : \mathbf{V}_{21}\mathbf{M}) \text{ for some } \mathbf{A} \in \mathbb{R}^{q\times f}\},\\
\mathscr{S}(\mathbf{X}_*\boldsymbol{\beta}) &= \{\mathbf{Fy} : \mathbf{AF}(\mathbf{X} : \mathbf{VM}) = (\mathbf{X}_* : \mathbf{0}) \text{ for some } \mathbf{A} \in \mathbb{R}^{q\times f}\},\\
\mathscr{S}(\boldsymbol{\varepsilon}_*) &= \{\mathbf{Fy} : \mathbf{AF}(\mathbf{X} : \mathbf{VM}) = (\mathbf{0} : \mathbf{V}_{21}\mathbf{M}) \text{ for some } \mathbf{A} \in \mathbb{R}^{q\times f}\}.
\end{aligned}$$

As Kala et al. [51, Remark 2] points out, the notation of the above type is merely symbolic and it is not meant to refer to a set containing only one element which is a single fixed vector resulting from transformation of an observed vector $\mathbf{y}$, or is a single random vector variable being a specific linear transformation of the random vector $\mathbf{y}$. We are, of course, actually interested in the matrices $\mathbf{F}$ satisfying a certain property.

Gourieroux and Monfort [29, Sect. 3F] and Baksalary and Kala [9] considered corresponding problems without using the term "linear sufficiency" which is due to Drygas [26, Sect. 3]. Baksalary and Kala [9] used the phrase "linear transformations preserving best linear unbiased estimators". For further related references and concepts like minimal sufficiency and linear completeness, see Müller et al. [69], Baksalary and Mathew [11], Müller [68], Baksalary and Drygas [5], Groß [30], Isotalo and Puntanen [45], and Sengupta and Jammalamadaka [80, Sect. 11.1]. In this paper, we shall very briefly handle minimal sufficiency but skip over the linear completeness concept. The concept of linear minimal sufficiency, introduced by Drygas [26], is defined as follows:

Definition 11.3 A linear statistic $\mathbf{Fy}$ is called linearly minimal sufficient if for any other linearly sufficient statistics $\mathbf{Sy}$, there exists a matrix $\mathbf{A}$ such that $\mathbf{Fy} = \mathbf{ASy}$ almost surely. Notation $\mathbf{Fy} \in \mathscr{S}_0(\mathbf{X}\boldsymbol{\beta})$ indicates that $\mathbf{Fy}$ is linearly minimal sufficient for $\mathbf{X}\boldsymbol{\beta}$.

The minimal *prediction* sufficiency can be defined in an analogous way, see Isotalo and Puntanen [46, Def. 3.2].

Suppose that $\mathbf{F}_0\mathbf{y} \in \mathscr{S}_0(\mathbf{X}\boldsymbol{\beta})$ and that $\mathbf{Sy}$ is an arbitrary member of $\mathscr{S}(\mathbf{X}\boldsymbol{\beta})$. Then

$$\mathscr{C}(\mathbf{X}) = \mathscr{C}(\mathbf{WF}_0'), \text{ and } \mathscr{C}(\mathbf{X}) \subset \mathscr{C}(\mathbf{WS}'), \text{ where } \mathbf{W} \in \mathscr{W}.$$

By Definition 11.3, there exists a matrix $\mathbf{A}$ such that $\mathbf{F}_0\mathbf{y} = \mathbf{ASy}$ almost surely, i.e., $\mathbf{F}_0\mathbf{W} = \mathbf{ASW}$, which further means that

$$\mathscr{C}(\mathbf{X}) = \mathscr{C}(\mathbf{WF}_0') \subset \mathscr{C}(\mathbf{WS}'). \tag{11.24}$$

The column space of $\mathbf{W}$ can be termed as the general linear model subspace while $\mathscr{C}(\mathbf{X})$ determines the expectation subspace of $\mathscr{M}$. Postmultiplying $\mathbf{W}$ by any matrix does not increase its rank. Therefore, the relation (11.24) means that the transformation $\mathbf{F}_0$ leads to the maximum possible reduction of $\mathscr{C}(\mathbf{W}) = \mathscr{C}(\mathbf{X} : \mathbf{V})$ to the expectation subspace $\mathscr{C}(\mathbf{X})$ in which the $\mathrm{BLUE}(\mathbf{X}\boldsymbol{\beta})$ is contained. Simultaneously, $\mathbf{F}_0\mathbf{y}$ is the smallest amount of information necessary to reconstruct the $\mathrm{BLUE}(\mathbf{X}\boldsymbol{\beta})$.

Remark 11.1 (Linear error sufficiency) Groß [30] introduced the notion of linear error sufficiency while considering linear sufficient statistics for the prediction of the random error term $\boldsymbol{\varepsilon}$ in the general linear model. As pointed out by Isotalo et al. [44, Sect. 3], this is nothing but the BLUP-sufficiency of $\boldsymbol{\varepsilon}$. Namely, if we, instead of $\boldsymbol{\varepsilon}_*$, wish to find the BLUP for $\boldsymbol{\varepsilon}$, we have to minimize (in Löwner sense) $\mathrm{Cov}(\boldsymbol{\varepsilon} - \mathbf{Ay})$ subject to $\mathbf{Ay}$ being unbiased for $\boldsymbol{\varepsilon}$, that is, parallel to (11.14), the unbiased $\mathbf{Ay}$ must satisfy

$$\mathrm{Cov}(\boldsymbol{\varepsilon} - \mathbf{Ay}) \leq_{\mathrm{L}} \mathrm{Cov}(\boldsymbol{\varepsilon} - \mathbf{KMy}) \quad \text{for all } \mathbf{K} \in \mathbb{R}^{n\times n}.$$

Corresponding to (11.18), the matrix $\mathbf{A}$ is a solution to $\mathbf{A}(\mathbf{X} : \mathbf{VM}) = (\mathbf{0} : \mathbf{VM})$, and thus

$$\mathrm{BLUP}(\boldsymbol{\varepsilon}) = \mathbf{VM}(\mathbf{MVM})^{-}\mathbf{My},$$

and on account of (11.22) we have the decomposition

$$\mathbf{y} = \mathrm{BLUE}(\mathbf{X}\boldsymbol{\beta}) + \mathrm{BLUP}(\boldsymbol{\varepsilon}).$$

For the BLUP of $\boldsymbol{\varepsilon}$, see also Arendacká and Puntanen [2, Lemma 1]. □

In Proposition 11.3, we collect some well-known equivalent conditions for $\mathbf{Fy}$ being linearly sufficient for $\mathbf{X}\boldsymbol{\beta}$. For the proofs of parts (c) and (d), see Baksalary and Kala [9]. [Actually, according to Drygas [26, p. 92], (c) was originally proved by Baksalary and Kala [7].] For part (e), see Baksalary and Kala [10, Corollary 2]; and part (f), Müller [68, Prop. 3.1a]. For further related references, see Drygas [26], Baksalary and Mathew [11], Baksalary and Drygas [5], Groß [30], Isotalo and Puntanen [45, 46], Kornacki [57], Kala and Pordzik [53].

Proposition 11.3 *The statistic* $\mathbf{Fy}$ *is linearly sufficient for* $\boldsymbol{\mu} = \mathbf{X}\boldsymbol{\beta}$ *under the linear model* $\mathscr{M} = \{\mathbf{y}, \mathbf{X}\boldsymbol{\beta}, \mathbf{V}\}$ *if and only if any of the following equivalent statements holds:*

(a) $\mathscr{C}\begin{pmatrix}\mathbf{X}' \\ \mathbf{0}\end{pmatrix} \subset \mathscr{C}\begin{pmatrix}\mathbf{X}'\mathbf{F}' \\ \mathbf{MVF}'\end{pmatrix}$,
(b) $\mathscr{N}(\mathbf{FX} : \mathbf{FVM}) \subset \mathscr{N}(\mathbf{X} : \mathbf{0})$,
(c) $\mathscr{C}(\mathbf{X}) \subset \mathscr{C}(\mathbf{WF}')$, *where* $\mathbf{W} \in \mathscr{W}$,
(d) $\mathrm{r}(\mathbf{X} : \mathbf{VF}') = \mathrm{r}(\mathbf{WF}')$, *where* $\mathbf{W} \in \mathscr{W}$,
(e) $\mathscr{C}(\mathbf{X}'\mathbf{F}') = \mathscr{C}(\mathbf{X}')$ *and* $\mathscr{C}(\mathbf{FX}) \cap \mathscr{C}(\mathbf{FVM}) = \{\mathbf{0}\}$,
(f) $\mathscr{N}(\mathbf{F}) \cap \mathscr{C}(\mathbf{X} : \mathbf{V}) \subset \mathscr{C}(\mathbf{VM})$.

Moreover, $\mathbf{Fy}$ *is linearly minimal sufficient for* $\mathbf{X}\boldsymbol{\beta}$ *if and only if* $\mathscr{C}(\mathbf{X}) = \mathscr{C}(\mathbf{WF}')$, *or equivalently, the equality holds in* (a), (b) *or* (f).

Example 11.1 (When is $\mathbf{X}'\mathbf{y}$ linearly sufficient?) Following Kala et al. [52, Sect. 4], let us pose a question under which condition the statistic $\mathbf{X}'\mathbf{y}$ is linearly sufficient for $\mathbf{X}\boldsymbol{\beta}$ under $\mathscr{M}$. Now $\mathbf{X}'\mathbf{y} \in \mathscr{S}(\mathbf{X}\boldsymbol{\beta})$ if and only if $\mathscr{C}(\mathbf{X}) \subset \mathscr{C}(\mathbf{WX})$, where $\mathbf{W} = \mathbf{V} + \mathbf{XX}'$, i.e., $\mathscr{C}(\mathbf{X}) = \mathscr{C}(\mathbf{WX})$, which, noting that we always have $\mathrm{r}(\mathbf{WX}) = \mathrm{r}(\mathbf{X})$, further holds if and only if

$$\mathbf{P_X WX} = \mathbf{WX}. \tag{11.25}$$

Rewriting (11.25) yields $\mathbf{P_X}(\mathbf{V} + \mathbf{XX}')\mathbf{X} = (\mathbf{V} + \mathbf{XX}')\mathbf{X}$, i.e., $\mathbf{P_X VX} = \mathbf{VX}$, or in other words,

$$\mathscr{C}(\mathbf{VX}) \subset \mathscr{C}(\mathbf{X}). \tag{11.26}$$

The column space inclusion (11.26) is the well-known necessary and sufficient condition for the equality between the BLUE of $\mathbf{X}\boldsymbol{\beta}$ and the ordinary least-squares estimator, OLSE, of $\mathbf{X}\boldsymbol{\beta}$ under the model $\mathscr{M}$, see, e.g., Rao [73], Zyskind [87], Puntanen and Styan [71]. We can express our conclusion as follows:

Proposition 11.4 *The statistic* $\mathbf{X}'\mathbf{y}$ *is linearly sufficient for* $\boldsymbol{\mu} = \mathbf{X}\boldsymbol{\beta}$ *under the model* $\mathscr{M} = \{\mathbf{y}, \mathbf{X}\boldsymbol{\beta}, \mathbf{V}\}$ *if and only if* $\mathrm{OLSE}(\mathbf{X}\boldsymbol{\beta}) = \mathrm{BLUE}(\mathbf{X}\boldsymbol{\beta})$. *In this situation,* $\mathbf{X}'\mathbf{y}$ *is linearly minimal sufficient.*

The corresponding result as in Proposition 11.4, for a positive definite $\mathbf{V}$, appears also in Baksalary and Kala [9, p. 913]. We may further mention that requesting $\mathbf{P_X y} = \mathrm{OLSE}(\mathbf{X}\boldsymbol{\beta})$ to be linearly sufficient for $\mathbf{X}\boldsymbol{\beta}$ leads to the same condition as in Proposition 11.4, i.e.,

$$\mathbf{P_X y} \in \mathscr{S}(\mathbf{X}\boldsymbol{\beta}) \iff \mathbf{P_X y} = \mathrm{BLUE}(\mathbf{X}\boldsymbol{\beta}).$$

As a simple special case of the linear sufficiency of $\mathbf{X}'\mathbf{y}$, let us consider the model $\mathscr{A} = \{\mathbf{y}, \mathbf{1}\alpha, \mathbf{V}\}$, where $\mathbf{V}$ is positive definite and $\mathbf{1}$ is a vector of ones. We know that the BLUE of α is $\widetilde{\alpha} = (\mathbf{1}'\mathbf{V}^{-1}\mathbf{1})^{-1}\mathbf{1}'\mathbf{V}^{-1}\mathbf{y}$. Now $\mathbf{1}'\mathbf{y} \in \mathscr{S}(\alpha)$ if and only if there exists a scalar a such that $\widetilde{\alpha} = a\,\mathbf{1}'\mathbf{y}$ for all $\mathbf{y} \in \mathbb{R}^n$, i.e.,

$$(\mathbf{1}'\mathbf{V}^{-1}\mathbf{1})^{-1}\mathbf{1}'\mathbf{V}^{-1} = a\,\mathbf{1}'. \tag{11.27}$$

Equation (11.27) means that $\mathbf{1}$ is an eigenvector of $\mathbf{V}$, i.e.,

$$\mathbf{V1} = \lambda\mathbf{1} \quad \text{for some } \lambda \in \mathbb{R},$$

which corresponding to (11.26) can be expressed as that $\mathscr{C}(\mathbf{V1}) = \mathscr{C}(\mathbf{1})$. □

Baksalary and Kala [10] proved parts (a)–(e) of the following theorem. For other claims, see Kala et al. [51, 52].

Proposition 11.5 *Let* $\boldsymbol{\mu}_* = \mathbf{X}_*\boldsymbol{\beta}$ *be an estimable parametric function under* $\mathscr{M} = \{\mathbf{y}, \mathbf{X}\boldsymbol{\beta}, \mathbf{V}\}$, *i.e.,* $\mathscr{C}(\mathbf{X}_*') \subset \mathscr{C}(\mathbf{X}')$. *Then the following statements hold:*

(a) $\mathbf{Fy}$ *is linearly sufficient for* $\mathbf{X}_*\boldsymbol{\beta}$ *under* $\mathscr{M}$ *if and only if any of the following equivalent statements holds:*

(a$_1$) $\mathscr{C}\begin{pmatrix}\mathbf{X}_*' \\ \mathbf{0}\end{pmatrix} \subset \mathscr{C}\begin{pmatrix}\mathbf{X}'\mathbf{F}' \\ \mathbf{MVF}'\end{pmatrix}$,

(a_2) $\mathscr{N}(\mathbf{FX} : \mathbf{FVX}^{\perp}) \subset \mathscr{N}(\mathbf{X}_* : \mathbf{0})$,

(a_3) $\mathscr{C}[\mathbf{X}(\mathbf{X}'\mathbf{W}^-\mathbf{X})^-\mathbf{X}_*'] \subset \mathscr{C}(\mathbf{WF}')$, *where* $\mathbf{W} \in \mathscr{W}$;

(b) $\mathbf{Fy}$ *is linearly minimal sufficient for* $\mathbf{X}_*\boldsymbol{\beta}$ *if and only if equality holds in* (a$_1$), (a$_2$) *or equivalently in* (a$_3$);

(c) *If* $\mathscr{C}(\mathbf{X}'\mathbf{F}') = \mathscr{C}(\mathbf{X}_*')$, *then* $\mathbf{Fy} \in \mathscr{S}(\mathbf{X}_*\boldsymbol{\beta}) \iff \mathscr{C}(\mathbf{FX}) \cap \mathscr{C}(\mathbf{FVM}) = \{\mathbf{0}\}$;

(d) $\mathbf{Fy} \in \mathscr{S}(\mathbf{FX}\boldsymbol{\beta}) \iff \mathscr{C}(\mathbf{FX}) \cap \mathscr{C}(\mathbf{FVM}) = \{\mathbf{0}\}$;

(e) $\mathbf{Fy} \in \mathscr{S}(\mathbf{X}_*\boldsymbol{\beta})$ *for every estimable* $\mathbf{X}_*\boldsymbol{\beta}$ *if and only if* $\mathbf{Fy} \in \mathscr{S}(\mathbf{X}\boldsymbol{\beta})$;

(f) $\mathscr{C}(\mathbf{W}) \cap \mathscr{C}(\mathbf{F}')^{\perp} = \{\mathbf{0}\} \implies \mathbf{Fy} \in \mathscr{S}(\mathbf{X}_*\boldsymbol{\beta})$; *holds, e.g., if* $\mathbf{F}$ *is invertible;*

(g) $\mathrm{r}(\mathbf{LF}) = \mathrm{r}(\mathbf{F}) \implies \mathbf{LFy} \in \mathscr{S}(\mathbf{X}_*\boldsymbol{\beta})$;

(h) $\mathscr{C}(\mathbf{F}') \subset \mathscr{C}(\mathbf{F}_1') \implies \mathbf{F}_1\mathbf{y} \in \mathscr{S}(\mathbf{X}_*\boldsymbol{\beta})$;

(i) $(\mathbf{F}' : \mathbf{F}_2')'\mathbf{y}$ *is linearly sufficient for* $\mathbf{X}_*\boldsymbol{\beta}$ *for any conformable* $\mathbf{F}_2$.

As a curiosity, we may mention that Tian [81, Th. 3.1] and Tian and Puntanen [85, Th. 2.8] express (a$_1$) of Proposition 11.5 in the form

$$\mathscr{C}\begin{pmatrix}\mathbf{X}_*' \\ \mathbf{0}\end{pmatrix} \subset \mathscr{C}\begin{pmatrix}\mathbf{X}'\mathbf{F}' & \mathbf{0} \\ \mathbf{VF}' & \mathbf{X}\end{pmatrix}.$$

This follows at once from

$$\begin{pmatrix}\mathbf{X}'\mathbf{F}' \\ \mathbf{MVF}'\end{pmatrix} = \begin{pmatrix}\mathbf{X}'\mathbf{F}' & \mathbf{0} \\ \mathbf{VF}' & \mathbf{X}\end{pmatrix}\begin{pmatrix}\mathbf{I}_f \\ -\mathbf{X}^+\mathbf{VF}'\end{pmatrix}.$$

It is also noteworthy that in view of part (h) of Proposition 11.5, it is the basis of $\mathscr{C}(\mathbf{F}') \subset \mathbb{R}^n$ that matters, that is, we can choose the columns of $\mathbf{F}'$ linearly independent but spanning $\mathscr{C}(\mathbf{F}')$. In this context, we might ask: how many rows there are in $\mathbf{F}$, i.e., what is f? Suppose that $\mathbf{Fy} \in \mathscr{S}(\mathbf{X}_*\boldsymbol{\beta})$ and $\mathrm{r}(\mathbf{F}) = f$. Then by part (a$_3$) of Proposition 11.5,

$$\mathrm{r}[\mathbf{X}(\mathbf{X}'\mathbf{W}^-\mathbf{X})^-\mathbf{X}_*'] = \mathrm{r}(\mathbf{X}_*) \leq \mathrm{r}(\mathbf{WF}') \leq f \leq n. \tag{11.28}$$

For parts (a) and (b) of the following Proposition, see Isotalo and Puntanen [46] and Isotalo et al. [44], respectively.

Proposition 11.6 *Consider the linear model with new observations defined as $\mathscr{M}_*$ in (11.4), where $\mathscr{C}(\mathbf{X}_*') \subset \mathscr{C}(\mathbf{X}')$, i.e., $\mathbf{y}_*$ is predictable.*

(a) $\mathbf{Fy} \in \mathscr{S}(\mathbf{y}_*)$ *if and only if any of the following equivalent conditions holds:*

(i) $\mathscr{C}\begin{pmatrix} \mathbf{X}_*' \\ \mathbf{MV}_{12} \end{pmatrix} \subset \mathscr{C}\begin{pmatrix} \mathbf{X}'\mathbf{F}' \\ \mathbf{MVF}' \end{pmatrix}$,

(ii) $\mathscr{N}(\mathbf{FX} : \mathbf{FVM}) \subset \mathscr{N}(\mathbf{X}_* : \mathbf{V}_{21}\mathbf{M})$,

(iii) $\mathscr{N}(\mathbf{F}) \cap \mathscr{C}(\mathbf{W}) \subset \mathscr{C}\left((\mathbf{X} : \mathbf{VM}) \begin{pmatrix} \mathbf{X}_*' \\ \mathbf{MV}_{12} \end{pmatrix}^{\perp}\right)$.

(b) $\mathbf{Fy} \in \mathscr{S}(\boldsymbol{\varepsilon}_*)$ *if and only if any of the following equivalent conditions holds:*

(iv) $\mathscr{C}\begin{pmatrix} \mathbf{0} \\ \mathbf{MV}_{12} \end{pmatrix} \subset \mathscr{C}\begin{pmatrix} \mathbf{X}'\mathbf{F}' \\ \mathbf{MVF}' \end{pmatrix}$,

(v) $\mathscr{C}(\mathbf{MV}_{12}) \subset \mathscr{C}(\mathbf{MVF}'\mathbf{Q}_{\mathbf{FX}})$.

In particular, if $\mathbf{Fy} \in \mathscr{S}(\mathbf{X}\boldsymbol{\beta})$, *then* (iv) *becomes*

(vi) $\mathscr{C}(\mathbf{MV}_{12}) \subset \mathscr{C}(\mathbf{MVF}')$.

Moreover, the minimal prediction sufficiency above is obtained if and only if the corresponding inclusion is equality.

Remark 11.2 (Properties of $\mathscr{C}(\mathbf{WF}')$) Consider the following three questions (a), (b), and (c) related to the linear sufficiency condition (c) of Proposition 11.3:

$$\mathbf{Fy} \in \mathscr{S}(\mathbf{X}\boldsymbol{\beta}) \iff \mathscr{C}(\mathbf{X}) \subset \mathscr{C}(\mathbf{WF}'), \quad \text{where } \mathbf{W} \in \mathscr{W}. \tag{11.29}$$

(a) The matrix $\mathbf{W}$ in (11.29) belongs to the set $\mathscr{W}$ of (symmetric) nonnegative definite matrices. One question: is the column space $\mathscr{C}(\mathbf{WF}')$ unique, i.e., does it remain invariant for any choice of $\mathbf{W} \in \mathscr{W}$? It might be somewhat tempting to conjecture that for a given $\mathbf{F}$, the column space $\mathscr{C}(\mathbf{WF}')$ would be invariant. However, Kala et al. [52, Ex. 1] provide a counterexample showing that this is not the case.
(b) Kala et al. [52, Sect. 4] also studied whether the column space $\mathscr{C}(\mathbf{WF}')$ is invariant for any choice of $\mathbf{W}$ if $\mathbf{Fy} \in \mathscr{S}(\mathbf{X}\boldsymbol{\beta})$. Proposition 11.7 is a reply to this question. We formulate it in a more general setup using the set $\mathscr{W}^{\#}$ instead of $\mathscr{W}$.

Proposition 11.7 *Consider the linear model* $\mathscr{M} = \{\mathbf{y}, \mathbf{X}\boldsymbol{\beta}, \mathbf{V}\}$, *let* $\mathbf{W} \in \mathscr{W}^{\#}$ *and suppose that* $\mathscr{C}(\mathbf{X}) \subset \mathscr{C}(\mathbf{WF}')$. *Then the column space* $\mathscr{C}(\mathbf{WF}')$ *is invariant for any choice of* $\mathbf{W} \in \mathscr{W}^{\#}$ *and*

$$\mathscr{C}(\mathbf{WF}') = \mathscr{C}(\mathbf{X}) \oplus \mathscr{C}(\mathbf{MVF}') = \mathscr{C}(\mathbf{W}'\mathbf{F}').$$

(c) Kala et al. [52, Sect. 4] were wondering whether in (11.29) the set $\mathscr{W}$ can be replaced with the more general set $\mathscr{W}^{\#}$. The answer is interestingly (but not trivially) positive and can be expressed as follows:

Proposition 11.8 *Let* $\mathbf{W} \in \mathscr{W}^{\#}$. *Then the statistic* $\mathbf{Fy}$ *is linearly sufficient for* $\mathbf{X}\boldsymbol{\beta}$ *under the linear model* $\mathscr{M} = \{\mathbf{y}, \mathbf{X}\boldsymbol{\beta}, \mathbf{V}\}$ *if and only if*

$$\mathscr{C}(\mathbf{X}) \subset \mathscr{C}(\mathbf{WF}'), \quad \textit{or, equivalently,} \quad \mathscr{C}(\mathbf{X}) \subset \mathscr{C}(\mathbf{W}'\mathbf{F}').$$

According to our knowledge, in all linear sufficiency considerations appearing in literature, it is assumed that $\mathbf{W}$ is nonnegative definite. However, this is not necessary, and $\mathbf{W}$ can also be nonsymmetric. Of course, sometimes it can be simpler to have $\mathbf{W}$ from set $\mathscr{W}$. For detailed proofs of Propositions 11.7 and 11.8, see Kala et al. [52, Sect. 4]. □

Example 11.2 (Random walk and linear sufficiency) One interesting example of the BLUP is described by Isotalo and Puntanen [46, p. 1021] and Haslett et al. [37, Sect. 1]. They consider the model

$$y_t = \beta t + \varepsilon_t, \tag{11.30}$$

where t denotes discrete time variable, β is the unknown parameter, and ε_t is a random walk process, see, e.g., Davidson and MacKinnon [23, p. 606], with a form $\varepsilon_t = \varepsilon_{t-1} + u_t$, $\varepsilon_0 = 0$, $u_t \sim \mathrm{IID}(0, 1)$. Suppose that the time series y_t given in (11.30) is observable at times of $t \in \{1, 2, 3, 4\}$. Then putting $\mathbf{x} = (1, 2, 3, 4)'$, we have in matrix terms

$$\mathbf{y} = \mathbf{x}\beta + \boldsymbol{\varepsilon}, \quad \text{where } \boldsymbol{\varepsilon} = \mathbf{Du}, \ \mathbf{D} = \begin{pmatrix} 1 & 0 & 0 & 0 \\ 1 & 1 & 0 & 0 \\ 1 & 1 & 1 & 0 \\ 1 & 1 & 1 & 1 \end{pmatrix}, \ \mathrm{Cov}(\mathbf{u}) = \mathbf{I}_4.$$

Being interested in predicting the outcome of y_t at time of $t = 5$ based on observable random variables $y_1, \dots, y_4$, we can write the model as

$$\mathscr{M}_* = \left\{ \begin{pmatrix} \mathbf{y} \\ y_* \end{pmatrix}, \begin{pmatrix} \mathbf{x}\beta \\ x_*\beta \end{pmatrix}, \begin{pmatrix} \mathbf{V} & \mathbf{v}_{12} \\ \mathbf{v}_{12}' & v_{22} \end{pmatrix} \right\},$$

where $y_* = y_5$, $x_* = 5$, the variance of y_5 is $v_{22} = 5$, and

$$\mathrm{Cov}(\mathbf{y}) = \mathbf{V} = \mathbf{DD}' = \begin{pmatrix} 1 & 1 & 1 & 1 \\ 1 & 2 & 2 & 2 \\ 1 & 2 & 3 & 3 \\ 1 & 2 & 3 & 4 \end{pmatrix}, \quad \mathrm{Cov}(\mathbf{y}, y_5) = \mathbf{v}_{12} = \begin{pmatrix} 1 \\ 2 \\ 3 \\ 4 \end{pmatrix}.$$

Now $\mathbf{x} = \mathbf{v}_{12}$, and thereby $\mathscr{C}(\mathbf{x}^{\perp}) = \mathscr{C}(\mathbf{v}_{12}^{\perp})$ and

$$\mathbf{Q}_{\mathbf{x}} = \mathbf{I}_4 - \mathbf{v}_{12}\mathbf{v}_{12}'/\mathbf{v}_{12}'\mathbf{v}_{12} = \mathbf{I}_4 - \mathbf{V}\mathbf{f}\,\mathbf{f}'\mathbf{V}/\mathbf{f}'\mathbf{V}^2\mathbf{f},$$

where $\mathbf{f} = (0, 0, 0, 1)'$ and $\mathbf{v}_{12} = \mathbf{V}\mathbf{f}$. We further have

$$\mathbf{f}'\mathbf{x} = 4, \quad \mathbf{f}'\mathbf{V}\mathbf{Q}_{\mathbf{x}} = (0, 0, 0, 0), \quad x_* = 5, \quad \mathbf{v}_{12}'\mathbf{Q}_{\mathbf{x}} = (0, 0, 0, 0).$$

Thus

$$\mathscr{N}(\mathbf{f}'\mathbf{x} : \mathbf{f}'\mathbf{V}\mathbf{Q}_{\mathbf{x}}) = \mathscr{N}(x_* : \mathbf{v}_{12}'\mathbf{Q}_{\mathbf{x}}),$$

which by Proposition 11.6 means that $\mathbf{f}'\mathbf{y} = y_4$ is linearly minimal sufficient predictor for y_5. The BLUP for y_5 in this situation is simply $\frac{5}{4}\, y_4$. □

11.4 The Transformed Model

Consider the model $\mathscr{M} = \{\mathbf{y}, \mathbf{X}\boldsymbol{\beta}, \mathbf{V}\}$ and let $\mathbf{F} \in \mathbb{R}^{f \times n}$ be such a matrix that $\mathbf{F}\mathbf{y}$ is linearly sufficient for $\mathbf{X}\boldsymbol{\beta}$. Then the transformation $\mathbf{F}$ applied to $\mathbf{y}$ induces the transformed model

$$\mathscr{T} = \{\mathbf{F}\mathbf{y}, \mathbf{F}\mathbf{X}\boldsymbol{\beta}, \mathbf{F}\mathbf{V}\mathbf{F}'\}.$$

As the statistic $\mathbf{F}\mathbf{y}$ is linearly sufficient for $\mathbf{X}\boldsymbol{\beta}$, it sounds intuitively natural that both models provide the same starting point for obtaining the BLUE of $\mathbf{X}\boldsymbol{\beta}$. Indeed this is true as proved by Baksalary and Kala [9, 10]. We can also do the prediction of $\mathbf{y}_*$ (or $\boldsymbol{\varepsilon}_*$) based on the transformed model using the following setup:

$$\mathscr{T}_* = \left\{ \begin{pmatrix} \mathbf{F}\mathbf{y} \\ \mathbf{y}_* \end{pmatrix}, \begin{pmatrix} \mathbf{F}\mathbf{X} \\ \mathbf{X}_* \end{pmatrix} \boldsymbol{\beta}, \begin{pmatrix} \mathbf{F}\mathbf{V}\mathbf{F}' & \mathbf{F}\mathbf{V}_{12} \\ \mathbf{V}_{21}\mathbf{F}' & \mathbf{V}_{22} \end{pmatrix} \right\}.$$

Recall that trivially the BLUE-considerations are identical under $\mathscr{M}$ and $\mathscr{M}_*$, i.e., $\mathrm{BLUE}(\mathbf{X}_*\boldsymbol{\beta} \mid \mathscr{M}) = \mathrm{BLUE}(\mathbf{X}_*\boldsymbol{\beta} \mid \mathscr{M}_*)$, etc.

In order to find BLUEs and BLUPs, we need to have some estimability conditions.

Proposition 11.9 *Consider the models $\mathscr{M}$ and $\mathscr{T}$. Then the following estimability conditions hold:*

(a) $\mathbf{X}_*\boldsymbol{\beta}$ *is estimable under* $\mathscr{M} \iff \mathscr{C}(\mathbf{X}_*') \subset \mathscr{C}(\mathbf{X}')$,
(b) $\mathbf{X}_*\boldsymbol{\beta}$ *is estimable under* $\mathscr{T} \iff \mathscr{C}(\mathbf{X}_*') \subset \mathscr{C}(\mathbf{X}'\mathbf{F}')$,
(c) $\mathbf{X}\boldsymbol{\beta}$ *is estimable under* $\mathscr{T} \iff \mathscr{C}(\mathbf{X}') = \mathscr{C}(\mathbf{X}'\mathbf{F}')$, *i.e.*, $\mathrm{r}(\mathbf{X}) = \mathrm{r}(\mathbf{F}\mathbf{X})$.

Moreover,

(d) *the column space condition in* (b) *implies that one in* (a),
(e) *the column space conditions in* (a) *and* (c) *imply that one in* (b).

Suppose that $\mathbf{X}_*\boldsymbol{\beta}$ is estimable under the transformed model $\mathscr{T}$. Then $\mathbf{CFy}$ is the BLUE for $\mathbf{X}_*\boldsymbol{\beta}$ under $\mathscr{T}$ if and only if $\mathbf{C}$ satisfies the condition

$$\mathbf{C}(\mathbf{FX} : \mathbf{FVF}'\mathbf{Q}_{\mathbf{FX}}) = (\mathbf{X}_* : \mathbf{0}),$$

or equivalently

$$\mathbf{C}(\mathbf{FX} : \mathbf{FVMF}'\mathbf{Q}_{\mathbf{FX}}) = (\mathbf{X}_* : \mathbf{0}), \text{ i.e., } \mathbf{C}(\mathbf{FX} : \mathbf{FVMN}) = (\mathbf{X}_* : \mathbf{0}),$$

where $\mathbf{Q}_{\mathbf{FX}} = \mathbf{I}_f - \mathbf{P}_{\mathbf{FX}}$, and $\mathbf{N} = \mathbf{P}_{\mathscr{C}(\mathbf{M})\cap\mathscr{C}(\mathbf{F}')}$, see Lemma 11.4. We use the notation

$$\mathbf{C} \in \{\mathbf{P}_{\mathbf{X}_*|\mathscr{T}}\} \iff \mathbf{C}(\mathbf{FX} : \mathbf{FVF}'\mathbf{Q}_{\mathbf{FX}}) = (\mathbf{X}_* : \mathbf{0}).$$

The set of products $\mathbf{CF}$, where $\mathbf{C} \in \{\mathbf{P}_{\mathbf{X}_*|\mathscr{T}}\}$, will be denoted as $\{\mathbf{P}_{\mathbf{X}_*|\mathscr{T}}\mathbf{F}\}$. It means that each matrix $\mathbf{D} \in \{\mathbf{P}_{\mathbf{X}_*|\mathscr{T}}\mathbf{F}\}$ applied to $\mathbf{y}$ provides the BLUE for $\mathbf{X}_*\boldsymbol{\beta}$ under the transformed model $\mathscr{T}$, i.e.,

$$\mathbf{D} \in \{\mathbf{P}_{\mathbf{X}_*|\mathscr{T}}\mathbf{F}\} \iff \mathbf{D} = \mathbf{CF}, \text{ where } \mathbf{C} \in \{\mathbf{P}_{\mathbf{X}_*|\mathscr{T}}\}.$$

Recall, by (11.19), that the set $\{\mathbf{P}_{\mathbf{X}_*|\mathscr{M}}\}$ providing the BLUEs for $\mathbf{X}_*\boldsymbol{\beta}$ under the original model $\mathscr{M}$ is defined as

$$\mathbf{B} \in \{\mathbf{P}_{\mathbf{X}_*|\mathscr{M}}\} \iff \mathbf{B}(\mathbf{X} : \mathbf{VM}) = (\mathbf{X}_* : \mathbf{0}).$$

Baksalary and Kala [10, Th. 1] consider $\mathbf{X}_*\boldsymbol{\beta}$, which is estimable under $\mathscr{M}$. They assume that $\mathbf{Fy}$ is linearly sufficient for $\mathbf{X}_*\boldsymbol{\beta}$ so that there exists some matrix $\mathbf{A}$ such that $\mathbf{AFy}$ is the BLUE of $\mathbf{X}_*\boldsymbol{\beta}$ in $\mathscr{M}$, i.e., $\mathbf{AF} \in \{\mathbf{P}_{\mathbf{X}_*|\mathscr{M}}\}$ so that

$$\mathbf{AF}(\mathbf{X} : \mathbf{VM}) = (\mathbf{X}_* : \mathbf{0}). \tag{11.31}$$

They show that for each $\mathbf{C} \in \{\mathbf{P}_{\mathbf{X}_*|\mathscr{T}}\}$ satisfying the equation

$$\mathbf{C}(\mathbf{FX} : \mathbf{FVF}'\mathbf{Q}_{\mathbf{FX}}) = (\mathbf{X}_* : \mathbf{0}), \tag{11.32}$$

the equality

$$\mathbf{AF}(\mathbf{X} : \mathbf{V}) = \mathbf{CF}(\mathbf{X} : \mathbf{V}) \tag{11.33}$$

holds and thereby $\mathbf{AFy}$ equals $\mathbf{CFy}$ with probability 1. In other words,

$$\text{BLUE}(\mathbf{X}_*\boldsymbol{\beta} \mid \mathscr{M}) = \text{BLUE}(\mathbf{X}_*\boldsymbol{\beta} \mid \mathscr{T}) \quad \text{with probability 1.}$$

It is worth emphasizing that (11.33) holds for *every* $\mathbf{C} \in \{\mathbf{P}_{\mathbf{X}_*|\mathscr{T}}\}$ and *at least for one matrix* $\mathbf{A}$.

Equality (11.33) also implies that

$$\mathbf{AF}(\mathbf{X} : \mathbf{VM}) = \mathbf{CF}(\mathbf{X} : \mathbf{VM}) = (\mathbf{X}_* : \mathbf{0}). \tag{11.34}$$

Now (11.34) means that every multiplier $\mathbf{CF}$ of $\mathbf{y}$ (where $\mathbf{C} \in \{\mathbf{P}_{\mathbf{X}_*|\mathscr{T}}\}$) provides the BLUE under the original model $\mathscr{M}$. In other notation,

$$\mathbf{Fy} \in \mathscr{S}(\mathbf{X}_*\boldsymbol{\beta}) \implies \{\mathbf{P}_{\mathbf{X}_*|\mathscr{T}}\mathbf{F}\} \subset \{\mathbf{P}_{\mathbf{X}_*|\mathscr{M}}\}. \tag{11.35}$$

It is obvious that the implication in (11.35) holds also in the reverse direction.

Moreover, we can conclude the following: there exists at least one representation of BLUE under $\mathscr{M}$ [this is $\mathbf{AFy}$, where $\mathbf{A}$ satisfies (11.34)] which is BLUE also under the transformed model $\mathscr{T}$. To confirm this, we have to show that

$$\mathbf{AF}(\mathbf{X} : \mathbf{VM}) = \mathbf{A}(\mathbf{FX} : \mathbf{FVM}) = (\mathbf{X}_* : \mathbf{0}) \tag{11.36}$$

implies

$$\mathbf{A}(\mathbf{FX} : \mathbf{FVF}'\mathbf{Q}_{\mathbf{FX}}) = \mathbf{A}(\mathbf{FX} : \mathbf{FVMF}'\mathbf{Q}_{\mathbf{FX}}) = (\mathbf{X}_* : \mathbf{0}), \tag{11.37}$$

where we have used Lemma 11.4. Now it is obvious that (11.36) implies (11.37). The feature that there exists at least one representation of BLUE under $\mathscr{M}$ which is BLUE also under the transformed model $\mathscr{T}$ can be denoted as

$$\{\mathbf{P}_{\mathbf{X}_*|\mathscr{T}}\mathbf{F}\} \cap \{\mathbf{P}_{\mathbf{X}_*|\mathscr{M}}\} \neq \{\emptyset\}. \tag{11.38}$$

As a matter of fact, the above proof provides the following result, where the concept of linear sufficiency is not explicitly present:

Proposition 11.10 *Suppose that* $\mathbf{X}_*\boldsymbol{\beta}$ *is estimable under* $\mathscr{M}$. *Then*

(a) $\mathbf{AF} \in \{\mathbf{P}_{\mathbf{X}_*|\mathscr{M}}\} \implies \mathbf{A} \in \{\mathbf{P}_{\mathbf{X}_*|\mathscr{T}}\}$,
(b) $\mathbf{AF} \in \{\mathbf{P}_{\boldsymbol{\varepsilon}_*|\mathscr{M}_*}\} \implies \mathbf{A} \in \{\mathbf{P}_{\boldsymbol{\varepsilon}_*|\mathscr{T}_*}\}$,
(c) $\mathbf{AF} \in \{\mathbf{P}_{\mathbf{y}_*|\mathscr{M}_*}\} \implies \mathbf{A} \in \{\mathbf{P}_{\mathbf{y}_*|\mathscr{T}_*}\}$.

Notice that if (11.31) is solvable for $\mathbf{A}$ then (11.32) is solvable for $\mathbf{C}$ and $\mathbf{X}_*\boldsymbol{\beta}$ is estimable under the transformed model $\mathscr{T}$. Another noteworthy fact is that (11.38) implies that $\mathbf{Fy} \in \mathscr{S}(\mathbf{X}_*\boldsymbol{\beta})$.

The matrix $\mathbf{C} \in \{\mathbf{P}_{\mathbf{X}_*|\mathscr{T}}\}$ satisfying the equation (11.32) is unique if and only if

$$\mathrm{r}(\mathbf{FX} : \mathbf{FVF}'\mathbf{Q}_{\mathbf{FX}}) = f, \quad \text{i.e.,} \quad \mathrm{r}(\mathbf{FW}) = f.$$

The rank rule of the matrix product of Marsaglia and Styan [66, Corollary 6.2] gives

$$r(\mathbf{FW}) = r(\mathbf{F}) - \dim \mathscr{C}(\mathbf{F}') \cap \mathscr{C}(\mathbf{W})^{\perp},$$

and thus the matrix $\mathbf{C} \in \{\mathbf{P}_{\mathbf{X}_*|\mathscr{T}}\}$ is unique if and only if

$$r(\mathbf{F}) = f \quad \text{and} \quad \mathscr{C}(\mathbf{Q}_{\mathbf{F}'} : \mathbf{W}) = \mathbb{R}^n.$$

Kala et al. [51, Th. 1] represented the parts (a)–(f) of the following proposition dealing with statistical implications of the linear sufficiency.

Proposition 11.11 *Suppose that* $\mathbf{Fy}$ *is linearly sufficient for estimable* $\mathbf{X}_*\boldsymbol{\beta}$ *under the model* $\mathscr{M} = \{\mathbf{y}, \mathbf{X}\boldsymbol{\beta}, \mathbf{V}\}$ *and let* $\mathscr{T} = \{\mathbf{Fy}, \mathbf{FX}\boldsymbol{\beta}, \mathbf{FVF}'\}$ *denote the transformed model and* $\mathbf{W} \in \mathscr{W}$. *Then the following statements hold:*

(a) $\mathbf{X}_*\boldsymbol{\beta}$ *is estimable also under the transformed model* $\mathscr{T}$, *i.e.,* $\mathscr{C}(\mathbf{X}_*') \subset \mathscr{C}(\mathbf{X}'\mathbf{F}')$;
(b) $\text{BLUE}(\mathbf{X}_*\boldsymbol{\beta} \mid \mathscr{M}) = \text{BLUE}(\mathbf{X}_*\boldsymbol{\beta} \mid \mathscr{T})$ *with probability* 1, *in other notation,* $\mathbf{BW} = \mathbf{CFW}$, *where* $\mathbf{B} \in \{\mathbf{P}_{\mathbf{X}_*|\mathscr{M}}\}$, *and* $\mathbf{C} \in \{\mathbf{P}_{\mathbf{X}_*|\mathscr{T}}\}$;
(c) *Every representation of the* BLUE *of* $\mathbf{X}_*\boldsymbol{\beta}$ *under* $\mathscr{T}$ *is* BLUE *also under* $\mathscr{M}$, *i.e.,* $\{\mathbf{P}_{\mathbf{X}_*|\mathscr{T}}\mathbf{F}\} \subset \{\mathbf{P}_{\mathbf{X}_*|\mathscr{M}}\}$;
(d) *There exists at least one representation of* BLUE *of* $\mathbf{X}_*\boldsymbol{\beta}$ *under* $\mathscr{M}$ *which is* BLUE *also under the transformed model* $\mathscr{T}$, *i.e., there exists at least one matrix in the set* $\{\mathbf{P}_{\mathbf{X}_*|\mathscr{M}}\}$ *which belongs also to* $\{\mathbf{P}_{\mathbf{X}_*|\mathscr{T}}\mathbf{F}\}$;
(e) *If* $\mathbf{By}$ *is an arbitrary* BLUE *for* $\mathbf{X}_*\boldsymbol{\beta}$ *under* $\mathscr{M}$, *then there exists* $\mathbf{C}$ *such that* $\mathbf{By} = \mathbf{CFy}$ *for all* $\mathbf{y} \in \mathscr{C}(\mathbf{X} : \mathbf{V})$, *and* $\mathbf{CFy}$ *is the* BLUE *for* $\mathbf{X}_*\boldsymbol{\beta}$ *under* $\mathscr{T}$;
(f) *If* $\mathscr{C}(\mathbf{X}_*') = \mathscr{C}(\mathbf{X}'\mathbf{F}')$, *then* $\mathbf{Fy} \in \mathscr{S}(\mathbf{X}_*\boldsymbol{\beta}) \iff \mathbf{Fy} \in \mathscr{S}(\mathbf{FX}\boldsymbol{\beta})$;
(g) *There is only one matrix in the class* $\{\mathbf{P}_{\mathbf{X}_*|\mathscr{T}}\mathbf{F}\}$ *if and only if* $r(\mathbf{F}) = f$ *and* $\mathscr{C}(\mathbf{Q}_{\mathbf{F}'} : \mathbf{W}) = \mathbb{R}^n$.

Moreover, if $\mathbf{Fy}$ *is linearly sufficient for* $\mathbf{X}\boldsymbol{\beta}$ *then the statements* (a)–(e) *hold also when* $\mathbf{X}_*$ *is replaced with* $\mathbf{X}$.

Proof [of Part (e)]. Part (e) of the above Proposition 11.11 was given, without a proof, in Kala et al. [51, Th. 1]. For completeness, we briefly go through the proof. Suppose that $\mathbf{Fy} \in \mathscr{S}(\mathbf{X}_*\boldsymbol{\beta})$ and $\mathbf{B} \in \{\mathbf{P}_{\mathbf{X}_*|\mathscr{M}}\}$. Then the general representation of $\mathbf{B}$ is

$$\mathbf{B}_0 = \mathbf{X}_*(\mathbf{X}'\mathbf{W}^-\mathbf{X})^-\mathbf{X}'\mathbf{W}^- + \mathbf{E}\mathbf{Q}_{\mathbf{W}}, \tag{11.39}$$

where $\mathbf{E} \in \mathbb{R}^{q \times n}$ is arbitrary. Postmultiplying (11.39) by $\mathbf{W}$ yields

$$\mathbf{B}_0\mathbf{W} = \mathbf{X}_*(\mathbf{X}'\mathbf{W}^-\mathbf{X})^-\mathbf{X}' = \mathbf{CFW}, \tag{11.40}$$

for some $\mathbf{C} \in \mathbb{R}^{q \times f}$; here we have used part (a_3) of Proposition 11.5, which says that under linear sufficiency, $\mathbf{X}(\mathbf{X}'\mathbf{W}^-\mathbf{X})^-\mathbf{X}_*' = \mathbf{WF}'\mathbf{C}'$ for some $\mathbf{C}$. Equality (11.40) means that $\mathbf{B}_0\mathbf{y} = \mathbf{CFy}$ for all $\mathbf{y} \in \mathscr{C}(\mathbf{W})$. It remains to show that $\mathbf{C} \in \{\mathbf{P}_{\mathbf{X}_*|\mathscr{T}}\}$, which immediately follows from Proposition 11.10. □

Next, we consider the explicit expression for the BLUE of $\mathbf{X}\boldsymbol{\beta}$ under the transformed model $\mathscr{T}$. To do this, we have to assume that $\mathbf{X}\boldsymbol{\beta}$ is estimable under $\mathscr{T}$. Then the statistic $\mathbf{CFy}$ is the BLUE for $\mathbf{X}\boldsymbol{\beta}$ under $\mathscr{T}$ if and only if $\mathbf{C}$ satisfies

$$\mathbf{C}(\mathbf{FX} : \mathbf{FVF}'\mathbf{Q}_{\mathbf{FX}}) = (\mathbf{X} : \mathbf{0}), \quad \text{or shortly,} \quad \mathbf{C} \in \{\mathbf{P}_{\mathbf{X}|\mathscr{T}}\}. \tag{11.41}$$

One solution for $\mathbf{C}$ in (11.41) is

$$\mathbf{C}_1 := \mathbf{X}[\mathbf{X}'\mathbf{F}'(\mathbf{FWF}')^{-}\mathbf{FX}]^{-}\mathbf{X}'\mathbf{F}'(\mathbf{FWF}')^{-} \in \{\mathbf{P}_{\mathbf{X}|\mathscr{T}}\}.$$

Notice that by (11.7), $\mathbf{FWF}'$ is a $\mathbf{W}$-matrix under the model $\mathscr{T}$. Thus, see Kala et al. [52, Sect. 6] and Markiewicz and Puntanen [64, Sect. 3], the BLUE of $\mathbf{X}\boldsymbol{\beta}$ under $\mathscr{T}$ has, for example, the representation

$$\mathrm{BLUE}(\mathbf{X}\boldsymbol{\beta} \mid \mathscr{T}) = \mathbf{C}_1\mathbf{F}\mathbf{y} = \mathbf{G}_t\mathbf{y},$$

where

$$\mathbf{G}_t = \mathbf{C}_1\mathbf{F} = \mathbf{X}[\mathbf{X}'\mathbf{F}'(\mathbf{FWF}')^{-}\mathbf{FX}]^{-}\mathbf{X}'\mathbf{F}'(\mathbf{FWF}')^{-}\mathbf{F} \in \{\mathbf{P}_{\mathbf{X}|\mathscr{T}}\mathbf{F}\}. \tag{11.42}$$

Correspondingly, if $\mathbf{X}_*\boldsymbol{\beta}$ is estimable under $\mathscr{T}$, then $\mathbf{X}_* = \mathbf{LFX}$ for some $\mathbf{L} \in \mathbb{R}^{q\times f}$ and $\mathbf{CFy}$ is the BLUE for $\mathbf{X}_*\boldsymbol{\beta}$ if $\mathbf{C} \in \{\mathbf{P}_{\mathbf{X}_*|\mathscr{T}}\}$. One such a $\mathbf{C}$ is

$$\mathbf{C}_2 := \mathbf{X}_*[\mathbf{X}'\mathbf{F}'(\mathbf{FWF}')^{-}\mathbf{FX}]^{-}\mathbf{X}'\mathbf{F}'(\mathbf{FWF}')^{-} \in \{\mathbf{P}_{\mathbf{X}_*|\mathscr{T}}\},$$

and the BLUE of $\mathbf{X}_*\boldsymbol{\beta}$ under $\mathscr{T}$ can be expressed as

$$\mathrm{BLUE}(\mathbf{X}_*\boldsymbol{\beta} \mid \mathscr{T}) = \mathbf{C}_2\mathbf{F}\mathbf{y} = \mathbf{LF}\,\mathrm{BLUE}(\mathbf{X}\boldsymbol{\beta} \mid \mathscr{T}) = \mathbf{LFG}_t\mathbf{y}.$$

The general representations of matrices in $\{\mathbf{P}_{\mathbf{X}|\mathscr{T}}\mathbf{F}\}$ and $\{\mathbf{P}_{\mathbf{X}_*|\mathscr{T}}\mathbf{F}\}$ providing the BLUE for $\mathbf{X}\boldsymbol{\beta}$ and $\mathbf{X}_*\boldsymbol{\beta}$, under $\mathscr{T}$, can be expressed as

$$\mathbf{P}_{\mathbf{X}|\mathscr{T}}\mathbf{F} = (\mathbf{C}_1 + \mathbf{E}_2\mathbf{Q}_{\mathbf{FW}})\mathbf{F}, \quad \mathbf{P}_{\mathbf{X}_*|\mathscr{T}}\mathbf{F} = (\mathbf{C}_2 + \mathbf{E}_3\mathbf{Q}_{\mathbf{FW}})\mathbf{F},$$

respectively; here $\mathbf{E}_2 \in \mathbb{R}^{n\times f}$ and $\mathbf{E}_3 \in \mathbb{R}^{q\times f}$ are free to vary.

The statistic $\mathbf{Dy}$ is the BLUP for $\boldsymbol{\varepsilon}_*$ under the transformed model $\mathscr{T}_*$ if and only if $\mathbf{D} = \mathbf{CF}$, where $\mathbf{C}$ satisfies

$$\mathbf{C}(\mathbf{FX} : \mathbf{FVF}'\mathbf{Q}_{\mathbf{FX}}) = (\mathbf{0} : \mathbf{V}_{21}\mathbf{F}'\mathbf{Q}_{\mathbf{FX}}), \quad \text{or shortly,} \quad \mathbf{C} \in \{\mathbf{P}_{\boldsymbol{\varepsilon}_*|\mathscr{T}_*}\}.$$

Thus, the BLUP of $\boldsymbol{\varepsilon}_*$ under $\mathscr{T}_*$ can be expressed as

$$\mathrm{BLUP}(\boldsymbol{\varepsilon}_* \mid \mathscr{T}_*) = \mathbf{V}_{21}\mathbf{F}'\mathbf{Q}_{\mathbf{FX}}(\mathbf{Q}_{\mathbf{FX}}\mathbf{FVF}'\mathbf{Q}_{\mathbf{FX}})^{-}\mathbf{Q}_{\mathbf{FX}}\mathbf{Fy}. \tag{11.43}$$

Recall that in (11.43), $\mathbf{F}'\mathbf{Q}_{\mathbf{FX}} = \mathbf{MF}'\mathbf{Q}_{\mathbf{FX}}$ and hence

$$\mathbf{Q}_{\mathbf{FX}}\mathbf{FVF}'\mathbf{Q}_{\mathbf{FX}} = \mathbf{Q}_{\mathbf{FX}}\mathbf{FWF}'\mathbf{Q}_{\mathbf{FX}}.$$

It can further be shown that (11.43) can be expressed also as

$$\text{BLUP}(\boldsymbol{\varepsilon}_* \mid \mathscr{T}_*) = \mathbf{V}_{21}\mathbf{N}(\mathbf{NVN})^-\mathbf{Ny} = \widetilde{\boldsymbol{\varepsilon}}_{t*},$$

where $\mathbf{N} = \mathbf{P}_{\mathbf{F}'\mathbf{Q}_{\mathbf{FX}}}$. This is based on the fact that if $\mathscr{C}(\mathbf{A}) = \mathscr{C}(\mathbf{B})$, then

$$\mathbf{P_W A(A'WA)^- A'P_W} = \mathbf{P_W B(B'WB)^- B'P_W}.$$

Thus, see also Isotalo et al. [44, Sect. 4], we have the following proposition.

Proposition 11.12 *Let $\mathbf{y}_*$ be predictable under $\mathscr{M}_*$, so that $\mathbf{X}_* = \mathbf{KX}$ for some $\mathbf{K} \in \mathbb{R}^{q\times n}$, and $\mathbf{G} = \mathbf{X}(\mathbf{X}'\mathbf{W}^-\mathbf{X}')^-\mathbf{X}'\mathbf{W}^-$. Then the* BLUP($\mathbf{y}_*$) *under $\mathscr{M}_*$ can be written as*

$$\begin{aligned}\text{BLUP}(\mathbf{y}_* \mid \mathscr{M}_*) &= \text{BLUE}(\boldsymbol{\mu}_* \mid \mathscr{M}) + \mathbf{V}_{21}\mathbf{V}^-[\mathbf{y} - \text{BLUE}(\boldsymbol{\mu} \mid \mathscr{M})]\\ &= \mathbf{KGy} + \mathbf{V}_{21}\mathbf{V}^-(\mathbf{I}_n - \mathbf{G})\mathbf{y}\\ &= \mathbf{KGy} + \mathbf{V}_{21}\mathbf{W}^-(\mathbf{I}_n - \mathbf{G})\mathbf{y}\\ &= \mathbf{KGy} + \mathbf{V}_{21}\mathbf{M}(\mathbf{MVM})^-\mathbf{My}\\ &= \text{BLUE}(\boldsymbol{\mu}_* \mid \mathscr{M}) + \text{BLUP}(\boldsymbol{\varepsilon}_* \mid \mathscr{M}_*),\end{aligned}$$

or shortly,

$$\widetilde{\mathbf{y}}_* = \widetilde{\boldsymbol{\mu}}_* + \widetilde{\boldsymbol{\varepsilon}}_*.$$

Let $\mathbf{y}_*$ be predictable under $\mathscr{T}_*$, so that $\mathbf{X}_* = \mathbf{LFX}$ for some $\mathbf{L} \in \mathbb{R}^{q\times f}$, and $\mathbf{G}_t$ is defined as in (11.42). *Then the* BLUP($\mathbf{y}_*$) *under $\mathscr{T}_*$ can be written as*

$$\begin{aligned}\text{BLUP}(\mathbf{y}_* \mid \mathscr{T}_*) &= \text{BLUE}(\boldsymbol{\mu}_* \mid \mathscr{T}) + \mathbf{V}_{21}\mathbf{F}'(\mathbf{FVF}')^-\mathbf{F}[\mathbf{y} - \text{BLUE}(\boldsymbol{\mu} \mid \mathscr{T})]\\ &= \mathbf{LFG}_t\mathbf{y} + \mathbf{V}_{21}\mathbf{F}'(\mathbf{FVF}')^-\mathbf{F}(\mathbf{I}_n - \mathbf{G}_t)\mathbf{y}\\ &= \mathbf{LFG}_t\mathbf{y} + \mathbf{V}_{21}\mathbf{N}(\mathbf{NVN})^-\mathbf{Ny}\\ &= \text{BLUE}(\boldsymbol{\mu}_* \mid \mathscr{T}) + \text{BLUP}(\boldsymbol{\varepsilon}_* \mid \mathscr{T}_*),\end{aligned}$$

or shortly,

$$\widetilde{\mathbf{y}}_{t*} = \widetilde{\boldsymbol{\mu}}_{t*} + \widetilde{\boldsymbol{\varepsilon}}_{t*},$$

where $\mathbf{N} = \mathbf{P}_{\mathbf{F}'\mathbf{Q}_{\mathbf{FX}}} = \mathbf{P}_{\mathscr{C}(\mathbf{F}')\cap\mathscr{C}(\mathbf{M})}$.

For Proposition 11.13, representing four equivalent statements, see, e.g., Baksalary and Kala [9, 10], Drygas [26], Tian and Puntanen [85, Th. 2.8], and Kala et al. [51, Th. 2]. To the properties of the covariance matrices of the BLUEs we return in Propositions 11.14 and 11.21.

Proposition 11.13 *Consider the model $\mathscr{M} = \{\mathbf{y}, \mathbf{X}\boldsymbol{\beta}, \mathbf{V}\}$ and its transformed version $\mathscr{T} = \{\mathbf{Fy}, \mathbf{FX}\boldsymbol{\beta}, \mathbf{FVF}'\}$, and let $\boldsymbol{\mu}_* = \mathbf{X}_*\boldsymbol{\beta}$ be estimable under $\mathscr{T}$ (and thereby under $\mathscr{M}$). Then the following five statements are equivalent:*

(a) $\mathbf{Fy}$ *is* BLUE-*sufficient for* $\boldsymbol{\mu}_* = \mathbf{X}_*\boldsymbol{\beta}$;
(b) $\widetilde{\boldsymbol{\mu}}_* = \widetilde{\boldsymbol{\mu}}_{t*}$ *with probability* 1;
(c) $\text{Cov}\,(\widetilde{\boldsymbol{\mu}}_*) = \text{Cov}\left(\widetilde{\boldsymbol{\mu}}_{t*}\right)$;
(d) $\{\mathbf{P}_{\mathbf{X}_*|\mathscr{T}}\mathbf{F}\} \cap \{\mathbf{P}_{\mathbf{X}_*|\mathscr{M}}\} \neq \{\emptyset\}$;

(e) $\{\mathbf{P}_{\mathbf{X}_*|\mathscr{T}}\mathbf{F}\} \subset \{\mathbf{P}_{\mathbf{X}_*|\mathscr{M}}\}$.

Example 11.3 (Centering the model) Consider the partitioned linear model

$$\mathscr{M}_{12} = \{\mathbf{y},\ \mathbf{1}\alpha + \mathbf{X}_0\boldsymbol{\beta}_{\mathbf{x}},\ \mathbf{I}_n\} = \{\mathbf{y},\ (\mathbf{1} : \mathbf{X}_0)\begin{pmatrix}\alpha \\ \boldsymbol{\beta}_{\mathbf{x}}\end{pmatrix},\ \mathbf{I}_n\},$$

where $\mathbf{1} \in \mathbb{R}^n$ is a vector of ones. Assume that $\mathbf{X} = (\mathbf{1} : \mathbf{X}_0)$ has full column rank. Premultiplying the model $\mathscr{M}_{12}$ by the centering matrix $\mathbf{Q}_1 = \mathbf{I}_n - \mathbf{P}_1$ yields the centered model

$$\mathscr{M}_{12\cdot 1} = \{\mathbf{Q}_1\mathbf{y},\ \mathbf{Q}_1\mathbf{X}_0\boldsymbol{\beta}_{\mathbf{x}},\ \mathbf{Q}_1\}.$$

In this centered model, we have a singular covariance matrix and hence it may seem that finding a BLUE would be problematic. However, between the covariance matrix $\mathbf{Q}_1$ and the model matrix $\mathbf{Q}_1\mathbf{X}_0$ we have the relation

$$\mathscr{C}(\mathbf{Q}_1 \cdot \mathbf{Q}_1\mathbf{X}_0) = \mathscr{C}(\mathbf{Q}_1\mathbf{X}_0).$$

Thus, corresponding to (11.26), we have the equality between OLSE($\mathbf{Q}_1\mathbf{X}_0\boldsymbol{\beta}_{\mathbf{x}}$) and BLUE($\mathbf{Q}_1\mathbf{X}_0\boldsymbol{\beta}_{\mathbf{x}}$), and thus

$$\mathrm{BLUE}(\boldsymbol{\beta}_{\mathbf{x}} \mid \mathscr{M}_{12\cdot 1}) = \mathrm{OLSE}(\boldsymbol{\beta}_{\mathbf{x}} \mid \mathscr{M}_{12\cdot 1}) = (\mathbf{X}_0'\mathbf{Q}_1\mathbf{X}_0)^{-1}\mathbf{X}_0'\mathbf{Q}_1\mathbf{y}. \tag{11.44}$$

On the other hand, it is well known that

$$\mathrm{BLUE}(\boldsymbol{\beta}_{\mathbf{x}} \mid \mathscr{M}_{12}) = (\mathbf{X}_0'\mathbf{Q}_1\mathbf{X}_0)^{-1}\mathbf{X}_0'\mathbf{Q}_1\mathbf{y}. \tag{11.45}$$

Now the equality of the BLUEs in (11.44) and (11.45) means that $\mathbf{Q}_1\mathbf{y}$ is linearly sufficient for $\boldsymbol{\beta}_{\mathbf{x}}$. This is a simple example of the Frisch–Waugh–Lovell theorem, into which we return in Sect. 11.7. □

Example 11.4 (Linear sufficiency of $\mathbf{V}^{-1/2}\mathbf{y}$ and $\mathbf{V}^{+1/2}\mathbf{y}$) For a positive definite $\mathbf{V}$, the linear sufficiency condition (c) of Proposition 11.3 becomes simply $\mathscr{C}(\mathbf{X}) \subset \mathscr{C}(\mathbf{V}\mathbf{F}')$. Thus, $\mathbf{V}^{-1/2}\mathbf{y}$ is linearly sufficient for $\mathbf{X}\boldsymbol{\beta}$ and the BLUE of $\mathbf{X}\boldsymbol{\beta}$ under the transformed model

$$\mathscr{T} = \{\mathbf{V}^{-1/2}\mathbf{y},\ \mathbf{V}^{-1/2}\mathbf{X}\boldsymbol{\beta},\ \mathbf{I}_n\}$$

is the same as in the original model $\mathscr{M} = \{\mathbf{y}, \mathbf{X}\boldsymbol{\beta}, \mathbf{V}\}$, i.e., the BLUE($\mathbf{X}\boldsymbol{\beta}$) under $\mathscr{M}$ equals the BLUE under $\mathscr{T}$:

$$\mathrm{BLUE}(\mathbf{X}\boldsymbol{\beta} \mid \mathscr{M}) = \mathrm{BLUE}(\mathbf{X}\boldsymbol{\beta} \mid \mathscr{T}) = \mathrm{OLSE}(\mathbf{X}\boldsymbol{\beta} \mid \mathscr{T}).$$

This technique, sometimes referred to as the Aitken approach, see Aitken [1], is well known in statistical textbooks. However, as remarked by Kala et al. [52, Sect. 3], usually these textbooks do not mention anything about linear sufficiency feature of this transformation.

Consider then so-called weakly singular model $\mathscr{M}_{ws}$, say, which means that $\mathscr{C}(\mathbf{X}) \subset \mathscr{C}(\mathbf{V})$. Transforming $\mathscr{M}$ by $\mathbf{V}^{+1/2}$ leads to the transformed model

$$\mathscr{T}_{ws} = \{\mathbf{V}^{+1/2}\mathbf{y},\ \mathbf{V}^{+1/2}\mathbf{X}\boldsymbol{\beta},\ \mathbf{P_V}\},$$

because $\mathbf{V}^{+1/2}\mathbf{V}\mathbf{V}^{+1/2} = \mathbf{P_V}$. Under the model $\mathscr{M}_{ws}$, the statistic $\mathbf{V}^{+1/2}\mathbf{y}$ is linearly sufficient for $\mathbf{X}\boldsymbol{\beta}$; this is due to the fact that $\mathscr{C}(\mathbf{X}) \subset \mathscr{C}(\mathbf{V}\mathbf{V}^{+1/2}) = \mathscr{C}(\mathbf{V})$. The $\mathbf{W}$-matrix in $\mathscr{T}_{ws}$ is $\mathbf{P_V}$ whose Moore–Penrose inverse is $\mathbf{P_V}$ and hence under $\mathscr{T}_{ws}$, we have

$$\begin{aligned}
\mathrm{BLUE}(\mathbf{V}^{+1/2}\mathbf{X}\boldsymbol{\beta}) &= \mathbf{V}^{+1/2}\mathrm{BLUE}(\mathbf{X}\boldsymbol{\beta}) \\
&= \mathbf{V}^{+1/2}\mathbf{X}(\mathbf{X}'\mathbf{V}^{+1/2}\mathbf{P_V}\mathbf{V}^{+1/2}\mathbf{X})^{-}\mathbf{X}'\mathbf{V}^{+1/2}\mathbf{P_V}\mathbf{V}^{+1/2}\mathbf{y} \\
&= \mathbf{V}^{+1/2}\mathbf{X}(\mathbf{X}'\mathbf{V}^{+}\mathbf{X})^{-}\mathbf{X}'\mathbf{V}^{+}\mathbf{y}. \qquad (11.46)
\end{aligned}$$

Premultiplying (11.46) by $\mathbf{V}^{1/2}$ gives the following presentation for the BLUE of $\mathbf{X}\boldsymbol{\beta}$ under a weakly singular linear model:

$$\mathrm{BLUE}(\mathbf{X}\boldsymbol{\beta} \mid \mathscr{M}_{ws}) = \mathbf{X}(\mathbf{X}'\mathbf{V}^{+}\mathbf{X})^{-}\mathbf{X}'\mathbf{V}^{+}\mathbf{y},$$

where $\mathbf{V}^{+}$ can be replaced with any $\mathbf{V}^{-}$. For the weakly singular models, see Zyskind and Martin [88]. □

Choosing $\mathbf{W} = \mathbf{V} + \mathbf{XUU'X'} \in \mathscr{W}$, we have, in light of (11.23), the following representations for the covariance matrix of the BLUE for $\boldsymbol{\mu} = \mathbf{X}\boldsymbol{\beta}$:

$$\begin{aligned}
\mathrm{Cov}\,(\tilde{\boldsymbol{\mu}} \mid \mathscr{M}) &= \mathbf{V} - \mathbf{VM}(\mathbf{MVM})^{-}\mathbf{MV} = \mathbf{X}(\mathbf{X}'\mathbf{W}^{+}\mathbf{X})^{-}\mathbf{X}' - \mathbf{T} \\
&= \mathbf{X}(\mathbf{X}'\mathbf{W}^{+1/2}\mathbf{W}^{+1/2}\mathbf{X})^{-}\mathbf{X}' - \mathbf{T}, \qquad (11.47)
\end{aligned}$$

where $\mathbf{T} = \mathbf{XUU'X'}$; for further representations, see, e.g., Isotalo et al. [49]. Assuming that $\mathbf{X}\boldsymbol{\beta}$ is estimable under the transformed model $\mathscr{T}$, we have

$$\begin{aligned}
\mathrm{Cov}\,(\tilde{\boldsymbol{\mu}} \mid \mathscr{T}) &= \mathbf{X}[\mathbf{X}'\mathbf{F}'(\mathbf{FWF}')^{-}\mathbf{FX}]^{-}\mathbf{X}' - \mathbf{T} \\
&= \mathbf{X}(\mathbf{X}'\mathbf{W}^{+1/2}\mathbf{P}_{\mathbf{W}^{1/2}\mathbf{F}'}\mathbf{W}^{+1/2}\mathbf{X})^{-}\mathbf{X}' - \mathbf{T}. \qquad (11.48)
\end{aligned}$$

Using (11.47) and (11.48), Markiewicz and Puntanen [64, Th. 1] gave some characterizations of the linear sufficiency in terms of covariance matrices:

Proposition 11.14 *Let* $\boldsymbol{\mu} = \mathbf{X}\boldsymbol{\beta}$ *be estimable under* $\mathscr{T}$ *and let* $\mathbf{W} \in \mathscr{W}$. *Then*

$$\mathrm{Cov}\,(\tilde{\boldsymbol{\mu}} \mid \mathscr{M}) \leq_{\mathrm{L}} \mathrm{Cov}\,(\tilde{\boldsymbol{\mu}} \mid \mathscr{T})\,.$$

Moreover, the following statements are equivalent:

(a) $\mathbf{Fy}$ *is linearly sufficient for* $\boldsymbol{\mu} = \mathbf{X}\boldsymbol{\beta}$ *under* $\mathscr{M}$,
(b) $\mathrm{Cov}\,(\tilde{\boldsymbol{\mu}} \mid \mathscr{M}) = \mathrm{Cov}\,(\tilde{\boldsymbol{\mu}} \mid \mathscr{T})$,

(c) $\mathbf{X}(\mathbf{X}'\mathbf{W}^{+}\mathbf{X})^{-}\mathbf{X}' = \mathbf{X}(\mathbf{X}'\mathbf{W}^{+1/2}\mathbf{P}_{\mathbf{W}^{1/2}\mathbf{F}'}\mathbf{W}^{+1/2}\mathbf{X})^{-}\mathbf{X}$,
(d) $\mathbf{X}'\mathbf{W}^{+}\mathbf{X} = \mathbf{X}'\mathbf{W}^{+1/2}\mathbf{P}_{\mathbf{W}^{1/2}\mathbf{F}'}\mathbf{W}^{+1/2}\mathbf{X}$,
(e) $\mathscr{C}(\mathbf{W}^{+1/2}\mathbf{X}) \subset \mathscr{C}(\mathbf{W}^{1/2}\mathbf{F}')$.

The following proposition collects together some important properties of the linear prediction sufficiency. For further details, see Isotalo and Puntanen [46] and Isotalo et al. [44]. We will denote

$$\begin{aligned} \{\mathbf{P}_{\mathbf{y}_*|\mathscr{M}_*}\} &= \{\mathbf{A} : \mathbf{Ay} = \mathrm{BLUP}(\mathbf{y}_* \mid \mathscr{M}_*)\}\,, \\ \{\mathbf{P}_{\boldsymbol{\varepsilon}_*|\mathscr{M}_*}\} &= \{\mathbf{B} : \mathbf{By} = \mathrm{BLUP}(\boldsymbol{\varepsilon}_* \mid \mathscr{M}_*)\}\,. \end{aligned}$$

The classes $\{\mathbf{P}_{\mathbf{y}_*|\mathscr{T}_*}\}$ and $\{\mathbf{P}_{\boldsymbol{\varepsilon}_*|\mathscr{T}_*}\}$ are defined in the corresponding way.

Proposition 11.15 *Suppose that $\mathbf{y}_*$ is predictable under $\mathscr{T}_*$. Then the following properties hold.*

(a) *The following seven statements are equivalent:*

(i) $\mathbf{Fy} \in \mathscr{S}(\mathbf{y}_*)$.
(ii) $\widetilde{\mathbf{y}}_* = \widetilde{\mathbf{y}}_{t*}$ *with probability* 1.
(iii) $\mathrm{Cov}\,(\widetilde{\mathbf{y}}_* - \widetilde{\mathbf{y}}_{t*}) = \mathbf{0}$.
(iv) $\mathrm{Cov}\,(\mathbf{y}_* - \widetilde{\mathbf{y}}_*) = \mathrm{Cov}\,(\mathbf{y}_* - \widetilde{\mathbf{y}}_{t*})$.
(v) $\{\mathbf{P}_{\mathbf{y}_*|\mathscr{T}_*}\mathbf{F}\} \cap \{\mathbf{P}_{\mathbf{y}_*|\mathscr{M}_*}\} \neq \{\emptyset\}$.
(vi) $\{\mathbf{P}_{\mathbf{y}_*|\mathscr{T}_*}\mathbf{F}\} \subset \{\mathbf{P}_{\mathbf{y}_*|\mathscr{M}_*}\}$.
(vii) $\mathrm{Cov}\,(\widetilde{\mathbf{y}}_*) = \mathrm{Cov}\,(\widetilde{\mathbf{y}}_{t*}) = \mathrm{Cov}\,(\widetilde{\mathbf{y}}_*, \widetilde{\mathbf{y}}_{t*})$.

(b) *The following six statements are equivalent:*

(viii) $\mathbf{Fy} \in \mathscr{S}(\boldsymbol{\varepsilon}_*)$.
(ix) $\widetilde{\boldsymbol{\varepsilon}}_* = \widetilde{\boldsymbol{\varepsilon}}_{t*}$ *with probability* 1.
(x) $\mathrm{Cov}\,(\widetilde{\boldsymbol{\varepsilon}}_*) = \mathrm{Cov}\,(\widetilde{\boldsymbol{\varepsilon}}_{t*})$.
(xi) $\mathrm{Cov}\,(\boldsymbol{\varepsilon}_* - \widetilde{\boldsymbol{\varepsilon}}_*) = \mathrm{Cov}\,(\boldsymbol{\varepsilon}_* - \widetilde{\boldsymbol{\varepsilon}}_{t*})$.
(xii) $\{\mathbf{P}_{\boldsymbol{\varepsilon}_*|\mathscr{T}_*}\mathbf{F}\} \cap \{\mathbf{P}_{\boldsymbol{\varepsilon}_*|\mathscr{M}_*}\} \neq \{\emptyset\}$.
(xiii) $\{\mathbf{P}_{\boldsymbol{\varepsilon}_*|\mathscr{T}_*}\mathbf{F}\} \subset \{\mathbf{P}_{\boldsymbol{\varepsilon}_*|\mathscr{M}_*}\}$.

Let us prove the equivalence of (iii) and (vii). Denoting $\widetilde{\mathbf{y}}_* = \mathbf{Ay}$, $\widetilde{\mathbf{y}}_{t*} = \mathbf{By}$, we observe that

$$\begin{aligned} \mathrm{Cov}\,(\widetilde{\mathbf{y}}_* - \widetilde{\mathbf{y}}_{t*}) &= (\mathbf{A} - \mathbf{B})\mathbf{V}(\mathbf{A} - \mathbf{B})' \\ &= \mathbf{AVA}' + \mathbf{BVB}' - \mathbf{AVB}' - \mathbf{BVA}' \\ &= \mathrm{Cov}\,(\widetilde{\mathbf{y}}_*) + \mathrm{Cov}\,(\widetilde{\mathbf{y}}_{t*}) - \mathrm{Cov}\,(\widetilde{\mathbf{y}}_*, \widetilde{\mathbf{y}}_{*t}) - \mathrm{Cov}\,(\widetilde{\mathbf{y}}_{t*}, \widetilde{\mathbf{y}}_*) \\ &= \mathbf{0} \end{aligned} \tag{11.49}$$

holds if and only if $\mathbf{AV} = \mathbf{BV}$. Now $\mathbf{AV} = \mathbf{BV}$ immediately implies (vii). On the other hand, (vii) obviously implies (11.49).

Notice that in part (a) of Proposition 11.15 we do *not* have the covariance equality condition $\mathrm{Cov}\,(\widetilde{\mathbf{y}}_*) = \mathrm{Cov}\,(\widetilde{\mathbf{y}}_{t*})$; instead, we have $\mathrm{Cov}\,(\widetilde{\mathbf{y}}_* - \widetilde{\mathbf{y}}_{t*}) = \mathbf{0}$. In view of Lemma 11.6, the equality $\widetilde{\mathbf{y}}_* = \widetilde{\mathbf{y}}_{t*}$ holds with probability 1 if and only if $\mathrm{Cov}\,(\widetilde{\mathbf{y}}_* - \widetilde{\mathbf{y}}_{t*}) = \mathbf{0}$. We return into this feature in Sect. 11.8.

11.5 Relative Linear Sufficiency

When studying the relative efficiency of OLSE vs BLUE of $\boldsymbol{\beta}$ we are dealing with two linear models $\mathscr{M}_{\mathbf{I}} = \{\mathbf{y}, \mathbf{X}\boldsymbol{\beta}, \mathbf{I}_n\}$, and $\mathscr{M} = \{\mathbf{y}, \mathbf{X}\boldsymbol{\beta}, \mathbf{V}\}$, where $\mathbf{X}$ has full column rank and $\mathbf{V}$ is positive definite. The corresponding BLUEs are $\widehat{\boldsymbol{\beta}} = (\mathbf{X}'\mathbf{X})^{-1}\mathbf{X}'\mathbf{y}$ and $\widetilde{\boldsymbol{\beta}} = (\mathbf{X}'\mathbf{V}^{-1}\mathbf{X})^{-1}\mathbf{X}'\mathbf{V}^{-1}\mathbf{y}$. Assuming that the model $\mathscr{M}$ is the correct one, the relative goodness of $\widehat{\boldsymbol{\beta}}$ with respect to $\widetilde{\boldsymbol{\beta}}$ can be measured by various means. The most common measure is the Watson efficiency, see Watson [86] and Bloomfield and Watson [19],

$$\phi = \frac{|\mathrm{Cov}\,(\widetilde{\boldsymbol{\beta}})|}{|\mathrm{Cov}\,(\widehat{\boldsymbol{\beta}})|} = \frac{|(\mathbf{X}'\mathbf{V}^{-1}\mathbf{X})^{-1}|}{|(\mathbf{X}'\mathbf{X})^{-1}\mathbf{X}'\mathbf{V}\mathbf{X}(\mathbf{X}'\mathbf{X})^{-1}|} = \frac{|\mathbf{X}'\mathbf{X}|^2}{|\mathbf{X}'\mathbf{V}\mathbf{X}| \cdot |\mathbf{X}'\mathbf{V}^{-1}\mathbf{X}|},$$

where $|\cdot|$ refers to the determinant. Obviously, $0 < \phi \leq 1$ and the upper bound is attained when $\widetilde{\boldsymbol{\beta}} = \widehat{\boldsymbol{\beta}}$.

Kala et al. [52, Sect. 5] consider the models $\mathscr{M} = \{\mathbf{y}, \mathbf{X}\boldsymbol{\beta}, \mathbf{V}\}$ and $\mathscr{T} = \{\mathbf{F}\mathbf{y}, \mathbf{F}\mathbf{X}\boldsymbol{\beta}, \mathbf{F}\mathbf{V}\mathbf{F}'\}$, and aim to do something similar with

$$\begin{aligned} \mathrm{BLUE}(\boldsymbol{\beta} \mid \mathscr{T}) = \widetilde{\boldsymbol{\beta}}_t &= [\mathbf{X}'\mathbf{F}'(\mathbf{F}\mathbf{V}\mathbf{F}')^{-}\mathbf{F}\mathbf{X}]^{-1}\mathbf{X}'\mathbf{F}'(\mathbf{F}\mathbf{V}\mathbf{F}')^{-}\mathbf{F}\mathbf{y}, \\ \mathrm{Cov}\,(\widetilde{\boldsymbol{\beta}}_t) &= [\mathbf{X}'\mathbf{F}'(\mathbf{F}\mathbf{V}\mathbf{F}')^{-}\mathbf{F}\mathbf{X}]^{-1}, \end{aligned}$$

where it is assumed that $\mathbf{F}\mathbf{X}$ is estimable under $\mathscr{T}$. Corresponding to the Watson efficiency, we could consider the ratio

$$\gamma = \frac{|\mathrm{Cov}\,(\widetilde{\boldsymbol{\beta}})|}{|\mathrm{Cov}\,(\widetilde{\boldsymbol{\beta}}_t)|} = \frac{|(\mathbf{X}'\mathbf{V}^{-1}\mathbf{X})^{-1}|}{|[\mathbf{X}'\mathbf{F}'(\mathbf{F}\mathbf{V}\mathbf{F}')^{-}\mathbf{F}\mathbf{X}]^{-1}|} = \frac{|\mathbf{X}'\mathbf{V}^{-1/2}\mathbf{P}_{\mathbf{V}^{1/2}\mathbf{F}'}\mathbf{V}^{-1/2}\mathbf{X}|}{|\mathbf{X}'\mathbf{V}^{-1/2}\mathbf{V}^{-1/2}\mathbf{X}|}.$$

Clearly $0 < \gamma \leq 1$, where the upper bound is attained if and only if $\mathbf{F}\mathbf{y}$ is linearly sufficient for $\boldsymbol{\beta}$. We could keep $\mathbf{X}$ and $\mathbf{V}$ given and try to figure out which $\mathbf{F}$ yields the minimum of γ subject to the condition $\mathrm{r}(\mathbf{X}) = \mathrm{r}(\mathbf{F}\mathbf{X})$. The lower bound for the Watson efficiency was found by Bloomfield and Watson [19]. However, it seems to be nontrivial to find the lower bound for γ. The (attainable) lower bound zero does not make sense, of course.

Kala et al. [52, Sect. 5] consider also another measure of the relative linear sufficiency based on the linear sufficiency condition $\mathscr{C}(\mathbf{X}) \subset \mathscr{C}(\mathbf{W}\mathbf{F}')$ which is equivalent to

$$\mathbf{P}_{\mathbf{W}\mathbf{F}'}\mathbf{X} = \mathbf{X}. \tag{11.50}$$

Hence, one can wonder how "badly" (11.50) is satisfied by considering the difference $\mathbf{D} = \mathbf{X} - \mathbf{P}_{\mathbf{W}\mathbf{F}'}\mathbf{X}$. The "size" of $\mathbf{D}$ could be measured by the Frobenius norm as

$$\|\mathbf{D}\|_F^2 = \mathrm{Tr}\,(\mathbf{D}'\mathbf{D}) = \mathrm{Tr}\,(\mathbf{X}'\mathbf{X}) - \mathrm{Tr}\,(\mathbf{X}'\mathbf{P}_{\mathbf{W}\mathbf{F}'}\mathbf{X}).$$

Hence, the relative linear sufficiency of $\mathbf{Fy}$ could be defined as

$$\psi = \frac{\mathrm{Tr}\,(\mathbf{X}'\mathbf{P}_{\mathbf{WF}'}\mathbf{X})}{\mathrm{Tr}\,(\mathbf{X}'\mathbf{X})},$$

where $\mathrm{Tr}\,(\cdot)$ refers to the trace. Now $0 \le \psi \le 1$, where the lower bound is attained when $\mathscr{C}(\mathbf{X}) \subset \mathscr{C}(\mathbf{WF}')^{\perp}$ and the upper bound when $\mathscr{C}(\mathbf{X}) \subset \mathscr{C}(\mathbf{WF}')$, i.e., when $\mathbf{Fy}$ is linearly sufficient for $\mathbf{X}\boldsymbol{\beta}$.

Kala et al. [52, Remark 3] also consider two transformation matrices $\mathbf{F}_1$ and $\mathbf{F}_2$ and the corresponding transformed models

$$\mathscr{T}_i = \{\mathbf{F}_i\mathbf{y},\ \mathbf{F}_i\mathbf{X}\boldsymbol{\beta},\ \mathbf{F}_i\mathbf{V}\mathbf{F}_i'\},\quad i \in \{1,2\},$$

and assume that $\mathrm{r}(\mathbf{F}_1\mathbf{X}) = \mathrm{r}(\mathbf{F}_2\mathbf{X}) = \mathrm{r}(\mathbf{X}) = p$, so that $\boldsymbol{\beta}$ is estimable in both models. Then the Löwner ordering $\mathrm{Cov}\,(\widetilde{\boldsymbol{\beta}}_{t1}) \le_{\mathrm{L}} \mathrm{Cov}\,(\widetilde{\boldsymbol{\beta}}_{t2})$ holds if and only if

$$\mathbf{X}'\mathbf{V}^{-1/2}\mathbf{P}_{\mathbf{V}^{1/2}\mathbf{F}_2'}\mathbf{V}^{-1/2}\mathbf{X} \le_{\mathrm{L}} \mathbf{X}'\mathbf{V}^{-1/2}\mathbf{P}_{\mathbf{V}^{1/2}\mathbf{F}_1'}\mathbf{V}^{-1/2}\mathbf{X}, \tag{11.51}$$

i.e.,

$$\mathbf{X}'\mathbf{V}^{-1/2}\left(\mathbf{P}_{\mathbf{V}^{1/2}\mathbf{F}_1'} - \mathbf{P}_{\mathbf{V}^{1/2}\mathbf{F}_2'}\right)\mathbf{V}^{-1/2}\mathbf{X} \ge_{\mathrm{L}} \mathbf{0}.$$

The matrix $\mathbf{P}_{\mathbf{V}^{1/2}\mathbf{F}_1'} - \mathbf{P}_{\mathbf{V}^{1/2}\mathbf{F}_2'}$ is nonnegative definite if and only if (see, e.g., Puntanen et al. see, e.g., [72, p. 152])

$$\mathscr{C}\left(\mathbf{F}_2'\right) \subset \mathscr{C}\left(\mathbf{F}_1'\right).$$

Hence, (11.51) holds if $\mathscr{C}\left(\mathbf{F}_2'\right) \subset \mathscr{C}\left(\mathbf{F}_1'\right)$. In this case, we can say that in a sense $\mathbf{F}_1\mathbf{y}$ is "more than or equally linearly sufficient" than $\mathbf{F}_2\mathbf{y}$ even though neither of them need to be "fully linearly sufficient." However, it may well be that there is no Löwner ordering between the covariance matrices $\mathrm{Cov}\,(\widetilde{\boldsymbol{\beta}}_{t1})$ and $\mathrm{Cov}\,(\widetilde{\boldsymbol{\beta}}_{t2})$. Then some other criteria should be used to compare the "linear sufficiency" of $\mathbf{F}_1\mathbf{y}$ and $\mathbf{F}_2\mathbf{y}$. Using the matrix-rank method, Dong et al. [24] provide an extensive study of the relations between the covariance matrices of the BLUEs under $\mathscr{M}_{t1}$ and $\mathscr{M}_{t2}$.

We conclude this section with a curious result related to the Watson efficiency and linear sufficiency. Chu et al. [20, 21] showed that in the partitioned (weakly singular) linear model the Watson efficiency can be decomposed into the product

$$\mathrm{eff}\left(\widehat{\boldsymbol{\beta}} \mid \mathscr{M}_{12}\right) = \mathrm{eff}\left(\widehat{\boldsymbol{\beta}}_1 \mid \mathscr{M}_1\right) \cdot \mathrm{eff}\left(\widehat{\boldsymbol{\beta}}_2 \mid \mathscr{M}_{12}\right) \cdot \frac{1}{\mathrm{eff}\left(\widehat{\boldsymbol{\beta}}_1 \mid \mathscr{M}_{1H}\right)}, \tag{11.52}$$

where $\mathscr{M}_1 = \{\mathbf{y}, \mathbf{X}_1\boldsymbol{\beta}_1, \mathbf{V}\}$, and $\mathscr{M}_{1H} = \{\mathbf{Hy}, \mathbf{X}_1\boldsymbol{\beta}_1, \mathbf{HVH}\}$, and $\mathbf{H} = \mathbf{P}_{\mathbf{X}}$. Thus, $\mathscr{M}_{1H}$ is a transformed version of $\mathscr{M}_1$, transformation matrix being $\mathbf{F} = \mathbf{H}$. Chu et al. [21, Th. 2.2] proved that a necessary and sufficient condition for

$$\mathrm{eff}\left(\widehat{\boldsymbol{\beta}} \mid \mathscr{M}_{12}\right) = \mathrm{eff}\left(\widehat{\boldsymbol{\beta}}_2 \mid \mathscr{M}_{12}\right) \tag{11.53}$$

is that $\mathbf{Hy}$ is linearly sufficient for $\mathbf{X}_1\boldsymbol{\beta}_1$ under $\mathscr{M}_1$, i.e., $\mathscr{C}(\mathbf{X}_1) \subset \mathscr{C}(\mathbf{VH}) = \mathscr{C}(\mathbf{VX})$. Given that (11.52) holds, the claim (11.53) is easy to prove using Proposition 11.14, which says that under linear sufficiency,

$$\operatorname{Cov}\left(\tilde{\boldsymbol{\beta}}_1 \mid \mathscr{M}_1\right) = \operatorname{Cov}\left(\tilde{\boldsymbol{\beta}}_1 \mid \mathscr{M}_{1H}\right).$$

Of course, $\operatorname{Cov}\left(\widehat{\boldsymbol{\beta}}_1 \mid \mathscr{M}_1\right) = \operatorname{Cov}\left(\widehat{\boldsymbol{\beta}}_1 \mid \mathscr{M}_{1H}\right) = (\mathbf{X}_1'\mathbf{X}_1)^{-1}\mathbf{X}_1'\mathbf{V}\mathbf{X}_1(\mathbf{X}_1'\mathbf{X}_1)^{-1}$. We find the reduction of (11.52) into (11.53) somewhat unexpected and it is not obvious to find a "natural explanation" for it.

11.6 The "Vice Versa" Problem

Let us take a close look at Theorem of Baksalary and Kala [9, p. 914]. In part (ii) they write the following (in our notation):

(a) "If the condition $\mathscr{C}(\mathbf{X}) \subset \mathscr{C}(\mathbf{WF}')$ is satisfied, then each BLUE of $\mathbf{X}\boldsymbol{\beta}$ in the transformed model $\mathscr{T} = \{\mathbf{Fy}, \mathbf{FX}\boldsymbol{\beta}, \mathbf{FVF}'\}$ is also a BLUE of $\mathbf{X}\boldsymbol{\beta}$ in the original model $\mathscr{M} = \{\mathbf{y}, \mathbf{X}\boldsymbol{\beta}, \mathbf{V}\}$, and vice versa."

On the other hand, Baksalary and Kala [10, Th. 1] states the following:

(b) "If $\mathbf{Fy}$ is linearly sufficient for estimable $\mathbf{X}_*\boldsymbol{\beta}$, then every representation of the BLUE of $\mathbf{X}_*\boldsymbol{\beta}$ in the induced model $\mathscr{T}$ is also the BLUE of $\mathbf{X}_*\boldsymbol{\beta}$ in the original model $\mathscr{M}$."

It is the phrase *vice versa* that may cause some confusion as stated by Kala et al. [51, Sect. 4]. Let us discuss the meaning of the *vice versa* part.

Consider the multipliers of the response vector $\mathbf{y}$ when playing with the BLUEs under $\mathscr{M}$ and under $\mathscr{T}$. Let $\mathbf{X}_*\boldsymbol{\beta}$ be an estimable parametric function under the model $\mathscr{T}$ (and thereby also under $\mathscr{M}$) and denote

$$\begin{aligned}
\mathscr{A} &= \{\mathbf{A} : \mathbf{AFy} = \operatorname{BLUE}(\mathbf{X}_*\boldsymbol{\beta} \mid \mathscr{M})\} \\
&= \{\mathbf{A} : \mathbf{AF}(\mathbf{X} : \mathbf{VM}) = (\mathbf{X}_* : \mathbf{0})\}, \\
\mathscr{C} &= \{\mathbf{C} : \mathbf{CFy} = \operatorname{BLUE}(\mathbf{X}_*\boldsymbol{\beta} \mid \mathscr{T})\} \\
&= \left\{\mathbf{C} : \mathbf{C}(\mathbf{FX} : \mathbf{FVF}'\mathbf{Q}_{\mathbf{FX}}) = (\mathbf{X}_* : \mathbf{0})\right\} \\
&= \left\{\mathbf{C} : \mathbf{CF}(\mathbf{X} : \mathbf{VMF}'\mathbf{Q}_{\mathbf{FX}}) = (\mathbf{X}_* : \mathbf{0})\right\} \\
&= \{\mathbf{P}_{\mathbf{X}_*\mid\mathscr{T}}\},
\end{aligned}$$

where we have used Lemma 11.4. Assume further that $\mathbf{Fy}$ is linearly sufficient for $\mathbf{X}_*\boldsymbol{\beta}$, which actually guarantees that $\mathscr{A}$ and $\mathscr{C}$ are not empty.

By Proposition 11.10 we observe immediately that $\mathscr{A} \subset \mathscr{C}$. To go the other way, let us pick $\mathbf{C} \in \mathscr{C} = \{\mathbf{P}_{\mathbf{X}_*\mid\mathscr{T}}\}$ and utilize part (c) of Proposition 11.11 which says that (under linear sufficiency of $\mathbf{Fy}$) $\mathbf{CF} \in \{\mathbf{P}_{\mathbf{X}_*\mid\mathscr{M}}\}$, i.e., $\mathbf{CF}$ satisfies $\mathbf{CF}(\mathbf{X} : \mathbf{VM}) = (\mathbf{X}_* : \mathbf{0})$. Thus, we have confirmed the following, see Kala et al. [51, Th. 3].

Proposition 11.16 *Suppose that* $\mathbf{Fy}$ *is linearly sufficient for the estimable parametric function* $\mathbf{X}_*\boldsymbol{\beta}$ *under the linear model* $\mathscr{M}$, *and let the sets of matrices* $\mathscr{A}$ *and* $\mathscr{C}$ *be defined as above. Then* $\mathscr{A} = \mathscr{C} = \{\mathbf{P}_{\mathbf{X}_*|\mathscr{T}}\}$ *and*

$$\mathscr{B} = \big\{\mathbf{B} : \mathbf{B} = \mathbf{AF} \in \{\mathbf{P}_{\mathbf{X}_*|\mathscr{M}}\}\big\} = \{\mathbf{P}_{\mathbf{X}_*|\mathscr{T}}\mathbf{F}\}.$$

Assume that $\mathbf{Fy}$ is linearly sufficient for estimable $\boldsymbol{\mu}_*$ under $\mathscr{M}$. Then Proposition 11.16 means that $\mathbf{CFy}$ is the BLUE for $\boldsymbol{\mu}_*$ under $\mathscr{T}$ if and only if $\mathbf{CFy}$ is the BLUE for $\boldsymbol{\mu}_*$ under $\mathscr{M}$. In other words, for each matrix $\mathbf{C}$ such that $\mathbf{CFy}$ is the BLUE of $\boldsymbol{\mu}_*$ in the transformed model $\mathscr{T}$, the statistic $\mathbf{CFy}$ is also the BLUE of $\boldsymbol{\mu}_*$ in the original model $\mathscr{M}$, and vice versa. Notice that in this statement the "vice versa" means that we consider such $\mathbf{C}$ for which $\mathbf{CFy}$ is BLUE under $\mathscr{M}$, not the set of matrices $\mathbf{B} \in \{\mathbf{P}_{\mathbf{X}_*|\mathscr{M}}\}$ such that $\mathbf{By}$ is BLUE under $\mathscr{M}$.

It is noteworthy that we have the inclusion

$$\mathscr{B} = \big\{\mathbf{B} : \mathbf{B} = \mathbf{AF} \in \{\mathbf{P}_{\mathbf{X}_*|\mathscr{M}}\}\big\} \subset \{\mathbf{P}_{\mathbf{X}_*|\mathscr{M}}\},$$

which immediately implies part (c) of Proposition 11.11:

$$\{\mathbf{P}_{\mathbf{X}_*|\mathscr{T}}\mathbf{F}\} \subset \{\mathbf{P}_{\mathbf{X}_*|\mathscr{M}}\}. \tag{11.54}$$

It is clear that it is not necessary that every matrix from the set $\{\mathbf{P}_{\mathbf{X}_*|\mathscr{M}}\}$ belongs to set $\mathscr{B}$.

The total sets of multipliers of $\mathbf{y}$ for the BLUEs of $\mathbf{X}_*\boldsymbol{\beta}$ under $\mathscr{M}$ and $\mathscr{T}$ are $\{\mathbf{P}_{\mathbf{X}_*|\mathscr{T}}\mathbf{F}\}$ and $\{\mathbf{P}_{\mathbf{X}_*|\mathscr{M}}\}$, respectively. Kala et al. [51, Sect. 4] shows that the equality in (11.54) holds if $\mathbf{F}$ is a nonsingular square matrix. Using a different approach, Tian [81, Th. 3.1] has shown that the nonsingularity of $\mathbf{F}$ is also necessary for the equality in (11.54). However, Tian's assumptions are slightly different from those of ours and thus in our considerations the equality in (11.54) can appear without $\mathbf{F}$ being nonsingular. A simple example is obtained by considering model $\mathscr{M}$, where $\mathbf{X}$ has full column rank and $\mathbf{V}$ is positive definite. Then choosing $\mathbf{F} = \mathbf{X}'\mathbf{V}^{-1}$, we have $\mathbf{Fy} \in \mathscr{S}(\boldsymbol{\beta})$ and

$$\mathscr{T} = \{\mathbf{X}'\mathbf{V}^{-1}\mathbf{y},\ \mathbf{X}'\mathbf{V}^{-1}\mathbf{X}\boldsymbol{\beta},\ \mathbf{X}'\mathbf{V}^{-1}\mathbf{X}\},$$

yielding

$$\mathrm{BLUE}\,(\boldsymbol{\beta} \mid \mathscr{M}) = (\mathbf{X}'\mathbf{V}^{-1}\mathbf{X})^{-1}\mathbf{X}'\mathbf{V}^{-1}\mathbf{y} = \mathrm{BLUE}\,(\boldsymbol{\beta} \mid \mathscr{T})\,.$$

Thus, there is only one member, $(\mathbf{X}'\mathbf{V}^{-1}\mathbf{X})^{-1}\mathbf{X}'\mathbf{V}^{-1}$, in sets $\{\mathbf{P}_{\mathbf{X}_*|\mathscr{T}}\mathbf{F}\}$ and $\{\mathbf{P}_{\mathbf{X}_*|\mathscr{M}}\}$; yet $\mathbf{F} = \mathbf{X}'\mathbf{V}^{-1} \in \mathbb{R}^{p\times n}$ is not a square matrix and of course not invertible.

A deeper look at the possible equality in (11.54) yields Proposition 11.17 (which according to our knowledge is new). For this purpose, we denote

$$\begin{aligned}\{\mathbf{P}_{\boldsymbol{\varepsilon}_*|\mathscr{T}_*}\} &= \{\mathbf{D} : \mathbf{DFy} = \mathrm{BLUP}(\boldsymbol{\varepsilon}_* \mid \mathscr{T}_*)\}\\ &= \big\{\mathbf{D} : \mathbf{D}(\mathbf{FX} : \mathbf{FVF}'\mathbf{Q}_{\mathbf{FX}}) = (\mathbf{0} : \mathbf{V}_{21}\mathbf{M})\big\}\,.\end{aligned}$$

Proposition 11.17 *Let $\mathbf{X}_*\boldsymbol{\beta}$ be estimable under $\mathscr{T}$ and consider condition*

$$\mathbf{Q_W} = \mathbf{Q_W P_{F'}}, \quad i.e., \quad \mathscr{C}(\mathbf{W})^{\perp} \subset \mathscr{C}(\mathbf{F}'). \tag{11.55}$$

(a) *If* $\mathbf{Fy} \in \mathscr{S}(\mathbf{X}_*\boldsymbol{\beta})$, *then* $\{\mathbf{P}_{\mathbf{X}_*|\mathscr{M}}\} = \{\mathbf{P}_{\mathbf{X}_*|\mathscr{T}}\mathbf{F}\} \iff$ (11.55) *holds.*
(b) *If* $\mathbf{Fy} \in \mathscr{S}(\boldsymbol{\varepsilon}_*)$, *then* $\{\mathbf{P}_{\boldsymbol{\varepsilon}_*|\mathscr{M}_*}\} = \{\mathbf{P}_{\boldsymbol{\varepsilon}_*|\mathscr{T}_*}\mathbf{F}\} \iff$ (11.55) *holds.*
(c) *If* $\mathbf{Fy} \in \mathscr{S}(\mathbf{y}_*)$, *then* $\{\mathbf{P}_{\mathbf{y}_*|\mathscr{M}_*}\} = \{\mathbf{P}_{\mathbf{y}_*|\mathscr{T}_*}\mathbf{F}\} \iff$ (11.55) *holds.*

In other words, under the linear sufficiency and condition (11.55), *each representation of the* BLUE *of* $\mathbf{X}_*\boldsymbol{\beta}$ *under* $\mathscr{M}$ *is a representation of the* BLUE *under* $\mathscr{T}$ *and vice versa, and the same concerns the* BLUP *for* $\boldsymbol{\varepsilon}_*$ *and* $\mathbf{y}_*$.

Proof We know, by (11.54) and part (c) of Proposition 11.11, that linear sufficiency of $\mathbf{Fy}$ with respect to $\mathbf{X}_*\boldsymbol{\beta}$ implies the inclusion $\{\mathbf{P}_{\mathbf{X}_*|\mathscr{T}}\mathbf{F}\} \subset \{\mathbf{P}_{\mathbf{X}_*|\mathscr{M}}\}$. Thus, the claim (a) means that if $\mathbf{Fy} \in \mathscr{S}(\mathbf{X}_*\boldsymbol{\beta})$, then

$$(11.55) \iff \{\mathbf{P}_{\mathbf{X}_*|\mathscr{M}}\} \subset \{\mathbf{P}_{\mathbf{X}_*|\mathscr{T}}\mathbf{F}\}.$$

The general expression for $\mathbf{B}_0 \in \{\mathbf{P}_{\mathbf{X}_*|\mathscr{M}}\}$ is

$$\mathbf{B}_0 = \mathbf{B}_1 + \mathbf{E}\mathbf{Q_W},$$

where $\mathbf{B}_1$ is one matrix from the set $\{\mathbf{P}_{\mathbf{X}_*|\mathscr{M}}\}$ and $\mathbf{E} \in \mathbb{R}^{q\times n}$ is arbitrary. Because $\mathbf{Fy}$ is linearly sufficient for $\mathbf{X}_*\boldsymbol{\beta}$, we know that there exists a matrix $\mathbf{A} \in \mathbb{R}^{q\times f}$ such that $\mathbf{AF} \in \{\mathbf{P}_{\mathbf{X}_*|\mathscr{M}}\}$. Choosing $\mathbf{B}_1 = \mathbf{AF}$ gives

$$\mathbf{B}_0 = \mathbf{AF} + \mathbf{E}\mathbf{Q_W}. \tag{11.56}$$

Assume now that

$$\{\mathbf{P}_{\mathbf{X}_*|\mathscr{M}}\} \subset \{\mathbf{P}_{\mathbf{X}_*|\mathscr{T}}\mathbf{F}\}. \tag{11.57}$$

We observe that (11.57) implies that $\mathbf{B}_0$ is of the form $\mathbf{B}_0 = \mathbf{LF}$ for some $\mathbf{L} \in \mathbb{R}^{q\times f}$, which further means that

$$\mathbf{B}_0\mathbf{P_{F'}} = \mathbf{B}_0. \tag{11.58}$$

Substituting (11.58) into (11.56) gives

$$\mathbf{AF} + \mathbf{E}\mathbf{Q_W}\mathbf{P_{F'}} = \mathbf{AF} + \mathbf{E}\mathbf{Q_W},$$

i.e.,

$$\mathbf{E}(\mathbf{Q_W} - \mathbf{Q_W}\mathbf{P_{F'}}) = \mathbf{0}. \tag{11.59}$$

Equation (11.59) has to hold for any $\mathbf{E} \in \mathbb{R}^{q\times n}$ and hence we necessarily must have

$$\mathbf{P_{F'}}\mathbf{Q_W} = \mathbf{Q_W}, \quad \text{i.e.,} \quad \mathbf{P_{F'}}(\mathbf{I}_n - \mathbf{P_W}) = (\mathbf{I}_n - \mathbf{P_W}),$$

which is precisely the condition (11.55). Thus, we have shown that if $\mathbf{Fy} \in \mathscr{S}(\mathbf{X}_*\boldsymbol{\beta})$, then (11.57) implies (11.55), i.e.,

$$\{\mathbf{P}_{\mathbf{X}_*|\mathscr{M}}\} = \{\mathbf{P}_{\mathbf{X}_*|\mathscr{T}}\mathbf{F}\} \implies (11.55).$$

It remains to show that if $\mathbf{Fy} \in \mathscr{S}(\mathbf{X}_*\boldsymbol{\beta})$, then

$$(11.55) \implies \{\mathbf{P}_{\mathbf{X}_*|\mathscr{M}}\} \subset \{\mathbf{P}_{\mathbf{X}_*|\mathscr{T}}\mathbf{F}\}.$$

In view of (11.56), we know that there exists a matrix $\mathbf{A}$ such that the general expression for $\mathbf{B}_0 \in \{\mathbf{P}_{\mathbf{X}_*|\mathscr{M}}\}$ is

$$\mathbf{B}_0 = \mathbf{AF} + \mathbf{EQ}_\mathbf{W}, \tag{11.60}$$

where $\mathbf{E}$ is free to vary. Postmultiplying (11.60) by $\mathbf{P}_{\mathbf{F}'} = \mathbf{F}^+\mathbf{F}$ yields

$$\mathbf{B}_0\mathbf{P}_{\mathbf{F}'} = \mathbf{AF} + \mathbf{EQ}_\mathbf{W}\mathbf{P}_{\mathbf{F}'}. \tag{11.61}$$

Assuming that (11.55) holds, we have $\mathbf{Q}_\mathbf{W}\mathbf{P}_{\mathbf{F}'} = \mathbf{Q}_\mathbf{W}$ and thereby in light of (11.60) and (11.61), we have

$$\mathbf{B}_0 = \mathbf{B}_0\mathbf{P}_{\mathbf{F}'} = \mathbf{B}_0\mathbf{F}^+ \cdot \mathbf{F} := \mathbf{B}_2\mathbf{F},$$

where $\mathbf{B}_2 = \mathbf{B}_0\mathbf{F}^+$. It remains to show that $\mathbf{B}_2 = \mathbf{B}_0\mathbf{F}^+ \in \{\mathbf{P}_{\mathbf{X}_*|\mathscr{T}}\}$. This follows at once from Proposition 11.10 which says that

$$\mathbf{B}_0\mathbf{y} = \mathbf{B}_2\mathbf{Fy} = \mathrm{BLUE}(\mathbf{X}_*\boldsymbol{\beta} \mid \mathscr{M}) \implies \mathbf{B}_2 \in \{\mathbf{P}_{\mathbf{X}_*|\mathscr{T}}\}.$$

Thus, the proof of claim (a) is completed.

The proofs of parts (b) and (c) of Proposition 11.17 are parallel to that of (a). □

As an aside we may mention that if $\mathscr{C}(\mathbf{W}) = \mathbb{R}^n$, then $\{\mathbf{P}_{\mathbf{X}_*|\mathscr{M}}\} = \{\mathbf{P}_{\mathbf{X}_*|\mathscr{T}}\mathbf{F}\}$ holds for any $\mathbf{Fy} \in \mathscr{S}(\mathbf{X}_*\boldsymbol{\beta})$. Notice that in that situation the BLUE's representation is unique.

11.7 Partitioned Linear Model

Consider a partitioned linear model $\mathscr{M}_{12} = \{\mathbf{y}, \mathbf{X}_1\boldsymbol{\beta}_1 + \mathbf{X}_2\boldsymbol{\beta}_2, \mathbf{V}\}$, where $\mathbf{X}_i \in \mathbb{R}^{n \times p_i}$, $i \in \{1, 2\}$, $p_1 + p_2 = p$. We denote the transformed model as

$$\mathscr{T}_{12} = \{\mathbf{Fy},\ \mathbf{FX}_1\boldsymbol{\beta}_1 + \mathbf{FX}_2\boldsymbol{\beta}_2,\ \mathbf{FVF}'\}.$$

Let the transformation matrix be $\mathbf{M}_2 = \mathbf{I}_n - \mathbf{P}_{\mathbf{X}_2}$, so that the transformed model is

$$\mathscr{M}_{12\cdot 2} = \{\mathbf{M}_2\mathbf{y},\ \mathbf{M}_2\mathbf{X}_1\boldsymbol{\beta}_1,\ \mathbf{M}_2\mathbf{V}\mathbf{M}_2\}.$$

Then, in view of part (d) of Proposition 11.5, the statistic $\mathbf{M}_2\mathbf{y}$ is linearly sufficient for its expectation $\mathbf{M}_2\mathbf{X}_1\boldsymbol{\beta}_1$ under $\mathscr{M}_{12}$ if and only if

$$\mathscr{C}(\mathbf{M}_2\mathbf{X}_1) \cap \mathscr{C}(\mathbf{M}_2\mathbf{V}\mathbf{M}) = \{\mathbf{0}\}. \tag{11.62}$$

Using the decomposition $\mathbf{M} = \mathbf{M}_2\mathbf{Q}_{\mathbf{M}_2\mathbf{X}_1}$, (11.62) becomes

$$\mathscr{C}(\mathbf{M}_2\mathbf{X}_1) \cap \mathscr{C}(\mathbf{M}_2\mathbf{V}\mathbf{M}_2\mathbf{Q}_{\mathbf{M}_2\mathbf{X}_1}) = \{\mathbf{0}\},$$

which obviously, see (11.9), holds and thus $\mathbf{M}_2\mathbf{y}$ is linearly sufficient for $\mathbf{M}_2\mathbf{X}_1\boldsymbol{\beta}_1$, see also Kala and Pordzik [53, Th. 1].

Is $\mathbf{M}_2\mathbf{y}$ linearly minimal sufficient? By part (b) of Proposition 11.5, the answer is positive if

$$\mathrm{r}[\mathbf{X}(\mathbf{X}'\mathbf{W}^-\mathbf{X})^-\mathbf{X}_*'] = \mathrm{r}(\mathbf{W}\mathbf{M}_2),$$

where $\mathbf{X}_* = \mathbf{M}_2\mathbf{X} = (\mathbf{M}_2\mathbf{X}_1 : \mathbf{0})$. Now

$$\mathrm{r}[\mathbf{X}(\mathbf{X}'\mathbf{W}^-\mathbf{X})^-\mathbf{X}_*'] = \mathrm{r}(\mathbf{X}_*) = \mathrm{r}(\mathbf{M}_2\mathbf{X}).$$

Using the rank rule of Marsaglia and Styan [66, Corollary 6.2] it can be shown that $\mathrm{r}(\mathbf{M}_2\mathbf{W}) = \mathrm{r}(\mathbf{M}_2\mathbf{X})$ if and only if $\mathrm{r}(\mathbf{W}) = \mathrm{r}(\mathbf{X})$, i.e.,

$$\mathbf{M}_2\mathbf{y} \in \mathscr{S}_0(\mathbf{M}_2\mathbf{X}_1\boldsymbol{\beta}_1) \iff \mathscr{C}(\mathbf{X}) \subset \mathscr{C}(\mathbf{V}).$$

The following proposition characterizes the estimability of $\boldsymbol{\mu}_1 = \mathbf{X}_1\boldsymbol{\beta}_1$ under $\mathscr{M}_{12}$, $\mathscr{T}_{12}$ and under $\mathscr{M}_{12\cdot 2} = \{\mathbf{M}_2\mathbf{y}, \mathbf{M}_2\mathbf{X}_1\boldsymbol{\beta}_1, \mathbf{M}_2\mathbf{V}\mathbf{M}_2\}$.

Proposition 11.18 *Consider the models $\mathscr{M}_{12}$ and its transformed versions $\mathscr{T}_{12}$ and $\mathscr{M}_{12\cdot 2}$. Then the following statements hold:*

(a) $\mathbf{X}_1\boldsymbol{\beta}_1$ *is estimable under $\mathscr{M}_{12}$ if and only if $\mathscr{C}(\mathbf{X}_1') = \mathscr{C}(\mathbf{X}_1'\mathbf{M}_2)$;*
(b) $\mathbf{X}_1\boldsymbol{\beta}_1$ *is estimable under $\mathscr{T}_{12}$ if and only if $\mathscr{C}(\mathbf{X}_1') = \mathscr{C}(\mathbf{X}_1'\mathbf{F}'\mathbf{Q}_{\mathbf{F}\mathbf{X}_2})$, or, equivalently, if and only if $\mathscr{C}(\mathbf{X}_1') = \mathscr{C}(\mathbf{X}_1'\mathbf{F}')$ and $\mathscr{C}(\mathbf{F}\mathbf{X}_1) \cap \mathscr{C}(\mathbf{F}\mathbf{X}_2) = \{\mathbf{0}\}$;*
(c) $\mathbf{X}_1\boldsymbol{\beta}_1$ *is estimable under $\mathscr{M}_{12\cdot 2}$ if and only if $\mathscr{C}(\mathbf{X}_1') = \mathscr{C}(\mathbf{X}_1'\mathbf{M}_2)$.*

It is noteworthy that if $\mathscr{C}(\mathbf{X}_1') = \mathscr{C}(\mathbf{X}_1'\mathbf{M}_2)$, i.e., $\mathscr{C}(\mathbf{X}_1) \cap \mathscr{C}(\mathbf{X}_2) = \{\mathbf{0}\}$, which is a condition for the estimability of $\mathbf{X}_1\boldsymbol{\beta}_1$, then by part (f) of Proposition 11.11,

$$\mathbf{M}_2\mathbf{y} \in \mathscr{S}(\mathbf{M}_2\mathbf{X}\boldsymbol{\beta}) = \mathscr{S}(\mathbf{M}_2\mathbf{X}_1\boldsymbol{\beta}_1) \iff \mathbf{M}_2\mathbf{y} \in \mathscr{S}[(\mathbf{X}_1 : \mathbf{0})\boldsymbol{\beta}] = \mathscr{S}(\mathbf{X}_1\boldsymbol{\beta}_1),$$

i.e., assuming the estimability of $\mathbf{X}_1\boldsymbol{\beta}_1$, $\mathbf{M}_2\mathbf{y}$ is linearly sufficient for $\mathbf{M}_2\mathbf{X}_1\boldsymbol{\beta}_1$ if and only if $\mathbf{M}_2\mathbf{y}$ is linearly sufficient for $\mathbf{X}_1\boldsymbol{\beta}_1$.

The next result is an interesting corollary to Proposition 11.17.

Proposition 11.19 *Consider the partitioned linear model $\mathscr{M}_{12}$ and the reduced model $\mathscr{M}_{12\cdot 2} = \{\mathbf{M}_2\mathbf{y}, \mathbf{M}_2\mathbf{X}_1\boldsymbol{\beta}_1, \mathbf{M}_2\mathbf{V}\mathbf{M}_2\}$. Then*

$$\{\mathbf{P}_{\mathbf{M}_2\mathbf{X}|\mathscr{M}_{12}}\} = \{\mathbf{P}_{\mathbf{M}_2\mathbf{X}|\mathscr{M}_{12\cdot 2}}\mathbf{M}_2\}, \tag{11.63}$$

that is, each representation of the BLUE *of* $\mathbf{M}_2\mathbf{X}\boldsymbol{\beta} = \mathbf{M}_2\mathbf{X}_1\boldsymbol{\beta}_1$ *under* $\mathscr{M}_{12}$ *is a representation of the* BLUE *under* $\mathscr{M}_{12\cdot 2}$ *and vice versa. Moreover, if* $\mathbf{X}_1\boldsymbol{\beta}_1$ *is estimable under* $\mathscr{M}$*, i.e.,* $\mathscr{C}(\mathbf{X}_1) \cap \mathscr{C}(\mathbf{X}_2) = \{\mathbf{0}\}$*, then*

$$\{\mathbf{P}_{(\mathbf{X}_1:\mathbf{0})|\mathscr{M}_{12}}\} = \{\mathbf{P}_{(\mathbf{X}_1:\mathbf{0})|\mathscr{M}_{12\cdot 2}}\mathbf{M}_2\}.$$

Proof By Proposition 11.17, (11.63) holds if

$$\mathscr{C}(\mathbf{F}')^{\perp} \subset \mathscr{C}(\mathbf{W}), \tag{11.64}$$

where now $\mathbf{F}' = \mathbf{M}_2$. Trivially (11.64) in this case holds. The second part of Proposition 11.19 is obvious. □

An alternative proof of Proposition 11.19 appears in Puntanen et al. [72, Sect. 15.4]. Proposition 11.19 is the general formulation of the well-known Frisch–Waugh–Lovell theorem, see, for example, Frisch and Waugh [27], Lovell [55, 56], Groß and Puntanen [32, 33], Bhimasankaram and Sengupta [18, Th. 6.1], and Arendacká and Puntanen [2, Th. 1]. Actually, Groß and Puntanen [32, Th. 4] shows that $\{\mathbf{P}_{\mathbf{M}_2\mathbf{X}|\mathscr{M}_{12\cdot 2}}\mathbf{M}_2\} \subset \{\mathbf{P}_{\mathbf{M}_2\mathbf{X}|\mathscr{M}_{12}}\}$ but state "boldly" that "It is obvious that also the reverse relation holds."

Consider then the estimation of $\boldsymbol{\mu}_1 = \mathbf{X}_1\boldsymbol{\beta}_1 = (\mathbf{X}_1 : \mathbf{0})\boldsymbol{\beta}$ under $\mathscr{M}_{12}$, where $\mathscr{C}(\mathbf{X}_1) \cap \mathscr{C}(\mathbf{X}_2) = \{\mathbf{0}\}$, so that $\boldsymbol{\mu}_1$ is estimable. Now $\mathbf{X}_* = (\mathbf{X}_1 : \mathbf{0})$ and on account of (11.20a), one expression for the BLUE of $\boldsymbol{\mu}_1$ is

$$\mathbf{A}\mathbf{y} = \mathbf{X}_*(\mathbf{X}'\mathbf{W}^-\mathbf{X})^-\mathbf{X}'\mathbf{W}^-\mathbf{y}.$$

An alternative expression for the BLUE of $\boldsymbol{\mu}_1$, obtainable from $\mathscr{M}_{12\cdot 2}$, is

$$\begin{aligned}\mathbf{B}\mathbf{y} &= \mathbf{X}_1[\mathbf{X}_1'\mathbf{M}_2(\mathbf{M}_2\mathbf{W}\mathbf{M}_2)^-\mathbf{M}_2\mathbf{X}_1]^-\mathbf{X}_1'\mathbf{M}_2(\mathbf{M}_2\mathbf{W}\mathbf{M}_2)^-\mathbf{M}_2\mathbf{y}\\ &= \mathbf{X}_1(\mathbf{X}_1'\dot{\mathbf{M}}_{2W}\mathbf{X}_1)^-\mathbf{X}_1'\dot{\mathbf{M}}_{2W}\mathbf{y},\end{aligned}$$

where, on account of (11.7), $\mathbf{M}_2\mathbf{W}\mathbf{M}_2 \in \mathscr{W}_{\mathscr{M}_{12\cdot 2}}$ and we have denoted

$$\dot{\mathbf{M}}_{2W} = \mathbf{M}_2(\mathbf{M}_2\mathbf{W}\mathbf{M}_2)^-\mathbf{M}_2. \tag{11.65}$$

It is clear that in (11.65) $\mathbf{W}$ can be replaced with any matrix of the form $\mathbf{W}_1 = \mathbf{V} + \mathbf{X}_1\mathbf{U}_1\mathbf{U}_1'\mathbf{X}_1'$ such that $\mathscr{C}(\mathbf{W}_1) = \mathscr{C}(\mathbf{V} : \mathbf{X}_1)$.

Now we have $\mathbf{Ay} = \mathbf{By}$ for all $\mathbf{y} \in \mathscr{C}(\mathbf{W})$, i.e., $\mathbf{AW} = \mathbf{BW}$ which implies

$$\mathbf{W}\dot{\mathbf{M}}_{2W}\mathbf{X}_1(\mathbf{X}_1'\dot{\mathbf{M}}_{2W}\mathbf{X}_1)^-\mathbf{X}_1' = \mathbf{X}(\mathbf{X}'\mathbf{W}^-\mathbf{X})^-\mathbf{X}_*'. \tag{11.66}$$

It is easy to confirm that $\mathscr{C}[\mathbf{W}\dot{\mathbf{M}}_{2W}\mathbf{X}_1(\mathbf{X}_1'\dot{\mathbf{M}}_{2W}\mathbf{X}_1)^-\mathbf{X}_1'] = \mathscr{C}(\mathbf{W}\dot{\mathbf{M}}_{2W}\mathbf{X}_1)$. Thus, in view of ($a_3$) of Proposition 11.5,

$$\mathbf{Fy} \in \mathscr{S}(\mathbf{X}_1\boldsymbol{\beta}_1) \iff \mathscr{C}(\mathbf{W}\dot{\mathbf{M}}_{2W}\mathbf{X}_1) \subset \mathscr{C}(\mathbf{WF}'). \tag{11.67}$$

From (11.67), we immediately see that $\mathbf{X}_1'\dot{\mathbf{M}}_{2W}\mathbf{y}$ is linearly minimal sufficient for $\mathbf{X}_1\boldsymbol{\beta}_1$, as observed by Isotalo and Puntanen [45, Th. 2].

In Proposition 11.20, we collect some interesting properties of linearly sufficient statistics of $\boldsymbol{\mu}_1$. See Markiewicz and Puntanen [64, Th. 2].

Proposition 11.20 *Let $\boldsymbol{\mu}_1 = \mathbf{X}_1\boldsymbol{\beta}_1$ be estimable under $\mathscr{M}_{12}$ and let $\mathbf{W} \in \mathscr{W}$. Then the statistic $\mathbf{Fy}$ is linearly sufficient for $\boldsymbol{\mu}_1$ under $\mathscr{M}_{12}$ if and only if*

$$\mathscr{C}(\mathbf{W}\dot{\mathbf{M}}_{2W}\mathbf{X}_1) \subset \mathscr{C}(\mathbf{WF}'), \tag{11.68}$$

where $\dot{\mathbf{M}}_{2W} = \mathbf{M}_2(\mathbf{M}_2\mathbf{W}\mathbf{M}_2)^-\mathbf{M}_2$, or, equivalently, by Lemma 11.3,

$$\mathscr{C}\{[\mathbf{I}_n - \mathbf{X}_2(\mathbf{X}_2'\mathbf{W}^+\mathbf{X}_2)^-\mathbf{X}_2'\mathbf{W}^+]\mathbf{X}_1\} \subset \mathscr{C}(\mathbf{WF}').$$

Moreover, the following statements hold:

(a) $\mathbf{M}_2\mathbf{y}$ *is linearly sufficient for* $\boldsymbol{\mu}_1$*;*
(b) *If* $\mathscr{C}(\mathbf{A}') = \mathscr{C}(\mathbf{M}_2)$ *then* $\mathbf{Ay}$ *is linearly sufficient for* $\boldsymbol{\mu}_1$*;*
(c) $\dot{\mathbf{M}}_{2W}\mathbf{y}$ *is linearly sufficient for* $\boldsymbol{\mu}_1$*;*
(d) $\mathbf{M}_2\mathbf{y}$ *is linearly minimal sufficient for* $\boldsymbol{\mu}_1$ *if and only if* $\mathscr{C}(\mathbf{X}) \subset \mathscr{C}(\mathbf{V})$*;*
(e) $\mathbf{X}_1'\dot{\mathbf{M}}_{2W}\mathbf{y}$ *is linearly minimal sufficient for* $\boldsymbol{\mu}_1$*;*
(f) $\mathbf{Q}_{\mathbf{M}_2\mathbf{VM}}\mathbf{M}_2\mathbf{y}$ *is linearly sufficient for* $\boldsymbol{\mu}_1$*;*
(g) *If* $\mathscr{C}(\mathbf{X}_1) \subset \mathscr{C}(\mathbf{X}_2 : \mathbf{V})$, (11.68) *becomes*

$$\mathscr{C}(\mathbf{W}\dot{\mathbf{M}}_2\mathbf{X}_1) \subset \mathscr{C}(\mathbf{WF}'), \quad \text{where } \dot{\mathbf{M}}_2 = \mathbf{M}_2(\mathbf{M}_2\mathbf{V}\mathbf{M}_2)^-\mathbf{M}_2,$$

and $\mathbf{V}\mathbf{M}_2(\mathbf{M}_2\mathbf{V}\mathbf{M}_2)^-\mathbf{M}_2\mathbf{y} \in \mathscr{S}(\boldsymbol{\mu}_1)$*;*
(h) *If* $\mathbf{V}$ *is positive definite,* (11.68) *becomes* $\mathscr{C}(\dot{\mathbf{M}}_2\mathbf{X}_1) \subset \mathscr{C}(\mathbf{F}')$*;*
(i) *If* $\boldsymbol{\beta}_1$ *is estimable under* $\mathscr{M}_{12}$*, then*

$$\mathbf{Fy} \in \mathscr{S}(\mathbf{X}_1\boldsymbol{\beta}_1 \mid \mathscr{M}_{12}) \iff \mathbf{Fy} \in \mathscr{S}(\boldsymbol{\beta}_1 \mid \mathscr{M}_{12}).$$

Part (f) of Proposition 11.20 is due to Groß and Puntanen [32, Sect. 3]. They consider the following choice of $\mathbf{F}$:

$$\begin{aligned}\mathbf{F} &= (\mathbf{I}_n - \mathbf{P}_{\mathbf{M}_2\mathbf{VM}})\mathbf{M}_2 = \mathbf{M}_2 - \mathbf{P}_{\mathbf{M}_2\mathbf{VM}} \\ &= \mathbf{I}_n - \mathbf{P}_{\mathbf{X}_2} - \mathbf{P}_{\mathbf{M}_2\mathbf{VM}} = \mathbf{I}_n - \mathbf{P}_{(\mathbf{X}_2:\mathbf{VM})},\end{aligned}$$

where we have used part (a) of Lemma 11.5. For this $\mathbf{F}$ we have

$$\mathscr{C}(\mathbf{F}') = \mathscr{C}(\mathbf{F}) = \mathscr{C}(\mathbf{M}_2) \cap \mathscr{C}(\mathbf{VM})^{\perp} = \mathscr{C}(\mathbf{M}_2) \cap \mathscr{N}(\mathbf{MV}).$$

It is easy to observe that $\mathscr{C}(\dot{\mathbf{M}}_{2W}\mathbf{X}_1) \subset \mathscr{C}(\mathbf{F}')$, and thereby (11.68) holds.

The next proposition gives some further characterizations for $\mathbf{Fy}$ being linearly sufficient for $\boldsymbol{\mu}_1$. For more details, see Markiewicz and Puntanen [64], who show, for example, that the following properties hold:

$$\begin{aligned}
\operatorname{Cov}(\widetilde{\boldsymbol{\mu}}_1 \mid \mathscr{M}_{12}) &= \mathbf{X}_1(\mathbf{X}_1'\dot{\mathbf{M}}_{2W}\mathbf{X}_1)^{-}\mathbf{X}_1' - \mathbf{T}_1 \\
&= \mathbf{X}_1[\mathbf{X}_1'\mathbf{M}_2(\mathbf{M}_2\mathbf{W}\mathbf{M}_2)^{-}\mathbf{M}_2\mathbf{X}_1]^{-}\mathbf{X}_1' - \mathbf{T}_1 \\
&= \mathbf{X}_1[\mathbf{X}_1'\mathbf{W}^{+1/2}\mathbf{P}_{\mathbf{W}^{1/2}\mathbf{M}_2}\mathbf{W}^{+1/2}\mathbf{X}_1]^{-}\mathbf{X}_1' - \mathbf{T}_1, \\
\widetilde{\boldsymbol{\mu}}_1(\mathscr{T}_{12}) &= \mathbf{X}_1[\mathbf{X}_1'\mathbf{N}_2(\mathbf{N}_2\mathbf{W}\mathbf{N}_2)^{-}\mathbf{N}_2\mathbf{X}_1]^{-}\mathbf{X}_1'\mathbf{N}_2(\mathbf{N}_2\mathbf{W}\mathbf{N}_2)^{-}\mathbf{N}_2\mathbf{y}, \\
\operatorname{Cov}(\widetilde{\boldsymbol{\mu}}_1 \mid \mathscr{T}_{12}) &= \mathbf{X}_1[\mathbf{X}_1'\mathbf{N}_2(\mathbf{N}_2\mathbf{W}\mathbf{N}_2)^{-}\mathbf{N}_2\mathbf{X}_1]^{-}\mathbf{X}_1' - \mathbf{T}_1 \\
&= \mathbf{X}_1(\mathbf{X}_1'\mathbf{W}^{+1/2}\mathbf{P}_{\mathbf{W}^{1/2}\mathbf{N}_2}\mathbf{W}^{+1/2}\mathbf{X}_1)^{-}\mathbf{X}_1' - \mathbf{T}_1,
\end{aligned}$$

where $\mathbf{W} = \mathbf{V} + \mathbf{XUU}'\mathbf{X}' \in \mathscr{W}$, $\mathbf{U}' = (\mathbf{U}_1' : \mathbf{U}_2')$ and

$$\mathbf{T}_1 = \mathbf{X}_1\mathbf{U}_1\mathbf{U}_1'\mathbf{X}_1', \quad \mathbf{N}_2 = \mathbf{P}_{\mathbf{F}'\mathbf{Q}_{\mathbf{FX}_2}} = \mathbf{P}_{\mathscr{C}(\mathbf{F}')\cap\mathscr{C}(\mathbf{M}_2)}.$$

Proposition 11.21 *Let* $\boldsymbol{\mu}_1 = \mathbf{X}_1\boldsymbol{\beta}_1$ *be estimable under* $\mathscr{T}_{12}$ *(and hence under* $\mathscr{M}_{12}$*), let* $\mathbf{W} \in \mathscr{W}$ *and denote* $\mathbf{N}_2 = \mathbf{P}_{\mathbf{F}'\mathbf{Q}_{\mathbf{FX}_2}} = \mathbf{P}_{\mathscr{C}(\mathbf{F}')\cap\mathscr{C}(\mathbf{M}_2)}$*. Then*

$$\operatorname{Cov}(\widetilde{\boldsymbol{\mu}}_1 \mid \mathscr{M}_{12}) \leq_{\mathrm{L}} \operatorname{Cov}(\widetilde{\boldsymbol{\mu}}_1 \mid \mathscr{T}_{12}).$$

Moreover, the following statements are equivalent:

(a) $\operatorname{Cov}(\widetilde{\boldsymbol{\mu}}_1 \mid \mathscr{M}_{12}) = \operatorname{Cov}(\widetilde{\boldsymbol{\mu}}_1 \mid \mathscr{T}_{12})$.
(b) $\mathscr{C}(\mathbf{X}_1) \subset \mathscr{C}(\mathbf{X}_2 : \mathbf{WN}_2) = \mathscr{C}(\mathbf{X}_2 : \mathbf{M}_2\mathbf{WN}_2)$.
(c) $\mathscr{C}(\mathbf{X}_1) \subset \mathscr{C}(\mathbf{X}_2) \oplus [\mathscr{C}(\mathbf{WF}') \cap \mathscr{C}(\mathbf{WM}_2)]$.
(d) $\mathbf{WM}_2(\mathbf{M}_2\mathbf{WM}_2)^{-}\mathbf{M}_2\mathbf{X}_1 = \mathbf{WN}_2(\mathbf{N}_2\mathbf{WN}_2)^{-}\mathbf{N}_2\mathbf{X}_1$.
(e) *The statistic* $\mathbf{Fy}$ *is linearly sufficient for* $\mathbf{X}_1\boldsymbol{\beta}_1$ *under* $\mathscr{M}_{12}$.

Notice the correspondence between the notation $\mathbf{N}$ and $\mathbf{N}_2$:

$$\mathbf{N} = \mathbf{P}_{\mathbf{F}'\mathbf{Q}_{\mathbf{FX}}} = \mathbf{P}_{\mathscr{C}(\mathbf{F}')\cap\mathscr{C}(\mathbf{M})}, \quad \mathbf{N}_2 = \mathbf{P}_{\mathbf{F}'\mathbf{Q}_{\mathbf{FX}_2}} = \mathbf{P}_{\mathscr{C}(\mathbf{F}')\cap\mathscr{C}(\mathbf{M}_2)}.$$

Consider the small model $\mathscr{M}_1 = \{\mathbf{y}, \mathbf{X}_1\boldsymbol{\beta}_1, \mathbf{V}\}$ and full model $\mathscr{M}_{12}$. We may ask, for example, what is the condition that the following implication holds:

$$\mathbf{Fy} \in \mathscr{S}(\boldsymbol{\mu}_1 \mid \mathscr{M}_1) \implies \mathbf{Fy} \in \mathscr{S}(\boldsymbol{\mu}_1 \mid \mathscr{M}_{12}). \tag{11.69}$$

Markiewicz and Puntanen [64, Th. 4] provided the following solution to (11.69).

Proposition 11.22 *Consider the models $\mathscr{M}_{12}$ and $\mathscr{M}_1$ and suppose that $\boldsymbol{\mu}_1 = \mathbf{X}_1\boldsymbol{\beta}_1$ is estimable under $\mathscr{M}_{12}$ and $\mathscr{C}(\mathbf{X}_2) \subset \mathscr{C}(\mathbf{X}_1 : \mathbf{V})$. Then the following statements are equivalent:*

(a) $\mathbf{X}_1'\mathbf{W}_1^+\mathbf{X}_2 = \mathbf{0}$,
(b) $\mathrm{BLUE}(\boldsymbol{\mu}_1 \mid \mathscr{M}_1) = \mathrm{BLUE}(\boldsymbol{\mu}_1 \mid \mathscr{M}_{12})$ *with probability* 1,
(c) $\mathbf{Fy} \in \mathscr{S}(\boldsymbol{\mu}_1 \mid \mathscr{M}_1) \iff \mathbf{Fy} \in \mathscr{S}(\boldsymbol{\mu}_1 \mid \mathscr{M}_{12})$.

From a different angle, the linear sufficiency in a partitioned linear model has been considered, e.g., in Isotalo and Puntanen [45, 47], Markiewicz and Puntanen [60], and Kala and Pordzik [53]. Baksalary [3, 4, Sects. 3.3 and 5] considered linear sufficiency under $\mathscr{M}_{12}$ and $\mathscr{M}_1$ assuming that $\mathbf{V} = \mathbf{I}_n$.

11.8 Mutual Relations of Linear Sufficiencies

In this section, we explore the mutual relations of linear sufficiencies. In addition, we go through some interesting connections between the covariance matrices of the BLUPs and the linear sufficiencies. We also comment on the upper bounds of the Euclidean distance between the BLUPs when the prediction is based on the original model $\mathscr{M}$ and when it is based on the transformed model $\mathscr{T}$.

Following Markiewicz and Puntanen [62], let us take a closer look at $\widetilde{\mathbf{y}}_* = \widetilde{\boldsymbol{\mu}}_* + \widetilde{\boldsymbol{\varepsilon}}_*$ and $\widetilde{\mathbf{y}}_{t*} = \widetilde{\boldsymbol{\mu}}_{t*} + \widetilde{\boldsymbol{\varepsilon}}_{t*}$. It can be seen that $\widetilde{\boldsymbol{\mu}}_*$ and $\widetilde{\boldsymbol{\varepsilon}}_*$ are uncorrelated and the corresponding property holds also for $\widetilde{\boldsymbol{\mu}}_{t*}$ and $\widetilde{\boldsymbol{\varepsilon}}_{t*}$. Hence

$$\mathrm{Cov}\,(\widetilde{\mathbf{y}}_*) = \mathrm{Cov}\,(\widetilde{\boldsymbol{\mu}}_*) + \mathrm{Cov}\,(\widetilde{\boldsymbol{\varepsilon}}_*)\,, \quad \mathrm{Cov}\,(\widetilde{\mathbf{y}}_{t*}) = \mathrm{Cov}\,\left(\widetilde{\boldsymbol{\mu}}_{t*}\right) + \mathrm{Cov}\,(\widetilde{\boldsymbol{\varepsilon}}_{t*})\,.$$

Now, we have

$$\widetilde{\boldsymbol{\varepsilon}}_* = \mathbf{V}_{21}\mathbf{M}(\mathbf{MVM})^-\mathbf{My}, \quad \widetilde{\boldsymbol{\varepsilon}}_{t*} = \mathbf{V}_{21}\mathbf{N}(\mathbf{NVN})^-\mathbf{Ny}, \tag{11.70}$$

with covariance matrices

$$\mathrm{Cov}\,(\widetilde{\boldsymbol{\varepsilon}}_*) = \mathbf{V}_{21}\mathbf{M}(\mathbf{MVM})^-\mathbf{MV}_{12}, \quad \mathrm{Cov}\,(\widetilde{\boldsymbol{\varepsilon}}_{t*}) = \mathbf{V}_{21}\mathbf{N}(\mathbf{NVN})^-\mathbf{NV}_{12}. \tag{11.71}$$

Notice that in view of (11.8), the matrix products $\mathbf{M}(\mathbf{MVM})^-\mathbf{M}$ and $\mathbf{N}(\mathbf{NVN})^-\mathbf{N}$ in (11.70) and (11.71) could be replaced with $(\mathbf{MVM})^+$ and $(\mathbf{NVN})^+$, respectively.

Straightforward calculation shows that $\mathrm{Cov}\,(\widetilde{\boldsymbol{\varepsilon}}_*, \widetilde{\boldsymbol{\varepsilon}}_{t*}) = \mathrm{Cov}\,(\widetilde{\boldsymbol{\varepsilon}}_{t*})$, and

$$\mathrm{Cov}\,(\widetilde{\boldsymbol{\varepsilon}}_* - \widetilde{\boldsymbol{\varepsilon}}_{t*}) = \mathrm{Cov}\,(\widetilde{\boldsymbol{\varepsilon}}_*) - \mathrm{Cov}\,(\widetilde{\boldsymbol{\varepsilon}}_{t*})\,, \tag{11.72}$$

and thereby we have the Löwner ordering $\mathrm{Cov}\,(\widetilde{\boldsymbol{\varepsilon}}_*) \geq_{\mathrm{L}} \mathrm{Cov}\,(\widetilde{\boldsymbol{\varepsilon}}_{t*})$. Moreover, in view of Lemma 11.6 and (11.72), the equality $\widetilde{\boldsymbol{\varepsilon}}_* = \widetilde{\boldsymbol{\varepsilon}}_{t*}$ holds with probability 1 if and only if $\mathrm{Cov}\,(\widetilde{\boldsymbol{\varepsilon}}_*) = \mathrm{Cov}\,(\widetilde{\boldsymbol{\varepsilon}}_{t*})$. This confirms that (b) of Proposition 11.15 indeed holds and so:

$$\mathbf{Fy} \in \mathscr{S}(\boldsymbol{\varepsilon}_*) \iff \operatorname{Cov}(\widetilde{\boldsymbol{\varepsilon}}_*) = \operatorname{Cov}(\widetilde{\boldsymbol{\varepsilon}}_{t*}) .$$

Proposition 11.23 provides some further characterizations for $\mathbf{Fy}$ being linearly sufficient for $\boldsymbol{\varepsilon}_*$, see Markiewicz and Puntanen [62, Th. 2].

Proposition 11.23 *Denoting* $\mathbf{N} = \mathbf{P}_{\mathbf{F}'\mathbf{Q}_{\mathbf{FX}}}$, *the following statements are equivalent:*

(a) $\mathbf{V}_{21}\mathbf{M} = \mathbf{V}_{21}\mathbf{N}(\mathbf{NVN})^{-}\mathbf{NVM}$,
(b) $\mathscr{C}(\mathbf{MV}_{12}) \subset \mathscr{C}(\mathbf{MVN}) = \mathscr{C}(\mathbf{MVF}'\mathbf{Q}_{\mathbf{FX}})$,
(c) $\mathscr{C}(\mathbf{V}_{12}) \subset \mathscr{C}(\mathbf{VN} : \mathbf{X}) = \mathscr{C}(\mathbf{VF}'\mathbf{Q}_{\mathbf{FX}} : \mathbf{X})$,
(d) $\mathbf{V}_{21}\mathbf{M}(\mathbf{MVM})^{-}\mathbf{MV}_{12} = \mathbf{V}_{21}\mathbf{N}(\mathbf{NVN})^{-}\mathbf{NV}_{12}$.

Moreover, each of the above conditions is a necessary and sufficient condition for the statistic $\mathbf{Fy}$ *to be linearly sufficient for* $\boldsymbol{\varepsilon}_*$ *under* $\mathscr{M}_*$.

Recall that

$$\text{(i) } \boldsymbol{\Sigma}_{\varepsilon\varepsilon} := \operatorname{Cov}(\widetilde{\boldsymbol{\varepsilon}}_* - \widetilde{\boldsymbol{\varepsilon}}_{t*}) = \operatorname{Cov}(\widetilde{\boldsymbol{\varepsilon}}_*) - \operatorname{Cov}(\widetilde{\boldsymbol{\varepsilon}}_{t*}) , \quad \text{(ii) } \operatorname{Cov}(\widetilde{\boldsymbol{\varepsilon}}_{t*}) \leq_{\mathrm{L}} \operatorname{Cov}(\widetilde{\boldsymbol{\varepsilon}}_*) .$$

Moreover, Markiewicz and Puntanen [62, Sect. 5] showed that

$$\text{(i) } \boldsymbol{\Sigma}_{\mu\mu} := \operatorname{Cov}\left(\widetilde{\boldsymbol{\mu}}_* - \widetilde{\boldsymbol{\mu}}_{t*}\right) = \operatorname{Cov}\left(\widetilde{\boldsymbol{\mu}}_{t*}\right) - \operatorname{Cov}(\widetilde{\boldsymbol{\mu}}_*) , \quad \text{(ii) } \operatorname{Cov}\left(\widetilde{\boldsymbol{\mu}}_{t*}\right) \geq \operatorname{Cov}(\widetilde{\boldsymbol{\mu}}_*) .$$

In other words,

$$\begin{aligned} \boldsymbol{\Sigma}_{\mu\mu} &= \operatorname{Cov}\left(\widetilde{\boldsymbol{\mu}}_* - \widetilde{\boldsymbol{\mu}}_{t*}\right) = \operatorname{Cov}\left(\widetilde{\boldsymbol{\mu}}_{t*}\right) - \operatorname{Cov}(\widetilde{\boldsymbol{\mu}}_*) , \\ \boldsymbol{\Sigma}_{\varepsilon\varepsilon} &= \operatorname{Cov}(\widetilde{\boldsymbol{\varepsilon}}_* - \widetilde{\boldsymbol{\varepsilon}}_{t*}) = \operatorname{Cov}(\widetilde{\boldsymbol{\varepsilon}}_*) - \operatorname{Cov}(\widetilde{\boldsymbol{\varepsilon}}_{t*}) . \end{aligned}$$

However, the following does *not* necessarily hold:

$$\boldsymbol{\Sigma}_{\mathbf{yy}} := \operatorname{Cov}(\widetilde{\mathbf{y}}_* - \widetilde{\mathbf{y}}_{t*}) = \operatorname{Cov}(\widetilde{\mathbf{y}}_*) - \operatorname{Cov}(\widetilde{\mathbf{y}}_{t*}) .$$

In view of Lemma 11.6, the equality $\widetilde{\mathbf{y}}_* = \widetilde{\mathbf{y}}_{t*}$ holds with probability 1 if and only if $\operatorname{Cov}(\widetilde{\mathbf{y}}_* - \widetilde{\mathbf{y}}_{t*}) = \mathbf{0}$, which thereby is a condition for $\mathbf{Fy}$ being linearly sufficient for $\mathbf{y}_*$. Thus, in terms of covariance matrices, we have

$$\mathbf{Fy} \in \mathscr{S}(\boldsymbol{\mu}_*) \iff \operatorname{Cov}(\widetilde{\boldsymbol{\mu}}_*) = \operatorname{Cov}\left(\widetilde{\boldsymbol{\mu}}_{t*}\right) , \tag{11.73a}$$

$$\mathbf{Fy} \in \mathscr{S}(\boldsymbol{\varepsilon}_*) \iff \operatorname{Cov}(\widetilde{\boldsymbol{\varepsilon}}_*) = \operatorname{Cov}(\widetilde{\boldsymbol{\varepsilon}}_{t*}) , \tag{11.73b}$$

$$\mathbf{Fy} \in \mathscr{S}(\mathbf{y}_*) \iff \operatorname{Cov}(\widetilde{\mathbf{y}}_* - \widetilde{\mathbf{y}}_{t*}) = \mathbf{0}. \tag{11.73c}$$

The above statements (11.73a)–(11.73c) are all appearing in Propositions 11.13 and 11.15.

Denoting

$$\boldsymbol{\Sigma}_{\mu\varepsilon} = \operatorname{Cov}\left(\widetilde{\boldsymbol{\mu}}_* - \widetilde{\boldsymbol{\mu}}_{t*}, \widetilde{\boldsymbol{\varepsilon}}_* - \widetilde{\boldsymbol{\varepsilon}}_{t*}\right) ,$$

it can be shown that $\boldsymbol{\Sigma}_{\mu\varepsilon} = -\operatorname{Cov}\left(\widetilde{\boldsymbol{\mu}}_{t*}, \widetilde{\boldsymbol{\varepsilon}}_*\right)$ and

$$\boldsymbol{\Sigma}_{\mathbf{yy}} = \boldsymbol{\Sigma}_{\mu\mu} + \boldsymbol{\Sigma}_{\varepsilon\varepsilon} + \left(\boldsymbol{\Sigma}_{\mu\varepsilon} + \boldsymbol{\Sigma}'_{\mu\varepsilon}\right).$$

Markiewicz and Puntanen [62, Th. 3] expressed the linear sufficiency of $\mathbf{Fy}$ for $\mathbf{y}_*$ in terms of covariance matrices as follows.

Proposition 11.24 *Denoting* $\boldsymbol{\Sigma}_{\mu\varepsilon} = \mathrm{Cov}\left(\widetilde{\boldsymbol{\mu}}_* - \widetilde{\boldsymbol{\mu}}_{t*}, \widetilde{\boldsymbol{\varepsilon}}_* - \widetilde{\boldsymbol{\varepsilon}}_{t*}\right)$, *the following statements are equivalent:*

(a) $\mathbf{Fy}$ *is* BLUP-*sufficient for* $\mathbf{y}_*$,
(b) $\mathrm{Cov}\,(\widetilde{\mathbf{y}}_*) = \mathrm{Cov}\,(\widetilde{\mathbf{y}}_{t*})$ *and* $\mathrm{Cov}\,(\widetilde{\boldsymbol{\varepsilon}}_{t*}) - \mathrm{Cov}\,(\widetilde{\boldsymbol{\varepsilon}}_*) = \frac{1}{2}\left(\boldsymbol{\Sigma}_{\mu\varepsilon} + \boldsymbol{\Sigma}'_{\mu\varepsilon}\right)$.

The mutual relations of the linear sufficiency of $\mathbf{Fy}$ for $\mathbf{X}_*\boldsymbol{\beta}$, $\boldsymbol{\varepsilon}_*$, and $\mathbf{y}_*$, can be characterized as follows, see Markiewicz and Puntanen [62, Th. 5].

Proposition 11.25 *Consider the following three statements:*

(a) $\mathbf{Fy} \in \mathscr{S}(\mathbf{X}_*\boldsymbol{\beta})$,
(b) $\mathbf{Fy} \in \mathscr{S}(\boldsymbol{\varepsilon}_*)$,
(c) $\mathbf{Fy} \in \mathscr{S}(\mathbf{y}_*)$.

Then above, any two conditions together imply the third one. Moreover, the equality

$$\boldsymbol{\Sigma}_{\mu\varepsilon} = -\boldsymbol{\Sigma}'_{\mu\varepsilon}, \quad \textit{where } \boldsymbol{\Sigma}_{\mu\varepsilon} = -\mathrm{Cov}\left(\widetilde{\boldsymbol{\mu}}_{t*}, \widetilde{\boldsymbol{\varepsilon}}_*\right),$$

is a necessary and sufficient condition for the implication (c) $\Longrightarrow$ (a) *and* (b).

We end this section by some remarks concerning the upper bounds for the Euclidean distance between the BLUPs when the prediction is based on the original model $\mathscr{M}$ and when it is based on the transformed model $\mathscr{T}$.

Markiewicz and Puntanen [63] considered the Euclidean norms of differences

$$\widetilde{\boldsymbol{\varepsilon}}_* - \widetilde{\boldsymbol{\varepsilon}}_{t*}, \quad \widetilde{\boldsymbol{\mu}}_* - \widetilde{\boldsymbol{\mu}}_{t*} \quad \text{and} \quad \widetilde{\mathbf{y}}_* - \widetilde{\mathbf{y}}_{t*}.$$

For this purpose, let $\mathrm{ch}_1(\cdot)$ denote the largest eigenvalue of the matrix argument and let the matrix norm be defined as $\|\mathbf{A}\|_2 = \sqrt{\mathrm{ch}_1(\mathbf{AA}')}$. In particular, $\|\mathbf{a}\|_2^2 = \mathbf{a}'\mathbf{a}$, where $\mathbf{a} \in \mathbb{R}^n$. Without going into any further details we cite below the result of Markiewicz and Puntanen [63, Th. 4.1].

Proposition 11.26 *Consider the model* $\mathscr{M}_*$. *Then for all* $\mathbf{y} = \mathbf{Xa} + \mathbf{VMb}$,

$$\begin{aligned}\|\widetilde{\boldsymbol{\varepsilon}}_* - \widetilde{\boldsymbol{\varepsilon}}_{t*}\|_2^2 &= \|\mathrm{BLUP}(\boldsymbol{\varepsilon}_* \mid \mathscr{M}_*) - \mathrm{BLUP}(\boldsymbol{\varepsilon}_* \mid \mathscr{T}_*)\|_2^2 \\ &\leq \mathrm{ch}_1(\mathbf{AA}')\,\mathbf{b}'\mathbf{Mb} := \alpha_1,\end{aligned} \tag{11.74}$$

where

$$\mathbf{A} = \mathbf{V}_{21}\mathbf{M}(\mathbf{I}_n - \mathbf{B}) \in \mathbb{R}^{q\times n}, \quad \mathbf{B} = \mathbf{N}(\mathbf{NVN})^-\mathbf{NVM} \in \mathbb{R}^{n\times n},$$

and $\mathbf{N} = \mathbf{P}_{\mathbf{F}'\mathbf{Q}_{\mathbf{FX}}}$. *If* $\mathbf{b} \notin \mathscr{C}(\mathbf{X})$, *then the upper bound* α_1 *in* (11.74) *is equal to zero if and only if* $\mathbf{A} = \mathbf{0}$, *i.e.,* $\mathbf{Fy}$ *is linearly sufficient for* $\boldsymbol{\varepsilon}_*$.

Let us take a look at the Euclidean distance between the BLUEs of $\boldsymbol{\mu}_* = \mathbf{X}_*\boldsymbol{\beta}$ in the original and the transformed model, assuming that $\mathbf{X}_* = \mathbf{LFX}$ for some $\mathbf{L} \in \mathbb{R}^{q \times f}$. Then, for all $\mathbf{y} \in \mathscr{C}(\mathbf{W})$, and $\boldsymbol{\mu}_* = \mathbf{LFX}\boldsymbol{\beta}$, we have, using the consistency and multiplicativity of the matrix norm $\|\cdot\|_2$, see, for example, Ben-Israel and Greville [16, pp. 19–20],

$$\begin{aligned}
\|\widetilde{\boldsymbol{\mu}}_* - \widetilde{\boldsymbol{\mu}}_{t*}\|_2^2 &= \|\mathbf{LF}(\mathbf{G}_t - \mathbf{G})\mathbf{y}\|_2^2 \\
&= \|\mathbf{LFG}_t\mathbf{VM}(\mathbf{MVM})^-\mathbf{My}\|_2^2 \\
&\leq \|\mathbf{LFG}_t\mathbf{VM}\|_2^2 \; \|(\mathbf{MVM})^+\|_2^2 \; \|\mathbf{My}\|_2^2 \\
&= \|\mathbf{C}\|_2^2 \; \|(\mathbf{MVM})^+\|_2^2 \; \|\mathbf{My}\|_2^2 \\
&= \frac{a}{b^2}\, \mathbf{y}'\mathbf{My} := \alpha_2,
\end{aligned} \tag{11.75}$$

where $\mathbf{G}$ and $\mathbf{G}_t$ are defined as in Proposition 11.12, $\mathbf{C} = \mathbf{LFG}_t\mathbf{VM}$, the scalar a is the largest eigenvalue of $\mathbf{CC}'$, and b is the smallest nonzero eigenvalue of $\mathbf{MVM}$. A model with property $\mathbf{VM} = \mathbf{0}$ is called a degenerated model, see Groß [31, p. 317]. If $\mathscr{M}$ is not a degenerated model then α_2 is zero if and only if $\mathbf{Fy}$ is linearly sufficient for $\boldsymbol{\mu}_*$. For (11.75), see also Kala et al. [52, Th. 5].

For the upper bound of $\|\widetilde{\mathbf{y}}_* - \widetilde{\mathbf{y}}_{t*}\|_2^2$, we refer to Markiewicz and Puntanen [63, Th. 4.2]. The properties of the Euclidean norm of the difference $\mathrm{OLSE}(\mathbf{X}\boldsymbol{\beta}) - \mathrm{BLUE}(\mathbf{X}\boldsymbol{\beta})$ have been studied by Baksalary and Kala [6, 8] and for the BLUEs under two models by Hauke et al. [42], see also Pordzik [70], Baksalary et al. [15] and Haslett et al. [37].

11.9 Mixed Linear Model

In this section, we consider the linear mixed model in the spirit of Isotalo et al. [44] and Haslett et al. [38] defined as

$$\mathbf{y} = \mathbf{X}\boldsymbol{\beta} + \mathbf{Zu} + \mathbf{e}, \quad \text{denoted as} \quad \mathscr{L} = \{\mathbf{y}, \mathbf{X}\boldsymbol{\beta} + \mathbf{Zu}, \mathbf{D}, \mathbf{R}, \mathbf{S}\}. \tag{11.76}$$

Here $\mathbf{X}_{n\times p}$ and $\mathbf{Z}_{n\times q}$ are known matrices, $\boldsymbol{\beta} \in \mathbb{R}^p$ is a vector of unknown fixed effects, $\mathbf{u}$ is an unobservable vector (q elements) of random effects with $\mathrm{E}(\mathbf{u}) = \mathbf{0}$, $\mathrm{Cov}(\mathbf{u}) = \mathbf{D}_{q\times q}$, $\mathrm{Cov}(\mathbf{e}, \mathbf{u}) = \mathbf{S}_{n\times q}$, and $\mathrm{E}(\mathbf{e}) = \mathbf{0}$, $\mathrm{Cov}(\mathbf{e}) = \mathbf{R}_{n\times n}$. In this situation

$$\mathrm{Cov}\begin{pmatrix}\mathbf{e}\\ \mathbf{u}\end{pmatrix} = \begin{pmatrix}\mathbf{R} & \mathbf{S}\\ \mathbf{S}' & \mathbf{D}\end{pmatrix} =: \dot{\mathbf{V}}, \quad \mathrm{Cov}\begin{pmatrix}\mathbf{y}\\ \mathbf{u}\end{pmatrix} = \begin{pmatrix}\boldsymbol{\Sigma} & \mathbf{ZD} + \mathbf{S}\\ (\mathbf{ZD} + \mathbf{S})' & \mathbf{D}\end{pmatrix},$$

and denoting $\mathbf{v} = \begin{pmatrix}\mathbf{e}\\ \mathbf{u}\end{pmatrix}$,

$$\begin{aligned}\boldsymbol{\Sigma} &= \mathrm{Cov}(\mathbf{y}) = \mathrm{Cov}(\mathbf{e}+\mathbf{Zu}) = \mathrm{Cov}[(\mathbf{I}_n : \mathbf{Z})\mathbf{v}] \\ &= (\mathbf{I}_n : \mathbf{Z})\dot{\mathbf{V}}(\mathbf{I}_n : \mathbf{Z})' = \mathbf{ZDZ}' + \mathbf{R} + \mathbf{ZS}' + \mathbf{SZ}'.\end{aligned}$$

The mixed model can be expressed as a version of the model with "new observations," the new observations being, for example, in $\mathbf{g} = \mathbf{X}\boldsymbol{\beta} + \mathbf{Zu}$:

$$\mathscr{L}_* := \left\{ \begin{pmatrix}\mathbf{y}\\ \mathbf{g}\end{pmatrix}, \begin{pmatrix}\mathbf{X}\\ \mathbf{X}\end{pmatrix}\boldsymbol{\beta}, \begin{pmatrix}\boldsymbol{\Sigma} & (\mathbf{ZD}+\mathbf{S})\mathbf{Z}' \\ \mathbf{Z}(\mathbf{ZD}+\mathbf{S})' & \mathbf{ZDZ}'\end{pmatrix} \right\}.$$

Corresponding to (11.1) and (11.3), we have

$$\begin{aligned}&\mathbf{y} = \mathbf{X}\boldsymbol{\beta} + \boldsymbol{\varepsilon}, \quad \boldsymbol{\varepsilon} = \mathbf{Zu} + \mathbf{e}, \quad \mathrm{Cov}(\boldsymbol{\varepsilon}) = \mathrm{Cov}(\mathbf{y}) = \boldsymbol{\Sigma}, \\ &\mathbf{y}_* = \mathbf{g} = \mathbf{X}\boldsymbol{\beta} + \mathbf{Zu}, \quad \mathbf{X}_* = \mathbf{X}, \\ &\boldsymbol{\varepsilon}_* = \mathbf{Zu}, \quad \mathrm{Cov}(\boldsymbol{\varepsilon}_*) = \mathbf{ZDZ}', \quad \mathrm{Cov}(\boldsymbol{\varepsilon}, \boldsymbol{\varepsilon}_*) = (\mathbf{ZD}+\mathbf{S})\mathbf{Z}'.\end{aligned}$$

Choosing the "new observations" being as $\mathbf{u}$, we get

$$\left\{ \begin{pmatrix}\mathbf{y}\\ \mathbf{u}\end{pmatrix}, \begin{pmatrix}\mathbf{X}\\ \mathbf{0}\end{pmatrix}\boldsymbol{\beta}, \begin{pmatrix}\boldsymbol{\Sigma} & \mathbf{ZD}+\mathbf{S} \\ (\mathbf{ZD}+\mathbf{S})' & \mathbf{D}\end{pmatrix} \right\}.$$

Now, see, e.g., Haslett et al. [40, Lemma 2], under the mixed model $\mathscr{L}$, $\mathbf{B}_1\mathbf{y}$ is the BLUE for $\mathbf{X}\boldsymbol{\beta}$ and $\mathbf{B}_2\mathbf{y}$ is the BLUP for $\mathbf{Zu}$ if and only if

$$\begin{pmatrix}\mathbf{B}_1\\ \mathbf{B}_2\end{pmatrix}(\mathbf{X} : \boldsymbol{\Sigma}\mathbf{M}) = \begin{pmatrix}\mathbf{X} & \mathbf{0} \\ \mathbf{0} & \mathbf{Z}(\mathbf{ZD}+\mathbf{S})'\mathbf{M}\end{pmatrix} = \begin{pmatrix}\mathbf{X} & \mathbf{0} \\ \mathbf{0} & \mathrm{Cov}(\mathbf{g}, \mathbf{y})\mathbf{M}\end{pmatrix}.$$

Similarly, $\mathbf{B}_3\mathbf{y}$ is the BLUP for $\mathbf{g} = \mathbf{X}\boldsymbol{\beta} + \mathbf{Zu}$ if and only if

$$\mathbf{B}_3(\mathbf{X} : \boldsymbol{\Sigma}\mathbf{M}) = \left[\mathbf{X} : \mathbf{Z}(\mathbf{ZD}+\mathbf{S})'\mathbf{M}\right] = \left[\mathbf{X} : \mathrm{Cov}(\mathbf{g}, \mathbf{y})\mathbf{M}\right].$$

Hence $(\mathbf{B}_1 + \mathbf{B}_2)(\mathbf{X} : \boldsymbol{\Sigma}\mathbf{M}) = \mathbf{B}_3(\mathbf{X} : \boldsymbol{\Sigma}\mathbf{M})$ and the following holds:

$$\begin{aligned}\mathrm{BLUP}(\mathbf{X}\boldsymbol{\beta} + \mathbf{Zu}) &= \mathrm{BLUE}(\mathbf{X}\boldsymbol{\beta}) + \mathrm{BLUP}(\mathbf{Zu}) \\ &= \mathrm{BLUE}(\mathbf{X}\boldsymbol{\beta}) + \mathbf{Z}\,\mathrm{BLUP}(\mathbf{u}),\end{aligned}$$

which can be denoted as $\tilde{\mathbf{g}} = \tilde{\boldsymbol{\mu}} + \mathbf{Z}\tilde{\mathbf{u}}$, and we have the following representations for the BLUP of $\mathbf{g} = \mathbf{X}\boldsymbol{\beta} + \mathbf{Zu}$:

$$\begin{aligned}\mathrm{BLUP}(\mathbf{g}) = \tilde{\mathbf{g}} &= \mathbf{G}_m\mathbf{y} + \mathbf{Z}(\mathbf{ZD}+\mathbf{S})'\boldsymbol{\Sigma}^-(\mathbf{I}_n - \mathbf{G}_m)\mathbf{y} \\ &= \mathbf{G}_m\mathbf{y} + \mathbf{Z}(\mathbf{ZD}+\mathbf{S})'\mathbf{M}(\mathbf{M}\boldsymbol{\Sigma}\mathbf{M})^-\mathbf{My} \\ &= \tilde{\boldsymbol{\mu}} + \mathbf{Z}\tilde{\mathbf{u}},\end{aligned}$$

where $\mathbf{G}_m = \mathbf{X}(\mathbf{X}'\mathbf{W}_m^-\mathbf{X})^-\mathbf{X}'\mathbf{W}_m^-$ and $\mathbf{W}_m \in \mathscr{W}_{\mathscr{L}}$,

$$\mathscr{W}_{\mathscr{L}} = \left\{\mathbf{W}_m \in \mathbb{R}^{n\times n} : \mathbf{W}_m = \boldsymbol{\Sigma} + \mathbf{XUU}'\mathbf{X}',\ \mathscr{C}(\mathbf{W}_m) = \mathscr{C}(\mathbf{X} : \boldsymbol{\Sigma})\right\}. \quad (11.77)$$

For example, in the simple situation when $\mathbf{X}$ has full column rank, $\mathbf{S} = \mathbf{0}$ and $\boldsymbol{\Sigma}_0 = \mathbf{ZDZ}' + \mathbf{R}$ is positive definite, we have

$$\text{BLUP}(\mathbf{u}) = \mathbf{DZ}'\boldsymbol{\Sigma}_0^{-1}(\mathbf{y} - \mathbf{X}\widetilde{\boldsymbol{\beta}}_0), \quad \widetilde{\boldsymbol{\beta}}_0 = (\mathbf{X}'\boldsymbol{\Sigma}_0^{-1}\mathbf{X})^{-1}\mathbf{X}'\boldsymbol{\Sigma}_0^{-1}\mathbf{y}.$$

We futher note that $\mathbf{B}_4\mathbf{y}$ is the BLUP for $\boldsymbol{\eta} = \mathbf{K}\boldsymbol{\beta} + \mathbf{Lu}$ (where $\mathbf{K}\boldsymbol{\beta}$ is estimable) if and only if

$$\mathbf{B}_4(\mathbf{X} : \boldsymbol{\Sigma}\mathbf{M}) = \big[\mathbf{K} : \mathbf{L}(\mathbf{ZD} + \mathbf{S})'\mathbf{M}\big] = \big[\mathbf{K} : \text{Cov}(\boldsymbol{\eta}, \mathbf{y})\mathbf{M}\big].$$

Now obviously we get the following:

$$\begin{aligned}
\mathbf{Fy} \in \mathscr{S}(\mathbf{g} = \mathbf{X}\boldsymbol{\beta} + \mathbf{Zu}) &\iff \mathscr{C}\begin{pmatrix} \mathbf{X}' \\ \mathbf{M}(\mathbf{ZD} + \mathbf{S})\mathbf{Z}' \end{pmatrix} \subset \mathscr{C}\begin{pmatrix} \mathbf{X}'\mathbf{F}' \\ \mathbf{M}\boldsymbol{\Sigma}\mathbf{F}' \end{pmatrix}, \\
\mathbf{Fy} \in \mathscr{S}(\mathbf{X}\boldsymbol{\beta}) &\iff \mathscr{C}\begin{pmatrix} \mathbf{X}' \\ \mathbf{0} \end{pmatrix} \subset \mathscr{C}\begin{pmatrix} \mathbf{X}'\mathbf{F}' \\ \mathbf{M}\boldsymbol{\Sigma}\mathbf{F}' \end{pmatrix}, \\
\mathbf{Fy} \in \mathscr{S}(\mathbf{Zu}) &\iff \mathscr{C}\begin{pmatrix} \mathbf{0} \\ \mathbf{M}(\mathbf{ZD} + \mathbf{S})\mathbf{Z}' \end{pmatrix} \subset \mathscr{C}\begin{pmatrix} \mathbf{X}'\mathbf{F}' \\ \mathbf{M}\boldsymbol{\Sigma}\mathbf{F}' \end{pmatrix}, \qquad (11.78)
\end{aligned}$$

and thereby the next proposition holds.

Proposition 11.27 *Consider the mixed model* $\mathscr{L} = \{\mathbf{y}, \mathbf{X}\boldsymbol{\beta} + \mathbf{Zu}, \mathbf{D}, \mathbf{R}, \mathbf{S}\}$, *and the following statements:*

(a) $\mathbf{Fy} \in \mathscr{S}(\mathbf{X}\boldsymbol{\beta})$,
(b) $\mathbf{Fy} \in \mathscr{S}(\mathbf{Zu})$,
(c) $\mathbf{Fy} \in \mathscr{S}(\mathbf{X}\boldsymbol{\beta} + \mathbf{Zu})$.
Then any of the two conditions above imply the third one.

As Haslett et al. [37, Sect. 4] points out, premultiplying (11.76) by $\mathbf{M}$ produces a reduced model from which $\mathbf{X}\boldsymbol{\beta}$ has been eliminated. Actually, $\mathbf{My}$ is linearly sufficient for $\mathbf{Zu}$ under the mixed model because then condition (11.78) becomes

$$\mathscr{C}[\mathbf{M}(\mathbf{ZD} + \mathbf{S})\mathbf{Z}'] \subset \mathscr{C}(\mathbf{M}\boldsymbol{\Sigma}\mathbf{M}) = \mathscr{C}(\mathbf{M}\boldsymbol{\Sigma}),$$

which obviously holds in view of $\mathscr{C}[(\mathbf{ZD} + \mathbf{S})\mathbf{Z}'] \subset \mathscr{C}(\boldsymbol{\Sigma})$.

Liu et al. [58, p. 1511] has a slightly different definition for the linear sufficiency. According to them, the statistic $\mathbf{Fy}$ is BLUP-sufficient if for all predictable parametric functions $\boldsymbol{\eta} = \mathbf{K}\boldsymbol{\beta} + \mathbf{Lu}$ there exists a matrix $\mathbf{A}$ such that $\mathbf{AFy}$ is the BLUP for $\boldsymbol{\eta}$ in the original model. Since $\boldsymbol{\eta} = \mathbf{K}\boldsymbol{\beta} + \mathbf{Lu}$ is predictable if and only if $\mathbf{K} = \mathbf{JX}$ for some matrix $\mathbf{J}$ (while $\mathbf{L}$ can be any conformable matrix) we can re-express this definition as follows.

Definition 11.4 The statistic $\mathbf{Fy}$ is BLUP-sufficient under the model $\mathscr{L}$ if for all $\mathbf{J}$ and $\mathbf{L}$, there exists a matrix $\mathbf{A}$ such that $\mathbf{AFy}$ is the BLUP for $\boldsymbol{\eta} = \mathbf{JX}\boldsymbol{\beta} + \mathbf{Lu}$, and then we denote $\mathbf{Fy} \in \mathscr{S}_r(\boldsymbol{\beta}, \mathbf{u})$.

We see that the difference between Definitions 11.2 and 11.4 is that in Definition 11.2 our object of estimation/prediction is a given predictable combination of fixed parameters and random effect like $\mathbf{g} = \mathbf{X}\boldsymbol{\beta} + \mathbf{Zu}$ (where $\mathbf{X}$ and $\mathbf{Z}$ are given and fixed) while in Definition 11.4 we consider *all* predictable combinations of the type $\boldsymbol{\eta} = \mathbf{K}\boldsymbol{\beta} + \mathbf{Lu}$. Actually, Kala and Pordzik [53, p. 635] uses the linear sufficiency concept in the spirit of Definition 11.4 when saying that a statistic $\mathbf{Fy}$ is linearly sufficient if it is linearly sufficient for all estimable parametric functions of the model.

Haslett et al. [37, Th. 2] proved the following result.

Proposition 11.28 *Consider the mixed model* $\mathscr{L} = \{\mathbf{y}, \mathbf{X}\boldsymbol{\beta} + \mathbf{Zu}, \mathbf{D}, \mathbf{R}, \mathbf{S}\}$, *and denote* $\boldsymbol{\Sigma} = \mathrm{Cov}(\mathbf{y}) = \mathbf{ZDZ}' + \mathbf{R} + \mathbf{ZS}' + \mathbf{SZ}'$, *and let* $\mathbf{W} \in \mathscr{W}_{\mathscr{L}}$, *where the class* $\mathscr{W}_{\mathscr{L}}$ *of matrices is defined as in* (11.77). *Then the following statements are equivalent:*

(a) $\mathbf{Fy} \in \mathscr{S}(\mathbf{X}\boldsymbol{\beta}) \cap \mathscr{S}(\mathbf{u})$, *i.e.,* $\mathbf{Fy}$ *is linearly sufficient for* $\mathbf{X}\boldsymbol{\beta}$ *and for* $\mathbf{u}$.
(b) $\mathbf{Fy} \in \mathscr{S}_r(\boldsymbol{\beta}, \mathbf{u})$, *i.e.,* $\mathbf{Fy}$ *is linearly sufficient for every predictable* $\mathbf{K}\boldsymbol{\beta} + \mathbf{Lu}$.
(c) $\mathscr{C}\begin{pmatrix} \mathbf{X}' & \mathbf{0} \\ \mathbf{0} & \mathbf{M}(\mathbf{ZD} + \mathbf{S}) \end{pmatrix} \subset \mathscr{C}\begin{pmatrix} \mathbf{X}'\mathbf{F}' \\ \mathbf{M}\boldsymbol{\Sigma}\mathbf{F}' \end{pmatrix}$.
(d) $\mathscr{C}(\mathbf{X}) \subset \mathscr{C}(\mathbf{WF}')$ *and* $\mathscr{C}[\mathbf{M}(\mathbf{ZD} + \mathbf{S})] \subset \mathscr{C}(\mathbf{MWF}')$.
(e) $\mathscr{C}(\mathbf{X} : \mathbf{ZD} + \mathbf{S}) \subset \mathscr{C}(\mathbf{WF}')$.

As Isotalo et al. [44, Sect. 5] points out, here is one further interesting link between the mixed model and the following extended partitioned model:

$$\mathscr{A} = \{\dot{\mathbf{y}}, \dot{\mathbf{X}}\boldsymbol{\pi}, \dot{\mathbf{V}}\} = \left\{ \begin{pmatrix} \mathbf{y} \\ \mathbf{y}_0 \end{pmatrix}, \begin{pmatrix} \mathbf{X} & \mathbf{Z} \\ \mathbf{0} & -\mathbf{I}_q \end{pmatrix} \begin{pmatrix} \boldsymbol{\beta} \\ \boldsymbol{\gamma} \end{pmatrix}, \begin{pmatrix} \mathbf{R} & \mathbf{S} \\ \mathbf{S}' & \mathbf{D} \end{pmatrix} \right\},$$

where both $\boldsymbol{\beta}$ and $\boldsymbol{\gamma}$ are *fixed* effect parameters. Expressed in error terms we have

$$\begin{aligned} \mathbf{y} &= \mathbf{X}\boldsymbol{\beta} + \mathbf{Z}\boldsymbol{\gamma} + \boldsymbol{\varepsilon}, \\ \mathbf{y}_0 &= \quad\quad -\boldsymbol{\gamma} + \boldsymbol{\varepsilon}_0, \end{aligned}$$

where $\mathrm{Cov}\begin{pmatrix} \mathbf{y} \\ \mathbf{y}_0 \end{pmatrix} = \mathrm{Cov}\begin{pmatrix} \boldsymbol{\varepsilon} \\ \boldsymbol{\varepsilon}_0 \end{pmatrix} = \dot{\mathbf{V}}$. Premultiplying $\mathscr{A}$ by the matrix $\dot{\mathbf{F}} = (\mathbf{I}_n : \mathbf{Z})$, as in Arendacká and Puntanen [2, Sect. 2], yields the equation

$$\mathbf{y} + \mathbf{Z}\mathbf{y}_0 = \mathbf{X}\boldsymbol{\beta} + \mathbf{Z}\boldsymbol{\varepsilon}_0 + \boldsymbol{\varepsilon}, \tag{11.79}$$

and in matrix terms we get the transformed model

$$\mathscr{B} = \{\dot{\mathbf{F}}\dot{\mathbf{y}}, \dot{\mathbf{F}}\dot{\mathbf{X}}\boldsymbol{\pi}, \dot{\mathbf{F}}\dot{\mathbf{V}}\dot{\mathbf{F}}'\} = \{\mathbf{w}, \mathbf{X}\boldsymbol{\beta}, \boldsymbol{\Sigma}\},$$

where $\mathbf{w} = \mathbf{y} + \mathbf{Z}\mathbf{y}_0$ and

$$\boldsymbol{\Sigma} = \mathrm{Cov}(\mathbf{w}) = \mathbf{ZDZ}' + \mathbf{R} + \mathbf{ZS}' + \mathbf{SZ}'.$$

Now (11.79) can be interpreted as a mixed model where the observable response is $\mathbf{w} = \mathbf{y} + \mathbf{Z}\mathbf{y}_0$, $\boldsymbol{\varepsilon}_0$ is the unobservable random effect, and using the mixed model notation we have

$$\mathscr{B} = \{\mathbf{w},\ \mathbf{X}\boldsymbol{\beta} + \mathbf{Z}\boldsymbol{\varepsilon}_0,\ \mathbf{D},\ \mathbf{R},\ \mathbf{S}\}.$$

It would be now interesting to know whether the BLUEs of $\mathbf{X}\boldsymbol{\beta}$ under $\mathscr{A}$ and $\mathscr{B}$ are equal. Isotalo et al. [44] answers to this question using the linear sufficiency concept while Haslett et al. [40], Arendacká and Puntanen [2] solved this problem using different approach. Isotalo et al. [44] show that $\dot{\mathbf{F}}\dot{\mathbf{y}} = (\mathbf{I}_n : \mathbf{Z})\dot{\mathbf{y}}$ is linearly sufficient for $\mathbf{X}\boldsymbol{\beta}$ and thereby

$$\mathrm{BLUE}(\mathbf{X}\boldsymbol{\beta} \mid \mathscr{A}) = \mathrm{BLUE}(\mathbf{X}\boldsymbol{\beta} \mid \mathscr{B}).$$

Corresponding considerations appear also in Baksalary and Kala [10, Sect. 3] but without referring to the mixed model. The connection between the models $\mathscr{A}$ and $\mathscr{B}$ can be used as a tool to calculate the BLUEs and BLUPs in mixed model and it is often referred to as a Henderson's method, see, e.g., Henderson et al. [43] and McCulloch et al. [67, Chap. 8]. As a reference to rank and inertia formulas for covariance matrices of BLUPs in linear mixed models, we may mention Güler and Büyükkaya [35].

11.10 Linear Sufficiency in the Misspecified Linear Model

Consider the models $\mathscr{M}_*$ and $\underline{\mathscr{M}}_*$, where $\mathscr{C}(\mathbf{X}_*') \subset \mathscr{C}(\mathbf{X}')$:

$$\mathscr{M}_* = \left\{ \begin{pmatrix} \mathbf{y} \\ \mathbf{y}_* \end{pmatrix}, \begin{pmatrix} \mathbf{X} \\ \mathbf{X}_* \end{pmatrix} \boldsymbol{\beta}, \begin{pmatrix} \mathbf{V} & \mathbf{V}_{12} \\ \mathbf{V}_{21} & \mathbf{V}_{22} \end{pmatrix} \right\},$$
$$\underline{\mathscr{M}}_* = \left\{ \begin{pmatrix} \mathbf{y} \\ \mathbf{y}_* \end{pmatrix}, \begin{pmatrix} \mathbf{X} \\ \mathbf{X}_* \end{pmatrix} \boldsymbol{\beta}, \begin{pmatrix} \underline{\mathbf{V}} & \underline{\mathbf{V}}_{12} \\ \underline{\mathbf{V}}_{21} & \underline{\mathbf{V}}_{22} \end{pmatrix} \right\}.$$

Thus, the difference appears only in the covariance matrices

$$\boldsymbol{\Gamma} = \begin{pmatrix} \mathbf{V} & \mathbf{V}_{12} \\ \mathbf{V}_{21} & \mathbf{V}_{22} \end{pmatrix}, \quad \underline{\boldsymbol{\Gamma}} = \begin{pmatrix} \underline{\mathbf{V}} & \underline{\mathbf{V}}_{12} \\ \underline{\mathbf{V}}_{21} & \underline{\mathbf{V}}_{22} \end{pmatrix}.$$

By $\mathscr{M}$ and $\underline{\mathscr{M}}$ we of course mean the models $\mathscr{M} = \{\mathbf{y}, \mathbf{X}\boldsymbol{\beta}, \mathbf{V}\}$ and $\underline{\mathscr{M}} = \{\mathbf{y}, \mathbf{X}\boldsymbol{\beta}, \underline{\mathbf{V}}\}$.

Suppose that $\mathbf{Fy}$ is linearly sufficient for $\mathbf{X}_*\boldsymbol{\beta}$ under $\mathscr{M}_*$, i.e., $\mathbf{Fy} \in \mathscr{S}(\mathbf{X}_*\boldsymbol{\beta} \mid \mathscr{M}_*)$. Obviously, we have

$$\mathscr{S}(\mathbf{X}_*\boldsymbol{\beta} \mid \mathscr{M}_*) = \mathscr{S}(\mathbf{X}_*\boldsymbol{\beta} \mid \mathscr{M}). \tag{11.80}$$

We can now pose the following question: what is the condition that the same $\mathbf{Fy}$ continues to be linearly sufficient under the misspecified model $\underline{\mathscr{M}}_*$, i.e.,

$$\mathscr{S}(\mathbf{X}_*\boldsymbol{\beta} \mid \mathscr{M}) \subset \mathscr{S}(\mathbf{X}_*\boldsymbol{\beta} \mid \underline{\mathscr{M}})\ ? \tag{11.81}$$

Observe that in view of (11.80) we have above dropped off the subscript $*$ from $\mathscr{M}$ and $\underline{\mathscr{M}}$.

We use the following notations:

$$\mathscr{W} = \{\mathbf{W} \in \mathbb{R}^{n\times n} : \mathbf{W} = \mathbf{V} + \mathbf{XUU'X'},\ \mathscr{C}(\mathbf{W}) = \mathscr{C}(\mathbf{X} : \mathbf{V})\},$$
$$\underline{\mathscr{W}} = \{\underline{\mathbf{W}} \in \mathbb{R}^{n\times n} : \mathbf{W} = \underline{\mathbf{V}} + \mathbf{X}\underline{\mathbf{U}}\underline{\mathbf{U}}'\mathbf{X}',\ \mathscr{C}(\underline{\mathbf{W}}) = \mathscr{C}(\mathbf{X} : \underline{\mathbf{V}})\}.$$

Thus, in light of part (a$_3$) of Proposition 11.5, the claim (11.81) can be expressed as

$$\mathscr{C}(\mathbf{Z}) \subset \mathscr{C}(\mathbf{WF'}) \implies \mathscr{C}(\underline{\mathbf{Z}}) \subset \mathscr{C}(\underline{\mathbf{W}}\mathbf{F'}), \tag{11.82}$$

where $\mathbf{W} \in \mathscr{W}$, $\underline{\mathbf{W}} \in \underline{\mathscr{W}}$, and

$$\mathbf{Z} = \mathbf{X}(\mathbf{X'W^-X})^-\mathbf{X}'_*,\quad \underline{\mathbf{Z}} = \mathbf{X}(\mathbf{X}'\underline{\mathbf{W}}^-\mathbf{X})^-\mathbf{X}'_*.$$

Correspondingly, in view of part 11.6 of Proposition 11.6, $\mathscr{S}(\boldsymbol{\varepsilon}_* \mid \mathscr{M}) \subset \mathscr{S}(\boldsymbol{\varepsilon}_* \mid \underline{\mathscr{M}})$ can be expressed as

$$\mathscr{C}(\mathbf{MV}_{12}) \subset \mathscr{C}(\mathbf{MVF'Q_{FX}}) \implies \mathscr{C}(\mathbf{M}\underline{\mathbf{V}}_{12}) \subset \mathscr{C}(\mathbf{M}\underline{\mathbf{V}}\mathbf{F'Q_{FX}}). \tag{11.83}$$

In this section, we shortly present, following Markiewicz and Puntanen [65], solutions to implications (11.82) and (11.83). Baksalary and Mathew [11] allowed misspecification also in the $\mathbf{X}$-part and considered the following inclusion:

$$\mathscr{S}(\mathbf{X}\boldsymbol{\beta} \mid \mathscr{M}) \subset \mathscr{S}(\underline{\mathbf{X}}\boldsymbol{\beta} \mid \underline{\underline{\mathscr{M}}}), \tag{11.84}$$

where $\underline{\underline{\mathscr{M}}} = \{\mathbf{y}, \underline{\mathbf{X}}\boldsymbol{\beta}, \underline{\mathbf{V}}\}$.

Below is a solution to (11.81), see Markiewicz and Puntanen [65, Th. 3.1].

Proposition 11.29 *Consider the linear models* $\mathscr{M} = \{\mathbf{y}, \mathbf{X}\boldsymbol{\beta}, \mathbf{V}\}$ *and* $\underline{\mathscr{M}} = \{\mathbf{y}, \mathbf{X}\boldsymbol{\beta}, \underline{\mathbf{V}}\}$, *let* $\mathscr{C}(\mathbf{X}'_*) \subset \mathscr{C}(\mathbf{X}')$, *and denote*

$$\mathbf{Z} = \mathbf{X}(\mathbf{X'W^-X})^-\mathbf{X}'_*,\quad \underline{\mathbf{Z}} = \mathbf{X}(\mathbf{X}'\underline{\mathbf{W}}^-\mathbf{X})^-\mathbf{X}'_*.$$

Then the inclusion $\mathscr{S}(\mathbf{X}_*\boldsymbol{\beta} \mid \mathscr{M}) \subset \mathscr{S}(\mathbf{X}_*\boldsymbol{\beta} \mid \underline{\mathscr{M}})$ *holds if and only if the following two conditions hold:*

(a) $\mathscr{C}(\underline{\mathbf{Z}}) \subset \mathscr{C}(\underline{\mathbf{W}}\mathbf{W}^+\mathbf{Z})$, *i.e.*, $(\mathbf{W}^+\mathbf{Z})'\mathbf{y} \in \mathscr{S}(\mathbf{X}_*\boldsymbol{\beta} \mid \underline{\mathscr{M}})$,
(b) $\mathscr{C}(\underline{\mathbf{W}}) \subset \mathscr{C}(\mathbf{W})$.

Let us briefly sketch the main idea of the proof of Proposition 11.29. Assume now that the inclusion $\mathscr{S}(\mathbf{X}_*\boldsymbol{\beta} \mid \mathscr{M}) \subset \mathscr{S}(\mathbf{X}_*\boldsymbol{\beta} \mid \underline{\mathscr{M}})$ holds, i.e.,

$$\mathscr{C}(\mathbf{Z}) \subset \mathscr{C}(\mathbf{WF'}) \tag{11.85}$$

implies

$$\mathscr{C}(\underline{\mathbf{Z}}) \subset \mathscr{C}(\underline{\mathbf{W}}\mathbf{F}').$$

Choosing $\mathbf{F}' = \mathbf{W}^{-}\mathbf{Z}$, condition (11.85) is satisfied and thereby $(\mathbf{W}^{-}\mathbf{Z})'\mathbf{y}$ is linearly sufficient for $\mathbf{X}_{*}\boldsymbol{\beta}$ under $\mathscr{M}$ for any choice of $\mathbf{W}^{-}$. By assumption this same $\mathbf{Fy}$ (for any $\mathbf{W}^{-}$) now belongs to $\mathscr{S}(\mathbf{X}_{*}\boldsymbol{\beta} \mid \underline{\mathscr{M}})$, which means that we must have

$$\mathscr{C}(\underline{\mathbf{Z}}) \subset \mathscr{C}(\underline{\mathbf{W}}\mathbf{W}^{-}\mathbf{Z}). \tag{11.86}$$

Then the proof proceeds by utilizing some conditions under which (11.86) is holding for any choice of $\mathbf{W}^{-}$; for details, see Markiewicz and Puntanen [65, Th. 3.1].

The following proposition gives the condition for the equality in Proposition 11.29.

Proposition 11.30 *The equality* $\mathscr{S}(\mathbf{X}_{*}\boldsymbol{\beta} \mid \mathscr{M}) = \mathscr{S}(\mathbf{X}_{*}\boldsymbol{\beta} \mid \underline{\mathscr{M}})$ *holds if and only if the following two conditions hold:*

(a) $\mathscr{C}(\mathbf{W}^{+}\mathbf{Z}) = \mathscr{C}(\underline{\mathbf{W}}^{+}\underline{\mathbf{Z}})$,
(b) $\mathscr{C}(\underline{\mathbf{W}}) = \mathscr{C}(\mathbf{W})$.

Let us see how Proposition 11.29 changes when we put $\mathbf{X}_{*} = \mathbf{X}$. Using

$$\mathscr{C}[\mathbf{X}(\mathbf{X}'\mathbf{W}^{-}\mathbf{X})^{-}\mathbf{X}'] = \mathscr{C}[\mathbf{X}(\mathbf{X}'\underline{\mathbf{W}}^{-}\mathbf{X})^{-}\mathbf{X}'] = \mathscr{C}(\mathbf{X}),$$

the statement (a) in Proposition 11.29 becomes $\mathscr{C}(\mathbf{X}) \subset \mathscr{C}(\underline{\mathbf{W}}\mathbf{W}^{+}\mathbf{X})$, i.e., $\mathscr{C}(\mathbf{X}) = \mathscr{C}(\underline{\mathbf{W}}\mathbf{W}^{+}\mathbf{X})$. This yields the following result.

Proposition 11.31 *(Baksalary and Mathew [11, Th. 1]) The inclusion*

$$\mathscr{S}(\mathbf{X}\boldsymbol{\beta} \mid \mathscr{M}) \subset \mathscr{S}(\mathbf{X}\boldsymbol{\beta} \mid \underline{\mathscr{M}})$$

holds if and only if the following two conditions hold:

(a) $\mathscr{C}(\mathbf{X}) = \mathscr{C}(\underline{\mathbf{W}}\mathbf{W}^{+}\mathbf{X})$, *i.e.,* $(\mathbf{W}^{+}\mathbf{X})'\mathbf{y} \in \mathscr{S}(\mathbf{X}\boldsymbol{\beta} \mid \underline{\mathscr{M}})$,
(b) $\mathscr{C}(\underline{\mathbf{W}}) \subset \mathscr{C}(\mathbf{W})$.

Baksalary and Mathew [11, Th. 1] give Proposition 11.31 in the situation when also the $\mathbf{X}$-parts can be different, see (11.84).

Consider then a partitioned linear model $\mathscr{M}_{12}$. We know that $\mathbf{Fy}$ is linearly sufficient for $\boldsymbol{\mu}_1 = \mathbf{X}_1\boldsymbol{\beta}_1$ if and only if

$$\mathscr{C}[\mathbf{X}(\mathbf{X}'\mathbf{W}^{-}\mathbf{X})^{-}\mathbf{X}_{*}'] \subset \mathscr{C}(\mathbf{W}\mathbf{F}'),$$

where $\mathbf{X}_{*} = (\mathbf{X}_1 : \mathbf{0})$. In this situation, in view of (11.66), we can express the matrix $\mathbf{Z}$ as follows:

$$\mathbf{Z} = \mathbf{X}(\mathbf{X}'\mathbf{W}^{-}\mathbf{X})^{-}\mathbf{X}_{*}' = \mathbf{W}\dot{\mathbf{M}}_{2W}\mathbf{X}_1(\mathbf{X}_1'\dot{\mathbf{M}}_{2W}\mathbf{X}_1)^{-}\mathbf{X}_1',$$

where $\dot{\mathbf{M}}_{2W} = \mathbf{M}_2(\mathbf{M}_2\mathbf{W}\mathbf{M}_2)^-\mathbf{M}_2$ and column space of $\mathbf{Z}$ is

$$\mathscr{C}(\mathbf{Z}) = \mathscr{C}(\mathbf{W}\dot{\mathbf{M}}_{2W}\mathbf{X}_1). \tag{11.87}$$

Using (11.87) and Proposition 11.29, Markiewicz and Puntanen [65, Sect. 4] showed the following.

Proposition 11.32 *Let $\boldsymbol{\mu}_1 = \mathbf{X}_1\boldsymbol{\beta}_1$ be estimable under $\mathscr{M}_{12}$ and let $\mathbf{W} \in \mathscr{W}$ and $\underline{\mathbf{W}} \in \underline{\mathscr{W}}$. Then the inclusion $\mathscr{S}(\mathbf{X}_1\boldsymbol{\beta}_1 \mid \mathscr{M}_{12}) \subset \mathscr{S}(\mathbf{X}_1\boldsymbol{\beta}_1 \mid \underline{\mathscr{M}}_{12})$ holds if and only if the following two conditions hold:*

(a) $\mathscr{C}(\mathbf{M}_2\mathbf{X}_1) \subset \mathscr{C}[\mathbf{M}_2\underline{\mathbf{W}}\mathbf{M}_2(\mathbf{M}_2\mathbf{W}\mathbf{M}_2)^-\mathbf{M}_2\mathbf{X}_1]$,
(b) $\mathscr{C}(\underline{\mathbf{W}}) \subset \mathscr{C}(\mathbf{W})$.

Consider then the misspecification and the linear sufficiency with respect to the error term, that is, when is the following holding: $\mathscr{S}(\boldsymbol{\varepsilon}_* \mid \mathscr{M}_*) \subset \mathscr{S}(\boldsymbol{\varepsilon}_* \mid \underline{\mathscr{M}}_*)$; in other words,

$$\mathscr{C}(\mathbf{M}\mathbf{V}_{12}) \subset \mathscr{C}(\mathbf{M}\mathbf{V}\mathbf{F}'\mathbf{Q}_{\mathbf{FX}}) \implies \mathscr{C}(\mathbf{M}\underline{\mathbf{V}}_{12}) \subset \mathscr{C}(\mathbf{M}\underline{\mathbf{V}}\mathbf{F}'\mathbf{Q}_{\mathbf{FX}}).$$

The next proposition gives the result.

Proposition 11.33 *Consider the linear models (with new observations) $\mathscr{M}_*$ and $\underline{\mathscr{M}}_*$. Then the inclusion $\mathscr{S}(\boldsymbol{\varepsilon}_* \mid \mathscr{M}_*) \subset \mathscr{S}(\boldsymbol{\varepsilon}_* \mid \underline{\mathscr{M}}_*)$ holds if and only if the following two conditions hold:*

(a) $\mathscr{C}(\mathbf{M}\underline{\mathbf{V}}_{12}) \subset \mathscr{C}[\mathbf{M}\underline{\mathbf{V}}\mathbf{M}(\mathbf{M}\mathbf{V}\mathbf{M})^+\mathbf{M}\mathbf{V}_{12}]$,
(b) $\mathscr{C}(\underline{\mathbf{W}}) \subset \mathscr{C}(\mathbf{W})$,

where (a) *can be equivalently expressed in the following two forms:*

(c) $\mathbf{V}_{21}\mathbf{M}(\mathbf{M}\mathbf{V}\mathbf{M})^+\mathbf{M}\mathbf{y} \in \mathscr{S}(\boldsymbol{\varepsilon}_* \mid \underline{\mathscr{M}}_*)$, *i.e.*, $\mathrm{BLUP}(\boldsymbol{\varepsilon}_* \mid \mathscr{M}_*) \in \mathscr{S}(\boldsymbol{\varepsilon}_* \mid \underline{\mathscr{M}}_*)$,
(d) $\mathscr{C}(\underline{\mathbf{V}}_{12}) \subset \mathscr{C}[\mathbf{X} : \underline{\mathbf{V}}\mathbf{M}(\mathbf{M}\mathbf{V}\mathbf{M})^+\mathbf{M}\mathbf{V}_{12}]$.

Using Proposition 11.25 we can write the next result.

Proposition 11.34 *Consider the linear models (with new observations) $\mathscr{M}_*$ and $\underline{\mathscr{M}}_*$ and the following statements:*

(a) $\mathscr{S}(\mathbf{X}_*\boldsymbol{\beta} \mid \mathscr{M}_*) \subset \mathscr{S}(\mathbf{X}_*\boldsymbol{\beta} \mid \underline{\mathscr{M}}_*)$,
(b) $\mathscr{S}(\boldsymbol{\varepsilon}_* \mid \mathscr{M}_*) \subset \mathscr{S}(\boldsymbol{\varepsilon}_* \mid \underline{\mathscr{M}}_*)$,
(c) $\mathscr{S}(\mathbf{y}_* \mid \mathscr{M}_*) \subset \mathscr{S}(\mathbf{y}_* \mid \underline{\mathscr{M}}_*)$.

Then each of the two statements above imply the third one. In particular, (a) *and* (b) *imply* (c), *i.e.,* (c) *holds if*

(i) $\mathscr{C}(\mathbf{M}\underline{\mathbf{V}}_{12}) \subset \mathscr{C}[\mathbf{M}\underline{\mathbf{V}}\mathbf{M}(\mathbf{M}\mathbf{V}\mathbf{M})^+\mathbf{M}\mathbf{V}_{12}]$,
(ii) $\mathscr{C}(\underline{\mathbf{Z}}) \subset \mathscr{C}(\underline{\mathbf{W}}\mathbf{W}^+\mathbf{Z})$,
(iii) $\mathscr{C}(\underline{\mathbf{W}}) \subset \mathscr{C}(\mathbf{W})$.

It would be interesting to find a necessary and sufficient condition for (c) in Proposition 11.34. This does not seem to be easy to do, so this question remains a topic for future research.

We complete this section by briefly commenting the models

$$\mathscr{M} = \{\mathbf{y}, \mathbf{X}\boldsymbol{\beta}, \mathbf{V}\}, \quad \text{and} \quad \underline{\underline{\mathscr{M}}} = \{\mathbf{y}, \underline{\mathbf{X}}\boldsymbol{\beta}, \underline{\mathbf{V}}\}.$$

Thus, the difference appears not only in the covariance matrices but also in the **X**-part. We can now characterize the condition under which

$$\mathscr{S}(\mathbf{X}_*\boldsymbol{\beta} \mid \mathscr{M}) \subset \mathscr{S}(\underline{\mathbf{X}}_*\boldsymbol{\beta} \mid \underline{\underline{\mathscr{M}}}). \tag{11.88}$$

We assume that $\mathbf{X}_*\boldsymbol{\beta}$ is estimable under $\mathscr{M}$ and$\underline{\mathbf{X}}_*\boldsymbol{\beta}$ is estimable under $\underline{\underline{\mathscr{M}}}$, i.e.,

$$\mathscr{C}(\mathbf{X}_*') \subset \mathscr{C}(\mathbf{X}') \quad \text{and} \quad \mathscr{C}(\underline{\mathbf{X}}_*') \subset \mathscr{C}(\underline{\mathbf{X}}'). \tag{11.89}$$

Using the notation

$$\underline{\underline{\mathscr{W}}} = \{\underline{\underline{\mathbf{W}}} \uplus : \underline{\underline{\mathbf{W}}} = \underline{\mathbf{V}} + \underline{\mathbf{X}}\underline{\mathbf{U}}\underline{\mathbf{U}}'\underline{\mathbf{X}}', \ \mathscr{C}(\underline{\underline{\mathbf{W}}}) = \mathscr{C}(\underline{\mathbf{X}} : \underline{\mathbf{V}})\},$$

the claim (11.88) above can be expressed as

$$\mathscr{C}(\mathbf{Z}) \subset \mathscr{C}(\mathbf{W}\mathbf{F}') \implies \mathscr{C}(\underline{\underline{\mathbf{Z}}}) \subset \mathscr{C}(\underline{\underline{\mathbf{W}}}\mathbf{F}'),$$

where $\mathbf{W} \in \mathscr{W}, \underline{\underline{\mathbf{W}}} \in \underline{\underline{\mathscr{W}}}$, and

$$\mathbf{Z} = \mathbf{X}(\mathbf{X}'\mathbf{W}^-\mathbf{X})^-\mathbf{X}_*', \quad \underline{\underline{\mathbf{Z}}} = \underline{\mathbf{X}}(\underline{\mathbf{X}}'\underline{\mathbf{W}}^-\underline{\mathbf{X}})^-\underline{\mathbf{X}}_*'.$$

The proof of the following result is parallel to that of Proposition 11.29.

Proposition 11.35 *Consider the models* $\mathscr{M} = \{\mathbf{y}, \mathbf{X}\boldsymbol{\beta}, \mathbf{V}\}$ *and* $\underline{\underline{\mathscr{M}}} = \{\mathbf{y}, \underline{\mathbf{X}}\boldsymbol{\beta}, \underline{\mathbf{V}}\}$. *Assume that* (11.89) *holds and denotes*

$$\mathbf{Z} = \mathbf{X}(\mathbf{X}'\mathbf{W}^-\mathbf{X})^-\mathbf{X}_*', \quad \underline{\underline{\mathbf{Z}}} = \underline{\mathbf{X}}(\underline{\mathbf{X}}'\underline{\underline{\mathbf{W}}}^-\underline{\mathbf{X}})^-\underline{\mathbf{X}}_*',$$

where $\mathbf{W} \in \mathscr{W}, \underline{\underline{\mathbf{W}}} \in \underline{\underline{\mathscr{W}}}$. *Then the inclusion*

$$\mathscr{S}(\mathbf{X}_*\boldsymbol{\beta} \mid \mathscr{M}) \subset \mathscr{S}(\underline{\mathbf{X}}_*\boldsymbol{\beta} \mid \underline{\underline{\mathscr{M}}})$$

holds if and only if the following two conditions hold:

(a) $\mathscr{C}(\underline{\underline{\mathbf{Z}}}) \subset \mathscr{C}(\underline{\underline{\mathbf{W}}}\mathbf{W}^+\mathbf{Z})$, *i.e.*, $(\mathbf{W}^+\mathbf{Z})'\mathbf{y} \in \mathscr{S}(\underline{\mathbf{X}}_*\boldsymbol{\beta} \mid \underline{\underline{\mathscr{M}}})$,
(b) $\mathscr{C}(\bar{\bar{\underline{\underline{\mathbf{W}}}}}) \subset \mathscr{C}(\bar{\bar{\mathbf{W}}})$,

where the part (a) *can be replaced with any of the following equivalent conditions:*

(c) $\mathscr{C}(\underline{\underline{\mathbf{W}}}^{+}\underline{\underline{\mathbf{Z}}}) \subset \mathscr{C}(\mathbf{P}_{\underline{\underline{\mathbf{W}}}}\mathbf{W}^{+}\mathbf{Z})$,
(d) $\mathscr{C}(\mathbf{Q}_{\underline{\underline{\mathbf{W}}}\mathbf{Q}_{\underline{\underline{\mathbf{Z}}}}}\underline{\underline{\mathbf{Z}}}) \subset \mathscr{C}(\mathbf{P}_{\underline{\underline{\mathbf{W}}}}\mathbf{Q}_{\mathbf{W}\mathbf{Q}_{\mathbf{Z}}}\mathbf{Z})$.

11.11 Conclusions

The idea of transforming $\mathscr{M} = \{\mathbf{y}, \mathbf{X}\boldsymbol{\beta}, \mathbf{V}\}$ by a matrix $\mathbf{F}$ of order $f \times n$ follows from a desire of reduction of the initial information delivered by an observed value of a random vector variable $\mathbf{y}$ in such a way that it is still possible to obtain the BLUE of $\mathbf{X}\boldsymbol{\beta}$ from the transformed model $\mathscr{T} = \{\mathbf{Fy}, \mathbf{FX}\boldsymbol{\beta}, \mathbf{FVF}'\}$. Hence, the concept of the linear sufficiency has an essential role when studying the connection between $\mathscr{M}$ and its transformed version $\mathscr{T}$ and thereby in our paper a lot of attention has been paid to the properties of the transformed model. The same concerns the prediction of the new unknown $\mathbf{y}_*$ on the basis of observable $\mathbf{y}$: we say that $\mathbf{Fy}$ is linearly (prediction) sufficient for $\mathbf{y}_*$ if the BLUP of $\mathbf{y}_*$ is obtainable by $\mathbf{AFy}$ for some matrix $\mathbf{A}$.

So the key point is that we do not lose anything essential when doing estimation or prediction if instead of $\mathscr{M}$ we use $\mathscr{T}$. In these considerations, we believe that the new unobservable random vector $\mathbf{y}_*$ is coming from $\mathbf{y}_* = \mathbf{X}_*\boldsymbol{\beta} + \boldsymbol{\varepsilon}_*$, where the expectation of $\mathbf{y}_*$ is $\mathbf{X}_*\boldsymbol{\beta}$ and the covariance matrix of $\mathbf{y}_*$ is known as well as the cross-covariance matrix between $\mathbf{y}_*$ and $\mathbf{y}$. We denote the supplemented setup shortly as

$$\mathscr{M}_* = \left\{ \begin{pmatrix} \mathbf{y} \\ \mathbf{y}_* \end{pmatrix}, \begin{pmatrix} \mathbf{X} \\ \mathbf{X}_* \end{pmatrix} \boldsymbol{\beta}, \begin{pmatrix} \mathbf{V} & \mathbf{V}_{12} \\ \mathbf{V}_{21} & \mathbf{V}_{22} \end{pmatrix} \right\}.$$

In this paper, we have considered the linear sufficiency of $\mathbf{Fy}$ with respect to $\mathbf{y}_*$, $\mathbf{X}_*\boldsymbol{\beta}$, and $\boldsymbol{\varepsilon}_*$. We have also applied our results into the linear mixed model which can be seen as special case of the model with new observations. We omit some related concepts like linear completeness and discuss only briefly the minimal linear sufficiency. Regarding Sect. 11.10, we may mention that the issue of misspecified covariances in linear models also applies when covariances must be estimated (as is nearly always true in practice). This happens very frequently, indeed almost any time a linear mixed model is fitted. A recent review article by Haslett and Welsh [41] attempts to disentangle the literature on the effect of estimating covariances.

Dong et al. [24] and Tian [81, 84] study interesting connections between the BLUEs under two transformed models. Yongge Tian and his collaborators have presented a comprehensive investigation to the linear transformations using so-called matrix-rank method. This approach provides effective matrix algebraic tools and the results often appear in somewhat complicated forms. Of course, the questions that the matrix-rank group presents are often more complicated than those that we have been posing.

The concept of linear sufficiency was essentially introduced in early 1980s by Baksalary, Kala and Drygas. Drygas [26, p. 97] pointed out that "The concept of a linearly sufficient statistic is rather unknown in statistical literature. Besides the paper by Baksalary and Kala [7], who prove Drygas' Theorem 3.2 without using the concept of sufficiency, there is only the paper by Barnard [17, p. 232]." Recently, several papers providing further properties of the linear sufficiency have been published and a bunch is by the present authors. Some of these papers have been under the auspices of recent meetings of an International Research Group on Multivariate and Mixed Linear Models in Będlewo, Poland. Our aim in this article is to provide an easy-to-read, i.e., a self-readable review of recent results that "ordinary mortals can appreciate," cf. Lindley [54, p. 232], and while doing that, to go through some basic concepts related to linear sufficiency.

Acknowledgements Thanks for helpful discussions go to Barbora Arendacká, Tadeusz Caliński, Xu-Qing Liu, and Yongge Tian. We express special thanks to anonymous referees for constructive comments. Part of this research has been done during the biannual meetings of an International Research Group on Multivariate and Mixed Linear Models in the Mathematical Research and Conference Center, Będlewo, Poland, from November 2013 to November 2018, supported by the Stefan Banach International Mathematical Center.

References

1. Aitken, A.C.: On least squares and linear combination of observations. Proc. R. Soc. Edinb. Sect. A **55**, 42–49 (1935)
2. Arendacká, B., Puntanen, S.: Further remarks on the connection between fixed linear model and mixed linear model. Stat. Pap. **56**, 1235–1247 (2015)
3. Baksalary, J.K.: A study of the equivalence between a Gauss-Markoff model and its augmentation by nuisance parameters. Math. Operationsforsch. Stat. Ser. Stat. **15**, 2–35 (1984)
4. Baksalary, J.K.: Algebraic characterizations and statistical implications of the commutativity of orthogonal projectors. In: Pukkila, T., Puntanen, S. (eds.) Proceedings of the Second International Tampere Conference in Statistics, vol. A 184, pp. 113–142. Department of Mathematics Sciences/Statistics, University of Tampere (1987)
5. Baksalary, J.K., Drygas, H.: A note on the concepts of sufficiency in the general Gauss–Markov model: a coordinate-free approach. Forschungsbericht 92/2, Universität Dortmund, Fachbereich Statistik (1992)
6. Baksalary, J.K., Kala, R.: A bound for the Euclidean norm of the difference between the least squares and the best linear unbiased estimators. Ann. Stat. **6**, 1390–1393 (1978)
7. Baksalary, J.K., Kala, R.: Linear transformations preserving the best linear unbiased estimators in a general linear model. In: Paper presented at the 6th International Conference on Mathematical Statistics, Wisła, Poland, December (1978)
8. Baksalary, J.K., Kala, R.: A new bound for the Euclidean norm of the difference between the least squares and the best linear unbiased estimators. Ann. Stat. **8**, 679–681 (1980)
9. Baksalary, J.K., Kala, R.: Linear transformations preserving best linear unbiased estimators in a general Gauss-Markoff model. Ann. Stat. **9**, 913–916 (1981)
10. Baksalary, J.K., Kala, R.: Linear sufficiency with respect to a given vector of parametric functions. J. Stat. Plan. Inference **14**, 331–338 (1986)
11. Baksalary, J.K., Mathew, T.: Linear sufficiency and completeness in an incorrectly specified general Gauss-Markov model. Sankhyā A **48**, 169–180 (1986)
12. Baksalary, J.K., Mathew, T.: Rank invariance criterion and its application to the unified theory of least squares. Linear Algebra Appl. **127**, 393–401 (1990)

13. Baksalary, J.K., Puntanen, S., Styan, G.P.H.: A property of the dispersion matrix of the best linear unbiased estimator in the general Gauss-Markov model. Sankhyā A **52**, 279–296 (1990)
14. Baksalary, J.K., Rao, C.R., Markiewicz, A.: A study of the influence of the natural restrictions on estimation problems in the singular Gauss-Markov model. J. Stat. Plan. Inference **31**, 335–351 (1992)
15. Baksalary, O.M., Trenkler, G., Liski, E.: Let us do the twist again. Stat. Pap. **54**, 1109–1119 (2013)
16. Ben-Israel, A., Greville, T.N.E.: Generalized Inverses: Theory and Applications, 2nd edn. Springer, New York (2003)
17. Barnard, G.A.: The logic of least squares. J. R. Stat. Soc. Ser. B. Stat. Methodol. **25**, 12–127 (1963)
18. Bhimasankaram, P., Sengupta, D.: The linear zero functions approach to linear models. Sankhyā B **58**, 338–351 (1996)
19. Bloomfield, P., Watson, G.S.: The inefficiency of least squares. Biometrika **62**, 121–128 (1975)
20. Chu, K.L., Isotalo, J., Puntanen, S., Styan, G.P.H.: On decomposing the Watson efficiency of ordinary least squares in a partitioned weakly singular linear model. Sankhyā **66**, 634–651 (2004)
21. Chu, K.L., Isotalo, J., Puntanen, S., Styan, G.P.H.: Some further results concerning the decomposition of the Watson efficiency in partitioned linear models. Sankhyā **67**, 74–89 (2005)
22. Christensen, R.: Plane Answers to Complex Questions: the Theory of Linear Models, 4th edn. Springer, New York (2011)
23. Davidson, R., MacKinnon, J.G.: Econometric Theory and Methods. Oxford University Press, New York (2004)
24. Dong, B., Guo, W., Tian, Y.: On relations between BLUEs under two transformed linear models. J. Multivariate Anal. **13**, 279–292 (2014)
25. Drygas, H.: The Coordinate-Free Approach to Gauss-Markov Estimation. Springer, Berlin (1970)
26. Drygas, H.: Sufficiency and completeness in the general Gauss-Markov model. Sankhyā A **45**, 88–98 (1983)
27. Frisch, R., Waugh, F.V.: Partial time regressions as compared with individual trends. Econometrica **1**, 387–401 (1933)
28. Goldberger, A.S.: Best linear unbiased prediction in the generalized linear regression model. J. Am. Stat. Assoc. **58**, 369–375 (1962)
29. Gourieroux, C., Monfort, A.: Sufficient linear structures: econometric applications. Econometrica **48**, 1083–1097 (1980)
30. Groß, J.: A note on the concepts of linear and quadratic sufficiency. J. Stat. Plan. Inference **70**, 88–98 (1998)
31. Groß, J.: The general Gauss-Markov model with possibly singular dispersion matrix. Stat. Pap. **45**, 311–336 (2004)
32. Groß, J., Puntanen, S.: Estimation under a general partitioned linear model. Linear Algebra Appl. **321**, 131–144 (2000)
33. Groß, J., Puntanen, S.: Extensions of the Frisch-Waugh-Lovell Theorem. Discuss. Math. Probab. Stat. **25**, 39–49 (2005)
34. Groß, J., Trenkler, G.: On the equality of linear statistics in general Gauss-Markov Model. In: Mukherjee, S.P., Basu, S.K., Sinha, B.K. (eds.) Frontiers in Probability and Statistics, pp. 189–194. Narosa Publishing House, New Delhi (1998)
35. Güler, N., Büyükkaya, E.: Rank and inertia formulas for covariance matrices of BLUPs in general linear mixed models. Comm. Stat. Theory Methods (2019). https://doi.org/10.1080/03610926.2019.1599950
36. Harville, D.A.: Matrix Algebra From a Statistician's Perspective. Springer, New York (1997)
37. Haslett, S.J., Isotalo, J., Liu, Y., Puntanen, S.: Equalities between OLSE, BLUE and BLUP in the linear model. Stat. Pap. **55**, 543–561 (2014)
38. Haslett, S.J., Liu, X.-Q., Markiewicz, A., Puntanen, S.: Some properties of linear sufficiency and the BLUPs in the linear mixed model. Stat. Pap. **61**, 385–401 (2020)

39. Haslett, S.J., Puntanen, S.: Best linear unbiased prediction (BLUP). In: Davidian, M., Kenett, R., Longford, N.T., Molenberghs, G., Piegorsch, W.W., Ruggeri, F. (eds.) Wiley StatsRef: Statistics Reference Online. stat08120, 6 pp. Wiley, Chichester (2017)
40. Haslett, S.J., Puntanen, S., Arendacká, B.: The link between the mixed and fixed linear models revisited. Stat. Pap. **56**, 849–861 (2015)
41. Haslett, S.J., Welsh, A.H.: Empirical best linear unbiased prediction (EBLUP). In: Davidian, M., Kenett, R., Longford, N.T., Molenberghs, G., Piegorsch, W.W., Ruggeri, F. (eds.) Wiley StatsRef: Statistics Reference Online. stat08120, 10 pp. Wiley, Chichester (2019)
42. Hauke, J., Markiewicz, A., Puntanen, S.: Comparing the BLUEs under two linear models. Comm. Stat. Theory Methods **41**, 2405–2418 (2012)
43. Henderson, C.R., Kempthorne, O., Searle, S.R., von Krosigh, C.N.: The estimation of environmental and genetic trends from records subject to culling. Biometrics **15**, 192–218 (1959)
44. Isotalo, J., Markiewicz, A., Puntanen, S.: Some properties of linear prediction sufficiency in the linear model. In: Tez, M., von Rosen, D. (eds.) Trends and Perspectives in Linear Statistical Inference: LinStat, Istanbul, 2016, pp. 111–129. Springer (2018)
45. Isotalo, J., Puntanen, S.: Linear sufficiency and completeness in the partitioned linear model. Acta Comment. Univ. Tartu. **10**, 53–67 (2006)
46. Isotalo, J., Puntanen, S.: Linear prediction sufficiency for new observations in the general Gauss-Markov model. Comm. Stat. Theory Methods **35**, 1011–1023 (2006)
47. Isotalo, J., Puntanen, S.: Linear sufficiency and completeness in the context of estimating the parametric function in the general Gauss-Markov model. J. Stat. Plan. Inference **139**, 722–733 (2009)
48. Isotalo, J., Puntanen, S., Styan, G.P.H.: A useful matrix decomposition and its statistical applications in linear regression. Comm. Stat. Theory Methods **37**, 1436–1457 (2008)
49. Isotalo, J., Puntanen, S., Styan, G.P.H.: The BLUE's covariance matrix revisited: a review. J. Stat. Plan. Inference **138**, 2722–2737 (2008)
50. Kala, R.: Projectors and linear estimation in general linear models. Comm. Stat. Theory Methods **10**, 849–873 (1981)
51. Kala, R., Puntanen, S., Tian, Y.: Some notes on linear sufficiency. Stat. Pap. **58**, 1–17 (2017)
52. Kala, R., Markiewicz, A., Puntanen, S.: Some further remarks on the linear sufficiency in the linear model. In: Bebiano, N. (ed.) Applied and Computational Matrix Analysis: MatTriad, Coimbra, Portugal, September 2015, Selected, Revised Contributions, pp. 275–294. Springer Proceedings in Mathematics and Statistics, vol. 192 (2017)
53. Kala, R., Pordzik, P.R.: Estimation in singular partitioned, reduced or transformed linear models. Stat. Pap. **50**, 633–638 (2009)
54. Lindley, D.V.: Professor George A. Barnard, 1915–2002, Obituary. The Statistician **52**, 231–234 (2003)
55. Lovell, M.C.: Seasonal adjustment of economic time series and multiple regression analysis. J. Am. Stat. Assoc. **58**, 993–1010 (1963)
56. Lovell, M.C.: A simple proof of the FWL Theorem. J. Econ. Educ. **39**, 88–91 (2008)
57. Kornacki, A.: Different kinds of sufficiency in the general Gauss-Markov model. Math. Slovaca **57**, 389–392 (2007)
58. Liu, X.-Q., Rong, J.-Y., Liu, J.-Y.: Best linear unbiased prediction for linear combinations in general mixed linear models. J. Multivar. Anal. **99**, 1503–1517 (2008)
59. Markiewicz, A.: Comparison of linear restricted models with respect to the validity of admissible and linearly sufficient estimators. Stat. Probab. Lett. **38**, 347–354 (1998)
60. Markiewicz, A., Puntanen, S.: Admissibility and linear sufficiency in linear model with nuisance parameters. Stat. Pap. **50**, 847–854 (2009)
61. Markiewicz, A., Puntanen, S.: All about the $\perp$ with its applications in the linear statistical models. Open Mathematics **13**, 33–50 (2015)
62. Markiewicz, A., Puntanen, S.: Further properties of linear prediction sufficiency and the BLUPs in the linear model with new observations. Afr. Stat. **13**, 1511–1530 (2018)
63. Markiewicz, A., Puntanen, S.: Upper bounds for the Euclidean distances between the BLUPs. Special Matrices **6**, 249–261 (2018)

64. Markiewicz, A., Puntanen, S.: Further properties of the linear sufficiency in the partitioned linear model. In: Ahmed, S.E., Carvalho, F., Puntanen, S. (eds.) Matrices, Statistics and Big Data: Selected Contributions from IWMS 2016, pp. 1–22. Springer (2019)
65. Markiewicz, A., Puntanen, S.: Linear prediction sufficiency in the misspecified linear model. Comm. Statist. Theory Methods (2019). https://doi.org/10.1080/03610926.2019.1584311
66. Marsaglia, G., Styan, G.P.H.: Equalities and inequalities for ranks of matrices. Linear Multilinear Algebra **2**, 269–292 (1974)
67. McCulloch, C.E., Searle, S.R., Neuhaus, J.M.: Generalized, Linear, and Mixed Models, 2nd edn. Wiley, New York (2008)
68. Müller, J.: Sufficiency and completeness in the linear model. J. Multivariate Anal. **21**, 312–323 (1987)
69. Müller, J., Rao, C.R., Sinha, B.K.: Inference on parameters in a linear model: a review of recent results. In: Hinkelmann, K. (ed.) Experimental design, statistical models, and genetic statistics, pp. 277–295. Dekker, New York (1984)
70. Pordzik, P.R.: A bound for the Euclidean distance between restricted and unrestricted estimators of parametric functions in the general linear model. Stat. Pap. **53**, 299–304 (2012)
71. Puntanen, S., Styan, G.P.H.: The equality of the ordinary least squares estimator and the best linear unbiased estimator [with comments by O. Kempthorne and by S.R. Searle and with "Reply" by the authors]. Am. Stat. **43**, 153–164 (1989)
72. Puntanen, S., Styan, G.P.H., Isotalo, J.: Matrix Tricks for Linear Statistical Models: Our Personal Top Twenty. Springer, Heidelberg (2011)
73. Rao, C.R.: Least squares theory using an estimated dispersion matrix and its application to measurement of signals. In: Le Cam, L.M., Neyman, J. (eds.) Proceedings of the Fifth Berkeley Symposium on Mathematical Statistics and Probability: Berkeley, California, 1965/1966 vol. 1, pp. 355–372. University of California Press, Berkeley (1967)
74. Rao, C.R.: Representations of best linear estimators in the Gauss-Markoff model with a singular dispersion matrix. J. Multivariate Anal. **3**, 276–292 (1973)
75. Rao, C.R.: Projectors, generalized inverses and the BLUE's. J. R. Stat. Soc. Ser. B. Stat. Methodol. **36**, 442–448 (1974)
76. Rao, C.R., Mitra, S.K.: Generalized Inverse of Matrices and Its Applications. Wiley, New York (1971)
77. Robinson, G.K.: That BLUP is a good thing: the estimation of random effects (discussion: pp. 32–51). Stat. Sci. **6**, 15–51 (1991)
78. Rong, J.-Y., Liu, X.-Q.: On misspecification of the dispersion matrix in mixed linear models. Stat. Pap. **51**, 445–453 (2010)
79. Searle, S.R.: The matrix handling of BLUE and BLUP in the mixed linear model. Linear Algebra Appl. **264**, 291–311 (1997)
80. Sengupta, D., Jammalamadaka, S.R.: Linear Models: An Integrated Approach. World Scientific, River Edge (2003)
81. Tian, Y.: On properties of BLUEs under general linear regression models. J. Stat. Plan. Inference **143**, 771–782 (2013)
82. Tian, Y.: A matrix handling of predictions of new observations under a general random-effects model. Electron. J. Linear Algebra **29**, 30–45 (2015)
83. Tian, Y.: A new derivation of BLUPs under random-effects model. Metrika **78**, 905–918 (2015)
84. Tian, Y.: Transformation approaches of linear random-effects models. Stat. Methods Appl. **26**, 583–608 (2017)
85. Tian, Y., Puntanen, S.: On the equivalence of estimations under a general linear model and its transformed models. Linear Algebra Appl. **430**, 2622–2641 (2009)
86. Watson, G.S.: Serial correlation in regression analysis. I. Biometrika **42**, 327–341 (1955)
87. Zyskind, G.: On canonical forms, non-negative covariance matrices and best and simple least squares linear estimators in linear models. Ann. Math. Stat. **38**, 1092–1109 (1967)
88. Zyskind, G., Martin, F.B.: On best linear estimation and general Gauss-Markov theorem in linear models with arbitrary nonnegative covariance structure. SIAM J. Appl. Math. **17**, 1190–1202 (1969)

Chapter 12
Linear Mixed-Effects Model Using Penalized Spline Based on Data Transformation Methods

Syed Ejaz Ahmed, Dursun Aydın, and Ersin Yılmaz

Abstract In this paper, we discuss two different data transformation techniques for dealing with censored data: Kaplan-Meier weights and the k-nearest neighbor imputation method. The main objective of this paper is to find penalized spline estimates for the components of a linear mixed effect model with right-censored data. In the context of a mixed model setting, the estimation procedure is performed based on the modified or transformed dataset obtained via these transformation techniques. In order to compare the outcomes from a linear mixed model using these two approaches, a Monte Carlo simulation and two real data examples are presented. According to our results, the k-nearest neighbor imputation is very successful in dealing with censored observations.

12.1 Introduction

Right-censored data is phenomenon in many scientific application areas and especially common in medical research, clinical studies, and public health studies. This kind of data arises when the observation process is left uncompleted. Therefore, only partial information about the observation is obtained during the process.

It is important to note that this study focuses on randomly right-censored data, which is the most common censor type in medical studies. Randomness of this censor type comes from the different censoring times for each observation. For classical right-censored data, there is one limit value at the right hand and, if an observation is greater than this limit, then it is marked as censored by the researcher. However in

S. E. Ahmed (✉)
Department of Mathematics & Statistics, Brock University, St. Catharines, ON L2S 3A1, Canada
e-mail: sahmed5@brocku.ca

D. Aydın · E. Yılmaz
Department of Statistics, Mugla Sitki Kocman University, 48000 Mentese, Turkey
e-mail: duaydin@hotmail.com

E. Yılmaz
e-mail: ersinyilmaz@mu.edu.tr

K. Filipiak et al. (eds.), *Multivariate, Multilinear and Mixed Linear Models*, Contributions to Statistics, https://doi.org/10.1007/978-3-030-75494-5_12

the case of randomly right-censored data, each observation has its own limit and this limit arises randomly, such as a subject withdrawing from the study or dying from another reason (i.e., if a cancer patient dies in a car accident, then we cannot know when the patient would have succumbed to the cancer) and so on. In such an example, it is clear that the censorship procedure works randomly. Note that throughout the paper, "right-censored" and "randomly right-censored" denote the same meaning.

Censored values may have a significant effect on statistical inferences, since the quality of an analysis highly depends on the completeness of the data. The problem of censorship should therefore be carefully adjusted, because estimates may be biased otherwise. There are various ways of dealing with censored data. The most primitive method is to discard the censored values from the dataset. However, such methods are not suitable in many cases, as they lead to loss of information and biased estimates. There are several competing approaches to solve the problem of censorship, including synthetic data transformations methods (e.g., Koul et al. [11]) and Kaplan-Meier weights (Kaplan and Meier [10], Miller [16], Stute [19]). These methods also change the data structure and cause increments in magnitudes of the observed data points. Alternatively, imputation methods can be used, such that they are more reliable techniques to replace the right-censored observations with estimated values. See, for example, Batista and Monard [5]. Randomly right-censored data may arise in the analysis of linear mixed effect models (LMEM), as well as in other fields of analysis. LMEMs are commonly used in the analysis of variance and longitudinal data or survival data. Note that LMEMs are extended versions of general linear regression models, which incorporate both fixed and random effects. These models may also be stated in different but equivalent form. For example, using the hierarchal notation of Laird and Ware [12], the LMEM can be written as follows for fitting p fixed effects parameters (population coefficients) and K random-effects parameters (individual coefficients) with a matrix notation:

$$\mathbf{y} = \mathbf{X}\boldsymbol{\beta} + \mathbf{Z}\mathbf{u} + \boldsymbol{\varepsilon}, \tag{12.1}$$

where $\mathbf{y} = (y_1, y_2, \dots, y_n)'$ is the $(n \times 1)$ vector of responses, $\boldsymbol{\beta} = (\beta_1, \dots, \beta_p)'$ is a $(p \times 1)$-dimensional vector of the fixed effect parameters, $\mathbf{X} = (\mathbf{x}_1, \dots, \mathbf{x}_n)'$ is $(n \times p)$ dimensional design matrix with $\mathbf{x}_i = (x_{i1}, x_{i2}, \dots, x_{ip})$ representing the ith p-dimensional row vector of design matrix $\mathbf{X}$ for fixed effects, $\mathbf{Z} = (\mathbf{z}_1, \mathbf{z}_2, \dots, \mathbf{z}_K)$ is a second design matrix with dimension $(n \times K)$ giving the values of random effects, $\mathbf{u} = (u_1, \dots, u_K)'$ is a $(K \times 1)$ vector of the random effect coefficients, and $\boldsymbol{\varepsilon} = (\varepsilon_1, \varepsilon_2, \dots, \varepsilon_n)'$ is an $(n \times 1)$ vector of random error terms. Note also that the LMEM given in (12.1) has these assumptions:

- The random-effects vector and the error vector have the following distributions because of model (12.1) is a classical LMEM:

$$\mathbf{u} \sim N(\mathbf{0}, \mathbf{G}) \quad \text{and} \quad \boldsymbol{\varepsilon} \sim N(\mathbf{0}, \mathbf{C}),$$

  where $\mathbf{G} = \sigma_u^2 \mathbf{I}_K$ is a covariance matrix with $\mathbf{I}_K$ as a $(K \times K)$ dimensional identity matrix and $\mathbf{C} = \sigma_\varepsilon^2 \mathbf{I}_n$ is a positive definite symmetric covariance matrix.

- The random-effects vector $\mathbf{u}$ and the error vector $\boldsymbol{\varepsilon}$ are independent from each other.

LMEM models are estimated by parametric approaches in general; see, for example, McCulloch and Searle [14], Verbeke and Molenberghs [20], West et al. [22]. However, estimations with parametric methods may need more flexible nonparametric approaches. Wu and Pourahmedi [23] studied nonparametric modeling of LMEMs. Also, in the case of uncensored data, Ruppert et al. [18] shows the estimation procedure of the linear mixed effect model by using penalized splines in semiparametric context.

The penalized spline method is a powerful modeling technique based on penalized least squares, as proposed by Eilers and Marx [8]. It is commonly used as a smoothing method to estimate nonparametric or semiparametric regression models. Furthermore, penalized splines using truncated power basis functions may be easily extended to an LMEM by treating the basis functions as random variables. See Aydın and Memmedli [2] for the connections between penalized splines and LMEMs. In addition to those using uncensored data, there have been some studies on LMEMs which include censored data, especially for use in the medical field. For example, Bandyopadhyay et al. [4] discuss a Bayesian analysis of LMEMs with censored data replacing the Gaussian assumptions with skew-normal distributions, and Matos et al. [13] develop a new algorithm and tools, including influence diagnostics analyses, for computing both linear and nonlinear mixed effects models.

Significantly, there have been very few studies on LMEM under right-censored data in the literature. Pan and Louis [17] estimated the LMEM to multivariate failure time data using Buckley-James estimator. Vock et al. [21] are one of very few other examples of examining LMEMs under right-censored data. To the best of our knowledge, LMEM estimation based on penalized splines under randomly right-censored data has not been studied before. In this context, this paper mainly aims to provide appropriate nonparametric estimators based on two censorship solutions.

As indicated above, we are interested in estimating the components of model (12.1) when the values of the response variable are incompletely observed and randomly right-censored by a random censoring variable c_i, $i \in \{1, 2, \ldots, n\}$, but $\mathbf{x}_i$ and $\mathbf{z}_i$ are completely observed. Therefore, we now observe the pairs dataset (T_i, δ_i) such that

$$T_i = \min(y_i, c_i), \quad \delta_i = I(y_i < c_i), \quad i \in \{1, \ldots, n\}. \tag{12.2}$$

Here, T_i is the value of the updated response variable according to censorship and δ_i is the value of censor indicator associated with T_i. If the ith observation is censored then $\delta_i = 0$, otherwise $\delta_i = 1$. In this case, model (12.1) transforms into a linear effects model with randomly right-censored data, which can also be updated in terms the vectors of new response values. In order to understand the randomly right-censored data, a randomly right-censored dataset is generated and given in Fig. 12.1 which involves the T_i, y_i and c_i.

In Fig. 12.1, blue points denote completely observed data points (y_i), pink triangles denote the values of censoring variable c_i and the green line follows the incomplete observations T_i. Because in the real world, we cannot know the real values, we created a simulated dataset to illustrate the methods efficacy.

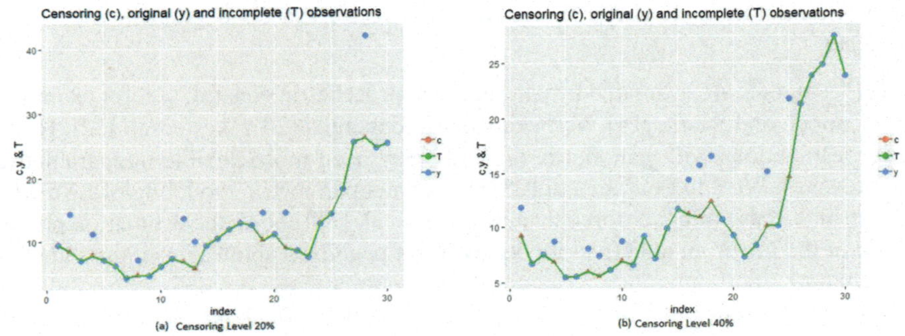

Fig. 12.1 Generated datasets for censoring levels 20 & 40%

Notice that since T_i is used in the estimation process and the effect of censorship is moved to the LMEM, $\mathbf{T} = \mathbf{X}\boldsymbol{\beta} + \mathbf{Z}\mathbf{u} + \boldsymbol{\varepsilon}_T$, where $\mathbf{u}$ and $\boldsymbol{\varepsilon}_T$ have unspecified distributions with $\mathrm{E}(\mathbf{u}) = \mathbf{0}$, $\mathrm{E}(\boldsymbol{\varepsilon}) = \mathbf{0}$ and finite variances $\sigma_{\mathbf{u}}^2 \mathbf{I}_K, \sigma_{\boldsymbol{\varepsilon}}^2 \mathbf{I}_n$, as mentioned before. In fact, due to the censorship effect, it would be more helpful to express the errors with an assumption in this model as follows.

It is assumed that heuristically $\mathrm{E}(\boldsymbol{\varepsilon}_T) \approx \mathbf{0}$ as $n \to \infty$. This heuristic argument assumption will help us to determine estimates for model (12.1). Note also that the accuracy of the estimation of model (12.1) is provided by two important assumptions, which are widely known in right-censored modeling literature (Koul et al. [11], Stute [19], Aydın and Yılmaz [3]):

A1 response variable y_i and censoring variable c_i are independent and,

A2 given time of failure (failure means death in clinical trials or an expected result from the corresponding experiment), explanatory variables cannot provide any further information whether dependent observation censored or not.

Here, A1 is the standard independence assumption to obtain an identifiable model under censored data. If A1 is violated, then additional information is needed about the censoring. In A2, "failure" describes death in clinical trials or an expected result from the corresponding experiment.

It should be noted that the estimation of random and fixed effects in an ordinary LMEM requires a dataset without any censored observations. The estimates of the components in the corresponding LMEM are biased and inconsistent when the response variable includes censored observations. Therefore, the censorship problem has to be solved appropriately. This paper introduces penalized spline estimator to estimate the components of LMEM with right-censored data based on two different data transformation approaches, the k-nearest neighbor (kNN) imputation method and Kaplan-Meier weights (KMW), which are commonly used to overcome right-censored data.

In summary, the main purpose of this paper is to estimate the components of an LMEM using the penalized spline method based on kNN imputation and KMW techniques for dealing with right-censored data. One should note that the penalized spline estimators based on kNN and KMW for LMEM introduced in this article provide useful contributions to the literature in terms of giving an estimate, especially for right-censored data. In addition, we compared the performance of these two different data transformation techniques under linear mixed model with censored data.

The remainder of this paper is arranged as follows. Section 12.2 contains the data transformation methods, the kNN imputation method with its algorithm, and the KMW method. The mixed model representation of the penalized spline using imputed response variables is explained in Sect. 12.3 and properties of the introduced estimators are given in Sect. 12.4. Section 12.5 includes evaluation criteria to measure the quality of the estimates. A simulation study and real-world data application are presented in Sects. 12.5, 12.6 and 12.7, respectively. Finally, discussion is given in Sect. 12.8.

12.2 Data Transformation Methods

12.2.1 KNN Imputation Technique

This paper considers a kNN imputation method to solve the problem of randomly right-censoring data in a fully nonparametric way. We would like to emphasize that the kNN method has both pros and cons. Imputed values are obtained using actual values. It does not transform data. Moreover, the kNN method provides additional information from the predictor variable. One of the most important properties of this method is that it is a fully nonparametric method.

The kNN imputation method uses an algorithm to match a point with its closest k-neighbors in a multi-dimensional space. This technique, which is quite suitable for dealing with all kinds of censored data, can be used for continuous, discrete, ordered, and categorical datasets. Given a positive integer k and an observation α_0, kNN imputation first identifies the k points in the data that are closest to α_0, represented by y_i. To optimally use the kNN imputation method, there are three main issues to consider:

1. *Number of nearest neighbors* k: For low values of k, results are highly affected by the variance of the data, possibly making the results inefficient. High values of k cause vogue structure of neighbors of the interested data point. In our experience (Ahmed et al. [1]), k should be determined as an odd number in the interval [2, 10].
2. *The aggregation method*: In this study, our interest is modeling of the randomly right-censored data which commonly comes from medical researches or clinical trials that are formed by continuous numeric variables. Therefore, the calculation of the arithmetic mean is decided as a suitable aggregation method.
3. *Distance metric*: The kNN is a similarity-based method which depends on the distance between data points and results may vary depending on the similarity measure used to assess the distance between the recipients (points of interests)

and the donors (neighbors). In this study, the Euclidean norm is used to evaluate the distances, which is the method generally used in the literature. The Euclidean norm can be calculated as follows:

$$D_E\left(\mathbf{x}_i, \mathbf{x}_j\right) = \left\|\mathbf{x}_i - \mathbf{x}_j\right\|, \quad j \in \{1, 2, \ldots, n_{uc}\}, \tag{12.3}$$

where $D_E(x_i, x_j)$ denotes the value of the distance measure and n_{uc} is the number of censored data points, which can be calculated as $n_{uc} = \sum_i^n \delta_i$. Note that $\mathbf{x}_i, \mathbf{x}_j$ are respective rows of the design matrix $\mathbf{X}$ given in model (12.1).

The kNN imputation procedure can be summarized thusly: to obtain $\mathbf{y}_{kNN}$, differences in (12.3) are obtained for each censored observation; then, the censored response value y_i is replaced by mean of k uncensored y_j's that have the smallest k difference value $D_E(\mathbf{x}_i, \mathbf{x}_j)$.

12.2.2 Kaplan-Meier Weights

The Kaplan-Meier weights method (KMW) is one of the most widely used methods for solving the right-censored data problem. Proposed by Miller [16], it was adopted to ordinary least squares, then Stute [19] discussed its statistical properties and extended its usage to nonlinear model estimation. In this paper, KMW is applied to linear mixed model estimation.

The fundamental idea of KMW is to estimate the distribution of the right-censored response variable T_i, the estimator proposed by Kaplan and Meier [10], and to add the effect of this distribution to the modeling process. If the component of the random effects $\mathbf{Zu} = \mathbf{0}$, then model (12.1) turns into a standard liner model and the vector of the random error terms (ω_i's) would be $\omega_i = \left(T_i - \mathbf{x}_i'\boldsymbol{\beta}\right)$, $i \in \{1, \ldots, n\}$. In this case, Miller's minimization criterion for right-censored ordinary least squares has to be minimized with respect to $\boldsymbol{\beta}$ and can be written as follows:

$$\min_{\boldsymbol{\beta}} \int \omega_i^2 d\widehat{F}(\omega_i; \boldsymbol{\beta}), \tag{12.4}$$

where $\widehat{F}(\omega_i; \boldsymbol{\beta})$ is the Kaplan-Meier estimator based on δ_i and residuals ω_i, $i \in \{1, \ldots, n\}$. It should be noted that expression given above is a function of $\boldsymbol{\beta}$, and $\widehat{F}(\omega_i; \boldsymbol{\beta})$ can be written more explicitly as follows:

$$1 - \widehat{F}(\omega_i; \boldsymbol{\beta}) = \prod_{\omega_{(i)} \leq s} \left(1 - \frac{h_{(i)}}{n_{(i)}}\right)^{\delta_{(i)}},$$

where s is a positive constant, $\omega_{(1)} < \omega_{(2)} < \cdots < \omega_{(n)}$ are ordered unique values of residuals, $n_{(i)}$'s are ordered numbers of subjects under failure risk, $h_{(i)}$'s are ordered numbers of failures, and $\delta_{(i)}$ denotes value of δ_i associated with ordered response $T_{(i)}$, where $T_{(1)} < T_{(2)} < \cdots < T_{(n)}$ are the ordered values of observed response variable T_i. It is obvious that (12.4) is a discontinuous function of $\boldsymbol{\beta}$ and this makes very hard to find its infimum. However, Miller [16] proposed the Kaplan-Meier weights as a result of an iterative process based on (12.4) to obtain $\widehat{\boldsymbol{\beta}}$; see Miller [16] for more detailed discussions.

Accordingly, to solve this minimization problem, an $(n \times n)$ diagonal weight matrix $\mathbf{W}$ is formed by the values of $\widehat{F}(\omega_i; \boldsymbol{\beta})$, which is called as Kaplan-Meier (K-M) weights associated to $T_{(i)}$.

The diagonal elements of the weight matrix $\mathbf{W}$ are computed by

$$w_{(i)} = \frac{\delta_{(i)}}{n - i + 1} \prod_{j=1}^{i-1} \left(\frac{n - j}{n - j + 1} \right)^{\delta_{(j)}}. \tag{12.5}$$

It should be emphasized that the K-M weights defined in (12.5) can also be computed as the contribution of the K-M estimator $\widehat{F}$ of the distribution function F of response observations y_i's at each ordered value $T_{(i)}$.

In this paper, the penalized spline method is modified by the KMW technique to handle censored observations. In the context of the penalized spline, the squared term in the penalized least criterion is multiplied by a matrix $\mathbf{W}$. Details are given in Sect. 12.4.

12.3 Penalized Spline as LMEM

This section will explain transition from a classical nonparametric regression model to the LMEM in the context of penalized spline. Accordingly, let us consider an ordinary nonparametric regression model with paired dataset (x_i, y_i), $i \in \{1, 2, \ldots, n\}$,

$$y_i = f(x_i) + \varepsilon_i,$$

where x_i and y_i are the same values as given in (12.1), $f(\cdot)$ is an unknown regression function to be estimated, and ε_i is an independent random error term having zero mean and a constant variance σ_ε^2. In a penalized spline context, $f(x_i)$ can be approximated by pth-degree splines with truncated polynomial basis as follows:

$$y = \beta_0 + \beta_1 x + \beta_2 x^2 + \cdots + \beta_p x^p + \sum_{j=1}^{K} u_j \left(x - r_j\right)_+^p, \tag{12.6}$$

where $(x - r_j)_+^p = (x - r_j)^p$ if $(X - r_j > 0)$ and $(x - r_j)_+^p = 0$ otherwise, $r_1 < \cdots < r_K$ is a set of determined knots that can be expressed as $\min(x) \le r_1 < r_2 < \cdots < r_K \le \max(x)$, $p \ge 1$ is a degree of polynomial functions, $\boldsymbol{\beta} = (\beta_0, \beta_1, \ldots, \beta_p)'$ consists of the coefficients of the polynomial functions that correspond to the fixed effects, and $\mathbf{u} = (u_1, u_2, \ldots, u_K)'$ consists of the coefficients of truncated functions that correspond to the random effect coefficients as defined in (12.1). Corresponding to these vectors, define

$$\mathbf{X} = \begin{pmatrix} 1 & x_1 & \ldots & x_1^p \\ 1 & x_2 & \ldots & x_2^p \\ \vdots & \vdots & \ddots & \vdots \\ 1 & x_n & \ldots & x_n^p \end{pmatrix} \text{ and } \mathbf{Z} = \begin{pmatrix} (x_1 - r_1)_+^p & (x_1 - r_2)_+^p & \ldots & (x_1 - r_K)_+^p \\ (x_2 - r_1)_+^p & (x_2 - r_2)_+^p & \ldots & (x_2 - r_K)_+^p \\ \vdots & \vdots & \ddots & \vdots \\ (x_n - r_1)_+^p & (x_n - r_2)_+^p & \ldots & (x_n - r_K)_+^p \end{pmatrix}.$$

Thus the LMEM based on penalized spline method can be obtained. Considering this model in terms of matrix and vector form, using **X** and **Z** defined above, the right-censored LMEM is written as follows:

$$\mathbf{y} = \mathbf{X}\boldsymbol{\beta} + \mathbf{Z}\mathbf{u} + \boldsymbol{\varepsilon}, \tag{12.7}$$

where **y** is an $(n \times 1)$ vector of values on the censored response variable, **X** and **Z** are design matrices for fixed and random effects, respectively, and are formed by correlated column vectors that are derived from power of x_i. $\boldsymbol{\beta}$ is a $(p \times 1)$ vector of unknown fixed effects, **u** is a $(K \times 1)$ vector of unknown random effects, and $\boldsymbol{\varepsilon}$ is an $(n \times 1)$ vector of random error terms. Also, we suppose that **u** and $\boldsymbol{\varepsilon}$ are independent.

Equation (12.7) is referred as a right-censored linear mixed effects model (RCLMEM). Note that the number of knots has to be determined carefully. In this manner, the full search algorithm introduced by Ruppert et al. [18] is used to determine knots successfully.

Full Search Algorithm: This algorithm scans all of the possible knot points. From (12.6) it can be seen that there are K possible knots. This algorithm computes GCV (generalized cross-validation) criterion according to penalized spline fit for a determined smoothing parameter λ, defined in (12.8), for each possible K. We consider the number of knots $K \in \{5, 6, \ldots, n\}$ to provide satisfying results. Accordingly, the full search algorithm obtains the GCV score with penalized spline fit for every K. Therefore, it has a computational cost. However, it provides a wider scale on the optimal number of knots. For details about determining the knots see Ruppert et al. [18]. The GCV score is written as follows:

$$\text{GCV}(\lambda) = \frac{n\|(\mathbf{I} - \mathbf{H})\mathbf{y}\|^2}{[\text{Tr}\,(\mathbf{I} - \mathbf{H})]^2},$$

where **H** (which is dependent on λ parameter) is given in (12.10) and (12.13), respectively, for kNN and KMW-based estimators.

For simplicity, we consider the LMEM (12.7) stated in matrix and vector form for the estimation and inference procedures. The key idea is to estimate the vectors $\boldsymbol{\beta}$, **u**, and vector of fitted values (i.e., $\widehat{\mathbf{y}}$). In this case, the penalized least squares estimates $(\widehat{\boldsymbol{\beta}}, \widehat{\mathbf{u}})$ of the vectors $(\boldsymbol{\beta}, \mathbf{u})$ are defined as minimizers to the following penalized least squares criterion:

$$\mathrm{PLS}(\boldsymbol{\beta}, \mathbf{u}) = \arg\min \left(\|\mathbf{T} - \mathbf{X}\boldsymbol{\beta} - \mathbf{Z}\mathbf{u}\|^2 + \lambda \mathbf{u}'\mathbf{D}\mathbf{u} \right), \tag{12.8}$$

where $\mathbf{T}$ is a vector with its entries determined by $T_i = \min(y_i, c_i)$, $\delta_i = I(y_i < c_i)$ is defined in (12.2), $\lambda > 0$ represents a smoothing parameter that controls the size of smoothness of the estimated curve, $\mathbf{D} = \mathrm{diag}\left(\mathbf{0}_{(p+1)}, \mathbf{1}_K\right)$ is a diagonal penalty matrix with K knots for the truncated polynomial function with vector $\mathbf{0}$ corresponding to the polynomial terms in (12.6), and vector $\mathbf{1}$ corresponding to the remaining entries in (12.6) (see Ruppert et al. [18]) and $\|\cdot\|$ denotes the Euclidean norm. Performing a little bit of algebra reveals that $(\widehat{\boldsymbol{\beta}}, \widehat{\mathbf{u}})$ estimates can be obtained as follows:

$$\left(\widehat{\boldsymbol{\beta}}', \widehat{\mathbf{u}}'\right)' = \left(\mathbf{R}'\mathbf{R} + \lambda\mathbf{D}\right)^{-1} \mathbf{R}'\mathbf{T}, \tag{12.9}$$

where $\mathbf{R} = (\mathbf{X} : \mathbf{Z})$ and $\mathbf{D}$ is the penalty matrix defined above. Note that (12.9) cannot be used without modification due to censorship. Also, necessary statistical properties of $\mathbf{T}$, such as expected value $\mathrm{E}(\mathbf{T})$ and variance $\mathrm{Var}(\mathbf{T})$, cannot be calculated. Consequently, usage of solution techniques are inevitable. Therefore in this paper $\mathbf{T}$ is replaced by $\mathbf{y}_{kNN}$ and KMW is added to estimation procedure.

As can be seen from (12.8), the penalized least squares approach imposes a penalty on the coefficients vector $\mathbf{u}$. When the coefficients in (12.6) are considered to optimize the fit, we need to avoid over-fitting the data. Moreover, if the estimated curve is assumed to be smooth, replicate knot points are not desirable. In this sense, the main idea in constructing a penalty is to penalize knot points that contain replicate and nearly replicate knots. Note also that the knot selection process is carried out by the full search algorithm explained above (Aydın and Yılmaz [3]).

In practice, however, the procedure for estimating $(\widehat{\boldsymbol{\beta}}, \widehat{\mathbf{u}})$ outlined in (12.9) cannot be applied directly here. This is because the values of $\mathbf{T}$ are the right-censored observations, and data transformation is necessary, as we mentioned earlier. Therefore, to obtain the estimates based on transformed data, we consider two alternative approaches that take censoring into account, kNN imputation and the KMW method, as expressed in the previous sections.

In the case of kNN imputation, the values of $\mathbf{T}$ defined in (12.9) are replaced by imputed values

$$\mathbf{y}_{kNN} = \left(y_{1_{kNN}}, y_{2_{kNN}}, \ldots, y_{n_{kNN}}\right)'.$$

The main reason for using imputed response values is to obtain estimates of components $(\widehat{\boldsymbol{\beta}}_{kNN}, \widehat{\mathbf{u}}_{kNN})$ that include the effect of censored data points. Note, that it is not a suitable way to use incomplete observations directly because they do not involve any effect of censorship. The kNN imputation opens a way for using censorship information for the estimated model. Thus, the penalized spline estimates based on kNN imputation can be calculated as follows:

$$\left(\widehat{\boldsymbol{\beta}}'_{kNN}, \widehat{\mathbf{u}}'_{kNN}\right)' = \left(\mathbf{R}'\mathbf{R} + \lambda\mathbf{D}\right)^{-1} \mathbf{R}'\mathbf{y}_{kNN}.$$

Accordingly, the fitted values based on kNN imputation method can be calculated as

$$\widehat{\mathbf{y}}_{kNN} = \widehat{\mathbf{f}}_{kNN} = \mathbf{X}\widehat{\boldsymbol{\beta}}_{kNN} + \mathbf{Z}\widehat{\mathbf{u}}_{kNN} = \mathbf{H}_{kNN}\mathbf{y}_{kNN}, \tag{12.10}$$

where $\mathbf{H}_{kNN} = \mathbf{R}\left(\mathbf{R}'\mathbf{R} + \lambda\mathbf{D}\right)^{-1}\mathbf{R}'$ is a smoothing matrix for the kNN imputation. Note that in the kNN imputation technique, determining k is an important task. From our experience about kNN imputation for right-censored data, the number of neighbors k is selected from interval [2, 10]. For different opinions see Batista and Monard [5].

In the case of the KMW method, the penalized least squares criterion stated in (12.8) is modified in the following way:

$$\text{PLSW}(\boldsymbol{\beta},\ \mathbf{u}) = \arg\min\left((\mathbf{T} - \mathbf{X}\boldsymbol{\beta} - \mathbf{Z}\mathbf{u})'\mathbf{W}(\mathbf{T} - \mathbf{X}\boldsymbol{\beta} - \mathbf{Z}\mathbf{u}) + \lambda\mathbf{u}'\mathbf{D}\mathbf{u}\right), \tag{12.11}$$

where diagonal weight matrix $\mathbf{W}$ involves Kaplan-Meier weights that are the mass attached to the uncensored observations via multiplication. With some algebra given in Appendix 12.8, it can be shown that the $(\widehat{\boldsymbol{\beta}},\ \widehat{\mathbf{u}})$ estimates based on the KMW method is equivalent to the following expression:

$$\left(\widehat{\boldsymbol{\beta}}'_{KMW},\ \widehat{\mathbf{u}}'_{KMW}\right)' = \left(\mathbf{R}'\mathbf{W}\mathbf{R} + \lambda\mathbf{D}\right)^{-1}\mathbf{R}'\mathbf{W}\mathbf{T}, \tag{12.12}$$

where $\left(\mathbf{R}'\mathbf{W}\mathbf{R} + \lambda\mathbf{D}\right)$ is assumed to be a non-singular matrix to provide an identifiability and computability of expression (12.12). Arguments similar to those given in (12.10) show that the fitted values of censored response variable based on the KMW method can be defined as

$$\widehat{\mathbf{y}}_{KMW} = \widehat{\mathbf{f}}_{KMW} = \mathbf{X}\widehat{\boldsymbol{\beta}}_{KMW} + \mathbf{Z}\widehat{\mathbf{u}}_{KMW} = \mathbf{H}_{KMW}\mathbf{T}, \tag{12.13}$$

where $\mathbf{H}_{KMW} = \mathbf{R}\left(\mathbf{R}'\mathbf{W}\mathbf{R} + \lambda\mathbf{D}\right)^{-1}\mathbf{R}'\mathbf{W}$ is a smoothing matrix for the KMW.

As previously indicated, the smoothing parameter λ plays a crucial role in estimating the components of an LMEM. For these purposes, the improved version of the Akaike information criterion (AIC$_c$) proposed by Hurvich et al. [9] is used, given by

$$\text{AIC}_c(\lambda) = \log\frac{\|(\mathbf{H} - \mathbf{I})\mathbf{y}\|^2}{n} + \frac{2(\text{Tr}\,(\mathbf{H}) + 1)}{n - \text{Tr}\,(\mathbf{H}) - 2}.$$

The value of λ that minimizes the AIC$_c$ is the optimum smoothing parameter. Note also that $\mathbf{H}$ and $\mathbf{y}$ are replaced by $\mathbf{H}_{kNN}$ and $\mathbf{y}_{kNN}$, respectively, as stated in (12.10), to get the values of AIC$_c(\lambda)$ for the kNN imputation. In a similar fashion, to obtain the values of AIC$_c(\lambda)$ for the KMW method, we use $\mathbf{H}_{KMW}$ and $\mathbf{T}$ as defined in (12.13).

12.4 Properties of Penalized Estimators Based on KMW and kNN

In this section, some properties of the penalized estimators of the fixed and random-effects coefficients in an LMEM are examined in terms of unbiasedness and covariance matrices. The penalized spline estimators based on the kNN and KMW methods will be examined.

By simple algebraic computations, it follows that the (12.11) is minimized when $\boldsymbol{\beta}$ and $\mathbf{u}$ provide the block matrix equation

$$\begin{pmatrix} \mathbf{X}'\mathbf{W}\mathbf{X} & \mathbf{X}'\mathbf{W}\mathbf{Z} \\ \mathbf{Z}'\mathbf{W}\mathbf{X} & \mathbf{Z}'\mathbf{W}\mathbf{Z} + \lambda\mathbf{D}_K \end{pmatrix} \begin{pmatrix} \boldsymbol{\beta} \\ \mathbf{u} \end{pmatrix} = \begin{pmatrix} \mathbf{X}'\mathbf{W}\mathbf{T} \\ \mathbf{Z}'\mathbf{W}\mathbf{T} \end{pmatrix}, \tag{12.14}$$

where $\mathbf{D}_K$ is a $(K \times K)$ penalty matrix similar to (12.8) but with different dimension. It can be expressed as $\mathbf{D} = \text{diag}\left(\mathbf{0}_{K-(\kappa+1)}, \mathbf{1}_{(\kappa+1)}\right)$ where κ is a determined number of knots. Note that if $\kappa = K$ then $\mathbf{D}$ will be the identity matrix. In (12.8) $\kappa = K$ is considered to simplify the definition. From (12.14), the estimates of $\boldsymbol{\beta}$ and $\mathbf{u}$ are computed, respectively, as

$$\widehat{\boldsymbol{\beta}}_{KMW} = \left(\mathbf{X}'\mathbf{W}\left(\mathbf{I}_n - \mathbf{S}_{KMW}\right)\mathbf{X}\right)^{-1}\mathbf{X}'\mathbf{W}\left(\mathbf{I}_n - \mathbf{S}_{KMW}\right)\mathbf{T}, \tag{12.15}$$

where $\mathbf{S}_{KMW} = \mathbf{Z}\left(\mathbf{Z}'\mathbf{W}\mathbf{Z} + \lambda\mathbf{D}\right)^{-1}\mathbf{Z}'\mathbf{W}$, and

$$\widehat{\mathbf{u}}_{KMW} = \left(\mathbf{Z}'\mathbf{W}\mathbf{Z} + \lambda\mathbf{D}\right)^{-1}\mathbf{Z}'\mathbf{W}\left(\mathbf{T} - \mathbf{X}\widehat{\boldsymbol{\beta}}_{KMW}\right). \tag{12.16}$$

Using (12.15) and (12.16), the vector of fitted values based on KMW is given by

$$\boldsymbol{\mu}_{KMW} = \left(\mathbf{X}\boldsymbol{\beta}_{KMW} + \mathbf{Z}\widehat{\mathbf{u}}_{KMW}\right) = \mathbf{T}_{KMW} = \mathrm{E}\left(\mathbf{T} \mid \mathbf{X},\ \mathbf{Z}\right). \tag{12.17}$$

Heuristically as $n \to \infty$, $\mathrm{E}\left(\widehat{\mathbf{T}}_{KMW} \mid \mathbf{T}\right) \cong \mathbf{E}(\mathbf{y})$, as discussed in Miller [16]. Given this expression, it means that the vector of the incomplete response $\mathbf{T}$ can be modeled to obtain fitted values of the real response vector $\mathbf{y}$, with help of Kaplan-Meier weights. It is proved by Stute [19] that the weighted least squares estimator (with Kaplan-Meier weights) is strongly consistent. The mentioned expression is heuristic, because it is dependent on several certain additional assumptions given in A1 and A2 (see Aydın and Yılmaz [3] for more detailed discussions on (12.15), (12.16) and (12.17)).

To obtain the kNN estimates of $\boldsymbol{\beta}$ and $\mathbf{u}$, the block matrix system stated in (12.14) can be updated as follows:

$$\begin{pmatrix} \mathbf{X}'\mathbf{X} & \mathbf{X}'\mathbf{Z} \\ \mathbf{Z}'\mathbf{X} & \mathbf{Z}'\mathbf{Z} + \lambda\mathbf{D} \end{pmatrix} \begin{pmatrix} \boldsymbol{\beta} \\ \mathbf{u} \end{pmatrix} = \begin{pmatrix} \mathbf{X}'\mathbf{T}_{kNN} \\ \mathbf{Z}'\mathbf{T}_{kNN} \end{pmatrix}. \tag{12.18}$$

After necessary algebraic operations, it follows from (12.18) that the estimates $\left(\widehat{\boldsymbol{\beta}}_{kNN}, \widehat{\mathbf{u}}_{kNN}\right)$ of $\boldsymbol{\beta}$ and $\mathbf{u}$ are obtained, respectively, as

$$\widehat{\boldsymbol{\beta}}_{kNN} = \left(\mathbf{X}'\left(\mathbf{I}_n - \mathbf{S}_{kNN}\right)\mathbf{X}\right)^{-1}\mathbf{X}'\left(\mathbf{I}_n - \mathbf{S}_{kNN}\right)\mathbf{T}, \tag{12.19}$$

where $\mathbf{S}_{kNN} = \mathbf{Z}\left(\mathbf{Z}'\mathbf{Z} + \lambda\mathbf{D}\right)^{-1}\mathbf{Z}$, and

$$\widehat{\mathbf{u}}_{kNN} = \left(\mathbf{Z}'\mathbf{Z} + \lambda\mathbf{D}\right)^{-1}\mathbf{Z}'\left(\mathbf{T} - \mathbf{X}\widehat{\boldsymbol{\beta}}_{kNN}\right). \tag{12.20}$$

Then, using (12.19) and (12.20), the vector of fitted values for the kNN method is written as follows:

$$\boldsymbol{\mu}_{kNN} = \left(\mathbf{X}\widehat{\boldsymbol{\beta}}_{kNN} + \mathbf{Z}\widehat{\mathbf{u}}_{kNN}\right) = \widehat{\mathbf{T}}_{kNN} = \mathrm{E}\left(\mathbf{T} \mid \mathbf{X}, \mathbf{Z}\right).$$

As mentioned before, to provide proper estimations for both the KMW and kNN methods, the smoothing parameter λ and number of knots are determined optimally, as it was presented in full search algorithm in Sect. 12.3.

12.5 Evaluation Criteria

In this section, some metrics are introduced to measure the performance of the considered data transformation methods, kNN and KMW.

In the regression setting, the most commonly used metric is the root mean square error (RMSE), given by

$$\mathrm{RMSE}(\widehat{\mathbf{f}}) = \sqrt{\frac{1}{n}\sum_{i=1}^{n}\left(\widehat{f}(x_i) - f(x_i)\right)^2},$$

where $\widehat{f}(x_i)$ denotes the estimator of the ith observation. If the estimated responses are very close to the real responses, the RMSE will be small. Thus the method that gives the lowest RMSE is preferred.

Another metric is the covariance matrix of the estimated fixed and random effects in the LMEM with censored data, which is given in Lemma 12.1.

Lemma 12.1 *Let* $(\widehat{\boldsymbol{\beta}}, \widehat{\mathbf{u}})$ *be the estimates of* $(\boldsymbol{\beta}, \mathbf{u})$ *for any of two methods. Accordingly, the covariance matrix can be calculated similarly to ordinary ridge regression using the sandwich formula*

$$\mathrm{Cov}(\widehat{\boldsymbol{\beta}}, \widehat{\mathbf{u}}) = \frac{\sigma^2}{n}\left(\mathbf{R}'\mathbf{R} + \lambda\mathbf{D}\right)^{-1}\left(\mathbf{R}'\mathbf{R}\right)\left(\mathbf{R}'\mathbf{R} + \lambda\mathbf{D}\right)^{-1}, \tag{12.21}$$

where σ^2 *is the variance of error terms. Note also that* $\mathbf{R}'\mathbf{R}$ *term given in* (12.21) *is replaced by* $\mathbf{R}'\mathbf{W}\mathbf{R}$ *to obtain the covariance matrix of the KMW method.*

The proof is given in Appendix 12.8.

It should be emphasized that, in practice, one needs to estimate the unknown variance σ^2 in (12.21) for both methods in a manner similar to this given in (12.12). In this sense, an unbiased estimate of this quantity can be described by the residual sum of squares (RSS)

$$\widehat{\sigma}^2 = \frac{\text{RSS}(\widehat{\mathbf{y}})}{n-(p+q+1)} = \frac{\mathbf{y}'(\mathbf{I}-\mathbf{H})^2\mathbf{y}}{\text{Tr}\,(\mathbf{I}-\mathbf{H})^2},$$

where $\mathbf{y}$ is replaced by the associated response vector $\mathbf{y}_{kNN}$ or $\mathbf{T}_{KMW}$. Notice that the RMSE is a reasonable measure of performance that measures the average squared difference between the estimate and the parameter. In general, the mean absolute error is a reasonable alternative to measure the suitability of an estimator and should be observed for every y_i. For our purposes, we use two measures, called averaged-bias (AvB) and inaccuracy measure (IA), given as follows.

Let n_c be the size of censored observations, y_i is the value of completely observed response variable and y_i^* is the response observation obtained from the KMW method or kNN imputation. Accordingly, AvB and IA can be defined as

$$\text{AvB} = \frac{1}{n_c}\sum_{i=1}^{n_c}\left|y_i - y_i^*\right| \quad \text{and} \quad \text{IA} = \frac{1}{n_c}\sum_{i=1}^{n_c}\frac{\left|y_i - y_i^*\right|}{y_i}. \tag{12.22}$$

Note that in the case of censored data, it is not possible to observe all the values of response variable y_i for the real data applications. Therefore, the metrics given in (12.22) are only used in the simulation study. In addition, (12.21) in Lemma 12.1 and the measures given by (12.22) are considered to evaluate the performance of the estimators in the simulation studies and the real data examples.

12.6 Simulation Study

In this section, we perform Monte Carlo simulations to compare the performance of the introduced methods, KMW and kNN imputation, for different sample sizes and censoring levels. To accomplish this, the process of creating data from model (12.1) and (12.2) is designed as follows:

1. an $(n \times 1)$ vector of the values $\mathbf{x}_i$ is sequentially constructed in the interval $[-1, 3]$;
2. smooth function $f(\cdot)$ is defined by

$$f(x_i) = -0.1812 - 0.3221x_i + 4\sin(x_i^2) + \exp(x_i);$$

3. completely observed real response observations are generated in the following way: $y_i = f(x_i) + \varepsilon_i, i \in \{1, \ldots, n\}$, where $\varepsilon_i \sim N\left(0, \sigma_\varepsilon^2\right)$;
4. the values of censoring variable c_i are generated from normal distribution with mean $\mu_{\mathbf{y}}$ and standard deviation $\sigma_{\mathbf{y}}$. This means that c_i and y_i are independently and identically distributed. Note that in practice, because of incomplete observations, real value of y_i cannot be known. Therefore, one can observe only uncensored y_i and censored c_i;
5. the censoring indicator $\delta_i = I(y_i \leq c_i)$ is generated from a Bernoulli distribution, with rates at 5%, 20%, and 40%. Accordingly, censoring levels (C.L.) are determined as 5%, 20%, and 40%;
6. the values of $T_i = min(y_i, c_i)$ are determined and considered as a new censored response variable;
7. 1000 random samples of size $n = 30$, 100 and 200 for each censoring rate in the simulation are generated.

All computations in the simulation experiments are provided by codes written in the R-language. The outcomes from the simulations are summarized in Figs. 12.2, 12.3 and Table 12.1.

Figure 12.2 shows the boxplots of the RMSE scores from the estimates of the RCLMEM using penalized spline based on the kNN and KMW methods, respectively, for all simulation configurations. In Fig. 12.2, K1, K2, and K3 denote the boxplots of the RMSE scores from the estimates using kNN imputation for samples of size $n = 30$, 100, and 200, respectively. Similarly, W1, W2, and W3 represent the boxplots for the KMW method. As can be seen in Fig. 12.2, although the results from both methods are similar to each other, the estimation performance of kNN seems better than KMW. When we look at this problem in more detail, it can be seen that as the sample size n gets larger, the vertical range of penalized spline estimates decreases, as expected. However, it should be noted that kNN imputation performs better for all simulation scenarios. This may be because of kNN nonparametric nature.

The curves estimated by penalized spline using both data transformation approaches are illustrated in Fig. 12.3. When the fitted curves are examined carefully, it clearly indicates that kNN imputation works better than KMW, especially for heavy censored data. To better understand the performance of the estimators, Table 12.1 is constructed based on the values of RMSEs, the AvB, and IA introduced in main text, and variances of coefficients from fixed and random effects of the RCLMEM using each of the two methods under each censoring level and sample size.

As stated above, Table 12.1 includes the results from empirical performances of the RCLMEM estimated by two solution techniques for all simulation configurations. A careful inspection of the findings from the RCLMEM presented in Table 12.1 shows that when the censorship ratio increases, kNN imputation slightly outperforms the KMW method in terms of AvB and IA, as in RMSE and variances. This means that KMW manipulates the original data structure more. In this sense, it can be said that kNN is better than KMW in completing censored observations.

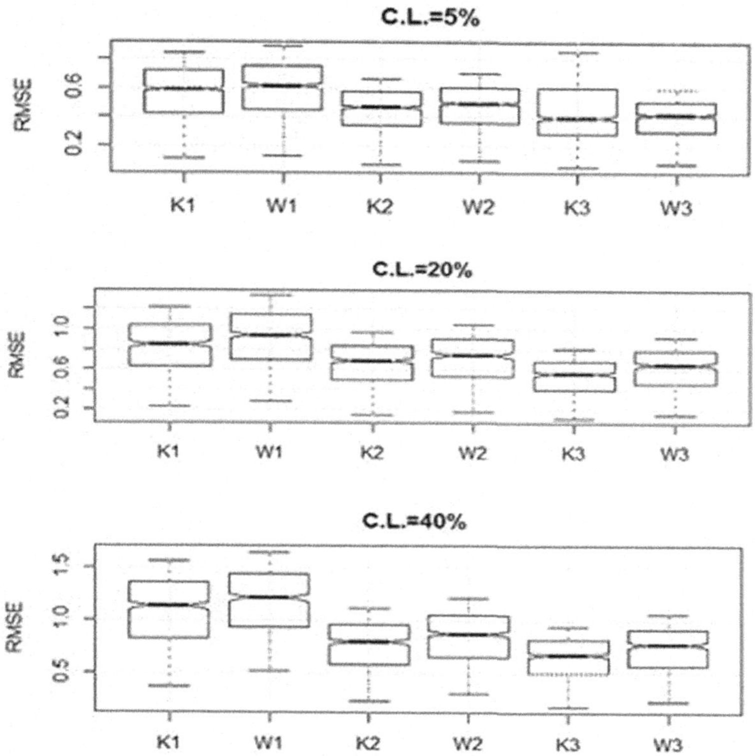

Fig. 12.2 Boxplots of RMSEs from 1000 runs for all censoring levels and samples

12.7 Real Data Examples

In this section, the penalized spline estimates in form of the LMEM with right-censored data is shown using two actual datasets, based on ovarian cancer survival times and kidney infection recurrence rates, for each of the two data transformation methods.

The ovarian cancer data is used as our first real data example. This data includes patients' survival time (time), age (age), and δ variable, which carries information from censored observations. The dataset is from Edmunson et al. [7]. Considering this information, the nonparametric model is written as follows for logarithm of survival time:

$$\log\left(\text{time}_i\right) = f\left(\text{age}_i\right) + \varepsilon_i, \qquad i \in \{1, \ldots, 26\}.$$

The results of the ovarian cancer data analysis are given in Tables 12.2 and 12.3 and Figs. 12.4 and 12.5.

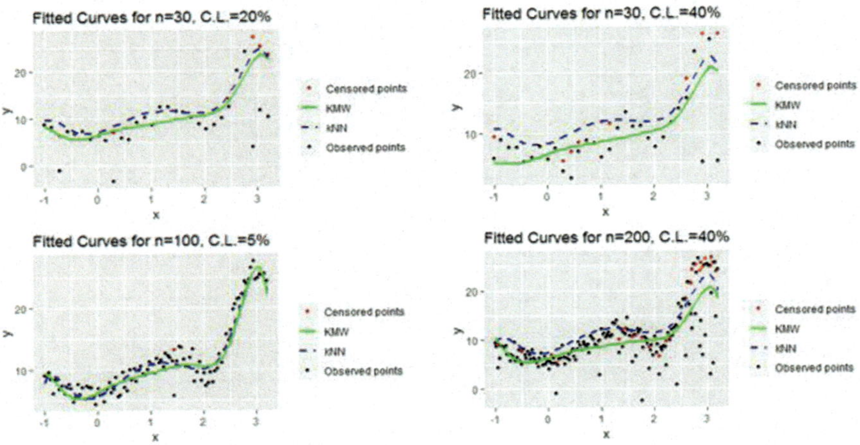

Fig. 12.3 Real observations, censored observations, and their curves fitted by kNN and KMW for determined combinations. Red points denote censored observations, black points are the observed ones, and the green and blue lines represent the curves fitted by KMW and kNN, respectively

Table 12.1 Finite sample performances of the proposed data transformation methods for each censoring levels (C.L.) and samples, with bold numbers denoting the best scores from each method

Method		kNN			KMW			kNN		KMW	
C.L.	n	RMSE	Var($\boldsymbol{\beta}_{FE}$)	Var($\boldsymbol{\beta}_{RE}$)	RMSE	Var($\boldsymbol{\beta}_{FE}$)	Var($\boldsymbol{\beta}_{RE}$)	AvB	IA	AvB	IA
5%	30	**0.877**	**8.409**	**0.160**	0.880	9.429	0.219	0.12	0.01	1.64	1.00
	100	**0.863**	**4.249**	**0.021**	0.865	5.147	0.181	**0.07**	**0.01**	1.60	1.01
	200	**0.859**	**3.419**	**0.012**	0.861	3.809	0.155	**0.27**	**0.02**	1.81	1.04
20%	30	**0.890**	**10.46**	**0.527**	0.898	13.81	0.493	**3.55**	**0.32**	6.12	1.53
	100	**0.868**	**5.139**	**0.030**	0.873	8.126	0.146	**1.59**	**0.11**	3.32	1.25
	200	**0.865**	**3.424**	**0.016**	0.869	4.606	0.095	**1.51**	**0.16**	3.43	1.28
40%	30	**0.908**	15.53	0.597	0.937	**14.14**	**0.381**	**4.62**	**0.40**	8.69	1.82
	100	**0.886**	**6.708**	**0.033**	0.898	10.99	0.295	**3.17**	**0.33**	6.25	1.61
	200	**0.878**	**5.409**	**0.022**	0.888	9.934	0.109	**2.93**	**0.26**	5.88	1.58

Table 12.2 The RMSE and variance values obtained from the transformation methods for the ovarian cancer data

Method	RMSE	Var($\boldsymbol{\beta}_{FE}$)	Var($\boldsymbol{\beta}_{RE}$)
kNN	**0.373**	**4.183**	**0.066**
KMW	0.544	10.893	0.132

Table 12.3 The RMSE and variance values obtained from the transformation methods for the ovarian cancer data involved prediction

Method	RMSE	Var(β_{FE})	Var(β_{RE})
kNN	**0.728**	**11.765**	**0.152**
KMW	0.812	22.258	0.479

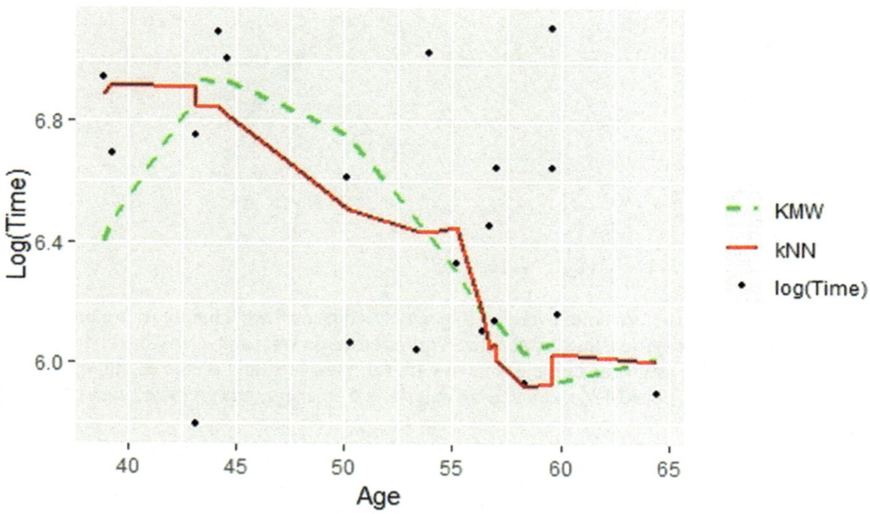

Fig. 12.4 Real log survival times and their curves fitted by penalized spline based on kNN (solid red line) and KMW (dashed green line) methods

Table 12.2 indicates that the kNN imputation method has a better quality in terms of the variance of coefficients and RMSE scores. Note that the RMSEs of the methods are similar, but the KMW method has large variances of coefficients. Here, AvB and IA measures are not shown, since completely observed responses are unknown.

Table 12.3 contains the performance scores the dataset with predicted values. Accordingly, it can be said that kNN performs better then KMW, as in Table 12.2. However, as expected, the magnitudes of the scores increase due to prediction of life times for the largest six values of age variable. As expected, the general trend of lifetime decreases for bigger ages. In addition, the results of Table 12.2 can be ensured from Fig. 12.5, which includes fitted curves for the predicted model. Better performance of kNN can also be seen from this figure, especially for the predicted three values.

Figures 12.4 and 12.5 illustrate the fitting performances of each of the methods for estimation and prediction. As can be seen in Fig. 12.4, the curve fitted by kNN imputation passes higher than KMW. This can be interpreted as the effect of censored observations completed by the kNN method. Furthermore, the KMW curve appears

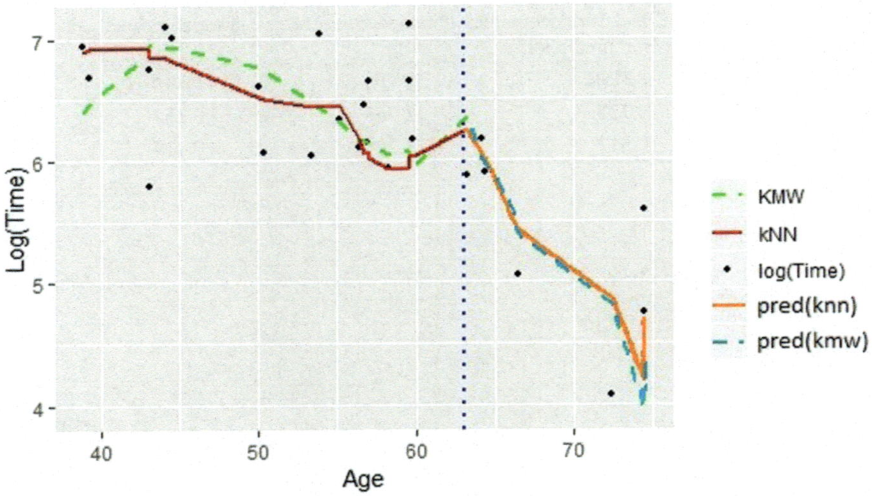

Fig. 12.5 Estimated (before dotted vertical line which indicates the beginning of prediction) and predicted curves (after vertical line) by penalized spline based on kNN and KMW methods. Regarding the kNN method, estimated curve is shown with solid red line, and predicted curve is shown with solid orange line. For KMW, estimated curve is dashed green line and predicted curve is dashed blue

to approach the x-axis, which is due to the fact that it assigns a zero value to the censored observations when calculating KMW. In addition, both prediction curves made an upward move. This is due to the small sample size ($n = 26$) and the over-scattered data structure, in our experience. It is seen that the curves estimated by both methods are easily affected by some extreme values. However, a similar situation is not seen in the kidney infection data estimates given below for the same two reasons. The kidney infection dataset has 76 data points and data structure is relatively less scattered then ovarian dataset.

In order to make clear the obtained results, another real data application is made with a kidney infection dataset. This dataset is collected to investigate how the frailties of patients affect the recurrence times of kidney infection. Here frailty values are estimated using a frailty distribution for each patient based on several risk variables about kidney disease (see McGilchrist and Aisbett [15]). The model for the kidney infection dataset is given as

$$\log\left(\text{time}_i\right) = f\left(\text{frailty}_i\right) + \varepsilon_i, \quad i \in \{1, \ldots, 76\}, \tag{12.23}$$

where time_i is the randomly right-censored response values denoting recurrence times and frailty_i denotes the kidney frailty values of 76 patients. This data is available in R-software and studied by McGilchrist and Aisbett [15]. We divided the dataset into two parts to make the prediction. 80% of the data is used for creating the model and 20% of the data is used for making the prediction. Results for the estimation of

Table 12.4 The RMSE and variance values obtained from the transformation methods for the kidney data

Method	RMSE	Var$(\boldsymbol{\beta}_{FE})$	Var$(\boldsymbol{\beta}_{RE})$
kNN	**1.259**	**16.020**	**0.163**
KMW	1.280	32.009	0.5824

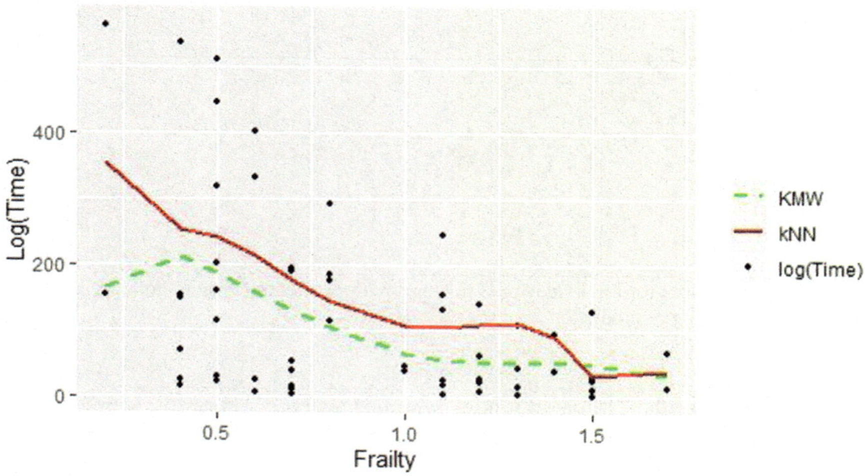

Fig. 12.6 Fitted curves for the kidney data

Table 12.5 Predicted RMSE and variance values obtained from the transformation methods for the kidney data

Method	RMSE	Var$(\boldsymbol{\beta}_{FE})$	Var$(\boldsymbol{\beta}_{RE})$
kNN	**3.464**	**23.402**	**1.9486**
KMW	3.580	36.473	2.4711

(12.23) are given in Table 12.4 and Fig. 12.6; outcomes of the prediction are given in Table 12.5 and Fig. 12.7.

As can be seen in Table 12.4 and Fig. 12.6, the kNN method provides better results than KMW, similarly as for the ovarian cancer dataset. Performances of fixed and random effects are measured by variances of $(\widehat{\boldsymbol{\beta}}, \widehat{\mathbf{u}})$. Moreover, from the obtained results, it can be said that RCLMEM estimation by penalized spline has reasonable results for both the KMW and kNN techniques, which can be counted as an important inference of this paper.

Figure 12.7 and Table 12.5 show the performance results from the predictions of each method. In Fig. 12.7, the predicted and estimated curves are presented together. The right side of the figure shows the predicted part and it can be said that these are reasonable in terms of fitting data. However, the performance scores are relatively

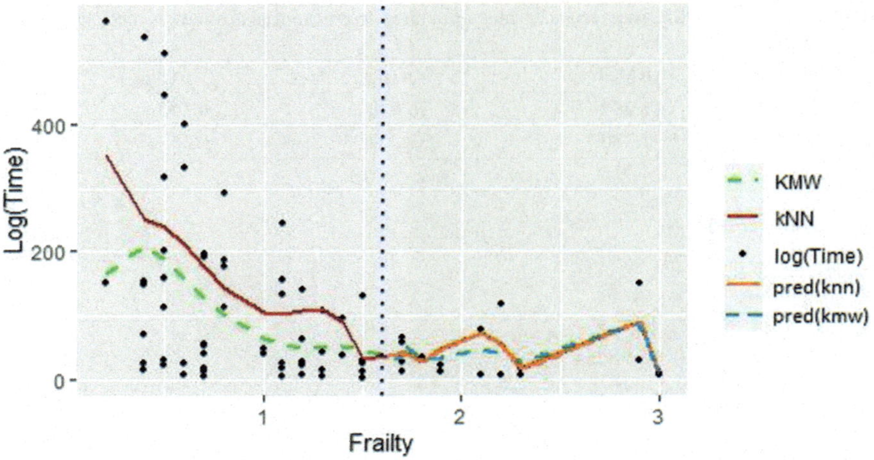

Fig. 12.7 Fitted curves for the kidney data together with predictions, where dotted vertical line indicates the beginning of prediction

worse than those of the estimated model, which is an expected result. Moreover, the kNN method performs better for both estimation and prediction when compared to KMW.

As a result, it can be said that the outcome of a real-world data and simulation study indicates that the estimation of the LMEM model based on kNN imputation gives better results in the case of right-censored observations.

12.8 Discussion

In this paper, right-censored data is modeled by a linear mixed-effects model (LMEM) based on two different censorship solution techniques: kNN imputation and KMW. In order to show how these methods work within an LMEM, a Monte Carlo simulation study and two real-world data examples are carried out. The results of these simulations are presented in Sects. 12.5 and 12.6. From the obtained results, it can be said that the kNN imputation technique is clearly more appropriate for solving the right-censored data problem when estimating the data via a nonparametric regression model based on an LMEM.

In terms of the simulation study, the kNN method performed better than KMW for nearly all combinations of censoring levels and sample sizes. As is expected, the general quality of the estimates obtained from both the KMW and kNN methods decreases under high censoring levels or small sample sizes; however, because of nonparametric nature of the kNN method, it was observed that kNN can give surprisingly satisfactory results even under heavy censorship or with small samples.

This was an important finding of this paper. It should be emphasized that the poor performance of the KMW method can be explained by its working procedure. KMW gives zero value to weight associated with the censored data points, which can be seen in (12.1); this affects data badly.

As a result, based on the outcomes of both our simulation study and the real-world examples of ovarian cancer survival and kidney infection recurrence, we recommend the kNN imputation method for the estimation of right-censored data modeling via an LMEM, especially under difficult conditions such as high censoring levels and small sample sizes. However, the KMW method is a very common tool for overcoming right-censored data, and our results show that it works satisfactory with lower censoring levels and that it has some theoretical bases that make it potentially more reliable than kNN.

Appendix A1—Proof of Equation (12.12)

One can see from (12.11), minimization criterion is given by

$$\text{PLSW}(\boldsymbol{\beta}, \mathbf{u}) = \arg\min \left((\mathbf{T} - \mathbf{X}\boldsymbol{\beta} - \mathbf{Z}\mathbf{u})'\mathbf{W}(\mathbf{T} - \mathbf{X}\boldsymbol{\beta} - \mathbf{Z}\mathbf{u}) + \lambda\mathbf{u}'\mathbf{D}\mathbf{u} \right).$$

From that equation, the system given below is obtained as

$$\begin{aligned}\begin{pmatrix} \widehat{\boldsymbol{\beta}}_{KMW} \\ \widehat{\mathbf{u}}_{KMW} \end{pmatrix} &= \begin{pmatrix} \mathbf{X}'\mathbf{W}\mathbf{X} & \mathbf{X}'\mathbf{W}\mathbf{Z} \\ \mathbf{Z}'\mathbf{W}\mathbf{X} & \mathbf{Z}'\mathbf{W}\mathbf{Z} + \lambda\mathbf{D} \end{pmatrix}^{-1} \begin{pmatrix} \mathbf{X}'\mathbf{W}\mathbf{T} \\ \mathbf{Z}'\mathbf{W}\mathbf{T} \end{pmatrix} \\ &= \left[(\mathbf{X} : \mathbf{Z})'\mathbf{W}(\mathbf{X} : \mathbf{Z}) + \lambda\mathbf{D} \right]^{-1} (\mathbf{X} : \mathbf{Z})'\mathbf{W}\mathbf{T}.\end{aligned}$$

Let $\mathbf{R} = (\mathbf{X} : \mathbf{Z})$. Accordingly, (12.12) is obtained as follows:

$$\left[\widehat{\boldsymbol{\beta}}'_{KMW}, \widehat{\mathbf{u}}'_{KMW} \right]' = \left(\mathbf{R}'\mathbf{W}\mathbf{R} + \lambda\mathbf{D} \right)^{-1} \mathbf{R}'\mathbf{W}\mathbf{T}.$$

Appendix A2—Proof of Lemma 12.1

Let $\boldsymbol{\theta} = (\boldsymbol{\beta}', \mathbf{u}')'$ and $\boldsymbol{\theta}$ can be written as the function of OLS estimator as

$$\begin{aligned}(\widehat{\boldsymbol{\beta}}', \widehat{\mathbf{u}}')' &= \left(\mathbf{R}'\mathbf{R} + \lambda\mathbf{D} \right)^{-1} \mathbf{R}'\mathbf{T} \\ &= \left(\mathbf{R}'\mathbf{R} + \lambda\mathbf{D} \right)^{-1} \mathbf{R}'\mathbf{R} \left(\mathbf{R}'\mathbf{R} \right)^{-1} \mathbf{R}'\mathbf{T} \\ &= \left(\mathbf{R}'\mathbf{R} + \lambda\mathbf{D} \right)^{-1} \mathbf{R}'\mathbf{R}\boldsymbol{\theta}.\end{aligned}$$

Thus,

$$\begin{aligned}\mathrm{Cov}(\widehat{\boldsymbol{\beta}},\widehat{\mathbf{u}}) = \mathrm{Cov}(\boldsymbol{\theta}) &= \left[\left(\mathbf{R}'\mathbf{R}+\lambda\mathbf{D}\right)^{-1}\mathbf{R}'\mathbf{R}\right]\mathrm{Cov}(\boldsymbol{\theta})\left[\left(\mathbf{R}'\mathbf{R}+\lambda\mathbf{D}\right)^{-1}\mathbf{R}'\mathbf{R}\right]' \\ &= \left(\mathbf{R}'\mathbf{R}+\lambda\mathbf{D}\right)^{-1}\left(\mathbf{R}'\mathbf{R}\right)\mathrm{Cov}(\boldsymbol{\theta})\left(\mathbf{R}'\mathbf{R}\right)\left(\mathbf{R}'\mathbf{R}+\lambda\mathbf{D}\right)^{-1} \\ &= \left(\mathbf{R}'\mathbf{R}+\lambda\mathbf{D}\right)^{-1}\left(\mathbf{R}'\mathbf{R}\right)\frac{\sigma^2}{n}\left(\mathbf{R}'\mathbf{R}\right)^{-1}\left(\mathbf{R}'\mathbf{R}\right)\left(\mathbf{R}'\mathbf{R}+\lambda\mathbf{D}\right)^{-1} \\ &= \frac{\sigma^2}{n}\left(\mathbf{R}'\mathbf{R}+\lambda\mathbf{D}\right)^{-1}\left(\mathbf{R}'\mathbf{R}\right)\left(\mathbf{R}'\mathbf{R}+\lambda\mathbf{D}\right)^{-1},\end{aligned}$$

where σ^2/n comes from the variance of $\widehat{\mathbf{u}}$, which is the property of the penalized spline method (see Claeskens et al. [6]). Thus, (12.21) would be obtained. It can be also obtained if $\mathbf{R}'\mathbf{R}$ is replaced by $\mathbf{R}'\mathbf{W}\mathbf{R}$ for KMW method.

References

1. Ahmed, S.E., Aydın, D., Yılmaz, E.: Nonparametric regression estimates based on imputation techniques for right-censored data. In: ICMSEM 2019: Proceeding of the Thirteenth International Conference on Management Science and Engineering Management, pp. 109–120. Springer (2020)
2. Aydın, D., Memmedli, M.: Optimum smoothing parameter selection for penazlied least squares in form of linear mixed effect models. Optimization **61**(4), 459–476 (2012)
3. Aydın, D., Yılmaz, E.: Modified spline regression based on randomly right-censored data: A comparative study. Comm. Stat. Simul. Comput. **47**(9), 2581–2611 (2018)
4. Bandyopadhyay, D., Lachos, V.H., Castro, L.M., Dey, D.K.: Skew-normal/independent linear mixed models for censored responses with applications to HIV viral loads. Biom. J. **54**(3), 405–425 (2012)
5. Batista, G., Monard, M.: An analysis of four missing data treatment methods for supervised learning. Appl. Artif. Intell. **17**, 519–533 (2002)
6. Claeskens, G., Krivobokova, T., Opsomer, J.D.: Asymptotic properties of penalized spline estimators. Biometrika **96**(3), 529–544 (2009)
7. Edmunson, J.H., Fleming, T.R., Decker, D.G., Malkasian, G.D., Jefferies, J.A., Webb, M.J., Kvols, L.K.: Different chemotherapeutic sensitivities and host factors affecting prognosis in advanced ovarian carcinoma vs. minimal residual disease. Cancer Treat. Rep. **63**, 241–247 (1979)
8. Eilers, P.H.C., Marx, B.D.: Flexible smoothing with B-splines and penalties. Stat. Sci. **11**(2), 89–102 (1996)
9. Hurvich, C.M., Simonoff, J.S., Tsai, C.-L.: Smoothing parameter selection in nonparametric regression using an improved Akaike information criterion. J. R. Stat. Soc. Ser. B. Stat. Methodol. **60**(2), 271–293 (1998)
10. Kaplan, E.L., Meier, P.: Nonparametric estimation from incomplete observations. J. Am. Stat. Assoc. **53**(282), 457–481 (1958)
11. Koul, H., Susarla, V., Van Ryzin, J.: Regression analysis with randomly right-censored data. Ann. Stat. **9**(6), 1276–1285 (1981)
12. Laird, N.M., Ware, J.H.: Random-effect models for longitudinal data. Biometrics **38**(4), 963–974 (1982)
13. Matos, L.A., Prates, M.O., Chen, M.-H., Lachos, V.H.: Likelihood-based inference for mixed effects models with censored response using the multivariate-t distribution. Stat. Sinica **23**, 1323–1345 (2013)
14. McCulloch, C.E., Searle, S.R.: Generalized Linear and Mixed Models. John Wiley and Sons, New York (2001)

15. McGilchrist, C.A., Aisbett, C.W.: Regression with frailty in survival analysis. Biometrics **47**, 461–466 (1991)
16. Miller, R.G.: Least squares regression with censored data. Biometrika **63**, 449–64 (1976)
17. Pan, W., Louis, T.A.: A linear mixed-effects model for multivariate censored data. Biometrics **56**, 160–166 (2000)
18. Ruppert, D., Wand, M.P., Carroll, R.J.: Semiparametric Regression. Cambridge University Press, New York (2003)
19. Stute, W.: Consistent estimation under random censorship when covariables are present. J. Multivar. Anal. **45**, 89–103 (1993)
20. Verbeke, G., Molenberghs, G.: Linear Mixed Models for Longitudinal Data. Springer Verlag, New York (2009)
21. Vock, D.M., Davidian, M., Tsiatis, A.A., Muir, A.J.: Mixed model analysis of censored longitudinal data with flexible random-effects density. Biostatistics **13**(1), 61–73 (2012)
22. West, B.T., Welch, K.B., Galecki, A.T.: Linear Mixed Models: A Practical Guide Using Statistical Software. CRC Press, New York (2015)
23. Wu, W.B., Pourahmedi, M.: Nonparametric estimation of large covariance matrices of longitudinal data. Biometrika **90**(4), 831–844 (2003)

MMLM Meetings—List of Publications

Books

1. von Rosen, D.: Bilinear Regression Analysis: An Introduction. Lecture Notes in Statistics, vol. 210, pp. 1–468. Springer, Cham (2018). ISBN 978-3-319-78784-8.

Articles

2010

1. Kollo, T., Pettere, G.: Parameter estimation and application of the multivariate skew t-copula. In: Jaworski, P., Durante, F., Härdle, W.K., Rychlik, T. (eds.) Copula Theory and Its Applications. In: Proceedings of the Workshop Held in Warsaw 25–26 Sept 2009, pp. 289–298. Lecture Notes in Statistics. Springer-Verlag, Berlin, Heidelberg.
2. Nahtman, T., von Rosen, D.: On a new class of singular nonsymmetric matrices with nonnegative integer spectra. In: Olshevsky, V., Tyrtyshnikov, E. (eds.) Matrix Methods: Theory, Algorithms and Applications: Dedicated to the Memory of Gene Golub, pp. 140–165. World Scientific Publishing.
3. Ohlson, M., von Rosen, D.: Explicit estimators of parameters in the growth curve model with linearly structured covariance matrices. J. Multivariate Anal. **101**, 1284–1295.
4. von Rosen, T., von Rosen, D.: On a class of singular nonsymmetric matrices with nonnegative integer spectra. Contemp. Math. Am. Math. Soc. **516**, 319–324.

2011

1. Kollo, T., von Rosen, T., von Rosen, D.: Estimation in high-dimensional analysis and multivariate linear models. Comm. Stat. Theory Methods **40**, 1241–1253.
2. Ohlson, M., Andrushchenko, Z., von Rosen, D.: Explicit estimators under m-dependence for a multivariate normal distribution. Ann. Inst. Stat. Math. **63**, 29–42.
3. Srivastava, M., Kollo, T., von Rosen, D.: Some tests for the covariance matrix with fewer observations than the dimension under non-normality. J. Multivariate Anal. **102**, 1090–1103.

K. Filipiak et al. (eds.), *Multivariate, Multilinear and Mixed Linear Models*, Contributions to Statistics, https://doi.org/10.1007/978-3-030-75494-5

2012

1. Filipiak, K., von Rosen, D.: On MLEs in an extended multivariate linear growth curve model. Metrika **75**, 1069–1092.
2. Li, Y., von Rosen, D.: Maximum likelihood estimators in a two step model for PLS. Comm. Stat. Theory Methods **41**, 2503–2511.
3. Nzabanita, J., Singull, M., von Rosen, D.: Estimation of parameters in the extended growth curve model with a linear structured covariance matrix. Acta Comment. Univ. Tartu. Math. *16*, 13–32.
4. Singull, M., Ahmad, R., von Rosen, D.: More on the Kronecker structured covariance matrix. Comm. Stat. Theory Methods **41**, 2512–2523.

2013

1. Kollo, T., Selart, A., Visk, H.: From multivariate skewed distributions to copulas. In: Bapat, R.B., Kirkland, S.J., Prasad, K. M., Puntanen, S. (eds.) Combinatorial Matrix Theory and Generalized Inverses of Matrices, pp. 63–72. New Delhi, Springer.
2. Li, Y., Udén, P., von Rosen, D.: A two-step PLS inspired method for linear prediction with group effect. Sankhyā A **75**, 96–117.

2015

1. Arendacká, B., Puntanen, S.: Further remarks on the connection between fixed linear model and mixed linear model. Stat. Pap. **56**, 1235–1247.
2. Haslett, S.J., Puntanen, S., Arendacká, B.: The link between the mixed and fixed linear models revisited. Stat. Pap. **56**, 849–861.
3. Klein, D., Žežula, I.: Maximum likelihood estimators for extended growth curve model with orthogonal between-individual design matrices. Stat. Methodol. **23**, 59–72.
4. Li, Y., Udén, P., von Rosen, D.: A two-step estimation method for grouped data with connections to the extended growth curve model and partial least squares regression. J. Multivariate Anal. **139**, 347–359.
5. Li, Y., Udén, P., von Rosen, D.: Erratum - A two-step PLS inspired method for linear prediction with group effect. Sankhyā A **77**, 433–436.
6. Liang, Y., von Rosen, D., von Rosen, T.: On estimation in hierarchical models with block circular covariance structures. Ann. Inst. Stat. Math. **67**, 773–791.
7. Markiewicz, A., Puntanen, S.: All about the $\perp$ with its applications in the linear statistical models. Open Math. **13**, 33–50. [Formerly Central European Journal of Mathematics].
8. Nzabanita, J., von Rosen, D., Singull, M.: Extended GMANOVA model with a linearly structured covariance matrix. Math. Methods Stat. **24**, 280–291.
9. Roy, A., Leiva, R., Žežula, I., Klein, D.: Testing the equality of mean vectors for paired doubly multivariate observations in blocked compound symmetric covariance matrix setup. J. Multivariate Anal. **137**, 50–60.
10. Rusnačko, R., Žežula, I.: The growth curve model with heterogeneous fractional decreasing correlation structure. Coll. Biom. **45**, 5–13.

2016

1. Filipiak, K., Klein, D., Roy, A.: Score test for a separable covariance structure with the first component as compound symmetric correlation matrix. J. Multivariate Anal. **150**, 105–124.
2. Hao, C., Liang, Y., Mathew, T.: Testing variance parameters in models with a Kronecker product covariance structure. Stat. Probab. Lett. **118**, 182–189.
3. Kollo, T., von Rosen, D., Valge, M.: Hypotheses testing on covariance structures: Comparison of likelihood ratio test, Rao's score test and Wald's score test. In: Bozeman, J.R., Oliveira, T., Skiadas, C.H. (eds.) Stochastic and Data Analysis Methods and Applications in Statistics and Demography, pp. 423–435. ISAST.
4. Kozioł, A. (2016). Best unbiased estimates for parameters of three-level multivariate data with doubly exchangeable covariance structure and structured mean vector. Discuss. Math. Probab. Stat. **36**, 93–113.
5. Roy, A., Zmy'slony, R., Fonseca, M., Leiva, R.: Optimal estimation for doubly multivariate data in blocked compound symmetric covariance structure. J. Multivariate Anal. **144**, 81–90.
6. Rusnačko, R., Žežula, I. (2016). Connection between uniform and serial correlation structure in the growth curve model. Metrika **79**, 149–164.

2017

1. Bailey, R.A., Cameron, P., Filipiak, K., Kunert, J., Markiewicz, A.: On optimality and construction of circular repeated-measurements designs. Stat. Sinica **27**, 1–22.
2. Filipiak, K., Klein, D.: Estimation of parameters under a generalized growth curve model. J. Multivariate Anal. **158**, 73–86.
3. Filipiak, K., Klein, D., Roy, A.: A comparison of likelihood ratio tests and Rao's score test for three separable covariance matrix structures. Biom. J. **59**, 192–215.
4. Filipiak, K., Markiewicz, A: Universally optimal designs under mixed interference models with and without block effects. Metrika **80**, 789–804.
5. Filipiak, K., Markiewicz, A.: Universally optimal designs under interference models with and without block effects. Comm. Stat. Theory Methods **46**, 1127–1143.
6. Fonseca, M., Volaufova, J., Puntanen, S.: Coordinate-free personal meetings with Professor Roman Zmyślony. Discuss. Math. Probab. Stat. **37**, 5–35.
7. Kala, R., Markiewicz, A., Puntanen, S.: Some further remarks on the linear sufficiency in the linear model. In: Bebiano, N. (ed.) Applied and Computational Matrix Analysis: MatTriad, Coimbra, Portugal, September 2015, Selected, Revised Contributions, pp. 275–294. Springer Proceedings in Mathematics and Statistics.
8. Kala, R., Puntanen, S., Tian, Y.: Some notes on linear sufficiency. Stat. Pap. **58**, 1–17.
9. Kozioł, A., Roy, A., Zmyślony, R., Leiva, R., Fonseca, M.: Best unbiased estimates for parameters of three-level multivariate data with doubly exchangeable covariance structure. Linear Algebra Appl. **535**, 87–104.

10. Ngaruye, I., Nzabanita, J., von Rosen, D., Singull, M.: Small area estimation under a multivariate linear model for repeated measures data. Comm. Stat. Theory Methods **46**, 1–22.
11. von Rosen, T., von Rosen, D.: On estimation in some reduced rank extended growth curve models. Math. Methods Stat. **26**, 299–310.
12. Rusnačko, R., Žežula, I.: Comparison of estimators of variance parameters in the growth curve model with a special variance structure. Acta Comment. Univ. Tartu. Math. **21(2)**, 171–184.
13. Szczepańska-Álvarez, A., Hao, C., Liang, Y., von Rosen, D.: Estimation equations for multivariate linear models with Kronecker structured covariance matrices. Comm. Stat. Theory Methods **46(16)**, 7902–7915.

2018

1. Filipiak, K., Klein, D.: Approximation with a Kronecker product structure with one component as compound symmetry or autoregression. Linear Algebra Appl. **559**, 11–33.
2. Filipiak, K., Klein, D. , Vojtková, E.: The properties of partial trace and block trace operators of partitioned matrices. Electron. J. Linear Algebra **33**, 3–15.
3. Filipiak, K., Klein, D., Mokrzycka, M.: Estimators comparison of separable covariance structure with one component as compound symmetry matrix. Electron. J. Linear Algebra **33**, 83–98.
4. Filipiak, K., Markiewicz, A., Mieldzioc, A., Sawikowska, A.: On projection of a positive definite matrix on a cone of non-negative definite Toeplitz matrices. Electron. J. Linear Algebra **33**, 74–82.
5. Fonseca, M., Kozioł, A., Zmyślony, R.: Testing hypotheses of covariance structure in multivariate data. Electron. J. Linear Algebra **33**, 53–62.
6. Isotalo, J., Markiewicz, A., Puntanen, S.: Some properties of linear prediction sufficiency in the linear model. In: Tez, M., von Rosen, D. (eds.) Trends and Perspectives in Linear Statistical Inference: LinStat, Istanbul, August 2016 (International Conference on Trends and Perspectives in Linear Statistical Inference, Istanbul, Turkey, 22–25 August 2016), pp. 111–129. Springer.
7. Kollo, T., Käärik, M., Selart, A.: Asymptotic normality of estimators for parameters of a multivariate skew-normal distribution. Comm. Stat. Theory Methods **47(15)**, 3640–3655.
8. Kozioł, A., Roy, A., Zmyślony, R., Leiva, R., Fonseca, M.: Free-coordinate estimation for doubly multivariate data. Linear Algebra Appl. **547**, 217–239.
9. Markiewicz, A., Puntanen, S.: Further properties of linear prediction sufficiency and the BLUPs in the linear model with new observations. Afr. Stat. **13**, 1511–1530.
10. Markiewicz, A., Puntanen, S.: Upper bounds for the Euclidean distances between the BLUPs. Spec. Matrices **6**, 249–261.
11. Roy, A., Filipiak, K., Klein, D.: Testing a block exchangeable covariance matrix. Statistics **52**(2), 393–408.
12. Žežula, I., Klein, D., Roy, A.: Testing of multivariate repeated measures data with block exchangeable covariance structure. Test **27(2)**, 360–378.

13. Zmyślony, R., Žežula, I., Kozioł, A.: Application of Jordan Algebra for testing hypotheses about structure of mean vector in model with block compound symmetric covariance structure. Electron. J. Linear Algebra **33**, 41–52.

2019

1. Markiewicz, A., Puntanen, S.: Further properties of the linear sufficiency in the partitioned linear model. In: Ahmed, S.E., Carvalho, F., Puntanen, S. (eds.) Matrices, Statistics and Big Data: Proceedings of the 25th International Workshop on Matrices and Statistics (IWMS-2016, Funchal, Madeira, Portugal, 6–9 June 2016), pp. 1–22. Springer, Cham.
2. Markiewicz, A., Puntanen, S.: Linear prediction sufficiency in the misspecified linear model. Comm. Stat. Theory Methods. https://doi.org/10.1080/03610926.2019.1584311
3. Mieldzioc A., Mokrzycka, M., Sawikowska, A.: Covariance regularization for metabolomic data on drought resistance of barley. Biom. Lett. **56(2)**, 165–181.
4. Ngaruye, I., von Rosen, D., Singull, M.: Mean-Squared errors of small area estimators under a multivariate linear model for repeated measures data. Comm. Stat. Theory Methods **48**, 2060–2073.

2020

1. Filipiak, K., John, M., Markiewicz, A.: Comments on maximum likelihood estimation and projections under multivariate statistical models. In: Holgersson, T., Singull, M. (eds.) Recent Developments in Multivariate and Random Matrix Analysis, pp. 51–66. Springer.
2. Haslett, S.J., Liu, X-Q., Markiewicz, A., Puntanen, S.: Some properties of linear sufficiency and the BLUPs in the linear mixed model. Stat. Pap **61**, 385–401.
3. Haslett, S.J., Markiewicz, A., Puntanen, S.: Properties of BLUEs and BLUPs in full vs. small linear models with new observations. In: Holgersson, T., Singull, M. (eds.) Recent Developments in Multivariate and Random Matrix Analysis, pp. 123–146. Springer.
4. Hauke, J., Markiewicz, A., Puntanen, S.: Beyond the MatTriad Conferences [a short history of the MatTriad conferences]. Appl. Math. **65**, 547–556.
5. Janiszewska, M., Markiewicz, A., Mokrzycka, M.: Block matrix approximation via entropy loss function. Appl. Math. **65(6)**, 829–844.
6. John, M., Mieldzioc, A.: The comparison of the estimators of banded Toeplitz covariance structure under the high-dimensional multivariate model. Comm. Stat. Simul. Comput. **49(3)**, 734–752.
7. Jurková, V., Žežula, I., Klein, D.: Testing in the Growth Curve Model with intraclass correlation structure. Statistics **54(5)**, 1124–1146.
8. Kollo, T., Valge, M.: Covariance structure tests for t-distribution. In: Holgersson, T., Singull, M. (eds.) Recent Developments in Multivariate and Random Matrix Analysis, pp. 199–217. Springer.
9. Kopčová, V., Žežula, I.: On intraclass structure estimation in the growth curve model. Stat. Pap. **61**(3), 1085–1106.

10. von Rosen, T., von Rosen, D.: Bilinear regression with rank restrictions on the mean and the dispersion matrix. Jpn. J. Stat. Data Sci. **3**, 63–72.
11. von Rosen, T., von Rosen, D., Volaufova, J.: A new method for obtaining explicit estimators in unbalanced mixed linear models. Stat. Pap. **61**, 371–383.
12. Roy, A., Klein, D.: Testing of mean interval for interval-valued data. Comm. Stat. Theory Methods **49**(20), 5028–5044.

2021

1. Filipiak, K., Klein, D., Markiewicz, A., Mokrzycka, M.: Approximation with a Kronecker product structure with one component as compound symmetry or autoregression via entropy loss function. Linear Algebra Appl. **610**, 625–646.
2. Jurková, V., Žežula, I., Klein, D, Hutník, O.: Unbiased estimator of correlation coefficient. Comm. Stat. Theory Methods. DOI: 10.1080/03610926.2020.1743314.
3. Klein, D., Žežula, I.: On drawbacks of least squares Lehmann-Scheffé estimation of variance components. Metron. Accepted.
4. Markiewicz, A., Puntanen, S., Styan, G.P.H.: The legend of the equality of OLSE and BLUE: highlighted by C.R. Rao in 1967. In: Arnold, B.C., Balakrishnan, N., Coelho, C.A. (eds.) Methodology and Applications of Statistics: A Volume in Honor of C.R. Rao on the Occasion of his 100th Birthday. Springer. To appear.

Index

K. Filipiak et al. (eds.), *Multivariate, Multilinear and Mixed Linear Models*, Contributions to Statistics, https://doi.org/10.1007/978-3-030-75494-5

Printed by Printforce, the Netherlands